Nuclear Magnetic Resonance Spectroscopy

Second Edition

Frank A. Bovey
AT&T Bell Laboratories
Murray Hill, New Jersey

Lynn Jelinski
(Chapter 2, Co-author of Chapter 9)

Peter A. Mirau
(Co-author of Chapter 6)

AT&T

Published by arrangement
with AT&T

Academic Press, Inc.
Harcourt Brace Jovanovich, Publishers

San Diego   New York   Berkeley   Boston
London   Sydney   Tokyo   Toronto

Academic Press Rapid Manuscript Reproduction

ACADEMIC PRESS, INC.
1250 Sixth Avenue
San Diego, California 92101

United Kingdom Edition published by
ACADEMIC PRESS INC. (LONDON) LTD.
24-28 Oval Road, London NW1 7DX

Library of Congress Cataloging-in-Publication Data

Bovey, Frank Alden, Date
    Nuclear magnetic resonance / Frank A. Bovey.
      p. cm.
    Includes bibliographies.
    ISBN 0-12-119752-2 (alk. paper)
    1. Nuclear magnetic resonance spectroscopy. I. Title.
QD96.N8B68 1987   87-30381
543′.0877—dc19   CIP

PRINTED IN THE UNITED STATES OF AMERICA
88 89 90 91   9 8 7 6 5 4 3 2 1

CONTENTS

Foreword by H. S. Gutowsky   ix
Preface to the First Edition   xi
Preface to the Second Edition   xiii

Chapter 1: Fundamental Principles

1.1   Introduction   1
1.2   Nuclear Spin   1
1.3   The Nuclear Magnetic Resonance Phenomenon   4
1.4   The Detection of the Resonance Phenomenon: CW and Pulse Fourier
      Transform Methods   7
1.5   Spin–Lattice Relaxation   12
1.6   Dipolar Broadening and Spin–Spin Relaxation   18
1.7   The Bloch Equations   23
1.8   Saturation   28
1.9   Nuclear Electric Quadrupole Relaxation   29
1.10 Magnetic Shielding and Chemical Shift   30
1.11 Electron-Mediated Coupling of Nuclear Spins   33
      References   36

Chapter 2: Experimental Methods

Lynn W. Jelinski

2.1   Introduction   39
2.2   Overview of the NMR Experiment   39
2.3   The Magnetic Field   41
2.4   Causing Transitions between the Nuclear Energy Levels   50
2.5   Detecting the Signal   60
2.6   Processing and Presenting the Data   67
2.7   Preparing the Sample   71
2.8   Resolution and Sensitivity Considerations   79
      References   85

Chapter 3: The Chemical Shift

3.1 Introduction 87
3.2 Atomic Shielding 87
3.3 Molecular Shielding 88
     References 142

Chapter 4: Coupling of Nuclear Spins

4.1 Introduction 147
4.2 Description and Analysis of Spin Systems 150
4.3 Dependence of J Coupling on Structure and Geometry 191
4.4 Spin Decoupling and Associated Techniques 216
     References 250

Chapter 5: Nuclear Relaxation and Chemical Rate Processes

5.1 Introduction 255
5.2 Spin Lattice Relaxation 255
5.3 Spin–Spin Relaxation 289
5.4 Rate Processes 291
     References 321

Chapter 6: Two-Dimensional Nuclear Magnetic Resonance Spectroscopy

Peter A. Mirau and Frank A. Bovey

6.1 Introduction 325
6.2 J-Resolved Spectroscopy 328
6.3 Correlated Spectroscopy 337
6.4 Multiple Quantum Spectroscopy 354
     References 355

Chapter 7: Macromolecules

7.1 Introduction 357
7.2 Chain Microstructure 357
7.3 Biopolymers 381
7.4 Chain Motion in Macromolecules 388
     References 396

Chapter 8: NMR of Solids

8.1 Introduction 399
8.2 Proton Dipolar Broadening 400
8.3 Chemical Shift Anisotropy 402
8.4 Magic Angle Spinning 404
8.5 Cross Polarization 406
8.6 Magic Angle Spectra 409
8.7 Multiple-Pulse Line Narrowing 422
8.8 Deuterium Quadruple Echo NMR 424
    References 434

Chapter 9: Special Topics

9.1 Less Common Nuclei 437
    F. A. Bovey
9.2 NMR Imaging 489
    L.W. Jelinski
9.3 Solvent Suppression 505
    P. A. Mirau and F. A. Bovey
9.4 Zero Field NMR 512
    F. A. Bovey
    References 515

Appendix A: Table of Nuclear Properties 523

Appendix B: Calculated Proton Spectra 529

Appendix C: Proton Spin-Spin Coupling Constants 604

Appendix D: Carbon-Proton Spin-Spin Coupling Constants 629

Author Index 635
Subject Index 642

FOREWORD

In 1980 Jiri Jonas and I wrote a review article titled "NMR in Chemistry — An Evergreen." We sought to capture in the title the continued vitality and growth of NMR spectroscopy, a never dormant field. Since its inception forty years ago NMR has been a cornucopia of exciting and extremely useful phenomena, with new and ever more powerful techniques for observing them — dipolar broadening and motional narrowing, chemical shifts, J-coupling, spin echoes, superconducting solenoids and high fields, Fourier transform methods, high resolution solid state NMR, two dimensional methods, and most recently, magnetic resonance imaging.

This book captures the always refreshing spirit of NMR today. Although billed as a second edition of Bovey's 1969 offspring, it is an accurate reflection of twenty more years of NMR — twice as big and twice as good. The chapter on experimental methods now stresses Fourier transform methods and superconducting magnets. Double resonance and spectral editing techniques are found in the chapter on coupling of nuclear spins, and an entire chapter is devoted to two-dimensional NMR and another to solids. The latest views on MR imaging, water suppression, and zero-field NMR are given in a special topics finale.

Today's seminar circuit style of learning has given new currency to Samuel Johnson's 1766 statement, "People have now-a-days got a strange opinion that everything should be taught by lectures. Now, I cannot see that lectures can do so much good as reading the books from which the lectures are taken." This is one of them.

H. S. Gutowsky

PREFACE TO THE FIRST EDITION

Following the appearance in 1959 of the first, and now classical, monograph on high resolution NMR by Pople, Schneider, and Bernstein,[1] no new comprehensive work appeared in this field for many years despite the fact that it was advancing very rapidly. Beginning in 1964, this vacuum began to be filled by an increasing flow of books of varying scope and technical level, including two annual review series. At the present time, one can no longer complain that there is a serious lack of coverage of the field, at least so far as the organic chemist is concerned.

In 1965, I published in *Chemical and Engineering News* an article describing in simplified terms the basis of high resolution NMR spectroscopy and its applications. The response to this article was large and widespread, and encouraged the idea that despite the other texts then available there was still a need which might be filled by expansion of this review to a book of moderate size, written in more or less the same spirit. This volume is the result.

This work is intended for the graduate student or research worker in industry who needs a text that is pitched at a modest, relatively nonmathematical level and that can be applied in solving problems related to the structure and behavior of organic molecules. The tone is empirical throughout; the theoretical basis of the more important effects being sketched in only lightly where it seemed to be appropriate. The text is supplemented by the following appendixes: Appendix A: Table of Nuclear Properties; Appendix B: Table of π-Values for Selected Organic Compounds in CCl_4 Solution; Appendix C: Calculated Shielding Effects Produced by Ring Currents in a Benzene Ring; Appendix D: Calculated Spectra; Appendix E: Proton Spin-Spin Coupling Constants.

A certain degree of repetition of mathematical developments given in other texts was unavoidable. For example, no worthwhile discussion of line shapes can be given without at least some consideration of the Bloch equations (Chapters I and VII), even though they are undoubtedly more authoritatively dealt with elsewhere. In Chapter IV, the discussion of the analysis of NMR spectra omits all except the most elementary quantum mechanical background, as this is amply treated in other sources. But tables of line positions and intensities are included, as these are necessary to support the text. In addition, some 300 computer-generated line spectra are given in Appendix D. These represent a fair proportion of the simpler types of "strong-coupled" spectra with which the reader may need to deal. This feature is believed to be unique in a text, although an earlier compilation was published as a book by Wiberg and Nist in 1962,[2] using a different type of pictorial display.

No effort has been made to discuss the spectral properties of classes of organic compounds in a systematic way, as, for example, in the comprehensive texts of Emsley, Feeney, and Sutcliffe[3] and Suhr.[4] Spectra are introduced only to illustrate specific points. This is in no sense a handbook or encyclopedia of data. For this purpose, the reader may wish to consult the spectral catalogue in two volumes distributed by Varian (Palo Alto, California) or my published tables.[5]

I wish to thank Mr. F. P. Hood and Mr. J. J. Ryan for running many of the spectra and assisting in the preparation of figures; Dr. W. P. Slichter and Dr. D. W. McCall for many helpful discussions; Dr. G. V. D. Tiers for permission to reproduce the data appearing in Appendix B; Dr. S. Meiboom for critically reading the entire manuscript; and the American Chemical Society for permission to reproduce figures from the *Chemical and Engineering News* article referred to above. The calculated spectra in the text and in Appendix D were produced by means of a digital computer program devised by Dr. L. C. Snyder and Mr. R. L. Kornegay of the Bell Telephone Laboratories. Grateful acknowledgment is due them for the use of the program and for many helpful discussions concerning it.

F. A. Bovey

Murray Hill, New Jersey
October, 1968

[1]J. A. Pople, W. G. Schneider, and H. J. Bernstein, "High Resolution Nuclear Magnetic Resonance." McGraw-Hill, New York, 1959.
[2]K. B. Wiberg and B. J. Nist, "The Interpretation of NMR Spectra." W. A. Benjamin, New York, 1962.
[3]J. W. Emsley, J. Feeney, and L. H. Sutcliffe, "High Resolution Nuclear Magnetic Resonance Spectroscopy" (2 vols.). Pergamon Press, London, 1965.
[4]H. Suhr, "Anwendungen der Kernmagnetischen Resonanz in der Organischen Chemie." Springer, Berlin, 1965.
[5]F. A. Bovey, "NMR Data Tables for Organic Compounds," Vol. 1. Wiley (Interscience), New York, 1967.

PREFACE TO THE SECOND EDITION

Professor Gutowsky vividly portrayed in the Foreword the dramatic progress that has been made in NMR since the First Edition of this book appeared in 1969. The present book is much altered and expanded compared to its predecessor. For example an entire section is devoted to Fourier transform instruments, which were in their infancy in 1969 and then only rated a page of attention. This book contains an entire chapter on two-dimensional NMR and another chapter on high resolution NMR of solids, fields that had not yet been discovered in 1969. The Special Topics chapter deals with techniques such as water suppression, multiple quantum spectroscopy, and NMR imaging, areas that have had a tremendous impact on fields as diverse as protein structure determinations, 2D NMR, and medicine.

Although the present book is much updated, it is still intended for the same audience: the graduate student or research worker who needs a discussion at an introductory level, accompanied by a substantial collection of "real" examples, including chemical shift and J-coupling data. As a data storehouse, however, it does not pretend to compete with the massive collections now available, many in multivolume and computer-searchable forms. Rather, it draws attention to the types of molecules that are encountered in an academic or industrial chemistry laboratory, and makes appropriate references to these extensive compilations of NMR data when necessary.

As in the First Edition, the level is purposely kept fairly non-mathematical. In this book there are no Hamiltonians and no matrices. The goal is to convey the basic fundamentals of NMR spectroscopy in a descriptive manner so that the concepts can be readily grasped and applied by the widest number of NMR users.

This edition is accompanied by a number of appendices that have proved extremely useful to graduate students and researchers over the years. These include Appendix A: Table of Nuclear Properties; Appendix B: Calculated Proton Spectra; and Appendix C: Proton Spin-Spin

Coupling Constants. However, Appendix D: Carbon-Proton and Carbon-Carbon Spin-Spin Coupling Constants, is completely new, and is indispensable for conformational analysis and structure determinations.

Throughout the book the reader will find a mix of early, low-field NMR spectra, complete with "ringing," and new high-field, Fourier transform spectra. These old spectra convey a sense of history and are useful for pedagogical purposes, whereas the high field spectra proclaim the incredible power of modern NMR methods. In either case, care has been taken to reproduce the spectra in a clear and large format. The same is true for the chemical formulas, which form an indispensable part of this volume.

It is most important to emphasize my indebtedness to my co-authors, Dr. Lynn W. Jelinski, who contributed Chapter 2 on experimental methods and Section 9.2 (of Chapter 9) on NMR Imaging; and Dr. Peter A. Mirau, who was co-author of Chapter 6 on two-dimensional NMR and of Section 9.3 (of Chapter 9) on Solvent Suppression. Dr. Mirau also obtained important data for Chapters 4 and 5, which is acknowledged in the text. There are also many references, particularly in Chapter 4, to unpublished work of Mr. Frederic Schilling, who played a most important role in the genesis of this work. Vital also were extensive discussions of NMR phenomena with all of these persons. Thanks are also due to Mary E. Flannelly and Susan A. Tarczynski, who "text-processed" the manuscript, and to Jean O'Bryan and Menju Parikh, who assembled the book in its final camera-ready form.

Frank A. Bovey
February 18, 1988

CHAPTER 1

FUNDAMENTAL PRINCIPLES

1.1 INTRODUCTION

In addition to charge and mass, which all nuclei have, many isotopes possess spin, or angular momentum. Since a spinning charge generates a magnetic field, there is associated with this angular momentum a magnetic moment. The magnetic properties of nuclei were first postulated by Pauli[1] to explain certain hyperfine structural features of atomic spectra. The phenomenon of nuclear magnetic resonance has long been known in molecular beams and has been effectively exploited by Rabi and his co-workers[2,3] to give much useful information concerning nuclear properties. The field of organic chemistry has for the most part been developed without any reference to these properties. Only since the discovery of nuclear magnetic resonance in bulk matter, i.e., in such substances as water and paraffin wax, by Purcell et al.[4] at Harvard and Bloch et al.[5] at Stanford has this phenomenon become of interest to the chemist.

1.2 NUCLEAR SPIN

According to a basic principle of quantum mechanics, the maximum experimentally observable component of the angular momentum of a nucleus possessing a spin (or of any particle or system having angular momentum) is a half-integral or integral multiple of $h/2\pi$, where h is Planck's constant. This maximum component is I, which is called the spin quantum number or simply "the spin." Each nuclear ground state is characterized by just one value of I. If $I = 0$, the nucleus has no magnetic moment. If I is not zero, the nucleus will possess a magnetic moment μ, which is always taken as parallel to the angular momentum vector. The permitted values of the vector moment along any chosen axis are described by means of a set of magnetic quantum numbers m given by the series

$$m = I, (I - 1), (I - 2)..., -I \qquad (1.1)$$

Thus, if I is 1/2, the possible magnetic quantum numbers are $+1/2$ and $-1/2$. If I is 1, m may take on the values 1, 0, -1, and so on. In general, then, there are $2I + 1$ possible orientations or states of the nucleus. In the absence of a magnetic field, these states all have the same energy. In the presence of a uniform magnetic field B_0, they correspond to states of different potential energy. For nuclei for which I is $1/2$, the two possible values of m, $+1/2$ and $-1/2$, describe states in which the nuclear moment is aligned with and against the field B_0, respectively, the latter state being of higher energy. The detection of transitions of magnetic nuclei (often themselves referred to as "spins") between the states is made possible by the nuclear magnetic resonance phenomenon.

The magnitudes of nuclear magnetic moments are often specified in terms of the ratio of the magnetic moment and angular momentum, or *magnetogyric ratio* γ, defined as

$$\gamma = \frac{2\pi\mu}{Ih} \tag{1.2}$$

A spinning spherical particle with mass M and charge e uniformly spread over its surface can be shown to give rise to a magnetic moment $eh/4\pi Mc$, where c is the velocity of light. For a particle with the charge, mass, and spin of the proton, the moment should be 5.0505×10^{-27} joules/tesla $(\text{JT}^{-1})^*$ (5.0505×10^{-24} erg-gauss^{-1} in cgs units) on this model. Actually, this approximation is not a good one even for the proton, which is observed to have a magnetic moment about 2.79 times as great as the oversimplified model predicts. No simple model can predict or explain the actual magnetic moments of nuclei. However, the predicted moment for the proton serves as a useful unit for expressing nuclear moments and is known as the *nuclear magneton*; it is the analog of the Bohr magneton for electron spin. Observed nuclear moments can be specified in terms of the nuclear magneton by

$$\mu = g\,\frac{ehI}{4\pi M_p c} \tag{1.3}$$

where M_p is the proton mass and g is an empirical parameter called the nuclear g factor. In units of nuclear magnetons, then

$$\mu = gI \tag{1.4}$$

In Table 1.1 and in Appendix A, nuclear moments are expressed in these units. It will be noted that some nuclei have negative moments. This is of

---

* Magnetic field strengths are expressed in kgauss or tesla T; 10 kgauss = 1T.

TABLE 1.1

Nuclei of Major Interest to NMR Spectroscopists

Isotope	Abundance (%)	Z	Spin	μ^a	$\gamma \times 10^{-8\,b}$	Relative[c] sensitivity	ν_0 at 1T (MHz)
^{1}H	99.9844	1	1/2	2.7927	2.6752	1.000	42.577
^{2}H	0.0156	1	1	0.8574	0.4107	0.00964	6.536
^{10}B	18.83	5	3	1.8006	0.2875	0.0199	4.575
^{11}B	81.17	5	3/2	2.6880	0.8583	0.165	13.660
^{13}C	1.108	6	1/2	0.7022	0.6726	0.0159	10.705
^{14}N	99.635	7	1	0.4036	0.1933	0.00101	3.076
^{15}N	0.365	7	1/2	−0.2830	−0.2711	0.00104	4.315
^{19}F	100	9	1/2	2.6273	2.5167	0.834	40.055
^{29}Si	4.70	14	1/2	−0.5548	−0.5316	0.0785	8.460
^{31}P	100	15	1/2	1.1305	1.0829	0.0664	17.235

[a] Magnetic moment in units of the nuclear magneton, $eh/(4\mu M_p c)$.
[b] Magnetogyric ratio in SI units.
[c] For equal numbers of nuclei at constant field.

some practical significance to NMR spectroscopists, as we shall see in Chapter 4, Section 4.4.4.2. Its theoretical meaning will be evident a little later. It should also be noted that the neutron, with no net charge, has a substantial magnetic moment. This is a particularly striking illustration of the failure of simple models to predict μ. Clearly, the neutron must contain separated charges (at least a part of the time) even though its total charge is zero.

Although magnetic moments cannot be predicted exactly, there are useful empirical rules relating the mass number A and atomic number Z to the nuclear spin properties:

1. If both the mass number A and the atomic number Z are even, $I = 0$.

2. If A is odd and Z is odd or even, I will have half-integral values 1/2, 3/2, 5/2, etc.

3. If A is even and Z is odd, I will have integral values 1, 2, 3, etc.

Thus, some very common isotopes, such as ^{12}C, ^{16}O, and ^{32}S, have no magnetic moment and cannot be observed by NMR. This is really a blessing, however, for if these nuclei did have magnetic moments, the spectra of organic molecules would be much more complex than they are.

1.3 THE NUCLEAR MAGNETIC RESONANCE PHENOMENON

Nuclei with spins of 1/2, such as protons, are often likened to tiny bar magnets. But because of their small size and because they spin, their behavior differs in some ways from the ordinary behavior of macroscopic bar magnets. When placed in a magnetic field, the spinning nuclei do not all obediently flip over and align their magnetic moments in the field direction. Instead, like gyroscopes in a gravitational field, their spin axes undergo *precession* about the field direction, as shown in Fig. 1.1. The frequency of this so-called Larmor precession is designated as ω_0 in radians per second or ν_0 in Hertz (Hz), cycles per second ($\omega_0 = 2\pi\nu_0$). If we try to force the nuclear moments to become aligned by increasing B_0, they only precess faster. They *can* be made to flip over, however, by applying a second magnetic field, designated as B_1, at right angles to B_0 and causing this second field to rotate at the precession frequency ν_0. This second field is represented by the horizontal vector in Fig. 1.1, although in practice (as we shall see in Chapter 2) it is actually very much smaller in relation to B_0 than this figure suggests. It can be seen that if B_1 rotates at a frequency close to but not exactly at the precession frequency, it will cause at most only some wobbling or *nutation* of the magnetic moment μ. If, however, B_1 is made to rotate exactly at the precession frequency, it will cause large

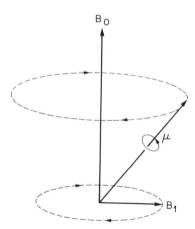

Fig. 1.1.   Nuclear moment in a magnetic field.

oscillations in the angle between μ and B_0. If we vary the rate of rotation of B_1 through this value, we will observe a resonance phenomenon as we pass through ν_0.

One might suppose that the Larmor precession of the nuclear moments could itself be detected by some means without the need to invoke a resonance phenomenon. This, however, is not possible because each nucleus precesses with a completely random phase with respect to that of its neighbors and there is therefore no macroscopic property of the system that changes at the Larmor frequency.

By a well-known relationship, the Larmor precession frequency is given by

$$\omega_0 = \gamma B_0 \tag{1.5}$$

or, from Eq. (1.2)

$$h\nu_0 = \frac{\mu B_0}{I} \tag{1.6}$$

The result of this classical treatment can also be obtained by a quantum mechanical description, which is in some ways a more convenient way of regarding the resonance phenomenon. It is best, however, not to try to adopt either approach to the exclusion of the other, since each provides valuable insights. From the quantum mechanical viewpoint, the quantity $h\nu_0$ is the energy separation ΔE between the magnetic energy levels (often termed Zeeman levels, after the investigator who first observed the corresponding splitting in atomic spectra) in a magnetic field B_0, as shown in Fig. 1.2. For a nucleus of spin 1/2, ΔE will be $2\mu B_0$, and as we have seen, only two energy levels are possible. For a nucleus of spin 1, there are three energy levels, as illustrated in Fig. 1.2. The quantum mechanical treatment gives us an additional result for such systems with $I \geqslant 1$, which the classical treatment does not: it tells us that only transitions between adjacent energy levels are allowed, i.e., that the magnetic quantum number can only change by ± 1. Thus, transitions between the $m = -1$ and $m = 0$ levels and between the $m = +1$ and $m = 0$ levels are possible, but transitions between the $m = -1$ and the $m = +1$ levels are not possible.

In Fig. 1.3 the separation of proton magnetic energy levels is shown as a function of magnetic field strength for a number of values of the latter employed in current spectrometers. The resonant proton rf field frequency is indicated (in megahertz, MHz, 10^6 Hz) and is a common way of

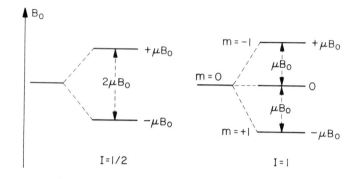

Fig. 1.2.   Magnetic energy levels for nuclei of spin 1/2 and 1.

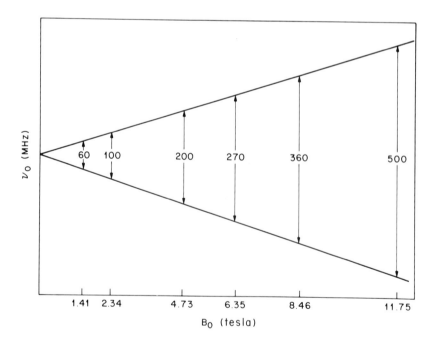

Fig. 1.3.   The splitting of magnetic energy levels of protons, expressed as resonance frequency ν_0 as a function of magnetic field strength, expressed in tesla (T).

designating the magnetic field strength.* Fields above *ca.* 2.5 T require superconducting solenoid magnets. For the ^{13}C nucleus, which has a magnetogyric ratio one-fourth that of the proton (Table 1.1), the resonant frequencies will be one-fourth of those shown in Fig. 1.3.

1.4 THE DETECTION OF THE RESONANCE PHENOMENON: CW AND PULSE FOURIER TRANSFORM METHODS

We have not yet said how the rotating magnetic field B_1 is to be provided or how the resonance phenomenon is to be detected. We give here a very simplified discussion, reserving a more detailed description for Chapter 2.

In the schematic diagram (Fig. 1.4), the sample tube is placed in the field of the magnet, which may be a permanent magnet, an electromagnet, or a superconducting solenoid. In Fig. 1.4 it is the last of these. The radio-frequency transmitter applies an rf field of appropriate frequency (Fig. 1.3) by means of the exciting coil indicated, which is of Helmholtz design. The magnetic field direction z is along the axis of the sample tube. The magnetic vector of the rf field oscillates along the y direction [i.e., in a direction perpendicular to the axis of the sample tube and to B_0 (see Fig. 1.5)]. This oscillating magnetic field provides what is required for flipping over the nuclear spins, for it may be thought of as being composed of two equal magnetic vectors rotating in phase with equal angular velocities, but in opposite directions. (An exact analog is the decomposition of plane polarized light into two equal and opposite circularly polarized vectors.) The precessing magnetic moments will pick out the appropriate rotating component of B_1 in accordance with Eq. (1.6) and the sign of μ (see Table 1.1 and Appendix A). The other rotating component is so far off resonance that it has no observable effect. (Note that if B_1 were circularly polarized, one would be able to determine the sign of μ.)

To display the resonance signal, the magnetic field may be increased slowly until Eq. (1.6) is satisfied and resonance occurs. At resonance, the nuclear magnetic dipoles in the lower energy state flip over and in so doing induce a voltage in the rf coil. This induced voltage is amplified and recorded. One may regard the rotating field B_1 as having given the precessing spins a degree of coherence, so that now there is a detectable *macroscopic* magnetic moment, precessing at a rate ν_0. The resonance phenomenon observed in this way is termed *nuclear induction* and is the

---

* The resonant frequency, ν_0, for a particular nucleus in a magnetic field of strength B_0 may be obtained from Eq. (1.5) and the relationship $\nu_0 = \omega_0/(2\pi)$. Some values of γ are listed in Table 1.1.

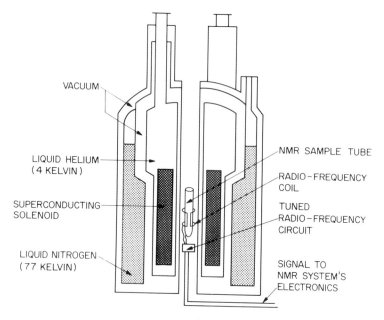

VACUUM

LIQUID HELIUM
(4 KELVIN)

SUPERCONDUCTING
SOLENOID

LIQUID NITROGEN
(77 KELVIN)

NMR SAMPLE TUBE

RADIO-FREQUENCY
COIL

TUNED
RADIO-FREQUENCY
CIRCUIT

SIGNAL TO
NMR SYSTEM'S
ELECTRONICS

Fig. 1.4.   Cross-section of a superconducting magnet for NMR spectroscopy. The magnet and sample tube are not drawn on the same scale. The diameter of the magnet assembly is approximately 70 cm, while that of the sample tube is approximately 1 cm.

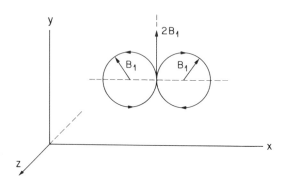

Fig. 1.5.   Decomposition of the magnetic vector B_1 into two counter-rotating vectors.

method originally used by the Bloch group. Alternatively, the nuclear flipping may be detected as an *absorption* of energy from the rf field.

The continuous wave (CW) method of detection just described is now employed only in old or very simple NMR spectrometers. Much greater sensitivity and flexibility is achieved in modern instruments by use of *pulse methods* with Fourier transformation of the resulting time-domain signal. In principle, the signal-to-noise ratio of any spectrum can be improved by extending the observing time, t. Background noise accumulates in proportion to $t^{1/2}$, whereas the strength of a coherent signal increases as t. Thus, one might obtain increased sensitivity by sweeping through the spectrum over a period of several hours instead of a few minutes, as is usual. This approach is of limited utility, however, as saturation can occur. It is somewhat more practical to sweep through the spectrum rapidly many times in succession, beginning each new sweep at the same point in the spectrum and then summing up the traces. This summation can be performed by a multichannel pulse height analyzer, a small computer that can store several thousand traces. The signal-to-noise ratio increases in proportion to the square root of the number of traces.

This method is still wasteful of time, however, as many spectra, particularly ^{13}C spectra, consist mostly of baseline. The time intervals spent between resonances are not profitable. It is much more efficient to excite all the nuclei of interest at the same time and to avoid sweeping entirely. This can be done by supplying the rf field in the form of *pulses*. These are commonly orders of magnitude stronger than the rf fields used in cw spectroscopy. Pulse spectroscopy utilizes the fact that a burst of resonant rf energy, although nominally monochromatic, with a frequency ν_0, actually contains a band of frequencies corresponding to the Fourier series of sine and cosine functions necessary to approximate a square wave. The band width may be estimated approximately as $\nu_0 \pm 1/t_p$, where t_p is the duration (or "width") of the pulse, generally a few microseconds. The use of pulsed rf fields in NMR spectroscopy was first suggested by Bloch.[5,6] It has long been standard for the measurement of proton relaxation in the solid state, particularly for polymers.[7,8] Its use with spectrum accumulation for observing high resolution spectra was first proposed by Ernst and Anderson.[9]

In CW spectroscopy, the rf field tips the macroscopic magnetic moment M_0 only very slightly. In pulsed spectroscopy, the angle of tipping, α, is much greater. This is a precession; α is proportional to the magnetogyric ratio of the nucleus and to the strength and duration of the pulse:

$$\alpha = \gamma B_1 t_p \qquad (1.7)$$

a variation of Eq. (1.5). It is common to employ a pulse that rotates the magnetic moment by $\pi/2$ radians or $90°$, i.e., just into the $x'y'$ plane, as shown in Fig. 1.6 (b) and (c). Such a pulse is termed a "$90°$ pulse." In Fig. 1.6, the magnetization M_0 is represented in a coordinate frame that rotates with B_1 and, at resonance, with M_0, and in which both are therefore static (Sec. 1.7). In Fig. 1.6(d) the magnetization vectors are shown fanning out, i.e., precessing at unequal rates, as a result of inhomogeneities in B_0 and in the sample. The resulting loss of phase coherence is manifested as a decay in the net magnetization and in the

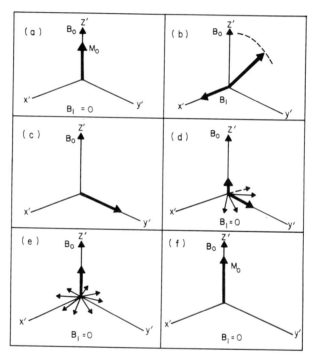

Fig. 1.6.   Rotating frame diagrams describing the pulsed NMR experiment. (The "primed" axes are used to indicate a rotating coordinate system.) (a) The net magnetization M_0 is aligned along the magnetic field direction, (b) and (c). An rf field B_1 is applied perpendicular to B_0. The duration of B_1 is sufficient to tip the net magnetization by $90°$. (d) and (e) The spins begin to relax in the x', y' plane by spin-spin (T_2) processes and in the z' direction by spin-lattice (T_1) processes; (f) the equilibrium magnetization is reestablished along B_0.

observed signal, occurring with a time constant T_2, as we shall see (Sections 1.6 and 1.7). At the same time, M_z, the longitudinal component of M_0, is growing back with a time constant T_1. The decay of the transverse magnetization (which provides the observed signal) is termed a *free induction decay* or FID, referring to the absence of the rf field. It may also be called the *time domain spectrum*. The FID following a single 90° pulse is shown in a schematic manner in Fig. 1.7(a). The FID is an interferogram with a simple exponential decay envelope; the beat pattern corresponds to the difference between the pulse carrier frequency ν_0 and the precession frequency of the nuclei ν_c. In Fig. 1.7(b) the FID is shown undergoing transformation to its Fourier inverse, appearing as a conventional or *frequency domain spectrum*.

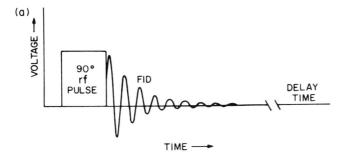

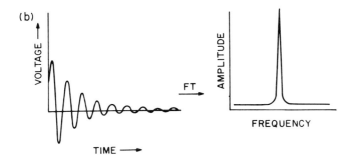

Fig. 1.7.   (a) Representation of a 90° rf pulse and the ensuing free induction decay or FID. (b) Fourier transformation of the time domain FID into the frequency domain signal.

1.5 SPIN-LATTICE RELAXATION

In the magnetic fields employed in NMR instruments, the separation of nuclear magnetic energy levels is very small. For example, for protons in a field of 11.75 tesla (the largest employed in commercial instruments) it is only about 0.5 joule. Even in the absence of the rf field there is usually a sufficiently rapid transfer of spins from the lower to the upper state and vice versa (for reasons to be explained shortly), so that an equilibrium population distribution is attained within a few seconds after B_0 is applied. If this distribution of spins is given by the Boltzmann factor $\exp(2\mu B_0/kT)$, this can be expressed with sufficient accuracy by

$$\frac{N_+}{N_-} = 1 + \frac{2\mu B_0}{kT} \tag{1.8}$$

where N_+ and N_- represent the spin populations of the lower and upper states, respectively, and T is the Boltzmann *spin temperature*. It can thus be seen that even for protons, the net degree of polarization of the nuclear moments in the magnetic field direction is small, for example, only about 100 ppm in an 11.75 tesla field; for all other nuclei (with the exception of tritium), it is even smaller. It is also clear that the nuclear magnetic energy cannot be expected to perturb the thermal energies of the molecules to an observable degree, except possibly at temperatures close to 0 K.

For nuclei with spins greater than 1/2, the $2I + 1$ equally spaced magnetic energy levels will be separated in energy by $\mu B_0/I$; the relative populations of adjacent levels will be given by expressions analogous to Eq. (1.8). To simplify the subsequent discussion, we shall confine our attention to nuclei with spins of 1/2, recognizing that what is said for such two-level systems will apply also to any pair of adjacent levels in systems of spin greater than 1/2.

We have seen that in the presence of the field B_1, there will be a net transfer of spins from the lower energy state to the upper energy state. In time, such a process would cause the populations of the levels to become equal, or nearly so, corresponding to a condition of *saturation* and to a very high Boltzmann spin temperature, unless there were some means by which the upper level spins could relax to the lower level. Such a process transfers energy from or *cools* the spin system. Similarly, when a system of spins is first thrust into a magnetic field, the populations of spins in the upper and lower energy states are equal, and the same relaxation must occur in order to establish an equilibrium spin population distribution and permit a resonance signal to be observed. It should be realized that since the energies involved are very small and the nuclei are, as we shall see, usually rather weakly coupled thermally to their surroundings, i.e., the thermal relaxation is a slow process, the spin temperature may readily be made very high with little or no effect on the actual temperature of the sample as ordinarily observed. We may say that the heat capacity of the nuclear spin system is very small.

The required relaxation can occur because each spin is not entirely isolated from the rest of the assembly of molecules, commonly referred to as the "lattice," a term which originated in dealing with solids but which has long been employed in discussing both liquids and solids. The spins and the lattice may be considered to be essentially separate coexisting systems with a very inefficient but nevertheless very important link by which thermal energy may be exchanged. This link is provided by *molecular motion*. Each nucleus sees a number of other nearby magnetic nuclei, both in the same molecule and in other molecules. These neighboring nuclei are in motion with respect to the observed nucleus, and this motion gives rise to fluctuating magnetic fields. The observed nuclear magnetic moment will be precessing about the direction of the applied field B_0, and will also be experiencing the fluctuating fields of its neighbors. Since the motions of each molecule and of its neighbors are random or nearly random, there will be a broad range of frequencies describing them. To the degree that the fluctuating local fields have components in the direction of B_1 and at the precession frequency ω_0 (expressed in radians-s^{-1} or $2\pi\nu_0$ to conform with the discussion in Chapter 5), they will induce transitions between energy levels because they provide fields equivalent to B_1. In solids or very viscous liquids, the molecular motions are relatively slow and so the component at ω_0 will be weak. The frequency spectrum will resemble curve (a) in Fig. 1.8. At the other extreme, in liquids of very low viscosity, the motional frequency spectrum may be very broad, and so no one component, in particular that at ω_0, can be very intense [curve (c) in Fig. 1.8]. We are then led to expect that at some intermediate condition, probably that of a moderately viscous liquid [curve (b)] the component at ω_0 will be at a maximum, and thermal relaxation of the spin system can occur with optimum efficiency.

The probability per unit time of a downward transition of a spin from the higher to the lower magnetic energy level, $W_{\downarrow}$, exceeds that of the reverse transition $W_{\uparrow}$ by the same factor as the *equilibrium* lower state population exceeds the upper state population [Eq. (1.8)]. We may then write

$$\frac{W_{\downarrow}}{W_{\uparrow}} = 1 + \frac{2\mu B_0}{kT} \tag{1.9}$$

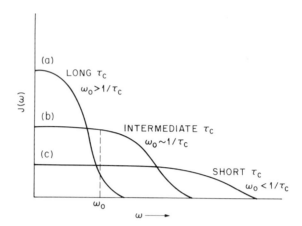

Fig. 1.8.   Motional   frequency   spectrum   at   (a) high   viscosity,
(b) moderate viscosity, and (c) low viscosity. The vertical
ordinate $J(\omega)$ represents the relative intensity of the motional
frequency ω. The observing frequency is ω_0.

This must be the case, or equilibrium could not be maintained. (For
isolated spins, i.e., those not in contact with the thermal reservoir provided
by the lattice, the transition probabilities are exactly the same in both
directions.) Let the spin population difference at any time t be given by n;
let the equilibrium population difference (i.e., in the presence of B_0, but in
the absence of the rf field) be given by n_{eq}. We shall also define more
completely the longitudinal relaxation time T_1 (Section 1.4), in units of
seconds, as the reciprocal of the sum of the probabilities (per second) of
upward and downward transitions, i.e., the total rate of spin transfer in
both directions:

$$1/T_1 = W_\downarrow + W_\uparrow \tag{1.10}$$

The approach to equilibrium will be described by

$$dn/dt = 2N - W_\downarrow - 2N_+ W_\uparrow \tag{1.11}$$

N_- and N_+ being the populations of the upper and lower levels,
respectively, as in Eq. (1.8). This represents the *net* transfer of spins from
the upper state. (The factors of 2 arise because each transfer of one spin
changes the population difference by 2.) If we designate the total spin
population $(N_+ + N_-)$ by N, we can express this rate in terms of N and n
as

$$dn/dt = (N - n)W_\downarrow - (N + n)W_\uparrow \qquad (1.12a)$$

$$= N(W_\downarrow - W_\uparrow) - n(W_\downarrow + W_\uparrow) \qquad (1.12b)$$

At equilibrium, we find from Eq. (1.12b) that

$$\frac{n_{eq}}{N} = \frac{W_\downarrow - W_\uparrow}{W_\downarrow + W_\uparrow} \qquad (1.13)$$

and so, using Eq. (1.10), we may rewrite Eq. (1.12b) as

$$dn/dt = \frac{1}{T_1}(n_{eq} - n) \qquad (1.14)$$

which gives upon integration

$$n = n_{eq}(1 - e^{-t/T_1}) \qquad (1.15)$$

We see then that, as might perhaps have been intuitively expected, the rate of approach to thermal equilibrium depends upon how far away from equilibrium we are. Here $1/T_1$ is the first-order rate constant (in the language of chemical kinetics) for this process. T_1 is called the *spin-lattice relaxation time*; it is the time required for the difference between the actual spin population and its equilibrium value to be reduced by the factor e.

Except in a static collection of nuclei, i.e., a solid near 0 K, spin-lattice relaxation through molecular motion always occurs, although it may be very slow. There are two other possible contributing causes to spin-lattice relaxation which may occur under some conditions and are of importance to the chemist:

1. Spin-lattice relaxation by interaction with unpaired electrons in paramagnetic substances. Fundamentally the mechanism is exactly the same as the one just discussed, but is much more effective because the magnetic moment of an unpaired electron is close to the magnitude of the Bohr magneton, $eh/4\pi M_e c$, and of the order of 10^3 times greater than the nuclear magneton owing to the small mass of the electron, M_e.

2. Spin-lattice relaxation by interaction of the *electric quadrupole moments* of nuclei of spin 1 or greater with electric fields within the tumbling molecule. This is a particularly potent cause of thermal relaxation and will be dealt with in Section 1.9 and Chapter 5, Section 5.2.4. (Other, less common causes of spin-lattice relaxation are also discussed in Chapter 5.)

Spin-lattice relaxation is often termed *longitudinal relaxation* because it involves changes of energy and therefore involves the component of the nuclear moment along the direction of the applied magnetic field. It also, of course, shortens the lifetimes of the spins in both the upper and lower energy states. This leads to an uncertainty in the energies of these states and to a broadening of the resonance lines. This can be estimated from the Heisenberg relation, $\delta E \cdot \delta t \approx h/2\pi$, from which the uncertainty in the frequency of absorption is given by

$$\delta\nu \simeq 1/(2\pi \cdot \delta t) \tag{1.16}$$

Under the present-day operating conditions, line widths are about 0.2 Hz (see Chapter 2), which, if arising from uncertainty broadening, would correspond to a spin lifetime of about 0.3 s. Since T_1 for protons and ^{13}C is usually of the order of 2-20 s in mobile liquids, spin-lattice relaxation ordinarily does not contribute observably to line broadening. As the temperature is lowered and viscosity increases, the component of the local magnetic noise spectrum at the Larmor frequency will increase, pass through a maximum, and decrease again. T_1 will correspondingly decrease, pass through a minimum, and increase again. It is useful to define a *correlation time*, τ_c, which we shall understand to be the average time between molecular collisions. The correlation time defines the length of time that the molecule can be considered to be in a particular state of motion. This time is sometimes referred to as the "phase memory" of the system. Thus, molecules with short correlation times are in rapid molecular motion. For rigid, approximately spherical molecules, τ_c is given to a good degree of approximation by

$$\tau_c = \tau_D/3 = 4\pi\eta a^3/3kT \tag{1.17}$$

τ_D being the correlation time used in the Debye theory of dielectric dispersion, η the viscosity of the liquid, and a the effective radius of the molecule. For most nonassociated molecules, T_1 is governed mainly by interactions of nuclei within the same molecule, rather than by motions of neighboring molecules. Under these conditions, it is found that the rate of spin-lattice relaxation for a spin-1/2 nucleus interacting with another nucleus of the same species is given by[10]

$$\frac{1}{T_1} = \frac{3}{10}\frac{\gamma^4\hbar^2}{r^6}\left[\frac{\tau_c}{1+\omega_0^2\tau_c^2} + \frac{4\tau_c}{1+4\omega_0^2\tau_c^2}\right] \tag{1.18}$$

where γ is the magnetogyric ratio of the interacting nuclei, $\hbar = h/2\pi$, r^6 is the internuclear distance, and ω_0 the resonant frequency in radians-s^{-1}. For mobile liquids near or above room temperature (or for moderate-sized molecules dissolved in mobile solvents), the correlation time is very short compared to $1/\omega_0$ and consequently *Eq.* (1.18) may be reduced to

$$\frac{1}{T_1} = \frac{3}{2} \frac{\gamma^4 \hbar^2 \tau_c}{r^6} \tag{1.19}$$

These circumstances are referred to as the "extreme narrowing condition." Combining *Eqs.* (1.17) and (1.19) and replacing $4/3\pi a^3$ by the molecular volume V we have

$$\frac{1}{T_1} = \frac{3}{2} \frac{\gamma^4 \hbar^2}{r^6} \cdot \frac{V\eta}{kT} \tag{1.20}$$

For proton relaxation, one will in general have to take account of more than one interaction with the observed nucleus and employ a summation over the inverse sixth powers of all the protons concerned. The spin-lattice relaxation of carbon-13 nuclei in organic molecules is more commonly studied than that of protons. It is somewhat easier to interpret because for the dipole-dipole mechanism just described the relaxation of the ^{13}C nucleus by directly bonded nuclei (if any) dominates, a consequence of the r^6 dependence. We find (see Chapter 5, Section 5.2.13)

$$\frac{1}{T_1} = \frac{N\gamma_H^2\gamma_c^2\hbar^2}{r_{C-H}^6} \cdot \tau_c \tag{1.21}$$

Here N is the number of directly bonded protons and the other terms are self-explanatory. In this case, other relaxation mechanisms, notably *spin rotation* and *chemical shift anisotropy* may make important contributions. These will be discussed in Chapter 5. Figure 1.9 shows the predicted dependence of the dipole-dipole contribution to T_1 as a function of τ_c for the relaxation of water protons measured at 200 MHz and 500 MHz. It will be observed that in the regime of short correlation times at the left of the plot, T_1 is independent of the observing magnetic field strength, whereas when molecular motion is slower spin-lattice relaxation becomes markedly slower as B_0 increases and the minimum moves to shorter τ_c values. Such plots [and Eqs. (1.17-(1.19)] tell us at least three things that have practical implications:

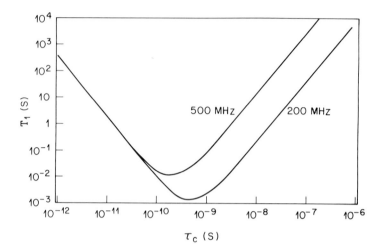

Fig. 1.9.   Spin-lattice relaxation time T_1 (s) corresponding to mutual relaxation of water protons (intramolecular only) as a function of correlation time τ_c, as observed at 200 MHz and 500 MHz.

1.   For a given liquid, T_1 will vary inversely with η/T over the usual temperature ranges employed in NMR spectroscopy, i.e., those over which the sample remains reasonably mobile. At any given temperature, narrower lines may be expected in solutions of low viscosity.

2.   Other things being equal, T_1 will vary inversely as a^3 or V. Thus for reasonably dilute solutions of a series of solutes in a given solvent of viscosity η, spin-lattice relaxation will increase in proportion to the molar volumes of the solutes and may for large rigid molecules become an important contributor to line broadening even though η remains small.

3.   Substances in which the magnetic nuclei are relatively far apart or have small moments will exhibit longer relaxation times, since T_1 is a sensitive function of both μ and r.

1.6 DIPOLAR BROADENING AND SPIN-SPIN RELAXATION

A cause of broadening of NMR spectral lines usually more important than spin-lattice relaxation is the so-called "dipolar" broadening. To understand this effect, let us first consider not the liquid samples that the chemist normally examines in the NMR spectrometer, but rather a proton-containing solid. If the protons are sufficiently removed from each other that they do not feel the effects of each other's magnetic fields, the resonant magnetic field at the nucleus will be essentially equal to B_0 (actually very slightly smaller because of the shielding effects of local electrons, to be discussed in Section 1.10). Therefore, if by careful design of the magnet (see Chapter 2) the field B_0 can be made very homogeneous over the volume occupied by the sample, the width of the absorption peak may be less than 10^{-4} g, i.e., of the order of a few tenths of a hertz, or 1 part in 10^9. In most such substances, the protons are actually near enough to each other so that each is appreciably influenced by the magnetic fields of its neighbors. Let us first imagine that we are dealing with a static system of isolated pairs of protons, i.e., each member of a pair experiences the field of the other member but not those of the other pairs. The field felt by each proton will be made up of the applied field B_0, plus this small additional field, B_{loc}. The sign and magnitude of this increment will depend upon the distance apart of the nuclei, r, and upon θ, the angle between the line joining the nuclei and the vector representing the direction of B_0. We suppose that all the pairs have the same values of these parameters. The equation expressing the functional dependence of the separation of the protons' magnetic energy levels upon r and θ is given by

$$\Delta E = 2\mu[B_0 \pm B_{loc}] = 2\mu[B_0 \pm \tfrac{1}{2}\mu r^{-3}(3\cos^2\theta - 1)] \qquad (1.22)$$

The $\pm$ sign corresponds to the fact that the local field may add to or subtract from B_0 depending upon whether the neighboring dipole is aligned with or against the direction of B_0. We recall that the net polarization of nuclear moments along B_0 is ordinarily only a few parts per million, and so in a large collection of such isolated pairs the probabilities of a neighboring dipole being aligned with or against B_0 are almost equal. Equation (1.22) then expresses the fact that the spectrum of this proton-pair system will consist of two equal lines [Fig. 1.10(a)] whose separation, at a fixed value of θ, will vary inversely as r^3. Only when the orientation of the pairs is such that $\cos^2\theta = 1/3$ ($\theta = 54.7°$) will the lines coincide to produce a single line. (We will see this angle reintroduced in Chapter 8 in conjunction with magic angle spinning.)

In many solids, for example, certain hydrated crystalline salts, we actually have pairs of protons in definite orientations and may expect to find a twofold NMR resonance. This is indeed observed.[11] The lines, however, are not narrow and isolated as in Fig. 1.10(a), but are broadened and partially overlapping as in Fig. 1.10(b). The separation of the maxima is found to depend on the orientation of the crystal in the magnetic field. In such substances, the magnetic interactions of the members of each pair dominate, but the interactions between pairs are not negligible. These many smaller interactions, varying with both r and θ, give rise in effect to a multiplicity of lines whose envelope is seen as the continuous curve of Fig. 1.10(b). [We should perhaps state that the simple picture suggested

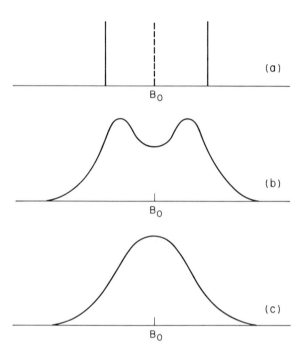

Fig. 1.10.   Schematic NMR spectra of static arrays of protons: (a) an
               isolated pair of protons; (b) a semi-isolated pair of protons;
               (c) a random or nearly random array of protons. In actual
               proton-containing solids, the widths of spectra (b) and (c)
               may be of the order of 50 kHz.

here, while qualitatively useful, does not correctly give the numbers and
intensities of the lines composing this envelope in a multi-spin system. The
solution of the problem actually involves quantum mechanical
considerations; in fact, the factor 3/2 in Eq. (1.22) is a result of this more
correct treatment.[12,13] In most organic solids, pairing of protons does not
occur to this degree, and instead we find a nearly Gaussian distribution of
magnetic interactions, such as one would expect if the protons were
randomly distributed throughout the substance. For such a complex array
of nuclei, the NMR spectrum is a single broad peak, as in Fig. 1.10(c). In
most rigid-lattice solids, where the protons are of the order of 1 Å apart,
the distribution of local field strengths about each proton is such that the
half-height width of the peak is of the order of 10 gauss (10^{-3} tesla), or *ca.*
0.5×10^5 Hz.

We have so far assumed that the nuclei are fixed in position. If *molecular motion* is allowed to take place, i.e., by raising the temperature of the solid, by melting it, or by dissolving it in a mobile solvent, the variables in Eq. (1.22) become functions of time. If we assume that r is constant and only θ varies with time, as would be true for pairs of protons within a rigid molecule, then the time-averaged local field will be given by

$$B_{\text{loc}} = \mu r^{-3} \cdot T_2^{-1} \int_0^{T_2} (3 \cos^2 \theta - 1) \, dt \qquad (1.23)$$

where T_2 is the time that the nucleus resides in a given spin state. If θ may vary rapidly over all values, this time average can be replaced by a space average[8]

$$B_{\text{loc}} = \mu r^{-3} \int_0^{\pi} (3 \cos^2 \theta - 1) \sin \theta \, d\theta = 0 \qquad (1.24)$$

Thus, if the correlation time τ_c [Eq. (1.17)] is so short that this space averaging is valid, then the net effect of the neighboring magnetic nuclei is effectively erased, and the line will be drastically narrowed (by a factor of 10^4 to 10^5) compared to its rigid-lattice value. If we begin with a solid at a temperature so low that it is virtually a rigid lattice, and then permit molecular motion (chiefly rotation) to occur with increasing frequency by raising the temperature, a narrowing of the resonance line will be observed when

$$1/\tau_c \geqslant 2\pi \, \delta\nu \qquad (1.25)$$

where $\delta\nu$ is the static line width (hertz). In liquids of ordinary viscosity, molecular motion is so rapid that Eq. (1.24) holds; the local variations in magnetic field strength have become so short lived that *motional averaging* is complete. But for large molecules and for viscous liquids this line narrowing effect may not be complete and broadening may still be observable. For high polymers, and even for molecules of 400-500 molecular weight (particularly if they are rigid), this broadening can remain conspicuous.

There is another aspect of the interaction of neighboring magnetic dipoles that is closely related to the effects we have just considered and must also be considered in relation to line broadening. We must recall that our nuclear spins are not merely small static magnetic dipoles, but that even in a rigid solid they are precessing about the field direction. We may resolve a precessing nuclear moment (Fig. 1.11) into a *static component a* along the direction of B_0, which we have considered so far, and into a *rotating component b*, whose effect we must now also consider. This component constitutes the right type of magnetic field, as we have seen (Section 1.4), to induce a transition in a neighboring nucleus if this is precessing at the same frequency. If this spin exchange or flip-flop occurs, the nuclei will exchange magnetic energy states with no overall change in the energy of the system but with a shortening of the lifetime of each. The magnitude of local field variations may be taken as μ/r^3 [an approximation of B_{loc} from Eq. (1.22)], and consequently the relative phases of the nuclei will change in a time of the order of hr^3/μ^2, the "phase memory time." From Eq. (1.16), we expect an uncertainty broadening of about μ^2/hr^3, i.e., of the same form and order of magnitude as that produced by the interaction of the static components of the nuclear moments. It has become customary to include both effects in the quantity T_2, which we defined as the spin lifetime. T_2 is an inverse measure of the broadening of the spectral lines:

$$T_2 = \frac{1}{\pi \, \delta\nu} \tag{1.26}$$

It is called the *spin-spin relaxation time* or *transverse relaxation time*. A detailed theory of its dependence on molecular correlation time has been

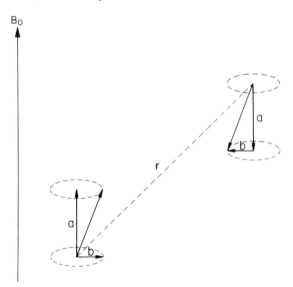

Fig. 1.11.   Static and rotating components of a precessing nuclear moment.

given by Bloembergen et al.[10] At short correlation times, as in mobile liquids (where "phase memory" is short), it becomes equal to T_1, but after T_1 passes through its minimum, T_2 continues to decrease as molecular motion becomes slower and finally levels out as the system begins to approach a rigid lattice.

Spin-spin exchange and dipolar broadening should not be considered as merely two alternative ways of looking at the same phenomenon, closely interrelated though they are. For example, in a lattice composed of magnetic nuclei α containing a dilute distribution of a different magnetic nuclear species β, spin exchange between nuclei α and β cannot occur since they precess at greatly different frequencies, but nevertheless, dipolar broadening will be present. An example is provided by solid-state ^{13}C spectra, which will be fully discussed in Chapter 8.

Spin-spin relaxation is associated with a decay of the macroscopic nuclear moment in the x, y plane; spin-lattice relaxation is associated with a decay of the macroscopic moment along the z direction. We may suspect that these relaxation rates, although they can be quite different in magnitude, are closely associated, and that both T_1 and T_2 must be considered in describing resonance signal shapes and intensities. This will be discussed in Section 1.7.

1.7 THE BLOCH EQUATIONS

In treating the experimental observation of nuclear magnetic resonance, it is convenient to adopt the approach of Bloch[14] and to consider the assembly of nuclei in macroscopic terms. We define a total moment M, which is the resultant sum, per unit volume, of all the individual nuclear moments in an assembly of identical nuclei, with magnetogyric ratio γ and $I = 1/2$. We consider that M is not collinear with any of the axes x, y, and z, as in Fig. 1.12. The static field B_0 is in the z direction, and M, like the individual moments, will precess about z with an angular frequency ω_0. In the absence of relaxation effects and the rotating field B_1, the projection of M on the z axis, M_z, would remain constant:

$$dM_z/dt = 0 \qquad (1.27)$$

The magnitudes of the x and y projections, M_x and M_y, will, however, vary with time as M precesses and as can be seen from Fig. 1.12, will be 180° out of phase, since when the projection of M along the x axis is at a maximum it will be zero along the y axis, and vice versa. This time dependence can be expressed by

$$dM_x/dt = \gamma M_y \cdot B_0 = \omega_0 M_y \qquad (1.28)$$

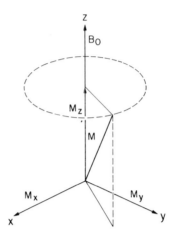

Fig. 1.12.   Precessing nuclear moment M in a magnetic field B_0 directed along the z axis of a fixed coordinate frame.

$$dM_y/dt = -\gamma M_x \cdot B_0 = -\omega_0 M_x \qquad (1.29)$$

(Note that the inclusion of γ in macroscopic expressions causes no difficulty if dimensions are consistent.) In addition to the fixed field B_0, we must now consider the rotating field B_1, which we recall from earlier discussion will be provided by one of the two counter-rotating magnetic vectors of a linearly polarized rf field. This field B_1 rotates in the xy plane, i.e., perpendicularly to B_0, with frequency ω (equal to ω_0 only at exact resonance, i.e., at the center of the absorption peak). Consideration of the effect of B_1 upon the magnitudes of M_x, M_y, and M_z, in accordance with the basic laws describing the tipping of magnetic vectors in magnetic fields, leads to these modifications of Eqs. (1.27)-(1.29):

$$dM_x/dt = \gamma[M_y B_0 - M_z (B_1)_y] \qquad (1.30)$$

$$dM_y/dt = -\gamma[M_x B_0 + M_z (B_1)_x] \qquad (1.31)$$

$$dM_z/dt = \gamma[M_x (B_1)_y - M_y (B_1)_x] \qquad (1.32)$$

where $(B_1)_x$ and $(B_1)_y$ are the components of B_1 along the x and y axes and are given by

$$(B_1)_x = B_1 \cos \omega t \qquad (1.33)$$

$$(B_1)_y = - B_1 \sin \omega t \qquad (1.34)$$

We have so far omitted any consideration of the relaxation of the components of M in the x, y, and z directions. By putting Eq. (1.14) in phenomenological terms, we may express the relaxation of the z component toward its equilibrium value M_0 as

$$dM_z/dt = - \frac{M_z - M_0}{T_1} \qquad (1.35)$$

The transverse relaxations may be expressed similarly, but with T_2 as the time constant:

$$dM_x/dt = - \frac{M_x}{T_2} \qquad (1.36)$$

$$dM_y/dt = - \frac{M_y}{T_2} \qquad (1.37)$$

The transverse relaxation processes differ in that they go to zero rather to equilibrium values. Upon adding these terms to Eqs. (1.30)-(1.34) and using Eqs. (1.33) and (1.34), we obtain the complete Bloch equations:

$$dM_x/dt = \gamma(M_y B_0 - M_z B_1 \sin \omega t) - \frac{M_x}{T_2} \qquad (1.38)$$

$$dM_y/dt = - \gamma(M_x B_0 - M_z B_1 \cos \omega t) - \frac{M_y}{T_2} \qquad (1.39)$$

$$dM_z/dt = - \gamma(M_x B_1 \sin \omega t + M_y B_1 \cos \omega t) - \frac{M_z - M_0}{T_1} \qquad (1.40)$$

A clearer insight into the significance of these equations in experimental terms is gained if the frame of reference is changed from fixed axes x, y, and z to the rotating frame (Fig. 1.6; see also Sec. 1.4). In the rotating frame, both B_0 and B_1 are fixed. We may then resolve the projection of M on the x, y plane into components u and v, which are along and perpendicular to B_1, respectively, and will accordingly be in-phase and out-of-phase, respectively, with B_1. (See Fig. 1.13). To transform to this new frame, we note that

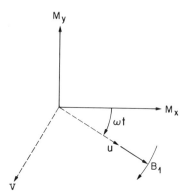

Fig. 1.13.   The components of the transverse macroscopic moment in a
fixed plane (M_x, M_y) and in a plane rotating at the Larmor
frequency about the static field B_0, taken as normal to the xy
plane (and upward); the rotating frame components u and v
are along and perpendicular to B_1, respectively.

$$M_x = u \cos \omega t - v \sin \omega t \tag{1.41}$$

$$M_y = -u \sin \omega t - v \cos \omega t \tag{1.42}$$

and also that $\gamma B_0 = \omega_0$. We then may replace Eqs. (1.38)-(1.40) with

$$du/dt + u/T_2 + (\omega_0 - \omega)v = 0 \tag{1.43}$$

$$dv/dt + v/T_2 - (\omega_0 - \omega)u + \gamma B_1 M_z = 0 \tag{1.44}$$

$$dM_z/dt\,(M_z - M_0)/T_1 - \gamma B_1 v = 0 \tag{1.45}$$

Here, the terms $(\omega_0 - \omega)$ are a measure of how far we are from exact
resonance. We see from Eq. (1.45) that changes of M_z, i.e., changes in
the energy of the spin system, are associated only with v, the out-of-phase
component of the macroscopic moment, and not with u. We may then
anticipate that absorption signals will be associated with the measurement
of v. The component u will be associated with "dispersion-mode" signals.
These methods of observation will be discussed in greater detail and in
specific instrumental terms in Chapter 2.

Under certain conditions of experimental observation, we are dealing with a steady state in which u, v, and M_z are constant in the rotating frame. We pass through the resonance peak varying ω (or in practice B_0) at a rate so slow that u, v, and M_z always have time to reach these steady values. Under such "slow-passage" conditions we obtain from Eqs. (1.43)-(1.45)

$$u = M_0 \frac{\gamma B_1 T_2^2 (\omega_0 - \omega)}{1 + T_2^2 (\omega_0 - \omega)^2 + \gamma^2 B_1^2 T_1 T_2} \qquad (1.46)$$

$$v = - M_0 \frac{\gamma B_1 T_2}{1 + T_2^2 (\omega_0 - \omega)^2 + \gamma^2 B_1^2 T_1 T_2} \qquad (1.47)$$

$$M_z = M_0 \frac{1 + T_2^2 (\omega_0 - \omega)^2}{1 + T_2^2 (\omega_0 - \omega)^2 + \gamma^2 B_1^2 T_1 T_2} \qquad (1.48)$$

We see from Eq. (1.47) that under conditions when $\gamma^2 B_1^2 T_1 T_2 \ll 1$, i.e., when B_1 is a few milligauss and T_1 and T_2 are no greater than a few seconds, the absorption or "v-mode" signal should be proportional to $\gamma B_1 T_2 / [1 + T_2^2 (\omega_0 - \omega)^2]$. This describes what is known as a Lorentzian line shape, as shown in Fig. 1.14(a), and is the type of signal normally used with high-resolution spectra. At the center, when the resonance condition is exactly fulfilled, $\omega_0 - \omega$ becomes zero and the signal height is

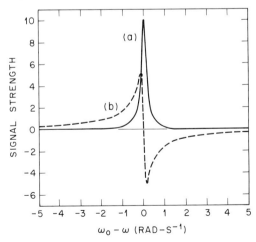

Fig. 1.14. v-Mode [absorption (a)] and u-mode [dispersion (b)] resonance signals.

proportional to $\gamma B_1 T_2$. It follows that the width must be inversely proportional to T_2, as we have already seen [Eq. (1.26)]. If we define the peak width $\delta\omega$ or δv in the customary way as its width at half the maximum height, it is easily shown that

$$\delta\omega = 2/T_2 \qquad (1.26a)$$

or

$$\delta v = 1/\pi T_2 \qquad (1.26b)$$

For some purposes, it is preferable to employ the dispersion or "u-mode" signal, which under the same conditions will be proportional to $\gamma B_1 T_2^2 (\omega_0 - \omega)/[1 + T_2^2 (\omega_0 - \omega)^2]$. This will give the line shape shown in Fig. 1.14(b). The maximum and minimum are separated by $1/\pi T_2$ for Lorentzian signals.

1.8 SATURATION

Equations (1.46)-(1.48) allow us to express the phenomenon of saturation, discussed qualitatively in Section 1.5, in more exact terms. Partial saturation will begin to become observable when the term $\gamma^2 B_1^2 T_1 T_2$ is no longer much less than unity, as assumed in the preceding discussion. The field B_1 is now beginning to cause a net transfer of lower state spins to the upper state at a rate that cannot be neglected in comparison to that of the spin-lattice relaxation. Equations (1.14) and (1.15) are no longer adequate to describe the relaxation. The new equilibrium population difference n'_{eq} will be given by

$$n'_{eq} = \frac{n_{eq}}{1 + \gamma^2 B_1^2 T_1 T_2} \qquad (1.49)$$

The ratio n'_{eq}/n_{eq} will be less than unity and has been termed the saturation factor Z_0:

$$Z_0 = \frac{1}{(1 + \gamma^2 B_1^2 T_1 T_2)} \qquad (1.50)$$

For ordinary mobile, proton-containing liquids under good experimental conditions (principally good field homogeneity; see Chapter 2), $T_1 T_2$ will be of the order of 10; γ^2 being approximately 10^8 for protons, we have

$$Z_0 = \frac{1}{(1 + 10^9 B_1^2)}$$

We thus find that observable saturation effects can be expected when B_1 is no greater than $10^{-8} - 10^{-9}$ tesla.

In pulsed NMR, the condition corresponding to Fig. 1.6c, following the 90° pulse, might be thought to correspond to saturation since the populations of the energy levels are now equal. However, this is not the case because the observable magnetization is here a maximum rather than zero. A condition of partial saturation may be induced by a train of pulses separated by intervals small compared to T_1 (see Chapter 5, Section 5.2.7.3).

1.9 NUCLEAR ELECTRIC QUADRUPOLE RELAXATION

Nuclei with spins of 1/2 possess a spherical distribution of the nuclear charge and are therefore not affected by the electric environment within the molecule. Nuclei with spins of 1 or greater, however, are found to have *electric quadrupole moments*, and their charge distribution can be regarded as spheroidal in form; the nucleus is to be regarded as spinning about the principal axis of the spheroid. The quadrupole moment may be *positive*, corresponding to a prolate spheroid [Fig. 1.15(a)] or *negative*, corresponding to an oblate spheroid [Fig. 1.15(b)]. The energies of spheroidal charges will depend upon their orientation in the molecular electric field gradient. In certain classes of molecules where an approximately spherical or tetrahedral charge distribution prevails, as for

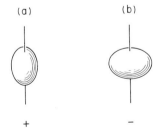

Fig. 1.15.   Nuclear charge distribution corresponding to (a) positive electric quadrupole moment (prolate spheroid); (b) negative electric quadrupole moment (oblate spheroid).

example in the ammonium ion NH_4^+, little or no electric field gradient will be present, and the nuclear quadrupole moment will not be disturbed by the tumbling of the molecule. But in most molecules, substantial electric field gradients are present and can interact with the nuclear quadrupoles so that the spin states of such nuclei may be rapidly changed by the tumbling of the molecular framework. This furnishes an additional pathway for energy exchange between the spins and the lattice, i.e., an important contribution to spin-lattice relaxation, and can broaden the resonance peaks very markedly. The resonance of nuclei such as ^2H or ^{14}N (Q positive; see Appendix A) or ^{17}O, ^{35}Cl, and ^{36}Cl (Q negative) may be so broad as to be difficult or impossible to observe. Nuclear quadrupole relaxation can also affect the resonance even of nuclei of spin 1/2 when these are in sufficiently close proximity to nuclei of spin 1 or greater. We shall discuss this further in Chapter 5 (Section 5.2.4).

1.10 MAGNETIC SHIELDING AND CHEMICAL SHIFT

From what we have said so far, it might be supposed that at any particular radio frequency v_0 all nuclei of a given species, for example, all protons, would resonate at the same value of B_0. Indeed this is true within the limits of the significant figures with which nuclear properties are expressed in Table 1.1 and in Appendix A. If this were *strictly* true, NMR would be of little interest to chemists. But about four years after the first demonstration of NMR in condensed phases, it was found that the characteristic resonant frequency of a nucleus depends to a very small, but measurable, extent upon its chemical environment. It was found that the protons of water do not absorb at quite the same frequency as those of mineral oil,[15,16] the difference being only a few parts per million. For heavier nuclei, much larger effects are noted — up to 2% for certain metals.[17] In 1950, Gutowsky and his students (see Jonas and Gutowsky[18]) observed different resonance frequencies for ^{19}F nuclei in an aromatic ring and on a CF_3 group in the same molecule. Similarly, Arnold, Dharmatti, and Packard[19] reported separate spectral lines for chemically different nuclei in alcohols. These discoveries opened a new era for the organic chemist.

The total range of variation of B_0 for protons is about 13 ppm or *ca.* 2600 Hz in a 4.73 T field. The proton spectrum of ethyl orthoformate, $HC(OCH_2CH_3)_3$, shown in Fig. 1.16(a), exhibits successive peaks (or groups of peaks; the cause of the splittings within the groups will be discussed shortly) for CH, CH_2, and CH_3 protons as we in effect increase the resonant magnetic field from left to right. The corresponding ^{13}C spectrum is shown in Fig. 1.16(b) (the much larger spacing of resonances

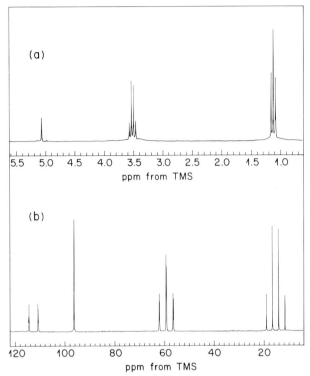

Fig. 1.16. (a) The 200 MHz proton spectrum of ethyl orthoformate, $HC(OCH_2CH_3)_3$, as a 15% solution in carbon tetrachloride; (b) 50.3 MHz carbon spectrum of the same solution. Tetramethylsilane reference appears at 0 ppm in both spectra. (F. C. Schilling, private communication.)

is noteworthy and will be more fully discussed in Chapter 3.) In both spectra the intensities of the peaks are proportional to the numbers of nuclei of each type.

The origin of this variation in resonant field strength is the cloud of electrons about each of the nuclei. When a molecule is placed in a magnetic field B_0, orbital currents are induced in the electron clouds, and these give rise to small, local magnetic fields, which, by a well-known physical principle, are always proportional to B_0 but opposite in direction.[*]

---

[*] It is assumed that all the electrons in the molecule are paired off, so that they have no net magnetic moment per se. If this is not the case, i.e., if the molecule is a free radical, the much stronger magnetic field arising from the unpaired electron will be *independent* of B_0.

Such behavior is common to all molecules and gives rise to the universally observed diamagnetic properties of matter. Each nucleus is, in effect, partially shielded from B_0 by the electrons and requires a slightly higher value of B_0 to achieve resonance. This can be expressed as

$$B_{loc} = B_0(1 - \sigma),\qquad\qquad(1.51)$$

where B_{loc} is the actual local field experienced by the nucleus and σ is the *screening constant*, expressing the reduction in effective field; σ is independent of B_0 but highly dependent upon chemical structure. Equation (1.6) can then be modified to

$$h\nu_0 = \mu B_0(1 - \sigma)/I\qquad\qquad(1.52)$$

The effect of nuclear screening is to decrease the spacing of the nuclear magnetic energy levels. It will be seen that at constant rf field frequency ν_0, an increase in σ, i.e., an increase in the magnetic shielding of the nucleus, means that B_0 will have to be *increased* to achieve resonance. Thus, if resonance peak positions are expressed on a scale of magnetic field strength increasing from left to right, as is now almost universally done, the peaks for the more shielded nuclei will appear on the right-hand side of the spectrum.

The causes of variations in nuclear shielding will be discussed in Chapter 3. In general, they may be thought of as arising from:

a. Variations in local electron density. Nuclei attached to or near electronegative groups or atoms such as O, OH, halogens, CO_2H, NH_3^+, and NO_2 experience a lower density of shielding electrons and resonate at lower values of B_0. Nuclei removed from such groups appear at higher values. Such inductive effects are the principle cause of shielding variations. The spectra of ethyl orthoformate (Fig. 1.16) furnishes an illustration of these effects. The methyl group carbons and protons, being furthest removed from the oxygen atoms, appear on the right-hand side of the spectrum; the methylene resonances are further "downfield"; while the formyl carbon and proton, because of the three attached oxygens, are the least shielded.

b. Special shielding effects produced by certain groups and structures that allow circulation of electrons only in certain preferred directions within the molecule, i.e., which exhibit diamagnetic anisotropy. Benzene rings, for example, show this behavior very strongly, and many other groups do so in varying degrees, including even carbon-carbon single bonds. These effects are in general smaller than inductive effects, but nonetheless can be vary marked and can provide valuable structural clues.

Protons, because of their low density of screening electrons, show smaller variations due to (a) than do other nuclei, but for the same reason exhibit relatively greater effects from (b) than do heavier nuclei.

Because nuclear shielding is proportional to the applied magnetic field, it of course follows that the spacing between peaks (or groups of peaks) corresponding to different types of nuclei is also proportional to the magnetic field. Thus, in NMR spectroscopy, in contrast to optical spectroscopy, there is no natural fundamental scale unit; the energies of transitions between quantum levels are proportional to the laboratory field. There is also no natural zero of reference. For practical purposes, these difficulties are evaded by these devices:

a. Using parts per million relative change in B_0 or ν_0 as the scale unit. (This convention is retained in Fourier transform spectra.)

b. Using an arbitrary reference substance dissolved in the sample and referring all displacements in resonance, called *chemical shifts*, to this "internal" reference.

The use of a dimensionless scale unit has the great advantage that the chemical shift values so expressed are independent of the value of B_0 of any particular spectrometer, and so a statement of chemical shift does not have to be accompanied by a statement of the frequency (or field) employed, as is the case when actual magnetic field strength or frequency are used as the scale unit. Tetramethylsilane, $(CH_3)_4Si$, was proposed many years ago by Tiers[20] as a reference for proton spectra because its resonance appears in a more shielded position than those of most organic compounds. For a similar reason, it is now also accepted as the reference for ^{13}C spectra. It is soluble in most solvents and is normally unreactive (see Chapter 2, Section 2.6). For both protons and ^{13}C, tetramethylsilane, commonly abbreviated TMS, is assigned a chemical shift of zero. (In older proton spectra, TMS appears at 10.0 ppm, as originally proposed by Tiers. This scale and the shielding both increase from left to right, but it has now been superseded.)

1.11 ELECTRON-MEDIATED COUPLING OF NUCLEAR SPINS

We have seen in Section 1.6 that nuclear spins may be directly coupled to each other through space, giving rise to the phenomenon of dipolar broadening. In addition to this direct coupling, magnetic nuclei may also transmit information to each other indirectly through the intervening chemical bonds. This interaction, discovered independently by Gutowsky, McCall, and Slichter[21,22] and Hahn and Maxwell,[23] occurs by slight polarizations of the spins and orbital motions of the valence electrons and, unlike the direct couplings, is not affected by the tumbling of the molecules and is independent of B_0. If two nuclei of spin 1/2 are so coupled, each will split the other's resonance to a doublet, for in a collection of many such pairs of nuclei there is an almost exactly equal probability of each finding the other's spin to be oriented with $(+1/2)$ or against $(-1/2)$ the applied field [Fig. 1.17(a)]. If one nucleus is coupled to a second group of two identical nuclei, the possible combinations of orientations of the latter will be as shown in Fig. 1.17(b). It is therefore to be expected that the resonance of the first nucleus will appear as a 1:2:1 triplet, while that of the group of two identical nuclei will be a doublet. Three equivalent neighboring spins would split the single nucleus's resonance to a 1:3:3:1 quartet [Fig. 1.17(c)]. Generalizing, we may say that if a nucleus of spin 1/2 has n equivalently coupled neighbors of spin 1/2, its resonance will be split into $n + 1$ peaks corresponding to the $n + 1$ spin states of the neighboring group of spins. The peak intensities are determined by simple statistical considerations and are proportional to the coefficients of the binomial expansion. A convenient mnemonic device for these is the Pascal triangle (Fig. 1.18).

The proton spectrum of the ethyl groups of ethyl orthoformate [Fig. 1.16(a)] illustrates these features of NMR spectra. The methyl resonance at 1.16 ppm is a triplet because of coupling to the CH_2 protons, which appear as a quartet centered at 3.52 ppm. The formyl proton is too distant to experience observable coupling. The strength of the coupling is denoted by J and is given by the spacing of the multiplets expressed in hertz. Although couplings through more than three intervening chemical

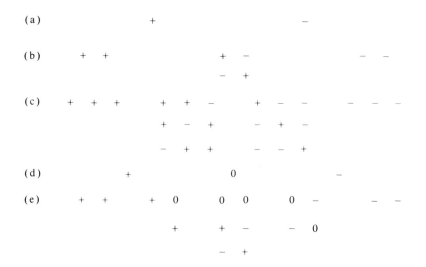

Fig. 1.17.   Spin states and statistical weights for: (a) one spin-1/2 nucleus; (b) two spin-1/2 nuclei; (c) three spin-1/2 nuclei; (d) one spin-1 nucleus; (e) two spin-1 nuclei.

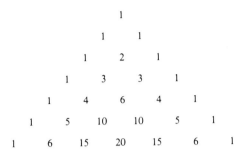

Fig. 1.18.   The Pascal triangle.

bonds tend to be weak or unobservable, as for the formyl proton in this case, this is not always true, as we shall see (Chapter 4, Section 4.3.5).

In the carbon spectrum [Fig. 1.16(b)], the multiplicity arises from the direct (one bond) couplings of the ^{13}C nuclei to the attached protons. These couplings are approximately 125 Hz. The $^{1}H-^{13}C$ couplings through two or more bonds are much smaller and are not resolved on the scale of Fig. 1.16(b). The numbers and intensities of the resonances are strictly binomial and can often serve to identify the carbons. (In practice, these couplings are usually removed and the carbon resonances reduced to singlets in order to increase the effective signal-to-noise ratio; see Chapter 2, Section 2.7, and Chapter 4, Section 4.4).

The coupling of nuclei with spins of 1 or greater is more rarely dealt with, deuterium (^{2}H) and ^{14}N being the most frequently encountered. For nuclei of spin 1, we have seen that three spin states are possible. One such nucleus will split another nucleus (or group of equivalent nuclei), of whatever spin, to a 1:1:1 triplet; two will give a 1:2:3:2:1 pentuplet, and so on [Fig. 1.17(d) and 1.17(e)]. In general, a nucleus coupled to n nuclei with spin I will give $2nI + 1$ lines.

It is particularly important to understand that the occurrence of spin-spin coupling does not depend upon the presence of the magnetic field B_0, and that the magnitude of J, unlike that of the chemical shift, is independent of the magnitude of B_0. Thus, the ratio $J/\Delta\nu$, where $\Delta\nu$ is the chemical shift difference between two given nuclei or two groups of equivalent nuclei, decreases as B_0 increases. The multiplets move further apart, on a field-strength or hertz scale (but not on a parts-per-million scale); however, the spacings within each multiplet (in hertz) remain the same.

It is observed that as $J/\Delta\nu$ becomes larger (as in a series of suitably chosen molecules or by observing a particular molecule at decreasing B_0), the spectrum begins to deviate noticeably from the simple "first-order" appearance which we have been describing. In the 200 MHz proton spectrum of the ethyl groups in ethyl orthoformate, for which $J/\Delta\nu$ is only about 0.01 and this tendency is not noticeable; but it can be clearly seen at 60 MHz (Chapter 7, Fig. 7.1). The members of the multiplets deviate from strict binomial intensities in such a way as to "lean" toward each other. As $J/\Delta\nu$ approaches and exceeds unity, these deviations become very severe, and chemical shifts and coupling constants can no longer be obtained by simple spacing measurements. We shall discuss these important effects in Chapter 4.

REFERENCES

1. W. Pauli, *Naturwissenschaften* **12**, 741 (1924).

2. I. Rabi, S. Millman, P. Kusch, and J. P. Zacharias, *Phys. Rev.* **55**, 526 (1939).

3. J. B. M. Kellogg, I. Rabi, N. F. Ramsey, and J. R. Zacharias, *Phys. Rev.* **56**, 728 (1939).

4. E. M. Purcell, H. C. Torrey, and R. V. Pound, *Phys. Rev.* **69**, 37 (1946).

5. F. Bloch, W. W. Hansen, and M. E. Packard, *Phys. Rev.* **69**, 127 (1946).

6. F. Bloch, *Phys. Rev.* **70**, 460 (1946).

7. W. P. Slichter, *Adv. Polym. Sci.* **1**, 35 (1958).

8. D. W. McCall, *Acc. Chem. Res.* **4**, 223 (1971).

9. R. R. Ernst and W. A. Anderson, *Rev. Sci. Instrum.* **37**, 93 (1966).

10. N. Bloembergen, E. M. Purcell, and R. V. Pound, *Phys. Rev.* **73**, 679 (1948). This classical paper on nuclear relaxation, written very early in the history of NMR when experimental methods were crude by present standards, should be read by anyone wishing real insight into this difficult and complex subject.

11. G. E. Pake, *J. Chem. Phys.* **16**, 327 (1948).

12. J. H. Van Vleck, *Phys. Rev.* **74**, 1168 (1948).

13. E. R. Andrew and R. Bersohn, *J. Chem. Phys.* **18**, 159 (1950).

14. F. Bloch, *Phys. Rev.* **70**, 460 (1946).

15. G. Lindström, *Phys. Rev.* **78**, 817 (1950).

16. H. A. Thomas, *Phys. Rev.* **80**, 910 (1950).

17. W. D. Knight, *Phys. Rev.* **76**, 1259 (1949).

18. J. Jonas and H. S. Gutowsky, *Ann. Rev. Phys. Chem.* **31**, 1 (1980).

19. J. T. Arnold, S. S. Dharmatti, and M. E. Packard, *J. Chem. Phys.* **19**, 507 (1951).

20. G. V. D. Tiers, *J. Phys. Chem.* **62**, 1151 (1958).

21. H. S. Gutowsky, D. W. McCall, and C. P. Slichter, *Phys. Rev.* **84**, 589 (1951).

22. H. S. Gutowsky, D. W. McCall, and C. P. Slichter, *J. Chem. Phys.* **21**, 279 (1953).

23. E. L. Hahn and D. E. Maxwell, *Phys. Rev.* **88**, 1070 (1952).

CHAPTER 2

EXPERIMENTAL METHODS

Lynn W. Jelinski

2.1 INTRODUCTION

In this chapter we describe in general terms how a nuclear magnetic resonance (NMR) spectrometer works. The essential ingredients — a magnet, a means to excite transitions between the nuclear energy levels, a method to detect the signal, and signal processing — are common to all spectrometers, regardless of their vintage or cost. We first look at a block diagram of an NMR spectrometer and at the NMR experiment in broad overview, and then in later sections, describe in more detail how each step works. At the end of this chapter we describe such practical matters as resolution, sensitivity, and how to prepare the sample for NMR measurement.

It is not our intent to set forth all of the experimental details necessary for obtaining an NMR spectrum, but rather to provide background material for the chapters that follow. Leading references are provided for the interested reader. Several excellent books have been published which describe the experimental details of NMR spectroscopy. The book by Fukushima and Roeder[1] gives a good description of NMR hardware, particularly as it concerns spectrometers for solid samples. The book by Martin, Delpuech, and Martin[2] contains a wealth of information on practical aspects of NMR spectroscopy, and the book by Derome[3] provides a very readable account of experimental techniques for solution NMR spectroscopy. The practical techniques for high resolution solid state NMR (Chapter 8) are described in a monograph by Jelinski and Melchior.[4]

2.2 OVERVIEW OF THE NMR EXPERIMENT

Figure 2.1 shows a block diagram of the major components of a pulsed Fourier transform NMR spectrometer. Using this figure, we can trace the main steps in obtaining an NMR signal. First, the sample is placed in a

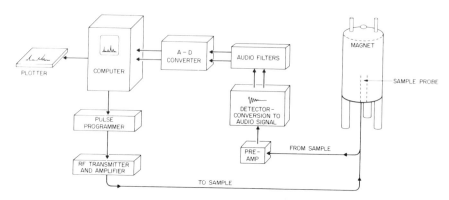

Fig. 2.1.   Block diagram of a pulsed Fourier transform NMR
             spectrometer. Only one channel, the observe channel, is
             shown. The lock and decoupler channels (not shown) are
             similar.

sample *probe* and this assembly is positioned so that the sample resides in
the most homogeneous part of the magnetic field (see Chapter 1, Fig. 1.4).
The computer then instructs the *pulse programmer* to begin the NMR
experiment. The pulse programmer is a timing device that sends out
precisely timed digital signals. The *radio frequency transmitter*
superimposes a signal of the correct frequency on the digital signals from
the pulse programmer. These radio frequency pulses are then amplified
and sent to the sample probe. (The sample probe has been previously
tuned to absorb the radio frequency pulse.) This pulse excites transitions
between the nuclear spin energy levels. The nuclei, having been disturbed
from their equilibrium energy states, undergo a *free induction decay* (FID)
as equilibrium is reestablished. This signal is very weak, of the order of
nanovolts. The *preamplifier* amplifies this weak signal. The amplified
radio frequency signal from the sample is then converted to an *audio
signal*. (The audio signal is obtained by "beating" the free induction
decay against an intermediate or carrier frequency.) The audio signal is
usually detected in *quadrature* (see Section 2.5.2) so that the frequencies
above and below the carrier frequency can be properly identified.
Quadrature detection creates two signals, each out of phase with the other
by $90°$. These audio signals are sent through a *band-pass filter* and are
then converted from their analog form to a digital representation using an
A-to-D (analog-to-digital) converter. These signals, now in digital form,
are stored in a computer for further processing. The processing involves
cosmetic manipulation of the raw data, Fourier transformation, and
presentation of the data.

In the foregoing description of the simple flow chart of Fig. 2.1, we described only the *observe channel*, which is the part of the spectrometer responsible for actually detecting the signal of interest. As we shall later see, there are also lock and decoupler channels that operate in a similar manner. Next, we will expand on these and other aspects of the NMR experiment in greater detail.

2.3 THE MAGNETIC FIELD

The magnetic field is a prerequisite for the NMR experiment because it removes the degeneracy of the nuclear spin energy levels (see Section 1.1). The magnetic field can be supplied by one of three types of magnets: permanent magnets, electromagnets, or superconducting magnets. The magnetic field for an NMR experiment must be stable and homogeneous, conditions that are met by *locking* and *shimming* the field.

2.3.1 Permanent Magnets

NMR spectrometers that are designed around permanent magnets are generally rugged and inexpensive. However, a field strength of 1.4 T [1 T (tesla) = 10 kG (kilogauss)] is about the maximum strength for such systems. A permanent magnet must be encased in some type of insulated housing to prevent temperature-induced drifts in the magnetic field. Because of field drift problems such systems are not ideal for long spectral accumulations. However, they are quite useful for teaching purposes and for analytical applications. Figure 2.2 shows a sketch of a spectrometer that employs a permanent magnet. The magnet is housed in the box standing behind the console; the box provides thermal isolation for the magnet system.

2.3.2 Electromagnets

Electromagnets are generally used to provide magnetic fields in the 1.4−2.4 T range. Electromagnets can take up considerable space. Because the forces between the magnetic pole pieces are enormous, a large yoke device is required to hold the pole pieces apart. These magnets also require a carefully regulated power supply, as well as a means to stabilize the flux of the magnetic field. The magnet is resistive, and thus the current used to provide the magnetic field causes considerable heating. Consequently, the pole pieces must be cooled with a continuous flow of chilled water. The cost of electricity (up to 10 kW/h) and chilled water (approximately 10 liters/h) can be great, although much has been done to produce economical systems. For these reasons, some manufacturers provide low-field superconducting magnets.

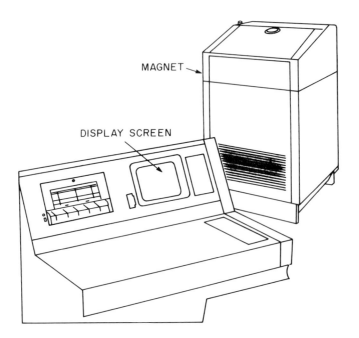

MAGNET

DISPLAY SCREEN

Fig. 2.2.   Sketch of an NMR spectrometer based on a permanent
               magnet. The sample tube is loaded from the top of the box
               containing the magnet. (Based on a photograph by Bruker
               Instruments.)

2.3.3 Superconducting Magnets

Certain metal alloys such as niobium—tin or niobium—tin—tantalum,
become superconducting at very low temperatures. When immersed in
liquid helium (4.2 K), solenoids wound from such alloys are able to
maintain sufficiently high current densities to provide persistent magnetic
fields of 12 T or greater. We have already seen (Fig. 1.4) a diagram of a
cross section of a superconducting magnet assembly. The solenoid, wound
from the superconducting wire, is bathed in liquid helium. This helium
dewar is thermally isolated by means of a vacuum chamber and thermal
baffles (not shown) from an outer dewar that contains liquid nitrogen
(77 K). Once energized, a persistent superconducting magnet requires no
additional current, as long as the magnet is held at liquid helium
temperature. Most superconducting magnets must be filled with liquid
nitrogen every 10 days and with liquid helium every 6 to 8 weeks. Recent
developments   in   superconducting   materials   may   make   possible
superconducting magnets that need only liquid nitrogen cooling.

The unintentional collapse of a superconducting field is called a *quench*. It is usually caused by accidentally transferring warm helium gas when intending to fill the magnet with liquid helium. The collapsing field generates considerable heat, the solenoid becomes resistive, and large amounts of helium evaporate rapidly. A quench can also be caused if the vacuum between the helium and nitrogen dewars is lost, or if a large metal object strikes the magnets and causes thermal contact between the dewars.

Superconducting solenoids have fringe fields, extending from the magnet by more than a meter, the intensity of which depends on the magnetic field strength and the bore size. Such stray fields have been known to capture paper clips, screw drivers, metal-edged rulers, and even gas tanks. Furthermore, they usually erase magnetically encoded information on credit cards. Copper—beryllium tools are non—magnetic and are especially handy when working near a superconducting magnet. (We note in passing that copper—beryllium is toxic, so the handles of these tools should be wrapped with electrical tape.)

The main advantages of a superconducting magnet are its stability, its low operating cost, and its simplicity and small size compared to an electromagnet. Figure 2.3 shows a sketch of an NMR spectrometer that employs a superconducting magnet. The location of the major components comprising this spectrometer are labelled in the figure.

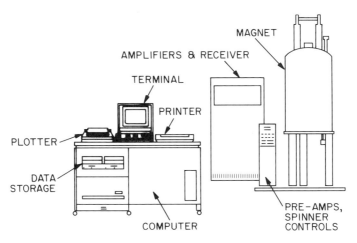

Fig. 2.3.   Sketch of an NMR spectrometer that uses a superconducting magnet. The sample tube is loaded from the top of the superconducting magnet. [Based on a drawing by JEOL (USA), Inc.]

2.3.4 Locking the Magnetic Field

Whether the magnet be permanent or superconducting or an electromagnet, the field must be locked to ensure magnetic field stability. Field stability is especially important for repeated signal accumulations, and several schemes are available for this purpose.

External lock systems are sometimes employed in low field, routine-use spectrometers. The lock material (usually a deuterium, proton, or fluorine-containing liquid) is permanently sealed in a vial that resides within the magnetic field, near the sample. The technique of audio modulation can be used to provide sidebands at known frequencies away from the center band. For example, 1 kHz modulation of the 60 MHz absorption signal for water will produce a series of upper and lower sidebands. These sidebands will be separated by 1 kHz surrounding the central absorption line. If the magnetic field strength were to drop, the position of the sidebands would change. The sideband position can be kept constant by varying the audio modulation frequency in a feedback circuit, thus keeping the field-to-frequency ratio constant.

Another method employs the dispersion signal of the external lock material. The point at which this signal changes sign will vary if the magnetic field changes. The field can be kept constant by using a feedback circuit that causes a compensating current to pass through the magnet stabilizing system.

A third scheme for an external lock system involves Fourier transform methods. The lock material is contained either in a sealed inaccessible vial within the sample probe or within a capillary placed concentrically within the NMR sample tube. The lock material is pulsed repeatedly, the free induction decay is detected, and a feedback circuit is used to keep the lock signal at constant frequency.

Solid-state NMR spectrometers that employ electromagnets use some type of an external lock system because it is not practical to place liquid lock materials in the solid samples. Usually the solid-state probe for an electromagnet contains a lock circuit and a sealed vial containing a lock solvent. The drift rate on superconducting magnets is so slow that solid samples can be run in the unlocked mode on these systems. Superconducting magnets are also run in the unlocked mode for biological tissue samples where the lines are broad or it is inconvenient to provide a lock signal.

Internal lock systems are most widely used, as they afford a more exact control of the field drift. In this method, the lock material is placed in the sample tube itself. Usually the lock material is deuterated and serves as the solvent (see Section 2.7.1). The actual manner in which the lock signal is detected depends upon the detection method used by the spectrometer. If the spectrometer uses a CW (continuous wave; see Section 2.4.2) detection scheme, some variation of the sideband modulation method is used. If the Fourier transform method is employed (see Section 2.4.3), the lock scheme is quite different. In the latter case, the lock circuit looks similar to the observe circuit shown in Fig. 2.1. A lock transmitter and receiver coil is wound so that it surrounds the sample. The sample, through this circuit, is pulsed repeatedly (several hundred times a minute) with pulses that have radio frequencies corresponding to the Larmor frequency of the deuterium-containing lock solvent. For example, the deuterium frequency would be 55.3 MHz when the proton frequency is 360 MHz. The free induction decay from the lock signal is amplified and detected. The operator makes small adjustments to the magnetic field so that the lock signal and the lock transmitter occur at identical frequencies. At this point, the feedback circuit is engaged. The frequency difference between the lock signal and the lock transmitter is monitored and kept as small as possible by applying a compensating current to the magnetic field.

2.3.5 Shimming and Shaping the Field

In addition to a stable field, it is also mandatory to have a homogeneous magnetic field. For permanent magnets and electromagnets, the pole pieces must be highly polished and precisely positioned. For superconducting systems the solenoid must be wound in a homogeneous manner. The goal is to have a magnetic field of uniform strength over the entire sample volume. Even with the most careful attention to detail, it is not possible to obtain a sufficiently homogeneous field without resorting to *field shaping* and *shimming*.

The fields of electromagnets are shaped by a process called *cycling*. When the magnet is turned off it can develop a dome-shaped field caused by the centers of the pole faces that have become more magnetized than their outer parts. Figure 2.4(a) shows a field plot of such a situation. The desired flat contour can be regenerated by cycling the magnet — that is, running the current up over its normal operating value and holding it there for several minutes. An over-cycled field develops a dished shape [Fig. 2.4(b)]. A domed field produces the type of signal distortion shown in Fig. 2.4(a). As the field is swept, the sample abruptly experiences a uniform field (producing the sharp part of the signal), followed by a field that falls off gradually. The gradual change in the field produces the tailing observed in the line shape. Conversely, the line shape that is

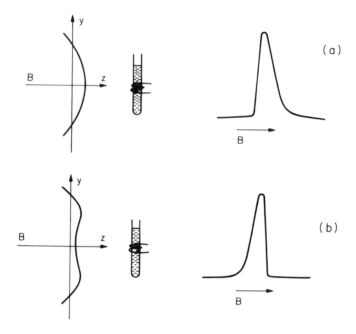

Fig. 2.4.   "Domed" (a) and "dished" (b) magnetic fields and the signal shapes which they yield. [From J. W. Emsley, J. Feeney, and L. H. Sutcliffe, *High Resolution Nuclear Magnetic Resonance Spectroscopy*, Vol. 1, Pergamon, Oxford (1965).]

diagnostic of a dished field [Fig. 2.4(b)] arises because the sample first experiences a gradually varying field, and then one that falls off abruptly.

The fields of superconducting magnets are shaped very differently from those of electromagnets. *Cryoshims* are used to shape superconducting fields. These are superconducting correction coils of various geometries that are energized with the proper amount of current necessary to produce a flat field over the sample. Once energized and adjusted, the cryoshims require no additional changes. Further homogeneity improvements on superconducting magnets are made through the use of *room temperature shims*.

In contrast to the field shaping methods just described, *shimming* refers to a method whereby small corrections are applied to the main field through the use of auxiliary coils of highly specific geometries. These coils are called *shim coils*, a holdover from the time when actual pieces of metal, or shims, were placed on the pole pieces to improve the magnetic field homogeneity.

The shim coils are wound to approximate the spherical harmonic functions listed in Table 2.1. Many of the shims interact with each other, and multiple iterations through the different shims are always necessary to arrive at a perfectly homogeneous field. The interested reader is directed to Refs. 5 and 6 for excellent step-by-step instructions on how to shim a magnet.

The letters X, Y, and Z in Table 2.1 refer to three orthogonal directions in the laboratory frame of reference. Z is taken as being along the magnetic field direction. At this point it is essential to recognize the difference in the direction of the main field between superconducting and electromagnets. For iron magnets, the direction of the main field is between the pole pieces, or parallel to the plane of the floor. For superconducting magnets, the Z direction is perpendicular to the winding of the main coil, which means that the main field runs along the direction of the center bore (from floor to ceiling in vertical magnets). These differences are important in understanding the effect of *sample spinning*, which is the additional method used to control the effective homogeneity of the field as felt by the sample. The solution samples are spun mechanically at $15-30$ revolutions per second about an axis that runs from floor to ceiling. Spinning the sample causes the magnetic field inhomogeneities experienced by the sample to be averaged out, as long as the spinning rate (in hertz) is fast compared to the field gradient across the sample. Spinning is particularly effective in electromagnets and in permanent magnets because the spinning axis is perpendicular to the main field direction. In superconducting magnets, sample spinning does not average out inhomogeneities in the Z direction (i.e., the main field), since the Z direction is coincident with the spinning axis. Sample spinning can produce *sidebands*, which can usually be minimized careful shimming and using high-precision sample tubes.

Essentially all NMR spectrometers have the first-order shim coils, X, Y, and Z, as well as the second-order gradient, $X^2 - Y^2$. This second-order gradient is sometimes called the *curvature*. Superconducting NMR spectrometers generally have more higher order shims, which include most of those listed in Table 2.1.

The lock signal (either the lock FID or the *lock level*) is usually used for the shimming process, although one can also shim on the FID of the sample of interest. The goal is to adjust the shims to obtain a symmetrical FID that persists for a long time, a situation that arises when the resonance line is very sharp. It can be shown empirically that the amplitude of the lock level corresponds to the sharpness of the lock signal,

TABLE 2.1
NMR Shims

Shim name	Function
Z^0	$1 - $ main field
Z^1	Z
Z^2	$2Z^2 - (X^2 + Y^2)$
Z^3	$Z[2Z^2 - 3(X^2 + Y^2)]$
Z^4	$8Z^2[Z^2 - 3(X^2 + Y^2)] + 3(X^2 + Y^2)^2$
Z^5	$48Z^3[Z^2 - 5(X^2 + Y^2)] + 90Z(X^2 + Y^2)^2$
X	X
Y	Y
ZX	ZX
ZY	ZY
XY	XY
$X^2 - Y^2$	$X^2 - Y^2$
Z^2X	$X[4Z^2 - (X^2 + Y^2)]$
Z^2Y	$Y[4Z^2 - (X^2 + Y^2)]$
ZXY	ZXY
$Z(X^2 - Y^2)$	$Z(X^2 - Y^2)$
X^3	$X(X^2 - 3Y^2)$
Y^3	$Y(3X^2 - Y^2)$

and hence to the homogeneity of the magnetic field. However, when using the lock level as the sole indicator of homogeneity, it is sometimes possible to get into local minima from which it is difficult to recover. Figure 2.5 shows the FIDs for a well-shimmed sample and two poorly shimmed ones. The FIDs in Figs. 2.5(b) and (c) are extreme examples of misadjustment of the shims. More subtle effects sometimes are best detected by obtaining a spectrum and looking at its Fourier transform. Figure 2.6 illustrates the effect of deliberate misadjustment of various superconducting shims.

Returning to Figs. 2.5 and (b), we see that the time required for the free induction decay signal to die away is related to the width of the line. The line width, or decay rate, is determined by two factors. One, the magnetic field inhomogeneity, can be controlled by shimming. The goal is

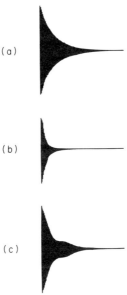

Fig. 2.5.   Free induction decays for (a) a well-shimmed magnetic field
and (b) and (c) poorly shimmed magnetic fields. [The rapid
decay (b) corresponds to a broad line.]

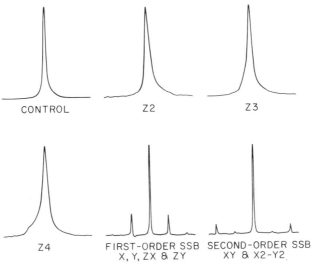

Fig. 2.6.   Deliberate misadjustment of various shims for a
superconducting magnet. (Adapted from Ref. 5.)

to shim out as much of this contribution as possible since it has a major effect on both sensitivity and resolution (see Section 2.8). The other contribution, the "natural" linewidth, is determined by the *spin−spin relaxation time*, or T_2 (see Section 1.6 and Chapter 5) and is inherent to the system under observation.

Many spectrometers have computer-controlled shims. The computer uses the lock level to iterate to optimum values for the shim settings. Once properly shimmed, either by the computer or the operator or both, the *auto shim* is engaged. This feature causes the computer to track the lock level, continually making corrections to the Z^1 shim setting, keeping the homogeneity at its optimal value. The auto shim feature is especially useful for overnight spectral accumulations where small temperature drifts may cause changes in the magnetic field homogeneity.

2.4 CAUSING TRANSITIONS BETWEEN THE NUCLEAR ENERGY LEVELS

We have already seen in Chapter 1 the concepts underlying the nuclear magnetic resonance phenomenon (Section 1.3) and the detection of resonance (Section 1.4). In this section we describe the experimental aspects of nuclear magnetic resonance and its detection.

2.4.1 The Concept of Resonance

Resonance occurs when the period, or frequency, of an excitation source matches the natural frequency of the system under investigation. For example, the frequency at which a radio circuit resonates, or absorbs a signal, occurs when there is an exact match between the impedance of the inductor, L, and the impedance of the capacitor, C, of the circuit; that is, when

$$\omega L = \frac{1}{\omega C} \tag{2.1}$$

which means that the resonant frequency, ω_o, of a simple LC circuit is given by

$$\omega_o = \left[\frac{1}{LC} \right]^{1/2} \tag{2.2}$$

The resonant frequency can be changed by varying either the capacitance or the inductance of the circuit. In practice it is usually easier to change the capacitance of the circuit, and this is exactly what is done when tuning in another radio station, or when tuning the NMR probe for another nucleus (see Section 2.5.1). The concept of resonance is not peculiar to radio frequency circuits; it is a phenomenon that occurs repeatedly in physics. For example, bells have characteristic resonant frequencies, and a pendulum exhibits its resonant frequency as it swings back and forth, trading kinetic for potential energy. Similarly, resonance in NMR occurs when the frequency of a second applied field, B_1, is exactly equal to the Larmor frequency of the nucleus under observation. We have already seen in Chapter 1 that this situation is given by

$$\omega_o = \gamma B_o \qquad (2.3)$$

Resonance occurs when the B_1 field of frequency ω exactly matches the energy difference between the nuclear spin energy levels (Fig. 1.2). This causes a flipping of the nuclear magnetic dipoles (see Section 1.4). As we have already seen in Section 1.7, the Bloch equations describe the motion of a magnetic moment in the fields B_o and B_1.

There are two main methods used to achieve a resonant condition in NMR spectroscopy — CW (continuous wave) and FT (Fourier transform). The actual method for achieving resonance is more complicated than it first appears. Because most samples generally have more than one type of magnetically equivalent group (perhaps a methyl and a methine), they have more than one distinct chemical shift (see Section 1.10 and Chapter 3). A single sample will generally have multiple Larmor frequencies, each corresponding to a different chemical shift. Any excitation method must cover all of the Larmor frequencies in the sample. In the CW method, the resonant condition is met for each magnetically equivalent group by either sweeping the B_1 frequency or the B_o field. As each group comes into resonance, it undergoes nuclear induction (Chapter 1), and a voltage is induced in the pickup coil. Rather than exciting one group at a time as in the CW method, the FT method provides a short but powerful pulse so that *all* of the groups in the sample are simultaneously excited. In either case the nuclei are upset from their equilibrium energy levels and immediately strive to reestablish their original Boltzmann distribution. We shall look more closely at these two excitation methods in the sections that follow.

2.4.2 Field Sweep and Frequency Sweep

Each group in a sample can successively be brought into resonance by sweeping either the frequency or the field. As we have already seen in Section 1.4 and in Fig. 1.5, the applied radio frequency field can be decomposed into two counter-rotating vectors in the $x'y'$ plane. One vector rotates in the same direction as the rotating frame, and if its frequency is ω_o, it will be stationary in the rotating frame. This component is the B_1 field. (The other vector has a negligible effect because it is rotating in the opposite direction, and in the rotating frame it is at twice the Larmor frequency.)

Since the radio frequency coil is wound so that it is perpendicular to the main field B_o, we must consider the effect that a slow sweep of either B_o or B_1 has on the x and y components of the magnetization. The magnetization vector, **M**, tries to follow the effective field, but its motion is balanced out by spin—spin and spin—lattice relaxation processes. We have already seen [Chapter 1, Eqs. (1.46)–(1.48)] that the Bloch equations describe the behavior of the magnetization in a field.

In a *crossed coil* system, the exciting and monitoring coils are perpendicular to each other. When a particular nucleus comes into resonance as a consequence of the exciting B_1 field, a voltage is induced in the perpendicular coil. This signal, microvolts or less in magnitude, is amplified and detected directly in the *frequency domain*. Although the CW method was the primary means for signal detection in the early days of NMR spectroscopy, it has largely been supplanted by FT methods in all but the most simple spectrometers.

Conceptually, FT NMR is rather different from CW spectroscopy. One of the main differences is that detection is performed in the *time domain* and the signal is Fourier transformed into the frequency domain (see Section 1.4 and Fig. 1.7).

2.4.3 Fourier Transform Methods

In the Fourier transform method, a short burst or pulse of radio frequency signal, nominally monochromatic at ν_o, is used to simultaneously excite all of the Larmor frequencies in a sample. This short pulse is ideally a square wave. A square wave can be approximated by orthogonal functions of higher and higher frequencies. This is illustrated in Fig. 2.7, where on the left a positive and a negative square wave are approximated by 1, 3, 5, and 7 orthogonal functions. (The error remaining after this approximation is shown on the right side of the diagram.) For this reason, the square wave, although monochromatic in frequency, can excite all of the frequencies in the spectrum if the pulse is powerful enough. The duration of a pulse necessary to tip the magnetization vector into the $x'y'$ plane, termed a *90° pulse*, is inversely related to the power of the radio frequency signal. The 90° pulse width, as it is called, is generally between

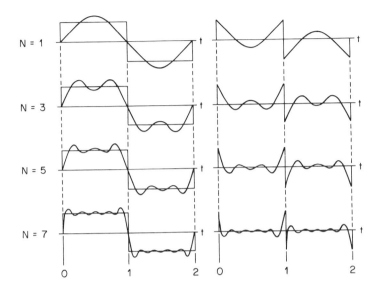

Fig. 2.7.   Approximation of two square waves by orthogonal functions (left) and the instantaneous error remaining after each approximation.

5 and 30 μs in duration. Its length depends upon the radio frequency power level, the frequency of the measurement, and the characteristics of the probe. As we have already seen (Section 1.4), the length of the pulse (t_p) is related to the range of frequencies that are excited:

$$\nu_o \pm \frac{1}{t_p} \tag{2.4}$$

where ν_o is the center frequency. The flip angle, θ, is related to the strength of the exciting field, B_1, and to t_p by

$$\theta = \frac{\gamma}{2\pi} B_1 t_p \tag{2.5}$$

In terms of energy level diagrams, we can visualize the effect of a 90° pulse as equalizing the populations of the nuclear spin energy levels (Fig. 1.2). Following the pulse the spins undergo a free induction decay (FID) as they reestablish their equilibrium populations. The effects of the 90° pulse and the ensuing T_2 and T_1 decay processes have already been illustrated in the rotating frame vector diagrams of Fig. 1.6. The FID is collected as a function of time and represents an interferogram in the time domain. The frequency domain information, or the normal spectrum, is obtained by Fourier transformation of the FID (see Fig. 1.7 and Section 2.6.2). The detector is positioned in the $x'y'$ plane, and hence it is desirable to perform a 90° pulse so that the maximum magnetization is detected. [In later discussions (Section 2.8.5) we shall see that a 90° pulse may not always be the optimum flip angle.] The effects of a 90° pulse and of a 45° pulse on the signal amplitudes are compared in Figs. 2.8(a) and (b). The detector samples only that projection of the magnetization that is in the $x'y'$ plane, so the signal obtained with a 45° pulse has 0.707 of the intensity of that observed with a 90° pulse. A 180° pulse should produce no signal, since there is no projection of the magnetization vector into the $x'y'$ plane [Fig. 2.8(c)]. A 270° pulse is 180° out of phase with a 90° pulse and produces the FID and spectrum shown in Fig. 2.8(d).

The 90° pulse width must be determined experimentally, and will vary somewhat with solvent. For example, samples prepared in ionic media such as physiological saline (0.15 M NaCl) will be lossy, and will usually

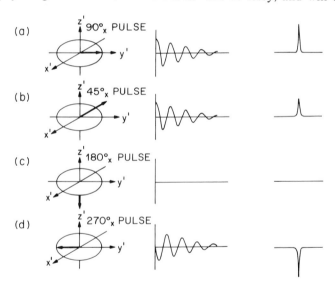

Fig. 2.8.   The magnetization vector (left column), the resultant FID (center column), and the transformed spectrum (right column) after a (a) 90° pulse, (b) 45° pulse, (c) 180° pulse, and (d) 270° pulse. In all cases, the B_1 field is applied along the x direction.

require a longer 90° pulse width than those prepared in an organic solvent. The length of the 90° pulse is best determined from a *null spectrum*, or one produced by a 180° pulse [see Fig. 2.8(c)]. The length of the 180° pulse is then halved to obtain the proper length for the 90° pulse.

In addition to having a magnitude or strength, the pulse in an FT experiment also has phase coherence. In all of the examples in Fig. 2.8, the radio frequency pulse was *along the x' direction*. A 90° pulse, phase shifted by 90°, would be along the $-y$ direction. Figure 2.9 shows the phases of the spectra resulting from 90° pulses along the x, $-x$, y, and $-y$ directions. There are situations, for example, for some forms of *two dimensional* NMR spectroscopy (see Chapter 6), where it is necessary to apply radio frequency pulses along directions other than x. Experimentally, a 90° phase shift is produced by introducing a delay equal to a quarter of the radio frequency cycle. As we shall later see, the ability to apply pulses of specified phases will let us manipulate the spins in highly specific ways.

2.4.4 Double and Triple Resonance

So far we have discussed in a generic sense the methods for obtaining an NMR signal but have deferred any discussion of methods, called scalar

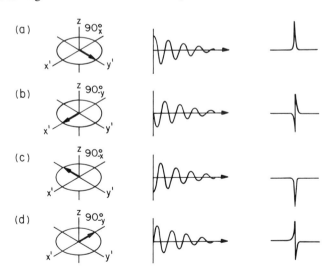

Fig. 2.9.   The magnetization vector (left column), the resultant FID (center column), and the transformed spectrum (right column) after a (a) 90°$_x$ pulse, (b) 90°$_{-y}$ pulse, (c) 90°$_{-x}$ pulse, and (d) 90°$_y$ pulse.

decoupling, which remove the splittings introduced by electron-mediated spin—spin couplings (see Sections 1.11 and 4.4). Furthermore, we have so far neglected the nuclear Overhauser enhancement (NOE; see Section 5.2.8), which arises from the direct magnetic coupling between nuclei and is observed as a change in the intensity of one resonance when the transitions of others nearby are irradiated. Both the scalar decoupling experiment and the NOE experiment require the technique of *double resonance*. Double resonance (Chapter 4) involves the simultaneous or sequential application of two different frequencies. The decoupling field, often called B_2, must be larger in frequency units than the interaction to be decoupled. The B_2 field "drives" the spins, causing their Z components to be flipped rapidly, effectively removing the interaction from the spectrum. Many different types of double resonance experiments are described in Section 4.4. Here we introduce the terminology and the basic concepts necessary for the ensuing discussion.

Homonuclear double resonance experiments for solution NMR spectroscopy are usually performed by irradiating the spectrum at a single frequency while observing the entire spectrum by pulse methods. This is accomplished experimentally by time-sharing the transmitter coil for the two purposes. Homonuclear double resonance is most commonly employed for protons. One can vary the power of the irradiating rf fields so that the spins are merely "tickled" or so they are completely saturated. When the spins are saturated, they are effectively *decoupled*, and the effects of spin—spin coupling to adjacent nuclei are removed. This usually produces substantial spectral simplification, and can be used to assign resonances or to determine coupling constants in proton NMR spectra. (In Chapter 6 we shall see *two-dimensional* NMR methods that can be used in place of single-frequency homonuclear decoupling and single-irradiation homonuclear NOE experiments. We shall defer discussion of homonuclear decoupling in solids until Section 8.7.)

Heteronuclear double resonance experiments are used in solution NMR spectroscopy to remove the multiplicities caused by spin—spin couplings between protons and the heteronucleus. (In solid state NMR, heteronuclear double resonance is used to remove the broadening caused by dipolar couplings. We shall defer discussion of this type of heteronuclear decoupling until Section 8.2.) In a heteronuclear double resonance experiment, the protons are saturated during the acquisition of the free induction decay signal. This causes the heteronuclear spin multiplets to collapse into a single peak. Continuous saturation of the proton spin system (as opposed to saturation only during acquisition) produces a signal enhancement caused by the NOE. A *proton-decoupled spectrum* is therefore simple and of enhanced signal-to-noise ratio. Figure 2.10 illustrates the effects of proton decoupling on the carbon-13 spectrum of dioxane. The spectral simplification and the increased sensitivity that decoupling produces are readily evident.

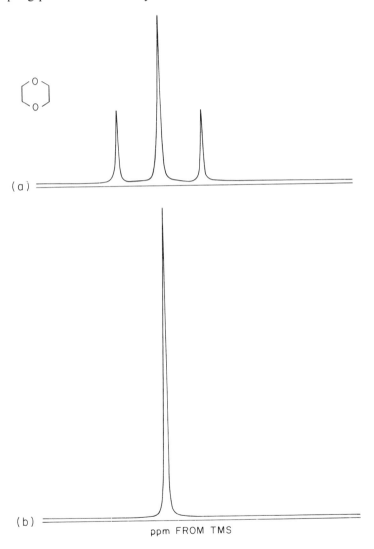

Fig. 2.10.   Computer simulation of a ^{13}C spectrum of dioxane with (a) proton coupling and (b) proton decoupling.

In some cases, particularly when observing carbon NMR spectra of molecules that contain both protons and fluorine, it is desirable to be able to perform a *triple resonance* experiment. Figure 2.11 illustrates the

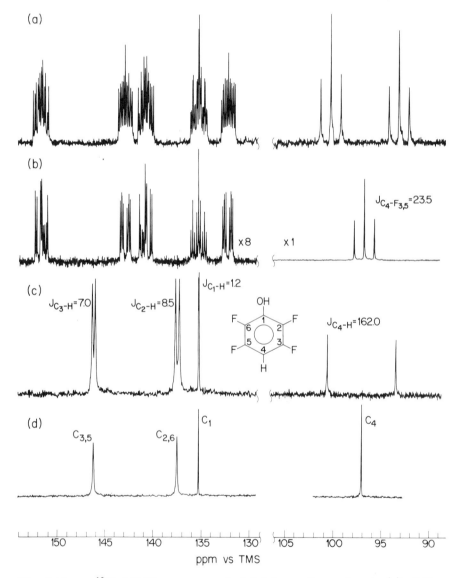

Fig. 2.11. ^{13}C NMR spectra of 2,3,5,6-tetrafluorophenol (a) fully coupled, (b) with proton decoupling, (c) with fluorine decoupling, and (d) with both proton and fluorine decoupling.

remarkable spectral simplification that can be obtained using triple resonance.[7] The spectra are all carbon-13 spectra of 2,3,5,6-tetrafluorophenol. The spectrum in Fig. 2.11(a) is fully coupled, and the carbons show splittings from both the protons and the fluorines. The spectrum in Fig. 2.11(b) was obtained with proton decoupling and thus shows only the carbon—fluorine splittings. The spectrum in Fig. 2.11(c) was obtained with fluorine-19 decoupling, and that of Fig. 2.11(d) with *both* proton and fluorine decoupling. The ability to decouple both protons and fluorine has been very useful in the analysis of carbon spectra of fluoropolymers.[8]

2.4.5 Pulse Sequences

Pulse sequences, described by *pulse sequence diagrams*, are used to manipulate the spins. In Chapter 1, Fig. 1.7, we have already seen a very simple pulse diagram, that for a one-pulse experiment. A pulse sequence consists of a precisely timed series of radio frequency pulses, whose frequencies, phases, and duration are used to produce the desired information. Figure 2.12 shows some more complicated examples, illustrating several variations on heteronuclear decoupling and NOE experiments. For example, the pulse sequence diagram in Fig. 2.12(a) is read as: a 90° pulse (of unspecified phase) is applied to the carbons, the carbon signal is detected, and there is an acquisition delay before repeating the sequence again. Simultaneously, the proton irradiating frequency is applied continuously during all three time periods. This produces a heteronuclear decoupled spectrum with NOE (see Section 5.2.8).

The pulse sequence diagram of Fig. 2.12(b) shows what is known as a *gated decoupling* experiment (see Chapter 4, Section 4.4). The resultant spectrum will be decoupled, but the signal intensities will not be perturbed by differential NOE's. Such an experiment is useful for obtaining quantitative carbon NMR spectra. This diagram shows that the proton irradiation frequency is turned on only during the time that the free induction decay signal is detected. This pulse sequence works because the spin—spin coupling is destroyed essentially instantaneously when the proton B_2 field is applied, but the NOE takes some time to build up to its maximum value (see Section 5.2.8). Furthermore, it is only those spins that are in the $x'y'$ plane that are being detected, and hence any Overhauser enhancement that builds up will be for those spins that have already relaxed back into the z' direction. This built-up Overhauser enhancement will be lost during the acquisition delay.

An additional variation of this straightforward double resonance experiment is to obtain a heteronuclear NMR spectrum that is fully

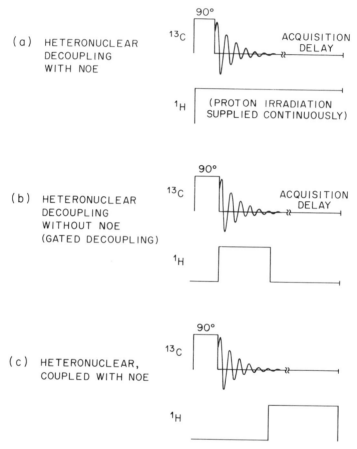

Fig. 2.12.   Heteronuclear pulse sequences: (a) ^{13}C decoupling with NOE, (b) gated decoupling, and (c) heteronuclear, coupled, with NOE.

coupled, yet has the enhanced signal caused by the NOE. The pulse sequence for such an experiment is shown in Fig. 2.12(c).

As we shall see in later chapters, pulse sequences can become substantially more complicated.

2.5 DETECTING THE SIGNAL

2.5.1 Sample Probe

The sample for NMR spectroscopy, whether it be a solid or a solution, is placed in the *probe* for measurement. In addition to the support structure and dewars for variable temperature control, the probe consists of at least one tuned radio frequency circuit. Hoult[9] has published a detailed analysis of NMR receivers, and the interested reader is directed to this paper for a rigorous description and design analysis.

Most high—resolution, solution—state probes have separate circuits for the lock, decoupler, and observe channels. Solid—state probes sometimes use single-coil, double-tuned circuits. The inductor, or coil, is wound and positioned so that the electromagnetic B_1 field that it creates is perpendicular to the direction of the main B_o field, although this is not strictly true for solids, where the coil may be tipped at the "magic angle" with respect to the field direction. Some typical coil designs are shown in Fig. 2.13. The Helmholtz coil [Fig. 2.13(a)] is generally used for high resolution applications with superconducting magnets. This coil provides a radio frequency field that is perpendicular to the main field, but allows the sample to be dropped into the bore from the top of the magnet. Solenoid coils [Fig. 2.13(b)] are often used for electromagnet geometries. These coils are also used for solid samples for both electromagnet and superconducting solenoid geometries. Figure 2.13(c) shows a topical coil, one of many possible coil designs used for surface NMR spectroscopy and for imaging.

These coils act as inductors in circuits that are tuned for maximum absorption of the desired frequency. Tuning is accomplished by adjusting the capacitors, which usually have extension rods so that they can be adjusted from outside the magnet. A probe is tuned when there is an exact

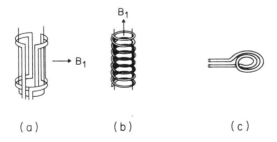

(a)   (b)   (c)

Fig. 2.13.   Some typical NMR coil designs. (a) A Helmholtz coil, generally used with superconducting magnets. (b) A solenoid coil, used for permanent and electromagnets, and also for solid—state NMR in superconducting magnets. (c) A surface coil, used for imaging applications. The B_1 field is perpendicular to the plane of the coil.

balance between the inductance and the capacitance of the circuit [see Eqs. (2.1) and (2.2)]. The exact tuning of a probe will be sample-dependent, as the sample is part of the effective circuit.

The probe coil can be characterized by a *quality factor*, Q. The Q, which is best determined experimentally, is given by $\omega L/r$, where L is the inductance of the coil and r is the resistance. In general, the higher the Q, the greater the sensitivity.

The NMR signal that comes from the probe is extremely weak, usually of the order of nanovolts. This signal must be amplified prior to detection, a job accomplished by the *preamplifier*. The preamplifier is normally placed as close to the probe as possible to minimize signal loss. A preamplifier can be either *broadbanded* (i.e., it amplifies signals over a very large frequency range) or *tuned* (i.e., it detects only a narrow band of frequencies and rejects all others). A tuned preamplifier or an associated band—pass (or high—pass) filter are methods to eliminate unwanted noise from the NMR spectrum.

2.5.2 Detecting the Signal and Digitizing the Data

The amplified signal from the sample is in the megahertz, or radio frequency regime. Because audio signals occur over a fairly narrow bandwidth and therefore are easier to work with than the radio signals at different frequencies, the high—frequency signal is converted to a common audio signal before detection. There are many methods for this conversion, all based on established radio technology. The basic goal is to strip off the radio frequency carrier, leaving only the frequency modulated signal. In this process the phase of the radio frequency carrier is used for *phase sensitive detection*. This audio signal FID then carries information about its offset (or frequency difference) from the carrier, or transmitter frequency.

The audio signal is then passed through audio band—pass filters to remove noise outside the spectral bandwidth. The audio filters attenuate the noise or unwanted signals that are outside of the region of interest. Although it is the audio signal that is actually filtered, it is easier to visualize the effects of filtering by looking at the time domain. This is illustrated in Fig. 2.14, in which the band pass of the audio filters is superimposed on the spectrum. The filters are always set slightly wider than the actual spectrum so that roll-off and end effects do not affect the information content of the spectrum. In most commercial spectrometers the width of the audio filters is set automatically when the desired *spectral width* (the total width of the spectrum, usually in hertz or kilohertz) is entered.

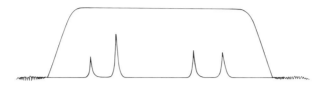

Fig. 2.14.   A filter cutoff function superimposed over a transformed
            spectrum. The filter prevents high-frequency noise from
            being folded into the spectrum.

This audio signal must now be digitized; that is, it must be converted
from its continuously varying analog format into a suitable digital
representation. This is performed by an A-to-D converter, which stores the
incoming signal in discrete time bins. Lindon and Ferrige[10] have written a
monograph on the subject of digitization and data processing in NMR, and
the interested reader is referred to this source for a complete description of
the important factors in this process. Three of the important factors to be
considered in the digitization process are the *digitization rate*, *digital
resolution*, and *dynamic range*.

First, we consider the rate that the data should be "sampled," or
converted into its digital representation. (For now, we shall assume that
all of the signal is acquired in a single channel; later we shall see that there
are many advantages in taking two channels of data.) According to the
sampling theorem, a sine wave must be described by at least two points in
a single period in order to be properly represented. A sampling rate of two
points per cycle, shown in Fig. 2.15(a) for a sine wave of frequency ν, is
known as the *Nyquist frequency*. If, instead of ν, the frequency of the sine
wave is somewhat greater, for example, $\nu + \Delta\nu$ [Fig. 2.15(b)], then that
frequency could not be accurately represented by sampling the data twice
per period ν, nor could that signal be distinguished from one whose
frequency was $\nu - \Delta\nu$ [Fig. 2.15(c)]. This means that a signal with a
frequency greater than the Nyquist frequency will appear in the spectrum
at $\nu - \Delta\nu$. This signal is *folded-in*, or *aliased*. In practical terms, the
Nyquist frequency dictates the rate at which the FID must be sampled.
For example, if a proton spectrum is 2 kHz wide, the signal must be
digitized at a 4-kHz rate, meaning that the data must be sampled at most
every 250 μs. This sampling time is called the *dwell time*.

The dwell time multiplied by the total number of points determines the
digital resolution. For example, if 8,192 points were used to acquire this

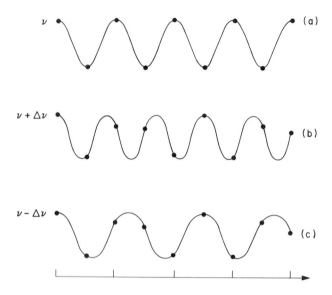

Fig. 2.15.   (a) To be properly represented, a signal of frequency ν must be sampled twice per cycle. (b) When sampled at this rate, a signal of higher frequency, $\nu + \Delta\nu$, cannot be distinguished from (c) a signal at the lower frequency, $\nu - \Delta\nu$.

hypothetical spectrum, the *acquisition time* would be 8,192 points × 250 μs/point = 2 × 10^6 μs = 2 s. The digital resolution would then be 2 kHz/8,192 points, or 0.24 Hz/point. [This calculation applies to a single channel of data. We shall see later that two channels of data are usually obtained, so in that the digital resolution would differ from this value (see Section 2.8.3)].

The final consideration is *dynamic range*. When the FID is split into time bins, the voltage gain must be adjusted so that the signal is properly represented by the number of available bits, or the *word size*. The strongest part of the signal must be represented without overflowing the computer word size, and the weakest part of the signal of interest must simultaneously be represented by at least one bit. A large word size (20, 24, or 32 bits) or the ability to perform double precision acquisition (i.e., linking together two words to represent each data point) is essential for "difficult" samples. An example of a difficult task would be the detection of end groups or branch points in the spectrum of a polymer that is dominated by very intense peaks from the main chain of the polymer. Another difficult sample would be the detection of small resonances from a protein in the presence of large amounts of water. In samples such as these, even the smallest part of the signal must be represented, otherwise it will be averaged away in the noise.

There are two ways to solve the dynamic range problem, and these are often used in conjunction with double precision acquisition of the data. The first of these is *block averaging*, where a number of individually accumulated spectra, none of which has exceeded the dynamic range of the digitizer, are added together. The other technique involves suppression of large water or solvent peaks (see Section 9.3). There are a number of methods available to suppress the solvent peak.[11] The most straightforward method simply involves presaturation of the undesirable resonance, which works well if the solvent protons do not exchange with other protons of interest in the spectrum. Another method is Redfield's 2-1-4 pulse sequence, which involves the application of "soft" or weak radio frequency pulses at the solvent resonance. Finally, there are a series of pulse sequences that are based on the idea of setting the transmitter frequency to the peak to be eliminated. That peak is stationary in the rotating frame, does not precess, and can be returned to its equilibrium magnetization along the z direction by the application of the sequence $90°_x - \tau - 90°_{-x}$. All other peaks will precess and will not return to the z axis. In reality, there are distortions with this simple sequence, and other pulses can be added to increase the effectiveness of this type of solvent suppression.

So far we have discussed the mechanics of detecting the signal in a single channel. However, in FT NMR the FID is usually acquired in *quadrature*, a technique that uses two phase sensitive detectors, each out of phase with the other by 90°. This produces *two* FIDs, also out of phase with each other by 90°. These FIDs can be digitized by a single A-to-D converter if the even time bins (0, 2, 4, 6, etc.) contain one of the FIDs and the odd ones contain the other. (Alternatively, two separate A-to-D converters could be used.) These FIDs are often called the "real" and the "imaginary" signals. (These FIDs are transformed in a complex Fourier transformation to produce the "absorption" and "dispersion" signals, again differing in phase by 90°. The absorption spectrum is the one that is normally displayed. We shall address the Fourier transform in Section 2.6.2.)

Quadrature detection offers several advantages over single channel detection. First, it allows one to distinguish between signals that are above and below the carrier frequency. If quadrature detection were not used and the transmitter, or carrier, were in the center of the spectrum, it would be impossible to distinguish between signals that were $+f$ hertz and $-f$ hertz from the carrier. Figure 2.16 illustrates this point. The alternative

(a)

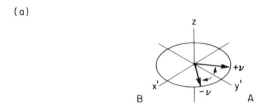

(b)

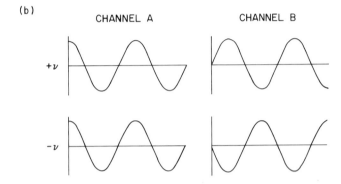

Fig. 2.16.   (a) Rotating frame vector diagram of two signals with frequencies $+\nu$ and $-\nu$, which have been flipped into the $x'y'$ plane for detection. The positions of the two quadrature detectors, A and B, are shown. (b) If the signal were detected only with detector A, the signals at $+\nu$ and $-\nu$ would appear identical. Although the signals appear identically with detector A, they appear differently with detector B.

then would be to place the carrier at one edge of the spectrum and make the effective bandwidth twice as large as the region of interest. The disadvantage here is that the noise from the unwanted part of the spectrum would be "folded" or "aliased" into the spectrum from the other side of the carrier, increasing the total noise in the spectrum. Furthermore, the power required to provide a suitable burst of rf signal would be much greater than if the carrier were in the center of the spectrum. Finally, with the carrier in the center of the spectrum, the data acquisition rate (the dwell time) can be twice as long, again because the bandwidth is half what it would otherwise be.

The *transmitter offset* is used to position the transmitter in the center of the spectrum, and the *spectral width* is used to define the entire width of the spectrum in frequency units. As we have already seen, the filter bandwidth is usually set to be only slightly greater than the width of the spectrum, so that extraneous noise outside of the observation window does not get folded into the spectrum.

Phase cycling is used to reduce *coherent noise* from a spectrum. This usually involves some permutation of cycling the phases of the receiver and the transmitter through 0, 90, 180, and 270°, while properly co-adding or co-subtracting the FIDs. This algorithm goes by many names, including CAPS, CYCLOPS, and ABC. The phase of the signal from the sample will vary as the transmitter phase is changed, whereas the phase of a signal from a coherent noise source will not be affected by the transmitter phase, thereby producing signals that do not co-add in a coherent fashion.

2.6 PROCESSING AND PRESENTING THE DATA

2.6.1 Digital Filtering and Weighting the Signal

Once obtained, the digitized signal is usually cosmetically conditioned — mathematically manipulated — to increase the apparent resolution, enhance the apparent signal-to-noise ratio, or emphasize one or another aspect of the spectrum. Of course one cannot increase the *actual* information content of the data beyond what is already there; one can only enhance some of the information at the expense of other information. The early points of the FID contain information about the broad components of the spectrum, whereas the highest frequency information is contained in the tail of the FID. Selective truncation or enhancement of various parts of the FID can selectively enhance or reduce these components.

If the signal has been acquired in quadrature, it is common practice to *baseline correct* the data, an operation that removes any residual effect from a DC offset in the two quadrature channels. The effect of baseline correction is shown in Fig. 2.17(a). The Fourier transform of an FID that has not been baseline corrected gives a sharp "glitch" at zero frequency.

Once baseline corrected, it is often desirable to *zero fill* the data, which consists of adding zeroes to the end of an FID. Often an FID of N data points is zero-filled to $2N$ points, or an FID that has been acquired with a number of points that is not an exact power of two is zero filled up to the next power of two in data points. Because the subsequent Fourier transformation uses more data points, zero filling results in an interpolation between data points in the spectrum and enables measurement of linewidths to better accuracy. Figure 2.17(b) shows the effect on the FID of zero filling.

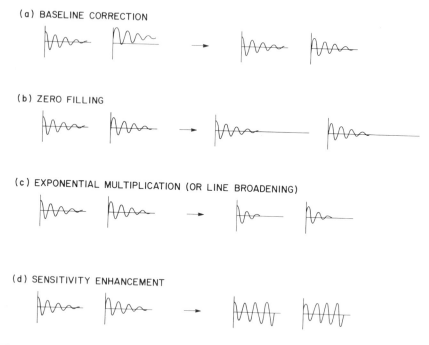

Fig. 2.17.   The effect of various weighting functions the FID and on the resultant spectrum.

There are a number of methods to cosmetically alter the FID that go under the general name of *digital filtering*. One of the most straightforward manipulations of the FID is to multiply it by a decaying exponential, $\exp(-t/k)$, where k is a time constant optimally matched for the signal. This process, called *exponential multiplication* or *line broadening*, optimally matches the natural T_2 decay of the FID with the decay of the exponential function. The resultant spectrum has an increased signal-to-noise ratio at the expense of resolution, where the amount of line broadening is given by $\pm 1/(\pi k)$. [When k is negative, the process is called *sensitivity enhancement*; (see Fig. 2.17(d).]

2.6.2 The Fourier Transform

The FID, a signal in the time domain, $f(t)$, is converted into a signal in the frequency domain, $F(\omega)$, by Fourier transformation. This process can be seen graphically in Fig. 2.18 and mathematically in what follows. The time $f(t)$ and frequency $F(\omega)$ functions are Fourier inverses:

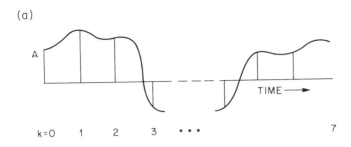

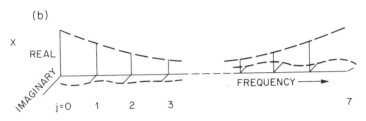

Fig. 2.18.   (a) A signal that has been digitized into eight discrete time bins. (b) The complex Fourier transform of the signal in (a). (Adapted from Reference 12.)

$$f(t) = \frac{1}{2\pi} \int_{-\infty}^{+\infty} F(\omega) e^{i\omega t} \, d\omega \tag{2.6}$$

and

$$F(\omega) = \int_{-\infty}^{+\infty} f(t) e^{-i\omega t} \, dt \tag{2.7}$$

We recall that

$$e^{ix} = \cos(x) + i \, \sin(x) \tag{2.8}$$

which means that Eq. (2.7) becomes

$$F(\omega) = \int_{-\infty}^{+\infty} f(t) \, \cos \omega_0 t - \int_{-\infty}^{+\infty} f(t) \, i \, \sin \omega_0 t \, dt \tag{2.9}$$

Because $f(t)$ can be expressed as a series of sines and cosines, Eq. (2.9) can be recast as

$$F(\omega) = \int [A_1 \cos(\omega_1 t) + A_2 \cos(2\omega_2 t) + \cdots$$

$$+ A_n \cos(n\omega_n t)] \cos(n\omega_o t) \, dt$$

$$-i \int [A_1 \cos(\omega_1 t) + A_2 \cos(2\omega_2 t) + \cdots$$

$$+ A_n \cos(n\omega_n t)] \sin(n\omega_o t) \, dt$$

(2.10)

$$-i \int [B_1 \sin(\omega_1 t) + B_2 \sin(2\omega_2 t) + \cdots$$

$$+ B_n \sin(n\omega_n t)] \cos(n\omega_o t) \, dt$$

$$-i^2 \int [B_1 \sin(\omega_1 t) + B_2 \sin(2\omega_2 t) + \cdots$$

$$+ B_n \sin(n\omega_n t)] \sin(n\omega_o t) \, dt$$

This equation applies to a continuous signal, but because we have digitized data, we must instead deal with discrete signals at discrete data points. The integrals must be replaced by sums, and the task now is to find the Fourier coefficients A_n so they can be summed to give $F(\omega)$. This is usually done with the fast Fourier transform (FFT) algorithm, an efficient method for computing the discrete Fourier transformation. The advantages of this method are severalfold. First, the transform can be performed "in place"; that is, there is no requirement for additional computer memory for storage of intermediate steps of the transformation. Second, the FFT reduces significantly the total number of computational steps. For example, for 1024 points, a standard Fourier transform would require approximately 1024^2 operations, whereas the FFT requires of order $n \log n$ operations and produces more than a 200-fold reduction in the number of computational steps.

The inner workings of the FFT can be understood by working through a very simple transform, such as an 8-point FID. The interested reader is referred to "A Guided Tour of the Fast Fourier Transform" by Bergland[12] for this exercise.

2.7 PREPARING THE SAMPLE

Specimens for NMR spectroscopy can range from soluble materials dissolved in an appropriate solvent (e.g., small molecules for photochemical studies, natural products, soluble polymers, and proteins) to solids (e.g., rocks, coal, polymers, and plants; see Chapter 8) to biological tissue (e.g., humans, organs from animals, and cells; see Section 9.2). Consequently there are a multitude of sample cells and sample preparation methods, each optimized for the particular material under study. In this section we outline some of the most critical considerations for sample preparation. Reference tables are provided for some of the most important NMR solvents and for NMR reference materials.

2.7.1 Soluble Samples

Samples for solution-state NMR measurements are generally dissolved in deuterated solvents, the deuterium providing the necessary signal for the lock channel. Table 2.2 lists some useful NMR solvents and their molecular formulas, melting and boiling points, proton and carbon chemical shifts, and multiplicities.[13] The most common NMR solvents are chloroform, benzene, and water, but other solvents are necessary if the spectroscopy involves poorly soluble materials, or low-temperature or high-temperature measurements. For example, it may be necessary to use methanol to record NMR spectra at very low temperatures in order to freeze out particular chemical exchange processes (see Chapter 5). On the other hand, it is sometimes useful to use mixed solvents for polymers; for example 1,2,4-trichlorobenzene (nondeuterated) plus dioxane-d_8 or benzene-d_6 as the lock material works well for observing chain branching in polyethylene samples at 110°C.

Most deuterated solvents are hygroscopic and pick up water from the atmosphere. This becomes a problem in proton NMR, particularly when recording spectra of material in low concentrations. Deuterated solvents are best stored in a dry box under a nitrogen atmosphere and opened and transferred only in the dry box. Contamination from residual water on glass surfaces can also be avoided if the sample tubes and pipettes are stored in the dry box.

The ideal sample for solution-state NMR is a fairly concentrated solution (> 3 mM for proton NMR and > 0.1 M for carbon NMR) of the material of interest dissolved in a volume that barely exceeds the dimensions of the NMR coil. The most common sample cells, or NMR tubes as they are usually called, have outer diameters of 5 and 10 mm, which require approximately 0.5 and 2 ml of solution, respectively. If the

TABLE 2.2

Properties of Some Common Deuterated NMR Solvents[a]

Compound	Molecular formula	mp^b	bp^b	δ_H (multiplicity)[c]	δ_C (multiplicity)[c]
Acetic Acid-d_4	CD_3CO_2D	17	118	11.53(1) 2.03(5)	178.4(br) 20.0(7)
Acetone-d_6	CD_3COCD_3	−94	57	2.04(5)	206.0(13) 29.8(7)
Acetonitrile-d_3	CD_3CN	−45	82	1.93(5)	118.2(br) 1.3(7)
Benzene-d_6	C_6D_6	5	80	7.15(br)	128.0(3)
Chloroform-d	CD_3Cl	−64	62	7.24(1)	77.0(3)
Cyclohexane-d_{12}	C_6D_{12}	6	81	1.38(br)	26.4(5)
Deuterium Oxide	D_2O	3.8	101.4	4.63(DDS) 4.67(TSP)	
1,2-Dichloroethane-d_4	CD_2ClCD_2Cl	−40	84	3.72(br)	43.6(5)
Diethyl-d_{10} Ether	$(CD_3CD_2)_2O$	−116	35	3.34(m) 1.07(m)	65.3(5) 14.5(7)
Dimethylformamide-d_7	$DCON(CD_3)_2$	−61	153	8.01(br) 2.91(5) 2.74(5)	162.7(3) 35.2(7) 30.1(7)

		mp (°C)	bp (°C)	^{1}H	^{13}C
Dimethyl-d_6 Sulfoxide	CD_3SOCD_3	18	189	2.49(5)	39.5(7)
p-Dioxane-d_8	$C_4D_8O_2$	12	101	3.53(m)	66.5(5)
Ethyl Alcohol-d_6 (anh)	CD_3CD_2OD	< -130	79	5.19(1) 3.55(br) 1.11(m)	56.8(5) 17.2(7)
HMPT-d_{18} (hexamethyl phosphoramide)	$C_6H_{18}N_3PO$	7	106(11)	2.53(2×5)	35.8(7)
Methyl Alcohol-d_4	CD_3OD	−98	65	4.78(1) 3.30(5)	49.0(7)
Methylene Chloride-d_2	CD_2Cl_2	−95	40	5.32(3)	53.8(5)
Nitrobenzene-d_5	$C_6D_5NO_2$	6	211	8.11(br) 7.67(br) 7.50(br)	148.6(1) 134.8(3) 129.5(3) 123.5(3)
Nitromethane-d_3	CD_3NO_2	−29	101	4.33(5)	62.8(7)
Isopropyl Alcohol-d_8	$(CD_3)_2CDOD$	−86	83	5.12(1) 3.89(br) 1.10(br)	62.9(3) 24.2(7)
Pyridine-d_5	C_5D_5N	−42	116	8.71(br) 7.55(br) 7.19(br)	149.9(3) 135.5(3) 123.5(3)

TABLE 2.2, (continued)
Properties of Some Common Deuterated NMR Solvents[a]

Compound	Molecular formula	mp[b]	bp[b]	δ_H (multiplicity)[c]	δ_C (multiplicity)[c]
Tetrahydrofuran-d_8	C_4D_8C	−109	66	3.58 (br) 1.73 (br)	67.4 (5) 25.3 (br)
Toluene-d_8	$C_6D_5CD_3$	−95	111	7.09 (m) 7.00 (br) 6.98 (m) 2.09 (5)	137.5 (1) 128.9 (3) 128.0 (3) 125.2 (3) 20.4 (7)
Trifluoroacetic Acid-d	CF_3CO_2D	−15	72	11.50 (1)	164.2 (4) 116.6 (4)
2,2,2-Trifluoroethyl Alcohol	CF_3CD_2OD	−44	75	5.02 (1) 3.88 (4×3)	126.3 (4) 61.5 (4×5)

[a] Adapted from a listing by Merck & Co., Inc., Isotopes.
[b] Physical property of protonated analog, except for D_2O.
[c] br, broad; m, multiplet.

amount of sample is limited, it is advantageous to use microcells, which require between 10−30 μl of solution. The rest of the volume in the coil, i.e., the area around the microcell, is filled with an index matching solvent. On the other hand, if the material is of limited solubility or is composed of biological cells (in which the compounds of interest are dilute), it is desirable to use sample cells that are larger than 10 mm in outer diameter. Finally, it is possible to obtain NMR spectra of air sensitive compounds by preparing the NMR sample on a vacuum line and sealing the sample tube under either vacuum or an inert atmosphere. It is essential to use NMR tubes of the highest possible quality for high-resolution, high-field applications. NMR tubes, although usually made of borosilicate glass, can also be made of plastic for corrosive fluorinated compounds, quartz for photochemical applications, and amber colored glass for photosensitive materials.

Vortex plugs are useful, particularly in the case of large sample tubes, to keep the liquid in the bottom of the tube while it is being spun in the probe. Figure 2.19 illustrates some typical sample cells, a microcell, a vortex plug, and a vortex plug puller.

Chemical shifts are almost always reported with respect to a reference material. A very small amount of the reference material is usually placed in the sample tube along with the solution. Tetramethylsilane (TMS) is the standard reference material for both proton and carbon NMR when organic solvents are used. DSS (4,4-dimethyl-4-silapentane sodium carboxylate) or similar compounds are used for aqueous solutions. Table

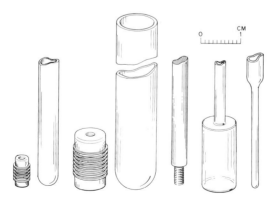

Fig. 2.19.   Sketches of sample cells, microcells, vortex plugs, and a vortex plug puller.

2.3 lists some common reference materials for solution NMR measurements.[14] A number of these can be purchased in partially deuterated form.

Solutions, particularly those of small molecules for relaxation studies, should be filtered through a plug of glass wool or a millipore filter and degassed prior to relaxation measurements. Dissolved oxygen must be removed because it is paramagnetic and competes with the dipole—dipole mechanism in inducing relaxation. Several freeze—pump—thaw cycles or degassing in a very fine stream of argon or nitrogen can accomplish this.

TABLE 2.3
Reference Compounds for Proton NMR[a]

Name	Chemical formula	Abbreviation	Boiling or melting point (°C)	δ_H (ppm TMS)
Tetramethylsilane	$(CH_3)_4Si$	TMS	BP = 26.3	0
Hexamethyldisilane	$(CH_3)_3Si-Si(CH_3)_3$		BP = 112.3	0.037
Hexamethyldisiloxane	$(CH_3)_3Si-O-Si(CH_3)_3$		BP = 100	0.055
Hexamethyldisilazane	$(CH_3)_3Si-NH-Si(CH_3)_3$		BP = 125	0.042
3-(trimethylsilyl)propane sulfonic acid Na salt		TSPSA		
4,4-dimethyl 4-silapentane sodium sulfonate	$(CH_3)_3Si(CH_2)_3SO_3Na$	DSS	MP = 200	0.015
3-(trimethylsilyl)propionic acid Na salt		TSPA		
4,4-dimethyl 4-silapentane sodium carboxylate	$(CH_3)_3Si(CH_2)_2COONa$	DSC	MP > 300	0.000
3-(trimethylsilyl) 3,3,2,2-tetra-deuteropropionic acid Na salt	$(CH_3)_3Si(CD_2)_2COONa$	d_4-TSPA	MP > 300	0.000
Octamethylcyclotetra-siloxane	$(CH_3)_2Si[O-Si(CH_3)_2]_3-O$	OCTS	BP = 175 MP = 16.8	0.085
11,33,55-hexakis-(trideutero-methyl)1,3,5-trisilacyclo-hexane	$(CD_3)_2Si-CH_2-Si(CD_3)_2$ $CH_2-Si(CD_3)_2-CH_2$	d_{18}-CS	BP = 208	−0.327
Tetrakis-(trimethylsilyl)-methane	$[(CH_3)_3Si]_4C$	TTSM	MP = 307	0.236

[a]Adapted from a listing published by Bruker Instruments.

It is sometimes desirable to add *relaxation reagents* to solution NMR samples to decrease the relaxation time and thereby decrease the amount of time necessary to obtain a spectrum. These relaxation reagents are ideally composed of paramagnetic materials that reduce the relaxation times but do not greatly increase the linewidths. They work by inducing electron–nucleus relaxation, which then competes with the dipole–dipole mechanism to cause relaxation (see Chapter 5, Section 5.2.6). At significantly high concentrations these materials shorten the transverse relaxation time (T_2). Examples of relaxation reagents used in solution NMR include chromium acetylacetonate [$Cr(acac)_3$] and the analogous iron complex. Nontoxic, site-selective relaxation reagents are useful as contrast enhancers for biomedical NMR imaging (see Section 9.2).

Shift reagents (see Section 3.3.7) are usually organometallic compounds containing a lanthanide. They act as Lewis acids and preferentially complex oxygen, nitrogen, or other Lewis base sites. They cause large shifts of groups adjacent to these centers, thereby making assignments easier. Shift reagents based on europium and ytterbium usually cause downfield shifts, and those based on praseodymium produce upfield shifts. Although shift reagents are not as essential as they once were, given high magnetic field strengths and the advent of powerful two-dimensional NMR methods, they nevertheless provide an additional assignment tool. In addition, *chiral shift reagents* provide the method of choice for determining the optical purity of organic materials when chiral impurities are present. Because the NMR method provides a direct measure of the enantiomeric excess of each component, it is preferable to polarimetric methods when the sample possibly contains other optically active impurities. The interested reader is referred to the monograph by Reuben[15] for a comprehensive review of lanthanide shift reagents in NMR.

2.7.2 Solid Samples

The techniques of solid-state NMR spectroscopy are described in Chapter 8, where it is shown that the solid sample must be spun mechanically about an axis which forms the "magic angle" (54.7°) with respect to the magnetic field direction. The required rate of spinning is usually greater than 2 kHz (120,000 rpm), which puts large stresses on the sample and on the sample cell. Because the sample must spin at this rate without wobbling for long periods of time, sample preparation for solids becomes particularly critical. The sample is usually pulverized so that the particles are uniform in size, and the powdered sample is distributed evenly and packed, sometimes with pressure, into appropriate sample cells. Examples of these sample cells, or *rotors*, are shown in Fig. 2.20. Most

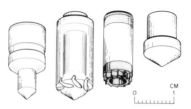

Fig. 2.20.   Sketches of various rotors for solid NMR samples.

rotors have a double bearing drive, which enhances stability. They can be made from polymers such as Kel-F [poly(chlorotrifluoroethylene)], Delrin [poly(oxymethylene)], machinable ceramics, aluminum oxides, or boron nitride. If one is obtaining a spectrum of a solid that can be machined, the sample cell can be made from the material itself.

In contrast to the situation in solution NMR spectroscopy, there is no single standard reference material for solid-state NMR. Furthermore, it is difficult to reference chemical shifts accurately in the solid state owing to anisotropic magnetic susceptibility problems. However, there are a number of secondary references that have proven useful. These are summarized in Table 2.4. Usually one can put a pressed wafer of the desired reference material in the sample rotor with the solid sample. The interested reader is directed to Earl and VanderHart[16] for a comprehensive examination of the problems and methods for referencing chemical shifts in the solid state.

An additional problem in the solid state is to know the sample temperature with accuracy. Haw and co-workers[17] have proposed an internal NMR thermometer for the solid state, samarium acetate. The chemical shifts of this material obey the Curie law over a large range in temperature, making it suitable for calibration of the sample temperature.

2.7.3 Tissue and Other Biological Samples

Biological samples that must be kept alive or functioning during the NMR measurement present great challenges to the NMR spectroscopist. Not only must the sample be confined to the coil of the NMR instrument, but there must be careful temperature control, proper oxygenation, pH monitoring, and often a flow system to perfuse the organ. Sample cells, and often sample probes, are specially constructed for this purpose. Figure 2.21 provides an illustration of some of the "plumbing" required to perform NMR measurements on perfused organs, in this case a hamster liver.

Experiments on biological tissues and whole cells are usually performed in the unlocked mode. It is essential to maximize the signal-to-noise ratio

TABLE 2.4
^{13}C **Chemical Shift References for Solid State NMR^a**

Compound	Chemical shifts (ppm)			
	1	2	3	4
Hexamethylbenzene	17.36	132.18	—	—
Adamantane	29.50	38.56	—	—
Delrin	89.1	—	—	—
Dodecamethylcyclohexasilane	−4.74	—	—	—
Polydimethylsilane	−1.96	—	—	—
Polydimethylsiloxane	1.50	—	—	—
Polyethylene	33.63	—	—	—
$(CH_3)_3SiCH_2CH_2CH_2SO_3Na$	0.40	18.53	20.44	57.39

^{a}From W. L. Earl and D. L. VanderHart, *J. Magn. Reson.*, **48**, 35 (1982)

of these types of experiments in order to reduce the amount of time required to perform the experiments.

2.8 RESOLUTION AND SENSITIVITY CONSIDERATIONS

Despite its enormous power to solve structural, dynamical, and analytical problems, NMR spectroscopy is inherently an insensitive method. A large number of parameters affect the resolution and particularly the sensitivity of an NMR experiment. In this section we discuss the effects that field strength, sample size, probe geometry, and other miscellaneous parameters have on the ultimate sensitivity of an NMR experiment.

2.8.1 Signals and Noise

In order to understand the effects of noise on an NMR spectrum, it is important to understand how signals differ from noise. Noise is generally random and unpredictable. It arises from the resistance of the coil and of the preamplifier. The noise *power*, N^2, in the frequency interval $\Delta\nu$, is given by[9]

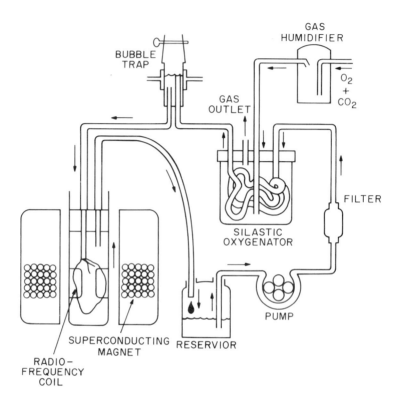

Fig. 2.21.   Apparatus for carbon-13 NMR spectroscopy on perfused
hamster livers. The hamster is anesthetized, and a cannula is
sutured into the portal vein of the liver. The liver, still in the
hamster, is then supplied with oxygen and nutrients by the
flow of perfusate into the vein. The liver is then surgically
removed from the animal and placed in a cylindrical sample
tube, which is positioned in the bore of a superconducting
magnet. The perfusate (a buffered saline solution containing
7% hemoglobin for increased oxygen-carrying ability) is
recycled during the experiment. (From J. R. Brainard, J. Y.
Hutson, R. E. London, N. A. Matwiyoff, *Los Alamos
Science Magazine*, p. 42, Summer 1983.)

$$\overline{N^2} = 4kTr \; \Delta\nu \qquad (2.11)$$

where k is Boltzmann's constant, T is the absolute temperature, and r is the resistance. Therefore, the root-mean-square noise over the bandwidth of interest is given by

$$N_{\text{RMS}} = \left[\overline{N^2} \right]^{1/2} = \sqrt{4ktr \; \Delta\nu} \qquad (2.12)$$

From Eq. (2.12) it can be seen that the noise does not depend on the spectrometer frequency; it depends only on the temperature and the resistance of the circuit. Consequently, there have been designs for NMR coils and preamplifiers that are cooled in liquid nitrogen or liquid helium to reduce the effects of thermal noise.

In contrast to the noise, the NMR signal, or voltage induced in the receiver coil, is predictable and can be made to reverse its polarity with the phase of the applied radio frequency pulses. The voltage, S, induced in the coil from the sample, is given by

$$S = \omega_o M_{xy} B_{1xy} \cos(\omega_o t + \phi) \qquad (2.13)$$

where M_{xy} is the magnetization in the xy plane, B_{1xy} is the component of the applied field, and ϕ is the phase. The voltage, S, adds linearly with the number of accumulations, whereas the noise *power* adds. This means that the signal-to-noise ratio, S/N after n accumulations, increases as the square root of the number of accumulations:

$$\left[\frac{S}{N} \right]_n = \sqrt{n} \left[\frac{S}{N} \right]_1 \qquad (2.14)$$

where $\left[S/N \right]_1$ is the signal-to-noise ratio after a single scan. In other words, in order to triple the signal-to-noise ratio of a spectrum, nine times as many scans must be accumulated.

2.8.2 Effect of the Magnetic Field

The strength of the magnetic field affects both the resolution and the sensitivity of NMR measurements. [In these and subsequent discussions, we shall assume that the magnetic field is properly shimmed (Section 2.3.5), which directly affects both the sensitivity and the resolution of the NMR spectrum.] We first look at the effect of the magnetic field on sensitivity. We have already seen that the NMR signal adds linearly with the voltage induced in the coil, and that the voltage depends on ω_o and on M_{xy} [Equation (2.13)]. We also know that $\omega_o = \gamma B_o$, that M_{xy} is proportional to M_o, and that M_o depends linearly on B_o. It then follows that the strength of the NMR signal increases as the square of the magnetic field strength. Since the signal–to–noise ratio increases as the square root of the number of accumulations, doubling the magnetic field strength will theoretically reduce the time required for spectral accumulation by a factor of 16. This means that an overnight experiment on a spectrometer operating at 200 MHz for protons could conceivably take less than an hour if the proton frequency were 400 MHz.

However, several factors complicate such predictions in actual practice. Depending on the correlation time, relaxation times (see Chapter 5) may vary with the magnetic field strength. In addition, the nuclear Overhauser enhancement (see Chapter 5), which provides enhanced intensity in some heteronuclear experiments, may not be as large at very high magnetic field strengths. It also becomes increasingly difficult to build high–sensitivity probes as the magnetic field strength increases, because of increased losses arising from the electronic leads.

We now turn to the effect of magnetic field strength on *resolution*. For solution NMR spectroscopy, in general, the higher the magnetic field strength, the higher the resolution. There are two main reasons for this. First, the chemical shift dispersion — the ability to distinguish between resonances whose chemical shifts are close together (see Section 1.10 and Chapter 3) — increases linearly with magnetic field strength. This means that a spectrum that may have been severely overlapping at lower magnetic field strength becomes substantially simplified. Second, the NMR spectra of complex spin systems (see Chapter 4) become simpler at high fields because the difference in frequency between two resonances becomes larger than the coupling constant, since the coupling constant is independent of the magnetic field.

Although it would seem that resolution must always increase with increasing magnetic field strength, this generalization does not hold if decoupling is inefficient. We have already seen that the spin–spin decoupling field (see Section 4.4) must be able to cover the entire frequency range of interest. For example, the carbon-13 frequency in an 11.7-T field is 125.7 MHz (500 MHz for protons), and in order to cover the entire carbon range ($\approx$ 200 ppm) the proton decoupler must cover over a 25-kHz range. Incomplete decoupling produces line broadening, which affects both the resolution and the sensitivity. Methods for efficient decoupling schemes are described in Section 4.4.

Although the highest possible magnetic field strength is advantageous for solutions, it is not necessarily the case for solids. We have already seen that the chemical shift dispersion increases with the strength of the magnetic field. The chemical shift anisotropy — the width of the solid—state NMR powder pattern — also increases with the strength of the magnetic field (see Section 8.3). This means that the width of the powder pattern that must be "spun out" increases, which means that the rate of sample spinning for solids must also increase if one is to obtain a spectrum that contains no sidebands. (It turns out that there are schemes for eliminating sidebands from magic angle spinning, so this is not as severe a problem as may be expected.) The decoupling argument also applies to solid-state experiments.

2.8.3 Digital Resolution

The digital resolution of an NMR spectrum can be calculated in the case of quadrature detection from

$$\text{digital resolution} = \frac{SW}{NP/2} = 2\,SW/NP \qquad (2.15)$$

where SW is the spectral width and NP is the total number of points (the number of data points used for the acquisition plus the additional points used for zero filling). The factor of two comes in because of quadrature detection; although NP represents the total number of points, the number of *real* data points is $NP/2$.

It is possible to obtain linewidths that are about 0.1 Hz wide at half-height when observing small molecules in a magnetic field that has been properly shimmed. Four to five points are required to properly define a peak, so in order to define a line that is 0.1 Hz wide, the digital resolution must be about 0.02 Hz/point. This means that a very large number of points would be required to obtain a spectrum over a large spectral width, and that the time to acquire the data would be very long. A long acquisition time means that more time is required to obtain each spectrum, and thus fewer FIDs can be averaged. Hence, increased resolution must be traded for decreased sensitivity.

In practice, for all but the most demanding solution NMR applications, a digital resolution of 0.2 to 0.5 Hz/point is sufficient. The digital resolution for solid-state NMR spectra (see Chapter 9) can be much less since the lines are almost always quite broad. In addition, for solids the decoupler must be on and at high power during the acquisition time for a solid-state heteronuclear experiment, and generally there are limits on this. Two-dimensional NMR spectroscopy of liquids (see Chapter 6) requires a rethinking of the concept of digital resolution. Part of this is because a digital resolution of 0.2 Hz/point in two dimensions would require a prohibitively large number of data points. Furthermore, it would take a very long time to acquire the data. Such incredible resolution in a two-dimensional spectrum is unnecessary, since the point of two-dimensional NMR is not to resolve the individual lines, but to determine the correlations *between* the peaks.

2.8.4 Sample Size and Probe Geometry

All things being equal, a probe made with a coil of solenoid design is two to three times more sensitive than one with a Helmholtz coil at low magnetic field strengths. However, the solenoid coil geometry is impractical for solution-state NMR applications in a superconducting magnet, and Helmholtz coils are usually used. In addition to the type of coil, the other factors that affect sensitivity are the Q of the coil (see Section 2.5.1); the filling factor, η; and the concentration of spins in the sample. The filling factor describes the fraction of the coil volume that is filled by the sample. For a solenoid coil, the filling factor is given by $\eta =$ (volume of sample)/2 (volume of coil). Table 2.5 illustrates the signal—to—noise ratios that can be realized for various combinations of tube or microcell size, filling factor, and quality factor. The data in this table illustrate that a microcell is ideal if the amount of material is limited, but that a large tube size is ideal is the material is of limited solubility or if there is a large amount of material.

2.8.5 Optimum Tip Angle

So far we have skirted around the idea that the tip angle could be other than 90°. For spin lattice relaxation experiments, two—dimensional NMR, and some solid—state NMR applications, this is indeed the case, as in these situations it is necessary to manipulate the spins in a precise way. However, in routine one—dimensional NMR, it is sometimes not practical to obtain fully quantitative spectra by applying 90° pulses and waiting five T_1's between each accumulation. Instead, one repeats each accumulation at a rate determined by the acquisition time, and sets the flip angle to a value less than 90°. The optimum flip angle is the one that maximizes the amount of z magnetization that returns to equilibrium before the next pulse begins. This tip angle is known as the *Ernst angle*,[18] and is given by

$$\cos \alpha_{\text{Ernst}} = e^{-AT/T_1} \tag{2.16}$$

where AT is the acquisition time and T_1 is the spin lattice relaxation time of the peak or peaks of interest.

TABLE 2.5
Relationship Between Signal-to-Noise Ratio (S/N) and Sample Size[a]

	Outer diameter of the sample tube (mm)			
	10	8	5	1.4
Internal diameter	8.95	7.0	4.2	1.0
Filling factor η	0.33	0.30	0.25	0.17
Quality factor Q	190	160	105	55
S/N for constant sample concentration	100	61.0	23.6	2.1
S/N for constant sample amount	53.5	57.0	63.9	100

[a]Adapted from Martin, Delpuech, and Martin, Ref. 2, p. 164.

If it is necessary to use 90° pulses, it can be shown that the optimum repetition time is given by $1.27 \times T_1$. Because the return of the z component of the magnetization to equilibrium is an exponential process, 99% of the magnetization will be regained by waiting $5 \times T_1$, 95% after $3 \times T_1$, and 72% after $1.27 \times T_1$ (see Chapter 5).

REFERENCES

1. E. Fukushima and S. B. W. Roeder, *Experimental Pulse NMR — A Nuts and Bolts Approach*, Addison-Wesley, Reading, Massachusetts, 1981.

2. M. L. Martin, J.-J. Delpuech, and G. J. Martin, *Practical NMR Spectroscopy*, Heyden, London, 1980.

3. A. E. Derome, *Modern Techniques for Chemistry Research*, Organic Chemistry Series, J. E. Baldwin, Ed., Vol. 6, Pergamon, Oxford, 1987.

4. L. W. Jelinski and M. T. Melchior in *NMR Spectroscopy Techniques*, C. Dybowski and R. L. Lichter, Eds., Dekker, New York, 1987, p. 253.

5. W. W. Conover, *Relaxation Times* (Nicolet Magnetics Corporation) **2**, 6 (1981).

6. W. W. Conover in *Topics in Carbon-13 NMR Spectroscopy*, G. C. Levy, Ed. Vol. **4**, 1984, p. 38.

7. F. C. Schilling, *J. Magn. Reson.* **47**, 61 (1982).

8. A. E. Tonelli, F. C. Schilling, and R. E. Cais, *Macromolecules* **14**, 560 (1981).

9. D. I. Hoult in *Progress in Nuclear Magnetic Resonance Spectroscopy*, J. W. Emsley, J. Feeney, and L. H. Sutcliffe, Eds., Vol. 12, Pergamon, Oxford, 1978, p. 41.

10. J. C. Lindon and A. G. Ferrige in *Progress in Nuclear Magnetic Resonance Spectroscopy*, J. W. Emsley, J. Feeney, and L. H. Sutcliffe, Eds., Vol. 14, Pergamon, Oxford, 1980, p. 27.

11. A. G. Redfield, S. D. Kuntz, and E. K. Ralph, *J. Mag. Res.* **19**, 114 (1977).

12. G. D. Bergland, *IEEE Spectrum*, 41 (July 1969).

13. Compiled from a data sheet by Merck Isotopes.

14. Adopted from a listing published by Bruker Instruments.

15. J. Reuben in *Progress in Nuclear Magnetic Resonance Spectroscopy*, J. W. Emsley, J. Feeney, and L. H. Sutcliffe, Eds., Vol. 9, Pergamon, Oxford, 1973, p. 1.

16. W. L. Earl and D. L. VanderHart, *J. Magn. Reson.* **48**, 35 (1982).

17. J. F. Haw, R. A. Crook, and R. C. Crosby, *J. Magn. Reson.* **66**, 551 (1986).

18. R. R. Ernst and W. A. Anderson, *Rev. Sci. Instrum.* **37**, 93 (1966).

CHAPTER 3

THE CHEMICAL SHIFT

3.1 INTRODUCTION

We have seen (Chapter 1, Section 1.10; Chapter 2, Section 2.6.2) that the position of the resonance of a particular nucleus on the NMR spectral scale depends on a number of influences. These include principally the bulk diamagnetic susceptibility of the medium (when an "external" reference is employed) and the inductive effects and magnetic anisotropies of substituent groups and atoms, as well as dispersion forces and solvent anisotropy. In this chapter we are concerned with shielding influences *within* molecules and will assume that a suitable "internal" reference is employed. Thus, variations in medium susceptibility will have no effect, and we will assume that other medium-effect terms are negligible or small.

3.2 ATOMIC SHIELDING

We consider first an atom in which the nucleus is surrounded by a spherical distribution of electronic charge, i.e., an atom in a ^1S state. In a magnetic field B_0, the entire system of electrons will undergo a Larmor precession about the field direction with angular frequency $eB_0/2M_e c$. The opposing magnetic field thus generated will have a strength at the center of charge, i.e., at the nucleus, which can be calculated as the summation of the effects of all circulating currents such as the one shown in Fig. 3.1. For the hydrogen atom, the local field at the nucleus is given by[1]

$$B_{\mathrm{loc}} = B_0 - \frac{4\pi e^2 B_0}{3 M_e c^2} \int_0^\infty r \rho(r) \, dr \qquad (3.1)$$

where $\rho(r)$ describes the electron density as a function of the distance r from the nucleus. For ^1S atoms of nuclear charge Z, we find from Eq. (3.1) and Eq. (1.51) that the nuclear *screening constant* σ is given by

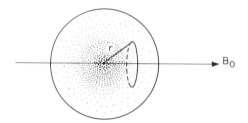

Fig. 3.1.   Electron currents in a magnetic field.

$$\sigma = \frac{4\pi Z e^2}{3 M_e c^2} \int_0^\infty r\rho(r)\ dr \qquad\qquad (3.2)$$

Calculation of σ for atoms from $Z = 1$ through $Z = 92$ by Dickinson,[2] using appropriate functions for the electron density, gave values of σ from 18×10^{-6} for hydrogen to values of the order of 10^{-2} for the heaviest atoms $(Z \geqslant 80)$. For ^{19}F, $\sigma = 133 \times 10^{-6}$ and for ^{31}P, $\sigma = 261 \times 10^{-6}$. Such calculations are very useful in suggesting the order of magnitude of shielding differences that one may expect for various nuclei, but when the atoms are incorporated into molecules, the electron distributions are in general no longer spherical, and there are also other perturbing effects of great importance.

In the remainder of this chapter, the discussion will be centered on the shielding of protons and carbon-13 nuclei in organic molecules, although the principles described are of general validity. Other nuclei will be dealt with in Chapter 9.

3.3 MOLECULAR SHIELDING

3.3.1 General Form

In molecular hydrogen and in hydrogen-containing molecules—organic molecules in particular—the electronic distribution about the nuclei cannot possess strict spherical symmetry, and so the electrons are not free to move in the circular paths of Fig. 3.1 for all orientations of the molecule with respect to B_0. The screening of the nuclei depends upon the molecular orientation in B_0. For liquids, the observed shielding is an average over all orientations of the tumbling molecule.

The effect of a departure from spherical symmetry can be understood qualitatively in the following way. In the development of Eq. (3.1), it was assumed that there was only one nucleus and that it was at the origin. In a molecule, we may choose a particular nucleus for observation and place this too at the origin of the molecular coordinate system. But the Larmor precession of the electron system about the field direction z is now impeded by the presence of the other nuclei. The simple model of circular electron paths is applicable only if the molecule can be so oriented that the nuclei and electrons have an arrangement that is axially symmetric along a line parallel to the z axis. For all other arrangements and even for axially symmetric molecules that tumble freely—in short, for all actual liquid systems—a Lamb-type term does not describe the screening correctly, even if averaged over all orientations, but instead predicts a screening that is too large. Another term is necessary to describe the hindering of electron circulation by the departure from axial symmetry, i.e., by the electric fields of the other nuclei. This second term turns out to be more complicated than one might at first expect.[3] It cannot be adequately dealt with classically, but requires a consideration of the excited electronic states of the molecule. The magnetic field causes a mixing of the ground state wave function with excited state wave functions. Because this second term corresponds to a magnetic moment opposing that of the diamagnetic Lamb-type term, it is customarily referred to as the paramagnetic term. These excited state wave functions are generally unknown except for very simple molecules, and so an approximation is employed in which an average is taken over all these wave functions and a summation replaced by a single term ΔE in which the average excitation energy of all upper states appears. The shielding may thus be expressed as

$$\sigma = \sigma_D + \sigma_P$$

where the diamagnetic term σ_D is of the form of Eq. (3.1) and the paramagnetic has the form

$$\sigma_P = \frac{e^2 h}{8\pi^2 M_e^2 c^2 \, \Delta E} \, \langle 1/r^3 \rangle_{2p} [Q_N + \sum Q_{ND}] \qquad (3.3)$$

Here, $\langle 1/r^3 \rangle_{2p}$ is the average value of the inverse cube of the distance of the $2p$ electrons from the nucleus, and the Q terms refer to terms in the matrix representation of the molecular orbital formalism for the unperturbed molecule (Q_N has a value of approximately 2; Q_{ND} is 0 for alkanes and acetylene, 0.4–0.6 for alkene carbons, and 0.4 for ketones).

For molecular hydrogen, an averaged Lamb term of 32.1 ppm has been calculated.[4] This is reduced to 26.7 when the paramagnetic term is included,[4,5] in excellent agreement with the experimental value of 26.6 ppm. For larger molecules, no similar success has been attained, because, with increase in molecular size and complexity, the paramagnetic term becomes larger and σ becomes a small difference between large numbers, neither of which can be precisely evaluated.

The molecular screening constant is actually anisotropic or *directional*: it depends on the orientation of the molecule with respect to the magnetic field. It is expressed as a tensor, a mathematical quantity having both direction and magnitude, and is composed of three principal values, σ_{ii}:

$$\sigma = \lambda_{11}^2 \sigma_{11} + \lambda_{22}^2 \sigma_{22} + \lambda_{33}^2 \sigma_{33} \qquad (3.4)$$

where λ_{ii} are the direction cosines of the principal axes of the screening constant with respect to the magnetic field. (Direction cosines represent the values of the cosines of the angles describing the orientation of the axes.) In Fig. 3.2 these mutually perpendicular axes are represented in the laboratory frame for a molecule containing a $^{13}C-X$ bond. One of these

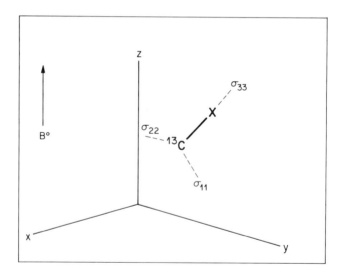

Fig. 3.2.   The principal values of the carbon-13 screening tensor, σ_{11}, σ_{22}, and σ_{33}, represented along three mutually perpendicular axes in a laboratory frame for a molecule containing a $^{13}C-X$ bond. One of these axes, corresponding to σ_{33}, is shown as coinciding with the $^{13}C-X$ bond, although in general it is not necessary that any of the axes coincide with chemical bonds.

axes, corresponding to σ_{33}, is shown as coinciding with the ^{13}C–X bond, although it is not necessarily the case that any of the axes coincide with chemical bonds. If the molecule is oriented so that the ^{13}C–X bond is along the field direction, the observed chemical shift will correspond to the screening constant σ_{33}; similar statements apply to σ_{11} and σ_{22}. For any arbitrary orientation, the chemical shift is prescribed by Eq. (3.4). In

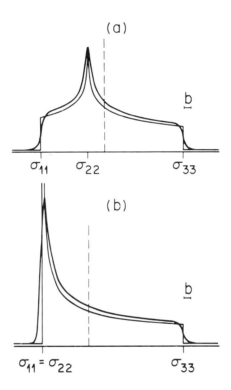

Fig. 3.3.   Powder line shapes for the shielding tensor: (a) general, $\sigma_{11} \neq \sigma_{22} \neq \sigma_{33}$; (b) having axial symmetry, $\sigma_{11} = \sigma_{22}$. Theoretical line shape is given by —; the lineshape with Lorentzian broadening equal to — (inset) is given by - - - -. (From M. Mehring, *High Resolution NMR in Solids*, 2nd ed., Springer-Verlag, Berlin and New York, 1983.)

solution, the screening constant is given by the isotropic average of this equation. Since this is $1/3$ for each λ_i^2 we have

$$\sigma = \tfrac{1}{3}(\sigma_{11}+\sigma_{22}+\sigma_{33}) \tag{3.5}$$

(The quantity in parentheses is the *trace* of the tensor, i.e., the sum of the diagonal elements of the matrix representing the tensor.) Such averaging also occurs when the solid is rotated at the "magic angle"; this is discussed in Chapter 8, Sections 8.4 and 8.6.

Most organic solids are polycrystalline, the crystals and their constituent molecules being oriented at all angles with equal probability. Under these circumstances, the screening constant—or chemical shift—takes on a continuum of values, forming the line shapes shown in Fig. 3.3. The dashed curves represent the theoretical line shape, and the solid curves represent the experimental line shape in which a Lorentzian broadening function (half-height width indicated) has been incorporated. Figure 3.3(a) shows a general, asymmetric powder line shape where the principal values of the screening tensor all differ. In Fig. 3.3(b) the chemical shift is axially symmetric and the screening constants σ_{11} and σ_{22} are equal and may be designated as $\sigma_\perp$; σ_{33} may then be designated as $\sigma_\parallel$. If $\sigma_{22} = \sigma_{33}$, as may also happen, the pattern will be reversed left to right. It should be noted that σ_{33} is customarily taken as corresponding to the largest shielding value, but actually the observation of such patterns does not tell us the orientations of the principal axis of the tensor with respect to the molecular framework. To decide this, one must study single crystals or consider symmetry questions and relationships to other known molecules (see Chapter 8, Section 8.3). In the remainder of this discussion, we shall deal only with the isotropic, motionally averaged value of the chemical shift, as observed in the molten or dissolved state.

It has been customary to separate electron currents in complex molecules into more manageable components.[6,7] Fixing our attention on a particular nucleus whose shielding is in question, we may recognize three contributions:

   a.   Local diamagnetic currents about this nucleus,

   b.   Local paramagnetic currents about this nucleus, and

   c.   Currents flowing in distant groups.

It must be emphasized that this division is arbitrary and to some extent fictional. Nevertheless, it is important because it enables one to attack the problem of shielding in organic molecules in a way which is intuitively reasonable and qualitatively useful, and which in some cases has been made the basis of successful quantitative treatment as well.

Local diamagnetic currents (a) account for the shielding of isolated nuclei as we have seen; in molecules, this term is still very significant, but its contribution is reduced by the deshielding term (b), which is absent in strictly spherical charge clouds. For protons, term (b) is not likely to be important because it depends upon the existence of low-lying excited states, and for hydrogen atoms there are no such states. However, because of their small local electron density, protons are particularly susceptible to the shielding and deshielding effects of currents in neighboring atoms and groups, component (c). Term (b) is very significant for ^{13}C and all nuclei other than ^1H and ^2H.

To these three terms should be added another contribution, (d) the intramolecular reaction field.[8] An atom having a spherical electron distribution will experience a distortion of the electron cloud when placed in an electric field, E. This distortion is proportional to E^2 and, as we have seen, will result in a reduction in the shielding of the nucleus. If the molecule has an electric dipole moment, the intramolecular electric field may produce significant shielding effects; in molecules with no overall electric dipole moment, local fields can still occur, and these may have a significant influence in some cases. Certain ^{13}C and ^{19}F shielding effects appear to arise from this cause.

Whereas proton chemical shifts cover a range of only about 15 ppm, all other nuclei exhibit much greater ranges. Carbon-13, for example, exhibits over 350 ppm or, if one includes the unusual case of CI_4, 650 ppm. These greater variabilities are to be attributed to the greater electron densities about such nuclei. These greater densities permit greater variations, under influences that we shall discuss in the remainder of this chapter.

3.3.2 Inductive Effects

As we have seen in Chapter 1, the most obvious correlation of chemical shift with molecular constitution is the observation that electronegative atoms and groups cause deshielding of nuclei attached to or near them, and that this deshielding effect falls off rapidly as the number of intervening bonds is increased. This correlation is by no means exact and can be seriously perturbed by other effects, but it is nevertheless one of the basic principles that the chemist relies on when interpreting an unfamiliar spectrum.

The influence of substituent negativity is illustrated by the general trends summarized for ^1H in Fig. 3.4 and ^{13}C in Fig. 3.5 (other nuclei are

described in Chapter 9, Section 9.1). Electron attracting groups such as
$-O-$, $-OH$, $-NO_2$, $-CO_2R$, and halogen atoms reduce the electron
density about nearby nuclei and thereby reduce their shielding. For
example, methyl halides show these values for the shielding of protons
(measured for dilute solutions in carbon tetrachloride and expressed in ppm
from tetramethylsilane): CH_3I, 2.16; CH_3Br, 2.68; CH_3Cl, 3.05; CH_3F,
4.26. For ^{13}C: $^{13}CH_3I$, -21.2; $^{13}CH_3Br$, 11.8; $^{13}CH_3Cl$, 19.8; $^{13}CH_3F$,
78.8. The organic chemist seeking to interpret the spectrum of a complex
compound is concerned with the propagation of such inductive effects
through a system of bonds. For protons in alkyl groups larger than methyl,
interpretation is unfortunately not so simple as one might like, although
some valuable correlations have been demonstrated. Shoolery[9,10] (see also
G. Klose[11] and R. Ettinger[12]) showed that for a series of ethyl compounds,
CH_3CH_2X, the *difference* in the proton chemical shift between the CH_3
and CH_2 groups is linearly related to the Pauling electronegativity of the

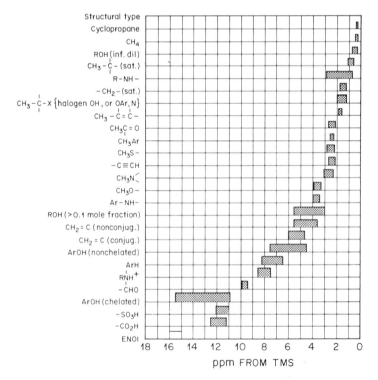

Fig. 3.4.   Chemical shifts for protons in various structures and
functional groups, expressed in ppm from tetramethylsilane
(TMS).

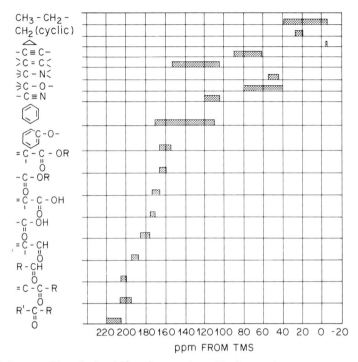

Fig. 3.5.   Chemical shifts for carbon-13 in various structures and functional groups, expressed in ppm from tetramethylsilane (TMS).

atom X. Spiesecke and Schneider[13] measured the CH_3 and CH_2 proton chemical shifts in a series of ethyl compounds and found that the shielding of the CH_2 groups decreased linearly with the electronegativity of X (but not so rapidly as in the corresponding methyl compounds) and that the CH_3 group showed the same trend but to a much weaker degree. But the shielding of the CH_3 protons in ethyl halides actually *increases* with increasing electronegativity. The same trends are found for the ^{13}C chemical shifts. This and similar trends have been attributed to the shielding anisotropies of neighboring groups (*vide infra*) but Cavanaugh and Dailey[14] have suggested that additional factors are involved.

In Fig. 3.6 are the proton spectra of a number of aliphatic compounds that exhibit the transmission of the inductive effect of electronegative groups in straight chains. Multiple substitutions of groups on the same carbon atom tend to produce at least approximately additive effects on the shielding of protons remaining on that carbon atom. It has been shown[15] that one may assign characteristic shielding constants to a number of

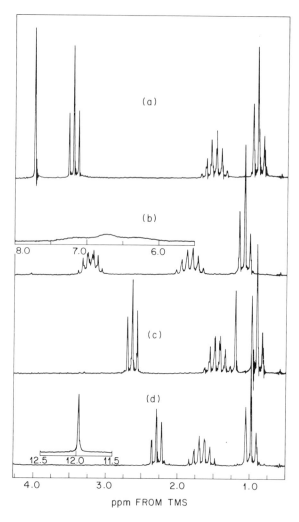

Fig. 3.6.   Proton spectra illustrating the transmission of inductive
effects in straight-chain aliphatic compounds, all observed at
100 MHz at 25° using 10% (wt./vol.) solutions in CCl$_4$
except for (b), which was observed in trifluoroacetic acid:
(a) *n*-propyl alcohol, $CH_{3(3)}CH_{2(2)}CH_2(1)OH$; (b) *n*-
propylammonium   ion, $CH_{3(3)}CH_{2(2)}CH_{2(1)}NH_3^+$;   (c)
propylamine, $CH_3(c)CH_{2(2)}CH_{2(1)}NH_2$; (d) butyric acid,
$CH_{3(3)}CH_{2(2)}CH_{2(1)}CO_2H$. In all spectra, H$_{(3)}$ appears as a
triplet near 1.0 ppm; H$_{(2)}$ is near 1.5 ppm but is appreciably

more deshielded by the NH_3^+ and CO_2H group than by OH
and NH_2; $H_{(1)}$ shows marked variation, the spectra being
arranged in order of increasing shielding of these protons. In
(a) the OH proton appears at *ca.* 4.0 ppm; in (b), the NH_3^+
protons appear as a broad triplet (inset) with 50 Hz spacing
arising from $^{14}N-H$ coupling; in (c) the NH_2 protons appear
at *ca.* 1.2 ppm, the deshielding effect of the positive charge in
(b) being thus shown; in (d) the carboxyl protons appear at
12 ppm (inset).

common groups, including halogens, $-OR$, $-NR_1R_2$, $-SR$,
$-CR = CR_1R_2$, $-C \equiv CH$, $-CH$, and CH_3, and thereby predict proton
chemical shifts in open-chain compounds of the types $CH_2X_1X_2$ or
$CHX_1X_2X_3$, more accurately for the former than the latter. Rules and
regularities tend to fail in cyclic compounds because their conformations
are more rigidly defined and anisotropic shielding effects are more
prominent.

The transmission of inductive effects in carbon-13 shielding is
illustrated by the schematic spectra in Fig. 3.7. Theory[16,17] suggests that
the deshielding influence of an electronegative element or group should be
propagated down the chain with alternating effect, decreasing as the third
power of the distance. This prediction is not borne out. It is true that the
shielding of C_1 shows the expected dependence on the electronegativity of
the attached atom X. Note particularly the effect of F and O. The
relative effect of the halogens is also what one expects. Iodine actually
causes C_1 to be more shielded than C_5. We observe, however, that the
shielding of C_2 is independent of electronegativity. The substitution of any
element for the H of pentane produces about the same deshielding effect at
C_2, i.e., 8 to 10 ppm. We also see that such substitution causes a *shielding*
of C_3 by *ca.* 2−6 ppm, the effect of carbon being at the low end of this
range. We shall deal with these effects when we discuss hydrocarbon
shieldings (*vide infra*). We may say at once, however, that they are not
understood theoretically.

3.3.3 Carbon-13 Shielding in Paraffinic Hydrocarbons; Steric Effects

Carbon-13 chemical shifts are very sensitive to molecular structure and
geometry, quite apart from the influence of substituent groups. This is
particularly clearly revealed in the ^{13}C spectra of paraffinic hydrocarbons.
In contrast to paraffinic protons, which embrace a range of only about
2 ppm (Fig. 3.4), carbon chemical shifts are spread over more than
50 ppm. This makes ^{13}C spectroscopy a particularly powerful means for

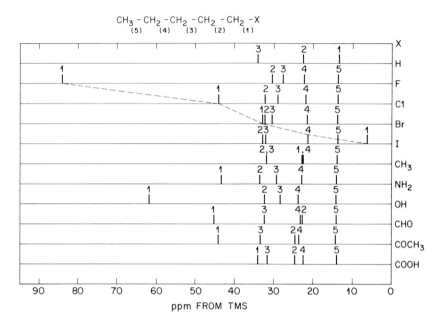

Fig. 3.7.   Schematic carbon-13 spectra illustrating the transmission of inductive effects in straight-chain aliphatic compounds. All spectra are represented as proton-decoupled so that carbon resonances appear as singlets.

the study of paraffins. Let us now consider the correlations and regularities that have been observed.[13,18-21] The empirical ordering of chemical shifts may be done in terms of α, β, and γ effects[19,20] (δ and ϵ effects are further refinements which we omit for the moment). Table 3.1 shows a series of data for simple hydrocarbons that illustrate the α effect. Here, we are to keep our eye on the °C carbon and ask what happens on adding carbons α to this one. We see a regular deshielding of about 9 ± 1 ppm for each added carbon, except in neopentane, where crowding apparently reduces the effect.

In Table 3.2 we see examples of the β effect, i.e., the effect of added carbons β to the one observed, °C. The effect is of similar magnitude to that produced by α carbons. Examples (a) through (d) illustrate the effect for terminal carbons. For non-terminal carbons [(e) through (l)], the effect is similar, but is reduced in magnitude if °C is a branch point.

Finally, we must consider the γ effect, which, although smaller than the α and β effects, is in some respects more significant. It is a *shielding*

TABLE 3.1 α-Effect in ^{13}C Shielding

	structure	ppm from TMS	α-effect ppm
(a)	$^\circ CH_3 \!-\! H$	-2.1	—
(b)	$^\circ CH_3 \!-\! {}^\alpha CH_3$	5.9	8.0
(c)	$^\circ CH_2 <{}^{{}^\alpha CH_3}_{{}_\alpha CH_3}$	16.1	10.2
(d)	$^\circ CH <{}^{{}^\alpha CH_3}_{{}^\alpha CH_3}\!\!-\!{}^\alpha CH_3$	25.2	9.1
(e)	$^\circ C$ with four ${}^\alpha CH_3$	27.9	2.7

rather than a deshielding effect, and it was suggested by Grant and Cheney[22] that it has a conformational origin, resulting from non-bonded steric interaction of the carbons. It was further suggested that the effect is mediated through their attached protons. We have seen, however (Fig. 3.7) that the γ effect is not limited to carbon–carbon interactions but is in fact larger for other elements and thus does not require proton mediation. It also appears to show a correlation with electronegativity.

For a clearer understanding of the γ effect, consider the staggered conformations of compounds (b), (c), and (d) in Table 3.3. (Scheme 1, p. 102)

In the case of butane, we divide the observed average γ shift by the gauche conformer content of butane near room temperature (ca. 0.45) and find a value of -5.3 ppm per gauche interaction. We shall later see that this quantity explains quite accurately, among other observations, the dependence of carbon chemical shifts on stereochemical configuration in many chiral compounds and macromolecules. For the more crowded compounds 2-methylbutane and 2,2-dimethylbutane, the effect appears somewhat smaller.

TABLE 3.2 β-Effect in ^{13}C Shielding

	structure	ppm from TMS	β-effect ppm
(a)	$^{\circ}CH_3 - {}^{\alpha}CH_3$	5.9	—
(b)	$^{\circ}CH_3 - {}^{\alpha}CH_2 - {}^{\beta}CH_3$	15.6	9.7
(c)	$^{\circ}CH_3 - {}^{\alpha}CH \bigg\langle {}^{\beta}CH_3 \atop {}^{\beta}CH_3$	24.3	8.7
(d)	$^{\circ}CH_3 - {}^{\alpha}C \Big\langle {}^{\beta}CH_3 , {}^{\beta}CH_3 , {}^{\beta}CH_3$	31.5	7.2
(e)	$^{\alpha}CH_3 - {}^{\circ}CH_2 - {}^{\alpha}CH_3$	16.1	—
(f)	$^{\alpha}CH_3 - {}^{\circ}CH_2 - {}^{\alpha}CH_2 - {}^{\beta}CH_3$	25.0	8.9
(g)	$^{\alpha}CH_3 - {}^{\circ}CH_2 - {}^{\alpha}CH \big\langle {}^{\beta}CH_3 , {}^{\beta}CH_3$	31.8	6.8
(h)	$^{\alpha}CH_3 - {}^{\circ}CH_2 - {}^{\alpha}C \Big\langle {}^{\beta}CH_3 , {}^{\beta}CH_3 , {}^{\beta}CH_3$	36.7	4.9
(i)	${}^{\alpha}CH_3 \atop {}^{\alpha}CH_3} {\Big\rangle} {}^{\circ}CH - {}^{\alpha}CH_3$	25.2	—
(j)	${}^{\alpha}CH_3 \atop {}^{\alpha}CH_3} {\Big\rangle} {}^{\circ}CH - {}^{\alpha}CH_2 - {}^{\beta}CH_3$	29.9	4.7
(k)	${}^{\alpha}CH_3 \atop {}^{\alpha}CH_3} {\Big\rangle} {}^{\circ}CH - {}^{\alpha}CH \big\langle {}^{\beta}CH_3 , {}^{\beta}CH_3$	34.1	4.2
(l)	${}^{\alpha}CH_3 \atop {}^{\alpha}CH_3} {\Big\rangle} {}^{\circ}CH - {}^{\alpha}CH \Big\langle {}^{\beta}CH_3 , {}^{\beta}CH_3 , {}^{\beta}CH_3$	38.1	4.0

TABLE 3.3 γ-Effect in ^{13}C Shielding

structure	ppm from TMS	γ-effect ppm
(a) $^{\circ}CH_3 \mathbin{\mathrm{\vert}} {^{\alpha}CH_2} - {^{\beta}CH_3}$	15.6	—
(b) $^{\circ}CH_3 \mathbin{\mathrm{\vert}} {^{\alpha}CH_2} - {^{\beta}CH_2} - {^{\gamma}CH_3}$	13.2	− 2.4
(c) $^{\circ}CH_3 \mathbin{\mathrm{\vert}} {^{\alpha}CH_2} - {^{\beta}CH}\!\!<\!{^{\gamma}CH_3} \atop {^{\gamma}CH_3}$	11.3	− 1.9
(d) $^{\circ}CH_3 \mathbin{\mathrm{\vert}} {^{\alpha}CH_2} - {^{\beta}C}\!\!<\!{^{\gamma}CH_3 \atop ^{\gamma}CH_3}\ {^{\gamma}CH_3}$	8.8	− 2.5
(e) $^{\alpha}CH_3 - {^{\circ}CH_2} \mathbin{\mathrm{\vert}} {^{\alpha}CH_2} - {^{\beta}CH_3}$	25.0	
(f) $^{\alpha}CH_3 - {^{\circ}CH_2} \mathbin{\mathrm{\vert}} {^{\alpha}CH_2} - {^{\beta}CH_2} - {^{\gamma}CH_3}$	22.6	− 2.4
(g) $^{\alpha}CH_3 - {^{\circ}CH_2} \mathbin{\mathrm{\vert}} {^{\alpha}CH_2} - {^{\beta}CH}\!\!<\!{^{\gamma}CH_3 \atop ^{\gamma}CH_3}$	20.7	− 1.9
(h) $^{\alpha}CH_3 - {^{\circ}CH_2} \mathbin{\mathrm{\vert}} {^{\alpha}CH_2} - {^{\beta}C}\!\!<\!{^{\gamma}CH_3 \atop ^{\gamma}CH_3}\ {^{\gamma}CH_3}$	18.8	− 1.9

Apart from specific considerations of conformation or stereochemistry, these correlations and regularities may be used to make approximate but very useful predictions of carbon-13 chemical shifts. To calculate the chemical shift of $^{\circ}C$ we proceed through four steps:

1. *Determine the extent of substitution of* $^{\circ}C$. This is represented by an index i, which is the number of carbons attached to $^{\circ}C$. Each type of carbon is associated with a base value (essentially the α effect value). These are listed in Table 3.4.

2. *Determine the extent of substitution of* $^{\alpha}C$. The number of carbons directly bonded to each $^{\alpha}C$ is designated by j. Using i and j one may choose a value from the "$\alpha - \beta$" shift parameters shown in Table 3.5. These incorporate corrections for branching and other effects observable in Tables 3.1–3.3. (There are no entries in the column for $j = 1$ since in this situation $^{\alpha}C$ is a methyl and there are no β–carbons to consider.) One selects from Table 3.5 an appropriate value for each $^{\alpha}C$, sums these values, and adds this quantity to the base value selected from Table 3.4.

Scheme 1

Butane:

*CH_3   *CH_3   *CH_3

H   H   H   CH_3   CH_3   H

H   H   H   H   H   H

CH_3   H   H

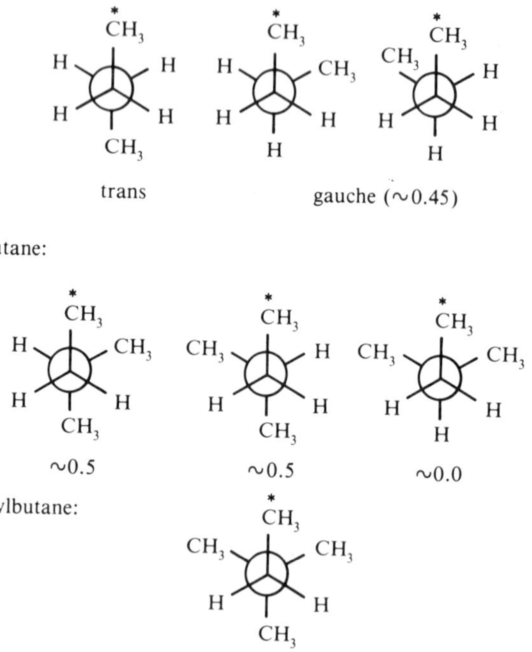

C* shielding
vs. propane

−2.4 ppm
or
5.3 ppm
per contact

trans   gauche (∼0.45)

2-Methylbutane:

*CH_3   *CH_3   *CH_3

H   CH_3   CH_3   H   CH_3   CH_3

H   H   H   H   H   H

CH_3   CH_3   H

∼0.5   ∼0.5   ∼0.0

−4.3 ppm
or
∼4.3 ppm
per contact

2,2-Dimethylbutane:

*CH_3

CH_3   CH_3

H   H

CH_3

−6.8 ppm
or
∼3.4 ppm
per contact

TABLE 3.4

°C	i	Base value
methyl	1	6.8
methylene	2	15.3
methine	3	23.5
quaternary	4	27.8

TABLE 3.5

$i \backslash j$	1	2	3	4
1	—	9.6	17.8	25.5
2	—	9.8	16.7	21.4
3	—	6.6	11.1	14.7
4	—	2.3	4.0	7.4

TABLE 3.6

i	γ effect
1	−3.0
2	−2.7
3	−2.1
4	+0.7

TABLE 3.7

i	δ effect
1	0.5
2	0.3
3	0.0
4	0.0

3. *Count the number of $^{\alpha}C$ carbons and choose a γ-effect parameter from Table 3.6 on the basis of the index i.*

4. *As a further refinement, count the number of $^{\delta}C$ carbons.* This is then multiplied by the appropriate number selected from Table 3.7.

Finally, one may include a small deshielding ϵ-*effect* of 0.1 ppm for each carbon atom five bonds removed from the observed carbon. This is principally useful in the prediction of chemical shift *differences* in closely related structures.

We now illustrate the application of these procedures to the prediction of the chemical shifts of two paraffinic hydrocarbons, one linear and one highly branched.

n-**Hexane**

$$C - C - C - C_3 - C_2 - C_1 \qquad \text{(calculated values}$$
$$31.8 \quad 22.9 \quad 13.7 \qquad \text{are in parentheses)}$$
$$(31.8) \ (22.5) \ (14.1)$$

For C_1, since it represents a methyl group, i is 1 and the base value is 6.8. There is one $^{\alpha}C$, C_2, and it is secondary; therefore $j = 2$ and the appropriate $\alpha - \beta$ factor is 9.8, making a total of 16.6. Since $i = 1$, the γ factor, chosen from Table 3.6, is −3.0 and the δ shift (Table 3.7) is 0.5, for a total of 14.1. The experimental value is 13.7.

For C_2, $i = 2$ and the base value is 15.3. Of the two $^\alpha C$'s, one is the methyl group, which is already accounted for by the base value. The other $^\alpha C$ is C_3 for which $j = 2$ and the $\alpha - \beta$ factor 9.6. Adding to this a γ effect of -2.7 and a δ effect of 0.3 we obtain a total of 22.5. The experimental value is 22.9.

For C_3, i is again 2, for a base value of 15.3. Since both $^\alpha C$'s are methylenes we have $j = 2$ and must add an $\alpha - \beta$ effect of 9.6 for each. There is one γ effect of -2.7 and no δ effect. The predicted value of 31.8 coincides with the experimental value.

We may see from these values that for a linear paraffinic chain of infinite length or for any methylene carbon 5 or more bonds removed from the end of a paraffinic chain, we predict a shielding of $15.3 + 2(9.6) - 2(2.7) + 2(0.3) + 2(0.1)$ — the last representing 2ϵ effects — totalling 29.9 ppm; this agrees with the observed value.[23,24]

2,2,4-trimethylpentane

$$\begin{array}{ccccc} & \overset{C_5}{|} & & \overset{C_1}{|} & \\ C_5 & - C_4 & - C_3 & - C & - C_1 \\ 24.7 & 25.3 & 53.3 & | \ 30.9 & 29.9 \\ (23.1) & (23.8) & (53.4) & \underset{|}{C} {}^{(31.5)} & (30.3) \end{array}$$

(calculated values are in parentheses)

Carbon-13 spectra of this molecule are shown in Chapter 4, Figs. 4.49 and 4.51. We note that all three C_1 carbons are made equivalent by fast rotation about the C_2–C_3 bond. Since they are methyls, $i = 1$ and we choose a base value of 6.8. There is one $^\alpha C$ and since it is quaternary $j = 4$, corresponding to an $\alpha - \beta$ factor of 25.5. So far we have a total of 32.3. We have a γ-effect from C_4 and we find from Table 3.6 that the appropriate value is -3.0. We also have δ-effects of 0.5 (Table 3.7) from each of the C_5 carbons. The predicted shift is 30.3 compared to an experimental value of 29.9.

For C_2, a quaternary carbon, the base value is 27.8. Three of the attached α carbons are methyls and so from Table 2, for $i = 4$ and $j = 1$, we find no contribution from these carbons since it has already been incorporated in the base value. The fourth $^\alpha C$ is a methylene ($j = 2$) and so we add an $\alpha - \beta$ shift of 2.3 ppm. There are two γ effects from the C_5 carbons, which have a value of $+ 0.7$ each. The predicted value is 31.5 compared to an observed value of 30.9.

C_3 has two α carbons and five β carbons, so we expect it to be highly deshielded. We select a value of 15.3 from Table 3.4, corresponding to a methylene carbon. Of the two $^\alpha C$'s, one is quaternary ($j = 4$) and so we choose an $\alpha - \beta$ factor of 21.4 from Table 3.5. The other $^\alpha$C is a methine, corresponding to an $\alpha - \beta$ factor of 16.7. The sum of these factors is 53.4 and there are no γ, δ, or ϵ effects. The experimental value is 53.3 ppm.

C_4 is a methine carbon so the base shift value is 23.5 ppm. There are three $^\alpha$C's bonded to it, one a methylene. For this, $i = 3$ and $j = 2$ and so the $\alpha - \beta$ factor is 6.6. But the C_1 carbons supply three γ interactions, each -2.1 ppm. There are no δ effects. The predicted shift is thus 23.8 compared to an experimental value of 25.3.

Finally, we consider the C_5 carbons; since these are methyls $i = 1$ and the base value is 6.8 ppm. There is one $^\alpha$C, which is methine, and one $^\beta$C, a methylene. Therefore $J = 3$, corresponding to a shift of 17.8 ppm (Table 3.5). We have one γ-effect from C_2, worth -3.0 since $i = 1$ (Table 3.6) and three δ effects from C_1 carbons, worth 0.5 each or 1.5 ppm. The total of these is 23.1 and the observed value is 24.7.

Again, as with C_1, the C_5 carbons are made equivalent by rapid rotation of the isopropyl group about the C_3–C_4 bond. However, it is to be noted that while *tert*-butyl methyl carbons are always made equivalent in this way, this is not necessarily so for isopropyl groups. Thus, in the isomeric 2,4-dimethylhexane

$$
\begin{array}{ccccc}
& \overset{\displaystyle C_7}{\underset{\displaystyle |}{}} & & \overset{\displaystyle C_1}{\underset{\displaystyle |}{}} & \\
C_6 - C_5 - & C_4 & - C_3 - & C_2 & - C_1
\end{array}
$$

the C_1 carbons give two distinct resonances separated by 1.0 ppm, while in 2,3,4-trimethylpentane

$$
\begin{array}{ccc}
\overset{\displaystyle C}{\underset{\displaystyle |}{}} \ \ \overset{\displaystyle C}{\underset{\displaystyle |}{}} \ \ \overset{\displaystyle C}{\underset{\displaystyle |}{}} \\
C_5 - C_4 - C_3 - C_2 - C_1
\end{array}
$$

the isopropyl carbons differ by 3.3 ppm. These molecules present circumstances that are not contemplated in the procedures described here. They require consideration of conformer populations and will be discussed further in Section 3.3.6.

We have discussed open chain paraffinic hydrocarbons in some detail, but the general principles of α, β, and γ effects are more broadly applicable. Relative chemical shifts in molecules with functional groups may be predicted. These considerations are particularly relevant in comparing the spectra of closely related isomers and homologs.

3.3.4 Shielding by Magnetically Anisotropic Groups

3.3.4.1 General

We have seen (Section 3.3.1) that it is convenient and useful to divide the shielding of a particular nucleus into terms arising from local diamagnetic and paramagnetic currents, and a term arising from electron currents flowing in neighboring parts of the molecule not necessarily directly associated with the observed nucleus. It is this latter effect which we now wish to examine.

Let us suppose first that the electron cloud in this neighboring group is spherically symmetrical, as might be nearly the case if it were simply a single atom. In this case, as we have seen, only the diamagnetic circulation of this atom's electrons needs to be considered, there being no paramagnetic term. The induced field will be independent of the direction of the applied field B_0. If the line joining the atomic group to the observed nucleus tumbles randomly and rapidly in all directions, as would be the case for a molecule in a liquid, it can be shown that the net averaged field produced by the group at the observed nucleus will be zero. Thus, remote shielding effects can be produced only by electron groups that are *not* spherically symmetrical and consequently are magnetically anisotropic. Such a group is represented in a generalized manner as g in Fig. 3.8. It has a principal axis A, along which the diamagnetic susceptibility is $\chi_{\parallel}$. What this means is that if all the molecules were aligned so that A were parallel to B_0 the observed magnetic susceptibility would be $\chi_{\parallel}$, and consequently the observed molecular diamagnetic moment would be $B_0\chi_{\parallel}$. For a group with a threefold or higher principal symmetry axis, there will be one other susceptibility, $\chi_{\perp}$, perpendicular to A, giving rise to a moment $B_0\chi_{\perp}$; if all of the molecules were to be aligned so that A were perpendicular to B_0, the observed diamagnetic susceptibility would be $\chi_{\perp}$. (For groups of lower symmetry, a third susceptibility must be defined, but we shall not consider this here.) Such a magnetically anisotropic group will influence the shielding of the "distant" nuclei $H_{(a)}$ or $H_{(b)}$ even when rapid, random tumbling of A occurs. McConnell[25] has shown that if these observed nuclei are sufficiently distant so that the anisotropic group g can be regarded as a point magnetic dipole, then the mean screening of these nuclei by the group g in the tumbling molecule will be expressed by

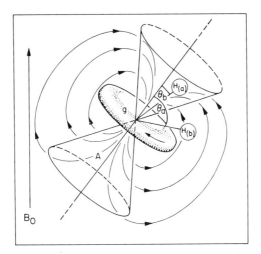

Fig. 3.8.   The shielding region about an electron grouping g, which has
cylindrical but not spherical symmetry. Protons $H_{(a)}$ and
$H_{(b)}$ are in the shielding and deshielding zones, respectively,
the nodal surface of zero shielding being the double cone.

$$\sigma_g = \frac{(\chi_\parallel - \chi_\perp)(1 - 3\cos^2\theta)}{3r^3} \tag{3.6}$$

where r is the distance from the observed nucleus to the point dipole and θ
is the angle between A and the line joining the nucleus to the point dipole.
Both θ and r will of course be fixed in a rigid molecule, even though
tumbling. Here, $\chi_\parallel - \chi_\perp$, or $\Delta\chi$, is the anisotropy of the diamagnetic
susceptibility. Since the susceptibilities are negative numbers, $\Delta\chi$ will be
negative when $\chi_\parallel$ is larger (more negative) than $\chi_\perp$.

As Eq. (3.6) and Fig. 3.8 indicate, the shielding of nuclei, protons in
particular, in the presence of such an electron group will depend on their
geometrical relationship to the symmetry axis. A proton located in or near
a plane passing perpendicularly to A through the group (such as $H_{(b)}$) will
experience deshielding, since the induced field reinforces B_0 here; σ_g will be
negative in this region, which may be termed the deshielding region. A
proton near the symmetry axis, such as H_a, will experience shielding since
here the induced field opposes B_0. Equation 3.6 delineates a cone-shaped
zone of excess shielding extending along the symmetry axis and making an
angle of 54.7° with it (Fig. 3.8). The surface of this zone is a nodal
surface in which σ_g is zero. Outside the cone there will be deshielding.

3.3.4.2 Acetylenes

A striking example of shielding arising from diamagnetic anisotropy is that occurring in acetylenes. By the usual criteria of bond hydridization and acidity, there should be a progressive deshielding of the protons as we pass from ethane to ethylene to acetylene. Instead, we find (for dilute solutions in carbon tetrachloride) the values 1.96, 5.84, and 2.88, respectively. The anomalously high degree of shielding in acetylene (observed also in substituted acetylenes) has been explained[26,27] as arising from paramagnetic circulation of electrons at the carbon atom occuring when the molecular axis is perpendicular to the applied field. An alternative and intuitively somewhat more appealing view of the magnetic anisotropy in acetylene follows from the previous discussion if we assume that electron currents flow preferentially in a direction perpendicular to the molecular axis (Fig. 3.9). The protons, being located on the symmetry axis, will clearly be expected to experience an increased shielding.

3.3.4.3 Aromatic Ring Currents

In contrast to acetylenic protons, protons on aromatic rings resonate at unexpectedly *low* fields, usually in the range 7–8 ppm (see Fig. 3.4). The known large diamagnetic anisotropies of aromatic molecules have been explained by Pauling[28] as being due to a circulation of π–electrons in the plane of the ring, giving rise to enhanced diamagnetic susceptibility in a direction perpendicular to the ring. This circulation of electrons may be considered as a superconducting current flowing in a ring having a radius equal to that of the aromatic ring. The current thus flows in a path

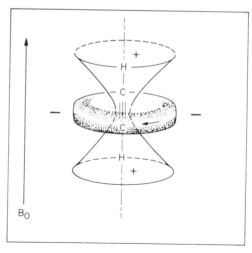

Fig. 3.9.   The shielding zone about an acetylenic bond.

embracing many nuclei, giving rise to a circular magnetic shell. This shell may be replaced by an equivalent point dipole[29] but a better approximation is to retain a circular path for the electrons. If n electrons circulate in a path of radius a, it can be shown[30-32] that the deshielding effect σ_g (expressed in ppm) of this current is given for the rapidly tumbling molecule by

$$\sigma_g = \frac{10^6 \times ne^2}{6\pi M_e c^2 a} \frac{1}{[(1+\rho)^2 + z^2]^{1/2}} \left[K + \frac{1-\rho^2-z^2}{(1-\rho)^2 + z^2} E \right] \quad (3.7)$$

where n is the number of circulating electrons of charge e and mass M_e, c is the velocity of light, ρ and z are the usual cylindrical coordinates (ρ being the perpendicular distance from the hexad axis and z the distance along the hexad axis from the origin at the center of the ring), and K and E the complete elliptic integrals whose modulus k is expressed by

$$k^2 = \frac{4\rho}{(1+\rho)^2 + z^2}$$

The calculation is somewhat complicated by the fact that the π-electron cloud in an aromatic ring, which we shall take to be a benzene ring in the remaining development, does not actually have its maximum density in the plane of the carbon atoms, but exists rather as two doughnutlike rings (probably somewhat hexagonal in form, as shown in Fig. 3.10), one on

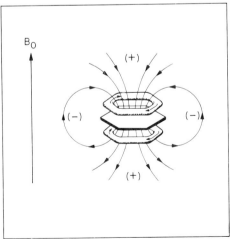

Fig. 3.10.   Electron density, ring currents, and magnetic lines of force about a benzene ring.

each side. The spacing of these rings may be taken as an adjustable parameter. Reasonably good agreement of predicted and calculated values of σ_g for benzene and simple alkylbenzenes can be obtained by assuming a separation of π clouds of *ca.* 0.92 *a* or 0.128 nm. Shielding values are then calculated by summing the two magnetic shells arising from these clouds. The results are presented in Figs. 3.11 and 3.12. In Fig. 3.11 "isoshielding" lines are shown over a region extending outward 5.0 ring radii (0.695 nm) in the plane of the carbon ring and 3.5 ring radii (0.487 nm) along the symmetry axis. This plot represents one quadrant of a plane passing normally through the center of the benzene ring. The line for $\sigma_g = 0$ represents the cross section of the nodal surface separating the shielding and deshielding regions, corresponding to the cone surface of Fig. 3.8, but distorted near the ring because a magnetic shell rather than a point dipole is assumed. The "isoshield" for $z = 0$ and $\rho = 1.78$, corresponding to a value of -1.50 ppm for σ_a, represents a proton attached directly to the ring.

The classical approach embodied in the Waugh–Fessenden and Johnson–Bovey ring current calculations is not the only method of dealing with the shielding effect of aromatic rings. Quantum mechanical calculations[33,34] produce a very similar result to that shown in Fig. 3.11 with the difference that this approach predicts less shielding in the "shielding cone" above the ring and less deshielding in the region in and

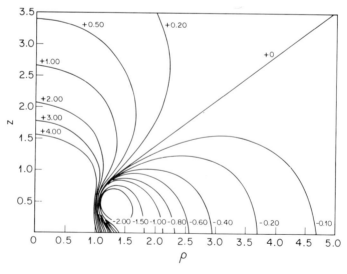

Fig. 3.11.   The shielding zone about a benzene ring. The "isoshielding" lines are calculated from the model represented in Fig. 3.10.

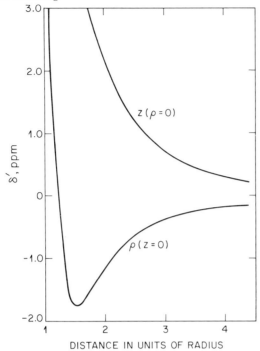

Fig. 3.12.   Shielding by a benzene ring as a function of distance along the hexad axis (upper curve) and in the plane of the ring, measured from its center (lower curve).

near the plane of the ring. The large literature on this subject has been reviewed by Haigh and Mallion.[35]

Most protons in aromatic molecules experience the deshielding influence of the ring current, but marked excess shielding ought to be observed if the molecular geometry is appropriate. Waugh and Fessenden[30,31] observed that in the 1,4-polymethylenebenzenes the central methylene protons, being held in a position directly above the ring, show such shielding. Thus, in 1,4-decamethylenebenzene or [10]paracyclophane, the four central (ϵ) methylene protons appear at *ca.* 0.3 ppm (CCl$_4$ solution), appreciably higher than the value of *ca.* 1.3 ppm generally observed for methylene

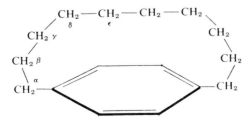

groups in paraffinic chains. Values have been reported for other paracyclophanes with shorter polymethylene chains.[36] For example, the protons of the two central methylene groups of [6]paracyclophane appear at −0.62 ppm, whereas one of those of the central methylene group in [5]paracyclophane appears at 0.01 ppm. This latter compound is unstable and was observed at −72°C, at which temperature the aliphatic chain is static. The unexpected lesser shielding is apparently not due to a reduced current in its ring, which is somewhat bent, because the aromatic protons are more deshielded than in the [6]paracyclophane.

For polynuclear aromatic hydrocarbons, a simple extension of the benzene ring current model, assuming a current in each ring equal to that in benzene and a point-dipole approximation for the magnetic moment, does not work well,[37] since the ring currents in each ring appear to vary markedly from that of benzene. (See Haigh and Mallion.[35]) Figure 3.13 shows the spectrum of phenanthrene. The 4,5 protons, which are expected to experience deshielding by all three rings, appear at lowest field. The 2,3,6,7 protons appear at highest field, being the most distant from the other rings. (The complexity of the spectrum is the result of couplings between protons on the same ring, interring couplings being very small.)

Larger conjugated rings may also show marked ring-current shifts. Porphyrins, in which 18 electrons may be regarded as circulating in a path

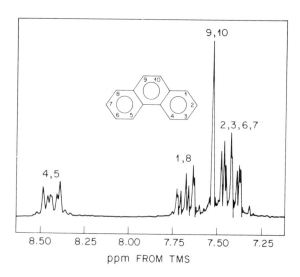

Fig. 3.13.   The proton spectrum of phenanthrene observed at 100 MHz, using a 10% (wt./vol.) concentration of CCl_4.

embracing 18 nuclei, show very strong shielding and deshielding of the ring protons. Becker and Bradley[38] and Becker et al.[39] observed that in copro-

$$M = CH_3; P = CH_2CH_2CO_2CH_3.$$

porphyrin I methyl ester the methine protons, which are "outside" the ring current, appear at 9.96 ppm (in $CDCl_3$), about 2.6–2.7 ppm more deshielded than expected in the absence of the current. Particularly striking is the resonance of the NH protons, which are in the same plane as the methine protons but "inside" the current. They appear at −3.89 ppm, ca. 13 ppm more shielded than the corresponding protons in pyrrole. Similar effects have been observed in a number of other porphyrins.

Many other large rings give evidence of ring currents. Thus [18]annulene shows chemical shifts of approximately 9.28 and −2.99 ppm

(in tetrahydrofuran) for the "outside" and "inside" protons, respectively.[40] Trans-15, 16-dimethyldihydropyrene[41] shows peaks at the remarkable position of −4.25 ppm for the methyl groups; in the absence of special shielding effects such groups would resonate at about 1 ppm.

It appears that ring currents may also be paramagnetic under appropriate circumstances. Quantum mechanical theory predicts that in conjugated monocyclic polyenes, the ring current will be diamagnetic for molecules with $4n + 2$ circulating π electrons, such as benzene and [14]- and [18]annulenes, but that it will be paramagnetic for molecules with $4n$ π electrons, such as [16]- and [24]annulenes. Calder and Sondheimer[42] report that at $-80°$ the outer protons appear at 5.73 ppm and the inner protons at *ca.* 12.9 to 11.2 ppm. Rapid bond shifts and rotations average all protons to a single peak at higher temperatures.

3.3.4.4 Alicyclic Ring Currents

It appears possible that there may be ring currents in saturated rings. Certainly both proton (Table 3.4) and carbon-13 (Table 3.5) shifts of cyclopropane have abnormally shielded values, and it is known that the diamagnetic susceptibilities of cyclopropanes are unusually large.[43] Patel *et al.*[44] have calculated proton chemical shifts for a number of substituted cyclopropanes by assuming circulations of electrons in the plane of the cyclopropane ring. For cyclopropane itself, agreement with observation was obtained by assuming four electrons circulating in a ring of 0.11 nm radius.

3.3.4.5 Double Bonds

The shielding anisotropy in the neighborhood of double bonds is not established with certainty, but appears to correspond to a substantial diamagnetism in a direction perpendicular to the double bond and to the plane containing the single bonds, as shown in Fig. 3.14. The following are two examples of many cases that support this hypothesis:

a. In $\alpha-$ and $\beta-$pinene, one of the gem methyl groups is much more shielded than the other. In $\alpha-$pinene, their chemical shifts are 1.27 ppm and 0.85 ppm (all values for CCl_4 solutions); in $\beta-$pinene, 1.23 ppm and 0.72 ppm. In pinene, the corresponding saturated molecule, these methyl protons appear at 1.17 and 1.01 ppm.[45] The most rational explanation of these marked shielding and deshielding effects is that the methyl groups labeled β are in the shielding zone of the double bond, as shown in Fig. 3.15, whereas the α methyl groups are in both cases within the deshielding zone. The allylic methyl group in $\alpha-$pinene appears at 1.63 ppm. This clearly lies in the deshielding zone, and is also subject to diminished shielding because of the double bond, both effects working in concert.

b. In bicyclo[3.1.0]hexene-2, one of the methylene protons of the cyclopropane ring

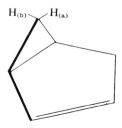

$H_{(b)}$ $H_{(a)}$

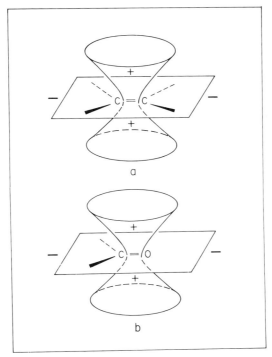

Fig. 3.14.   The shielding region about an olefinic double bond (a) and a carbonyl double bond (b).

actually appears upfield from the tetramethylsilane reference, at −0.17 ppm; the other appears at 0.83 ppm.[46] Cyclopropane itself resonates at 0.22 ppm (Fig. 3.4); thus the methylene protons in this compound are shielded and deshielded by 0.40 and 0.60 ppm, respectively. Again, the simplest explanation is in terms of the olefinic bond anisotropy shown in Fig. 3.14(a). There is ample qualitative evidence for shielding and deshielding effects by carbonyl groups. The very strong deshielding of aldehydic protons, which appears at 9−10 ppm (Fig. 3.4), is doubtless due in large measure to the anisotropy of the carbonyl group, as shown in Fig. 3.16.

Olefinic protons in $\alpha\beta$-unsaturated esters lie in the plane of the carbonyl group unless sterically prevented (Fig. 3.17). Both $H_{(a)}$ and $H_{(b)}$ will be deshielded by the carbonyl group, but one may expect the cis proton, $H_{(a)}$, to be more shielded because it is closer. This is confirmed for a large number of compounds. For example, in methyl acrylate, $H_{(a)}$ is at 6.38

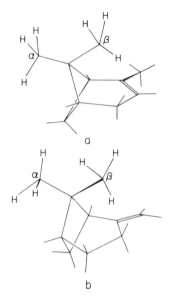

Fig. 3.15.   Orientation of the methyl groups in α-pinene (a) and β-pinene (b).

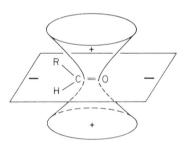

Fig. 3.16.   Shielding of the aldehydic proton.

Fig. 3.17.   Carbonyl conformation in α,β-unsaturated esters.

ppm and $H_{(b)}$ is at 5.82 ppm. In diethyl maleate the olefinic protons

appear at 6.28 ppm, while in diethyl fumarate they appear at 6.83 ppm.

However, there are some difficulties. For example, in formamide, in which

the two amide protons give separate resonances because of the high barrier to rotation about the C–N bond (Chapter 5, Section 5.4.4), $H_{(c)}$ appears at 6.44 ppm and $H_{(b)}$ at 6.65 ppm (see Chapter 4, Section 4.4.3.1), the reverse of expectation. The same trend is observed in a more marked degree by both the methyl protons[47] and carbons[48] of N,N'-dimethylformamide. Clearly, some other influence is at work here.

An example of carbonyl group anisotropy has been given by Williams et al.[49] who have observed that in 11-ketosteroids the equatorial $C_{(1)}$ protons are markedly shifted to lower field as compared to corresponding compounds with no carbonyl group in this position. This has been attributed to the fact that the equatorial $H_{(1)}$ is in the plane of the 11-keto carbonyl group (see Fig. 3.18).

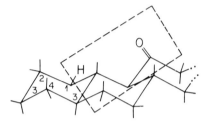

Fig. 3.18.   Carbonyl group shielding in 11-ketosteroids.

3.3.4.6 C − C and C − H Single Bonds

Anisotropic shielding by carbon−carbon (and possibly carbon−hydrogen) single bonds has been invoked to explain a number of commonly observed shielding correlations that do not seem to be readily explained in any other way. One of these may be cited: in cyclohexane, the equatorial protons are deshielded by about 0.48 ppm with respect to the axial protons. For cyclohexane itself, which is in rapid inversion at ordinary temperatures and gives only a single narrow peak, this can only be observed at very low temperature (see Chapter 5, Section 5.4.3). But substituted cyclohexanes with locked conformations show analogous behavior at normal temperatures. This is, in fact, one of the commonest and most useful of empirical correlations, and is widely applied in structural problems, particularly in the steroid field. It is usually explained in terms of the diamagnetic anisotropy of the carbon−carbon bonds (Fig. 3.19). The nearest carbon−carbon bonds (1,2 and 1,6 in the following structural formula) clearly affect both $H_{(ax)}$ and $H_{(eq)}$ equally because of symmetry;

but bonds 2,3 and 5,6 have different geometrical relationships to each. If it is assumed that the diamagnetic susceptibility of the C−C bond is greatest in a transverse direction, there will then be a deshielding zone extending out along the bond direction (Fig. 3.19). The anisotropy is thus opposite to that of the acetylenic bond. Jackman[50] showed that by using Eq. (3.6) and a reasonable value for $\chi_{\parallel} - \chi_{\perp}$, the observed value of the shielding difference for $H_{(ax)}$ and $H_{(eq)}$ in cyclohexane can be explained. It is evident that $H_{(ax)}$ will be in the shielding zones and $H_{(eq)}$ in the deshielding zones of the $C_{(2)(3)}$ and $C_{(5)(6)}$ bonds. (The $C_{(3)(4)}$ and $C_{(4)(5)}$ bonds have a shielding effect on both protons, but this constitutes only a small correction.)

3.3.5 Hydrogen Bonding

It has long been recognized that protons involved in hydrogen bonding exhibit a very marked deshielding. Arnold and Packard[51] observed that the hydroxyl peak of ethanol moved upfield by about 1.5 ppm (with respect to the methylene signal) over the temperature range $-117°$ to $78°$, while a number of investigators[52-55] have observed similar upfield shifts on dilution of alcohols and phenols with inert solvents. The appearance of the ethanol spectrum as a function of concentration in carbon tetrachloride is shown in Fig. 3.20. The positions of the CH_3 and CH_2 resonances are little affected by dilution. The OH peak,[*] in contrast, moves from 5.13 ppm in pure ethanol at $40°$ to *ca.* 1 ppm in 0.5% solution. This increased shielding is due to the breaking up of hydrogen bonded complexes, mainly dimers. The dependence of chemical shift upon dilution can be analyzed to give the free

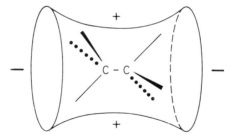

Fig. 3.19.   Shielding anisotropy in the neighborhood of a C—C single bond.

---

[*]

> For pure, dry ethanol, coupling between OH and CH_2 protons becomes evident, the OH resonance being now a triplet and the CH_2 resonance more complex than a quartet. This is a consequence of slower intermolecular proton exchange in the pure alcohol. In carbon tetrachloride solution, proton exchange is rapid enough to abolish this coupling, probably because of traces of water or acidic impurities. At high dilution, the OH and CH_2 peaks appear to be broadened [Fig. 3.26(d)] and upon still further dilution would again become as in pure ethanol, owing to slowing down of exchange by dilution. These effects are discussed further in Chapter 5, Section 5.4.5.

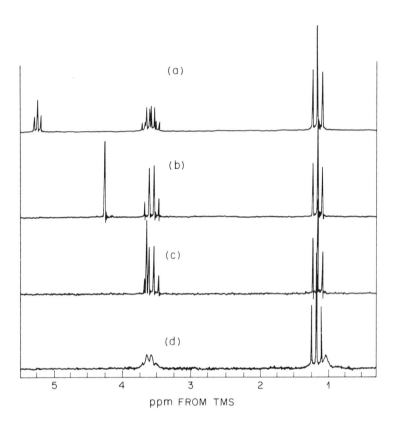

Fig. 3.20.   The spectrum of ethanol as a function of dilution: (a) neat;
(b) 10% (wt./vol.) in CCl$_4$; (c) 5% (wt./vol.) in CCl$_4$; (d)
0.5% (wt./vol.) in CCl$_4$. The CH$_3$ triplet at *ca.* 1 ppm and
the CH$_2$ resonance [a quartet except in spectrum (a)] at *ca.*
3.6 ppm are scarcely affected in chemical shift by dilution.
Observed at 60 MHz at 40°.

energy of dimer formation. The hydroxylic proton thus shows quite special
shielding behavior. In the absence of hydrogen bonding, its shielding is
comparable to that of the protons of a methyl group, probably because it
shares the electron cloud of the oxygen atom. When it participates in
hydrogen bonding, very marked deshielding occurs. This deshielding is
particularly marked when intramolecular hydrogen bonding may occur.
Acetylacetone is known to exist in both keto and enol forms:

$CH_3COCH_2COCH_3 \rightleftharpoons$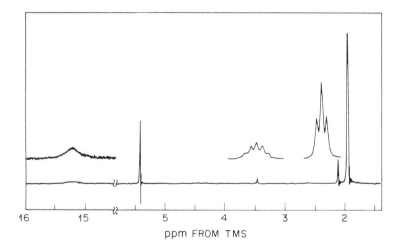

The keto—enol equilibration is slow enough (cf. Chapter 5, Section 5.4) so that spectra for both forms may be clearly seen (Fig. 3.21). The enol form is the major component (about 90 mole %), and its hydroxylic proton appears at the strikingly low field position of 15.18. Even greater shielding is observed for the chelated OH protons of phenols such as salicylaldehyde (18.1 ppm).

The deshielding of carboxyl protons, exemplified by the spectrum of butyric acid in carbon tetrachloride [Fig. 3.6(d)], no doubt has a similar origin. Here, NMR observation is complicated by the fact that the free energy of formation of the hydrogen-bonded dimers is quite large, and that higher polymers are present in which the CO_2H protons are more shielded. On dilution, the polymeric forms are disrupted but the dimer remains; the

ppm FROM TMS

Fig. 3.21.   Acetylacetone in keto and enol form. The major peaks correspond to the enol form: methyl at 1.98 ppm, olefinic at 5.40 ppm, and the OH at 15.2 ppm. The keto form shows a narrow methyl triplet at 2.16 ppm and a methylene septet at 3.46 ppm, both multiplets shown as insets and expanded 10×. Observed in 25% solution (wt./vol.) in CCl_4 at 100 MHz.

CO_2H resonance therefore moves first to *lower* field and then upfield only at high dilution.

Considering how large the effects are, theoretical understanding of the deshielding produced by hydrogen bonding cannot be said to be entirely satisfactory. We have seen that the electron clouds of the oxygen atoms are expected to cause marked shielding. There must therefore be some strong countervailing influence. Principal contributions may be:

a. Distortion of the electronic structure of the X-H bond by the presence of the donor molecule Y:

$$\overset{\delta+}{X}-\overset{\delta-}{H}...Y$$

This contribution is always negative. Qualitatively, it may be understood as arising because the electrostatic field of the hydrogen bond tends to draw H toward Y and repel the X—H bonding electrons toward X, resulting in a reduced electron density about H.

b. Magnetic anisotropy in group Y. This can have either sign, but will be positive (as we have seen in 3.3.4.2) if the principal symmetry axis of Y is along the hydrogen. A marked positive effect may be observed when Y is an aromatic ring, but generally contribution (a) appears to dominate.

The resonances of hydrogen bonded protons are often rather broad, as exemplified in Figs. 3.20 and 3.21. This may be in part the result of relatively slow exchange between hydrogen-bonded and non-bonded species.

3.3.6 Molecular Asymmetry and Chemical Shift Nonequivalence

Among spectroscopic methods, high—resolution NMR is uniquely powerful in providing information concerning the symmetry of molecules of uncertain structure. This information frequently permits a definite choice among several possible structures and may also define bond angles and conformational preferences. Observation of symmetry or lack of symmetry by NMR depends on the fact that otherwise similar nuclei that occupy geometrically nonequivalent sites in a molecule are very likely to be magnetically nonequivalent as well. These nuclei, therefore, will in general exhibit different chemical shifts and different couplings to other nuclei.

Environmental differences too subtle to detect otherwise are often very clear in NMR spectra. Some cases are quite obvious. For example, the spectra of $o-$, $m-$, and $p-$dinitrobenzenes are unambiguously assignable (Fig. 3.22). The ortho isomer gives an AA'BB' spectrum (a)[9] which is always symmetrical (see Chapter 4, Section 4.2.3.6); the meta compound gives an unsymmetrical A_2BC spectrum (b); and the para isomer shows a single peak, all the protons being equivalent (c). If the two substituents are unlike, only the para isomer gives a symmetrical AA'BB' spectrum.

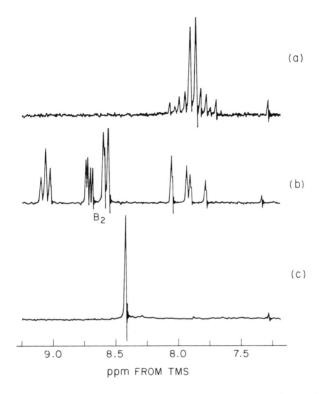

Fig. 3.22.   60 MHz proton spectra in CCl$_4$ solution of (a) *o*-dinitrobenzene; (b) *m*-dinitrobenzene; (c) *p*-dinitrobenzene.

But spectral symmetry alone cannot distinguish ortho from meta, both of which give complex ABCD spectra. However, couplings will distinguish these.

Substituted cyclobutanes (Fig. 3.23) provide similar problems. Thus, one can tell (a) from (b), since in (a) all B groups and all A groups are equivalent, while in (b) there are three different types of each. Spectral symmetry alone cannot distinguish (c) from (d), although their spectra will in general be different. Internal evidence will allow one to tell (e) from (f) since the C groups are all equivalent in (f) but not in (e).

More subtle problems are presented by molecules of the general type

$$X - CM_2 - Y^*$$

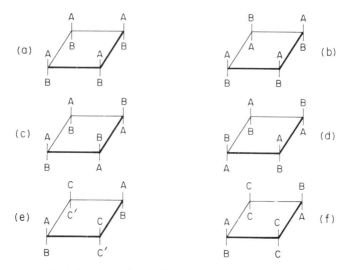

Fig. 3.23.   Cyclobutanes of varying symmetry.

where X is a group having a plane or axis of symmetry (such as CH_3, C_6H_5, halogen, etc.) and Y^* is a group having no plane or axis of symmetry. The groups M may be any observable atoms or groups such as H, F, CH_3, CF_3, C_6H_5, etc. In such a molecule, the groups M will be in nonequivalent environments, and will in general exhibit different couplings to X and Y. Such groups may be termed *diastereotopic*. If on the other hand Y^* is such that the M groups are equivalent, the M groups are termed *enantiotopic*.[56]

Let us consider some specific examples:

a. Y^* is an asymmetric carbon atom $CR_1R_2R_3$. For each optical isomer (not normally distinguishable by NMR, but see below) there will be three staggered conformers (Fig. 3.24), and in each the environments of $M_{(a)}$ and $M_{(b)}$ are non—equivalent. There is in fact no conformation in which they are equivalent. Under such circumstances, even when rotation about the C—C bond is fast, resulting in the averaging of the chemical shifts of the M group in each conformer (see Chapter 5, Section 5.4), $M_{(a)}$ and $M_{(b)}$ will remain non-equivalent. This was first observed by Drysdale and Phillips[57] for ^{19}F nuclei in acyclic compounds and confirmed by Nair and Roberts[58] and Shoolery and Crawford.[59] Waugh and Cotton[60] noted that such nonequivalence is not dependent on conformational preference but would persist even if all three staggered conformers were equally populated. Gutowsky[61] supported this conclusion but showed that nevertheless conformational preference was the main cause.

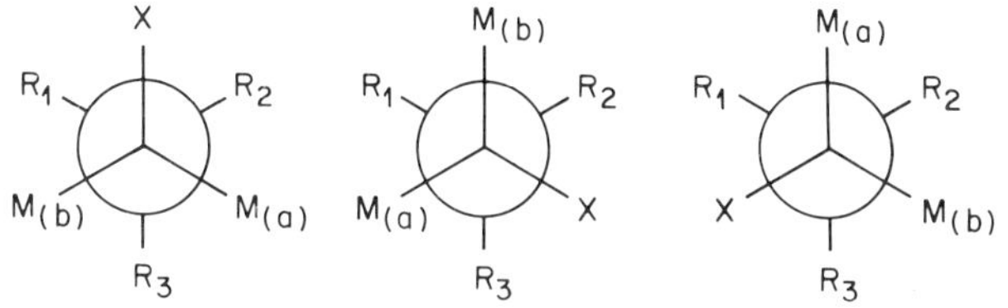

Fig. 3.24.   Staggered conformers of a substituted ethane $X-CM_2$ $-CR_1R_2R_3$.

A great many examples of this type of molecule have now been studied by NMR and the subject has been reviewed by van Gorkom and Hall[62] and Jennings.[63] $L-$valine is an example in which the M groups are methyls, the $\alpha-$carbon being asymmetric. Its proton spectrum in alkaline

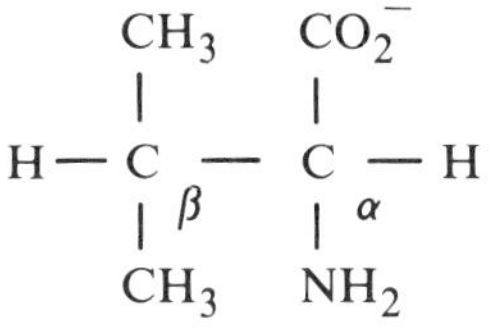

$$
\begin{array}{ccc}
CH_3 & & CO_2^- \\
| & & | \\
H-C & - & C-H \\
|\ \beta & & |\ \alpha \\
CH_3 & & NH_2
\end{array}
$$

D_2O is shown in Fig. 3.25. It exhibits separate doublets at 0.83 and 0.90 ppm for each of the methyl groups. The carbon-13 spectrum shows a similar differentiation.

In this example, a single enantiomer is observed. But even if a racemic mixture of d and l isomers were observed, the spectrum would be the same, as NMR cannot normally distinguish mirror image molecules. If a racemic mixture is dissolved in a chiral solvent, however, the d and l molecules experience on average different environments despite rapid molecular motion. Pirkle[64] observed that the ^{19}F resonance of the dl mixture of

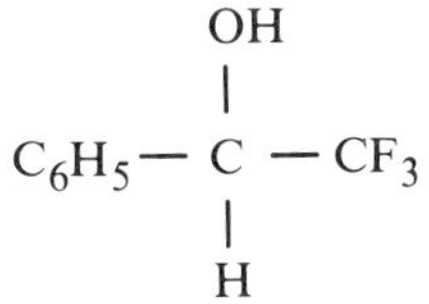

$$
\begin{array}{c}
OH \\
| \\
C_6H_5- C -CF_3 \\
| \\
H
\end{array}
$$

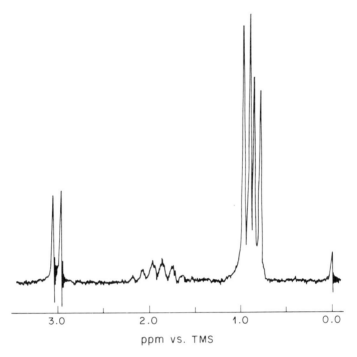

Fig. 3.25.   Proton spectrum of L-valine (15% solution in D_2O, 2M in NaOD, 60 MHz); separate doublets for each methyl group appear at 0.90 and 0.83 ppm.

appears as two sets of doublets separated by 2 Hz (at 56.4 MHz) when observed in optically active α–phenethylamine as solvent. Many such examples, for ^1H, ^{19}F, and ^{13}C, are now known, and the subject has been reviewed.[65] It is a useful means of determining the composition of *dl* mixtures that are not equimolar.

A similar example is provided by the methylene protons of an ethyl group attached to an asymmetric center; under these circumstances, a differentiation of these protons is observed:

$$
\begin{array}{ccc}
 & H_{(a)} & R_1 \\
 & | & | \\
CH_3\!-\! & C\!-\!C & \!-\!R_2 \\
 & | & | \\
 & H_{(b)} & R_3
\end{array}
$$

It is not necessary for the chiral center to be directly bonded to the carbon bearing the M groups, although the effect will generally be largest when this is the case. In principle it could be removed by several intervening bonds:

$$
\begin{array}{ccc}
M_{(a)} & & R_1 \\
| & & | \\
X - C - & \cdots - & C - R_2 \\
| & & | \\
M_{(b)} & & R_3
\end{array}
$$

b.　Y^* is　$-\overset{\displaystyle P}{\underset{\displaystyle Q}{C}}-CM_2X$

In this case the molecule has a plane of symmetry and does not have any asymmetric carbon atoms, but the CM_2 groups are nevertheless diastereotopic because the symmetry plane does not bisect the $M-C-M$ angle. The classic example of this type is provided by diethyl sulfite[66-69] (Fig. 3.26) in which the methylene protons of each ethyl group are made

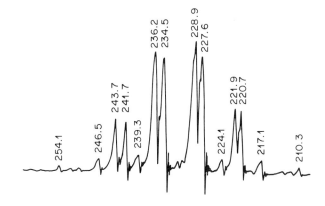

Fig. 3.26.　Methylene proton spectrum of diethyl sulfite, 50% solution in benzene, observed at 60 MHz. Peak positions are measured in hertz downfield from tetramethylsilane. [F. Kaplan and J. D. Roberts, *J. Am. Chem. Soc.* **83**, 4666 (1961)].

nonequivalent by the fact that in a sulfite group $P \neq Q$, so that the spin system is ABX_3 rather than A_2X_3. Other such systems are ethyl groups of sulfoxides (but not sulfones), diethylmethyl ammonium ion, acetaldehyde diethylacetal, phosphorous and phosphonic esters, sulfinic esters, and benzyl groups such as those of 3,3-dibenzylphthalide.[70]

Many examples have been reported of geminal nonequivalence in carbon-13 spectra. We have cited one such example in l−valine (p. 125) and mention has been made (p. 105) of differentiation of isopropyl methyl carbons falling into both class a and class b. It has been demonstrated[71] that this effect in hydrocarbons can be satisfactorily accounted for as arising from conformer preferences and the *gauche*−γ effect (Section 3.3.3). For example, in 2,4-dimethylhexane the conformer shown

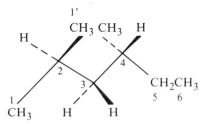

is favored; this has a gauche interaction between CH_3 group 1′ and C_4 (the chiral center), whereas CH_3 group 1 does not; CH_3 group 1′ is therefore the more shielded. The calculated and observed nonequivalence of isopropyl methyl carbons for this and a number of other hydrocarbons (data from Lindeman and Adams[19] and Carman *et al.*[21]) are shown in Table 3.8. It will be observed that if four or move bonds connect the isopropyl group to the chiral center (class a) or to the other isopropyl group (class b) the effect becomes vanishingly small.

c. It seems worthwhile to recognize a third classification: $X = Y^* = CR_1R_2R_3$. We now have a molecule with two similar asymmetric centers:

$$
\begin{array}{c}
R_1\ \ M_a\ \ R_1 \\
|\ \ \ \ |\ \ \ \ | \\
R_2-C-C-C-R_2 \\
|\ \ \ \ |\ \ \ \ | \\
R_3\ \ M_b\ \ R_3
\end{array}
$$

If the $CR_1R_2R_3$ groups are of the same handedness, we have the *racemic* diastereoisomer, which can be resolved into d and l enantiomers. As we have seen, these will in general be indistinguishable from each other and from the racemic mixture by NMR. The racemic diastereoisomer has

TABLE 3.8

Nonequilvalent ^{13}C NMR Chemical Shifts for the Isopropyl Methyl Carbons in Branched Alkanes

alkane	class	$\Delta\delta$ obsd[a]	calcd
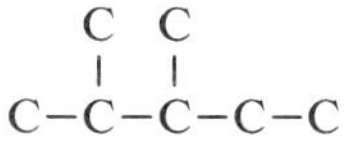 C C | | C−C−C−C−C	a	2.2,2.3	1.5
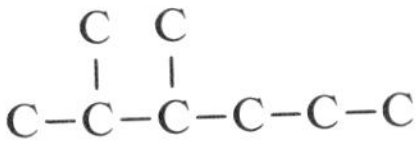 C C | | C−C−C−C−C−C	a	2.2	1.5
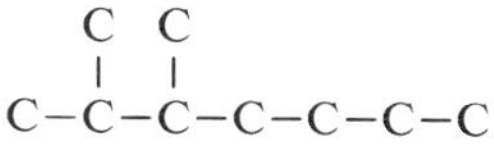 C C | | C−C−C−C−C−C−C	a	2.2	1.5
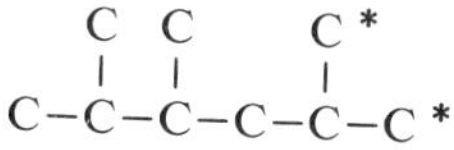 C C C* | | | C−C−C−C−C−C*	2.2(1.6*)	1.7(2.4*)	
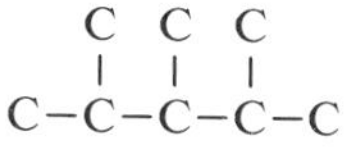 C C C | | | C−C−C−C−C	b	3.3	
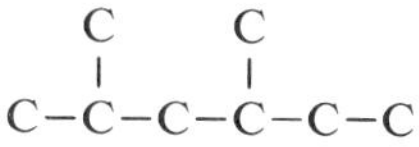 C C | | C−C−C−C−C−C	a	1.0(1.9, 1.1,0.9)[b]	1.6,1.1, 0.9
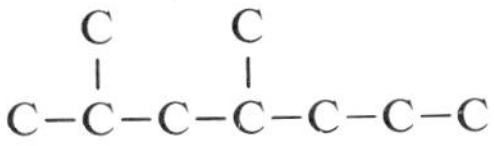 C C | | C−C−C−C−C−C−C	a	1.0	1.1
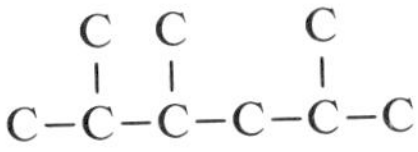 C C C | | | C−C−C−C−C−C	b	1.1	1.4
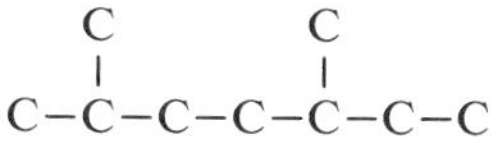 C C | | C−C−C−C−C−C−C	a	0.2	0.2
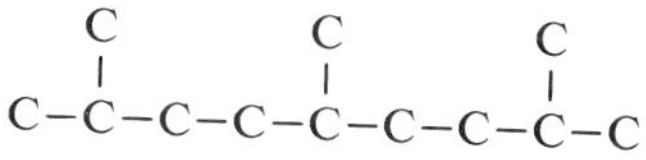 C C C | | | C−C−C−C−C−C−C−C−C	b	0.2	0.2
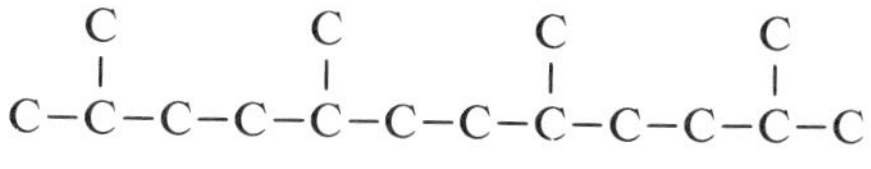 C C C C | | | | C−C−C−C−C−C−C−C−C−C−C−C b	0.2	0.2

(continues)

TABLE 3.8 (continued)

alkane	class	$\Delta\delta$ obsd[a]	calcd
C C | | C–C–C–C–C–C–C	b	0.0	0.0
C C | | C–C–C–C–C–C–C–C	a	0.1	0.04
C C | | C–C–C–C–C–C–C–C	b	0.0	0.0
C C | | C–C–C–C–C–C–C–C–C	a	0.0	0.0

[a] Observed between ambient temperature and 40°C.
[b] Observed at -120, 25, 90°C, Tonelli *et al.* (1984).

a twofold symmetry axis and so $M_{(a)}$ and $M_{(b)}$ are equivalent and the CM_2 group is enantiotopic. If the $CR_1R_2R_3$ groups are of opposite handedness, we have the *meso* diastereoisomer, which has a plane of symmetry and is optically inactive. However, unlike the *racemic* compound it has no twofold symmetry axis and so the CM_2 group is disastereotopic, $M_{(a)}$ and $M_{(b)}$ being nonequivalent.

Molecules of this type are closely related to vinyl polymers (Chapter 7, Section 7.2.1). The most carefully studied examples are *meso* and *racemic* 2,4-disubstituted pentanes:

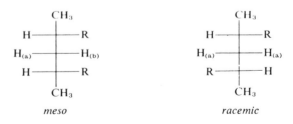

meso   *racemic*

Tiers and Bovey[72] showed that the fluorine nuclei of the central CF_2 group of *meso*–$CF_2Cl \cdot CFCl \cdot CF_2 \cdot CFCl \cdot CF_2Cl$ have differing chemical shifts and couplings, whereas those of the racemic isomer are identical.

Many other analogous cases have since been reported.[73,74] In Fig. 3.27 the 100-MHz spectra of the CH and CH_2 protons of racemic and meso 2,4-diphenylpentane (analogs of polystyrene) are shown. The racemic isomer's methylene resonance appears as a triplet (actually a doublet of doublets), corresponding to a single chemical shift, whereas the more complex CH_2 multiplet of the meso isomer corresponds to two chemical shifts separated by 0.21 ppm.

3.3.7 Chemical Shifts in Paramagnetic Compounds; Shift Reagents

It is usually difficult to obtain NMR spectra of paramagnetic compounds because of the very effective spin lattice relaxation and consequent line broadening caused by the unpaired electrons. For organic radicals, this broadening obliterates the spectrum completely. For certain complexes of transition metals, however, informative spectra can be obtained, thanks in part to the fact that extremely large chemical shifts may occur, which far exceed the line broadening in magnitude. Thus, for example, in the spectrum of the nickel complex:

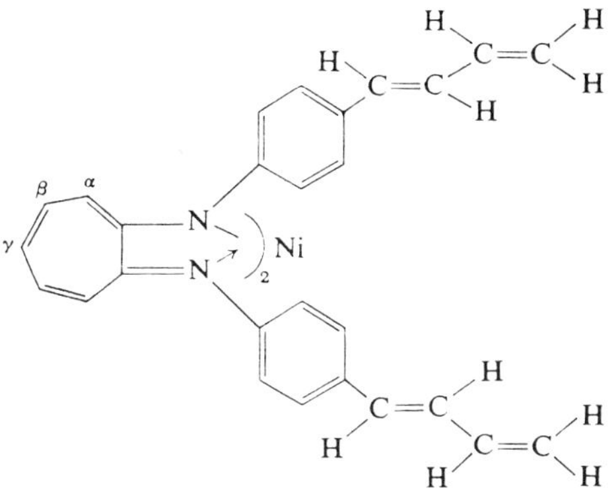

the α, β, and γ protons appear at -42, $+47$ ppm, and -64 ppm, respectively.[75,76] These large shifts are believed to arise because of a small donation of unpaired electron spin density from the paramagnetic ion to the bonding ligand atom (in this case, nitrogen) and its transmission throughout the rest of the molecule by the so-called *contact* interaction. This type of nucleus–electron interaction was first described by Fermi[77] and owes its name to the fact that it postulates a finite electron density at the nucleus. This interaction is also responsible for the hyperfine splitting in electron resonance spectra, its magnitude being given by a_N, which is characteristic of a given nucleus and its environment, and may be great as 1 MHz.

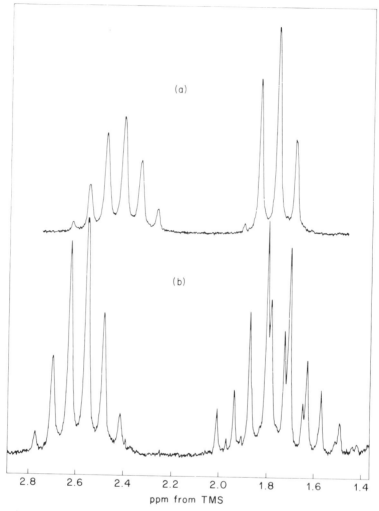

Fig. 3.27.   The 100-MHz proton spectra of (a) racemic and (b) meso-2,4-diphenylpentane, 10% (vol./vol.) in chlorobenzene at 35°. The CH_3 resonances, not shown, appear as a doublet centered at 1.50 ppm.

The origin of these very large chemical shift effects may be understood as follows. In many organic complexes with paramagnetic metals the electron spin relaxation times are very short — 10^{-10} to 10^{-13}s. The presence of a neighboring electron spin might be expected to split the resonance line of a nucleus, for example a proton, in the same way that a splitting is produced by a neighboring magnetic nucleus but to a much greater extent because of the larger electron magnetic moment. However, because of the very short electron relaxation time, a self-decoupling effect occurs and the split resonance is collapsed to a singlet. Because the self-decoupling many not be complete, the singlet resonance is likely to be broadened. In addition, the Boltzmann populations of the electron spin states differ markedly from equality (as compared to nuclear magnetic energy levels), and so the collapsed line deviates appreciably from a midpoint position between the split components. This results in the large contact shift effect. Its magnitude is given by

$$\Delta\sigma_c = -\frac{2\pi a_N g \mu_e S (S+1)}{3kT\gamma} \tag{3.8}$$

where g is the electron g–factor, assumed to be isotropic, S is the electron spin quantum number, μ_e the Bohr magneton, and γ the magnetogyric ratio of the observed nucleus; a_N is the electron-nucleus hyperfine splitting constant.

A second important effect causing shifts in paramagnetic compounds is the so-called *pseudocontact* interaction. This depends on the anisotropy of the electron g–factor, i.e. its effect as a magnetic dipole, and can be observed if the electron spin-lattice relaxation time is much shorter than the correlation time τ_c of the molecule. It is given by

$$\Delta\sigma_\psi = \frac{(g_{\parallel}^2 - g_{\perp}^2)\mu_e^2 S (S+1)(1-3\cos^2\theta)}{9kTr^3} \tag{3.9}$$

assuming that the g–factor has axial symmetry. This expression is closely parallel to Eq. 3.5. Again, r represents the distance between the observed nucleus and the paramagnetic nucleus, considered as a point magnetic dipole, and θ is the angle between the symmetry axis of g and the line joining the observed nucleus to the paramagnetic nucleus. $\Delta\sigma_\psi$ contains information concerning the geometry of the molecule (*vid. inf.*); it will be zero if the electronic g–factor is isotropic. Both $\Delta\sigma_c$ and $\Delta\sigma_\psi$ may be observed in the same molecule.

The large chemical shift effects produced by paramagnetic species are employed to advantage in *shift reagents*, which are complexes of lanthanide elements with chelating agents. Favored ligands are 2,2,6,6-tetramethylheptane-3,5-dionato, tmhd, also known as dipivaloylmethanato or dpm:

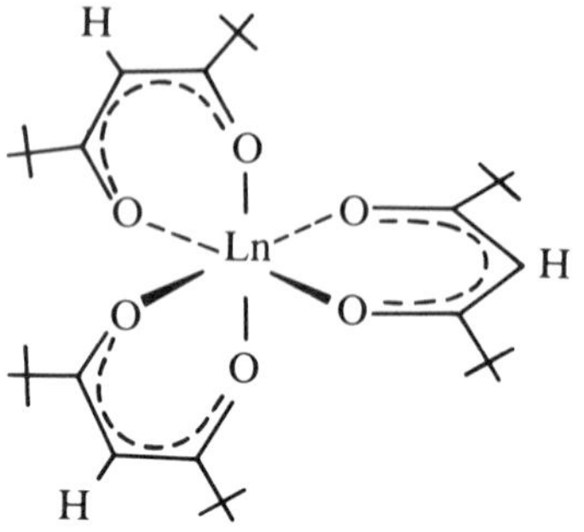

and 1,1,1,2,2,3,3,—heptafluoro-7,7-dimethyl-4,6-octadienato, known as fod:

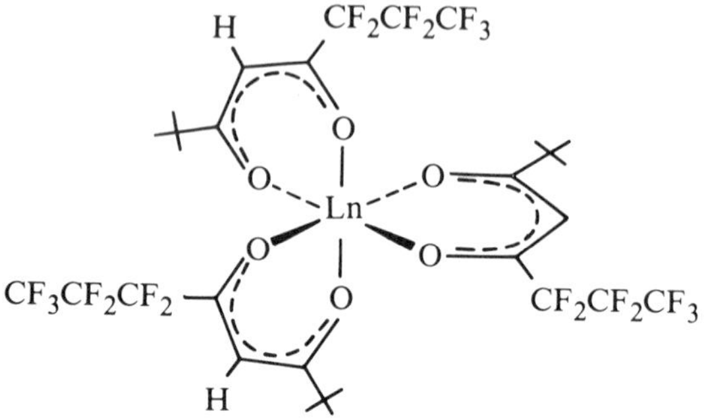

The lanthanides most commonly employed are europium (Eu) and praseodymium (Pr). Their complexes have very short electron spin relaxation times and therefore give sharp resonances with preservation of spin—spin coupling multiplet structure in proton spectra. These reagents exert their effect by forming rapidly equilibrating complexes—usually 1:1—with the molecule under study, which is therefore required to have functional groups capable of complexation with the ligands of the reagent. The value of such reagents lies in the fact that the shift effects are differential, resulting in a spreading out of the spectrum. In Fig. 3.28 100 MHz proton spectra of (a) n—hexanol in CCl_4 in the presence of Eu(dpm)$_3$ and (b) n—pentanol in CCl_4 in the presence of Pr(dpm)$_3$ are shown. In both cases the molar concentration of the reagent is substantially less than that of the substrate (see caption). (The reagent methyl protons appear at $ca.$ +0.9 ppm but are not shown.) It will be observed that Eu(dpm)$_3$ causes downfield shifts and Pr(dpm)$_3$ causes upfield shifts; this is consistently found for these two lanthanides regardless

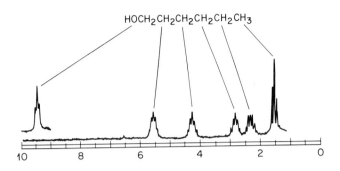

HOCH$_2$CH$_2$CH$_2$CH$_2$CH$_2$CH$_3$

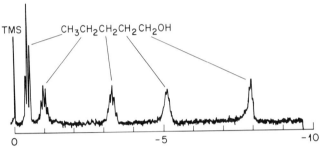

TMS   CH$_3$CH$_2$CH$_2$CH$_2$CH$_2$OH

Fig. 3.28.   (a) 100 MHz proton spectrum of *n*-hexanol in CCl$_4$ in the presence of Eu(dpm)$_3$. (J. K. M. Sanders and D. H. Williams, *Chem. Comm.* **1970**, 422.) (b) 100 MHz proton spectrum of *n*-pentanol in CCl$_4$ in the presence of Pr(dpm)$_3$ (0.053M). (J. Briggs, G. H. Frost, F. A. Hart, G. P. Moss, and M. L. Staniforth, *Chem. Comm.* **1970**, 749.)

of the nature of the reagent or the substrate. Since these shift effects are believed to arise exclusively or at least predominantly from a pseudo contact (i.e., dipolar) mechanism rather than by electron spin delocalization, the observed decrease in shift effect with distance along the chain from the hydroxyl group must be related primarily to the $1/r^3$ relationship in Eq. (3.9). The change in the sign of the influence of Eu compared to Pr must be attributed to the angular factor, the protons of the alcohols being in the deshielding zone of the Eu dipole and in the shielding zone of the Pr dipole. Inspection of molecular models of these reagents does not suggest that the geometry of the substrate complexes could differ substantially depending upon the nature of the lanthanide and that therefore the change in sign is most likely to be attributed to a change in orientation of the symmetry axis of the lanthanide g factor. In any event, the principal benefit conferred by the shift reagent in such cases is the spreading out of the chemical shifts, enabling the spectrum to be more readily comprehended and to be analyzed on a "first-order" or weak-coupled basis.

The dependence of $\Delta\sigma_\psi$ on molecular geometry has prompted many studies of the structure and conformation of substrates through the sign and magnitude of proton chemical shift effects. Unfortunately, such studies are fraught with difficulties. Unless a swamping excess of shift reagent is employed (often impractical), it is necessary to know the equilibrium constant of the reagent:substrate complex, and this is seldom available. This uncertainty is aggravated when more than one functional group capable of binding to the reagent is present. The orientation of the g—factor symmetry axis is often uncertain or unknowable, and it may be difficult to establish whether there is indeed a symmetry axis at all. There is the further question of whether the conformation of the substrate in the complex is the same as that of the free state—probably not a serious problem for rigid molecules. Finally, one must be sure that there is no significant participation of contact chemical shifts.

Shift reagents have also been employed in carbon-13 NMR and have proved useful in assignment of resonances,[78] mainly on the basis of proximity to the binding group. For carbon-13 the suspicion that the dipolar shift effect may be severely perturbed by contact interactions, which are not dependent on geometry, is better founded, and detailed structural studies are therefore even less meaningful than for protons.

In general, with the increasing availability of superconducting NMR spectrometers and the use of two-dimensional methods (Chapter 6), interest in shift reagents has very much diminished. A special application that retains utility, however, is that of *chiral* shift reagents in the investigation of mixtures of enantiomers.[79-81] The principle of action of these reagents is essentially the same as that of chiral solvents but the effects are larger. An effective class of reagents is represented by compound I:

I

Here, the presence of the d−camphor-based ligands permits differentiation of enantiomers and assessment of enantiomeric purity. In Fig. 3.29 the proton spectra of compounds II and III, prepared from the same sample of partially resolved 2-methyl-2-phenylbutanoate methyl ester, are shown.

$$\underset{II}{\underset{\displaystyle CH_3}{\underset{\displaystyle |}{CH_3OC}\overset{\displaystyle O}{\overset{\displaystyle \|}{-}}\overset{\displaystyle C_6H_5}{\overset{\displaystyle |}{C}}-CH_2CH_3}}$$

$$\underset{III}{\underset{\displaystyle CH_3}{\underset{\displaystyle |}{CH_3C}\overset{\displaystyle O}{\overset{\displaystyle \|}{-}}\overset{\displaystyle C_6H_5}{\overset{\displaystyle |}{C}}-CH_2CH_3}}$$

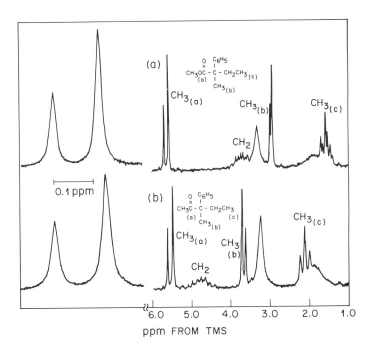

Fig. 3.29. Proton spectra of compounds II and III above in CCl$_4$ solution. The solutions are 0.9 M in II (a) and 0.6 M in III (b) in the presence of 0.5 M shift reagent I. Expanded traces of the CH$_3$O and CH$_3$CO resonances are shown downfield. Unassigned resonances originate from the shift reagent [H. L. Goering, J. N. Eikenberry, and G. S. Koermer, *J. Am. Chem. Soc.* **93**, 5913 (1971)].

The expanded spectra at the left show the methyl proton spectra of the CH_3O (a) and CH_3CO (b) groups, from which the ratio of enantiomers may be readily measured. It agrees with the ratio of enantiomers (ca. 3:1) obtained from the optical rotation of the starting acid. It may be noted that the singlets arising from the methyl groups at the quaternary carbons provide the same information but show a reversed sense of nonequivalence.

It is believed that these reagents function principally by providing different chiral environments for each enantiomer but with some contribution from differing binding equilibrium constants.

3.3.8 Shielding in Aromatic Rings

The shieldings of ring protons and carbon-13 nuclei in substituted benzene rings are of considerable interest and have been extensively investigated. The general behavior is in accordance with expectation (i.e., electron-withdrawing groups cause deshielding and electron-donating groups cause increased shielding), but resonance effects cause major perturbations. The principal effect is to be attributed to alterations in the π-electron density about the carbons. This concept has been put on a quantitative basis for protons by Schaefer and Schneider.[82] They compared chemical shifts in cyclopentadiene anion $C_5H_5^-$, benzene C_6H_6, tropylium cation $C_7H_7^+$, and cyclooctatetraene dianion $C_8H_8^=$, all of which have symmetrical structures, and deduced a shielding of -10.7 ppm per electron. For the carbon-13 nuclei, the much larger value of ca. 160 ppm per electron can be deduced[83,84]. This reflects the position of the protons at the periphery of the molecule and the fact that $1s$ electrons, unlike p electrons, are not excited in the magnetic field.

The chemical shifts of ortho, meta, and para protons in a number of monosubstituted benzenes are shown schematically in Fig. 3.30. In Fig. 3.31 the chemical shifts of the corresponding carbon atoms, together with those of the C_1 carbons, to which the R group is attached, are shown. In the proton spectra the ortho protons show the greatest excursions and the meta protons the least. In the carbon-13 spectra, the C_1 carbons show by far the widest range of chemical shifts — 67 ppm — and these appear in general to correlate with the electronegativity of the first bonded atom of the substituent group and with the shifts in substituted aliphatic chains (Fig. 3.7). They are thus responsive mainly to inductive effects. Both carbons and protons in the meta positions show the least influence of the nature of R, and appear to again reflect inductive effects but to a much attenuated extent.

The shielding effects for ortho and para protons and carbons are intermediate between those just discussed. They do not correlate well with

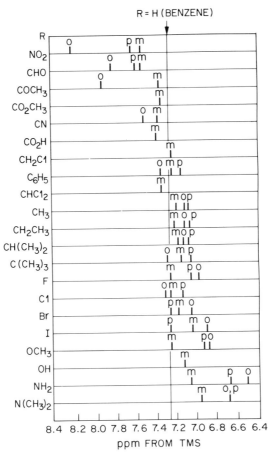

Fig. 3.30.   Chemical shifts of ortho, meta, and para ring protons in monosubstituted benzenes. The vertical line at 7.27 ppm corresponds to the chemical shift of benzene.

the effects of the same R groups on C_1 or in aliphatic chains. Thus, a methoxyl group strongly deshields the C_1 carbons and protons in aliphatic chains as well as the aromatic C_1 carbons, but shields the ortho and para positions. The effect of fluorine is closely parallel. (Iodine exerts a "heavy atom" shielding effect on all nearby nuclei.)

The ortho proton chemical shifts give some evidence of being influenced by special anisotropic shielding effects of the R group (e.g., NO_2 and *tert.-*butyl), but rationalization is hindered by the fact that the ortho carbons do not show parallel effects.

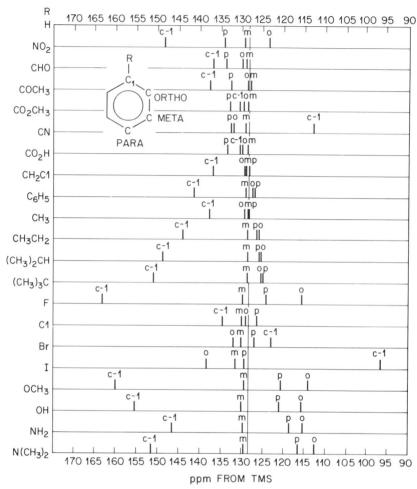

Fig. 3.31.   Chemical shifts of C_1, ortho, meta, and para ring carbons in monosubstituted benzenes. The vertical line at 128.5 ppm corresponds to the chemical shift of benzene.

It is natural to assume that the Hammett σ constants, which express the reactivity of substituted benzenes at meta and para positions, are related to the electron densities at these positions and should therefore be related to the carbon and proton chemical shifts. It has been shown that for both nuclei a strong correlation exists.[85-88]

The proton and carbon-13 chemical shifts of a few representative unsubstituted heterocyclic compounds are represented in Fig. 3.32. The shielding of the two nuclei is strongly correlated, as one might expect.

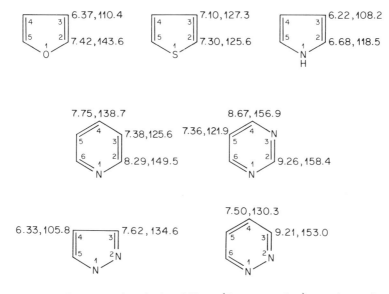

Fig. 3.32.   Proton chemical shifts (first number) and carbon-13 chemical shifts (second number) in some representative heterocyclic rings.

Inductive effects play a large role, but resonance contributions can be seen in, for example, the fact that shielding at the 3 position in pyridine is greater than at the 2 and 4 positions.

3.3.9 Shielding Effects of Isotopes

Protons, carbon-13 nuclei, and fluorine-19 nuclei are observed to exhibit small but measurable differences in chemical shift as a result of substitution of deuterium for hydrogen. The effect is of theoretical interest and can be large enough to cause interpretive errors. Tiers[89] found that the methyl group protons of α-d_1-toluene are 0.015 ± 0.002 ppm more shielded than those of toluene. Gutowsky[90] observed an upfield shift of 0.034 ± 0.001 ppm in d_5-acetone. A particularly marked shift is shown by d_{11}-cyclohexane, in which the lone proton is 0.060 ppm more shielded than those of cyclohexane.[91]

The effect of deuterium substitution on carbon-13 chemical shifts is considerably larger than on proton chemical shifts. Thus, the carbonyl carbon in acetone-d_6 is *less* shielded than in the protio compound by +0.32 ppm, while the methyl carbon is more shielded by −0.86 ppm.[92] Shifts of this order of magnitude are general and can sometimes be diagnostically useful.[93]

The effect of ^{13}C on proton chemical shifts is small but measurable. Thus, Tiers[94] observed the proton chemical shift of $^{13}CHCl_3$ to be 0.0059 ± 0.0012 ppm upfield from that of $^{12}CHCl_3$; a similar finding was reported for $^{13}CH_2Cl_2$. A somewhat smaller effect is observed for ^{13}C on a directly bonded ^{13}C: in tetrahydrofuran each ^{13}C is 0.0036 ppm upfield from its position in the molecules in which it has only ^{12}C neighbors (Chapter 4, Section 4.3.6.3).[95]

REFERENCES

1. W. E. Lamb, *Phys. Rev.* **60,** 817 (1941).

2. W. C. Dickinson, *Phys. Rev.* **80,** 563 (1951).

3. N. F. Ramsey, *Phys. Rev.* **78,** 699 (1950).

4. G. F. Newell, *Phys. Rev.* **80,** 476 (1950).

5. J. R. Hoyland and R. G. Parr, *J. Chem. Phys.* **38,** 2991 (1963).

6. A. Saika and C. P. Slichter, *J. Chem. Phys.* **22,** 26 (1954).

7. J. A. Pople, *Proc. Roy. Soc.* **A239,** 541, 550 (1957).

8. A. D. Buckingham, *Can. J. Chem.* **38,** 300 (1960).

9. J. N. Shoolery, *J. Chem. Phys.* **21,** 1899 (1953).

10. B. P. Dailey and J. N. Shoolery, *J. Am. Chem. Soc.* **77,** 3977 (1955).

11. G. Klose, *Ann. Phys.* **10,** 391 (1963).

12. R. Ettinger, *J. Phys. Chem.* **67,** 1558 (1963).

13. H. Spiesecke and W. G. Schneider, *J. Chem. Phys.* **35,** 722 (1961).

14. J. R. Cavanaugh and B. P. Dailey, *J. Chem. Phys.* **34,** 1094 (1961).

15. J. N. Shoolery, *Varian Tech. Inf. Bull.* **2** (3) (1959).

16. J. A. Pople and M. S. Gordon, *J. Am. Chem. Soc.* **89,** 4253 (1967).

17. R. Ditchfield and P. D. Ellis in *Topics in Carbon-13 NMR Spectroscopy*, Vol. 1, G. C. Levy, Ed., Academic Press, New York, 1974, p. 1.

18. D. M. Grant and E. G. Paul, *J. Am. Chem. Soc.* **86,** 2984 (1964).

19. L. P. Lindeman and J. Q. Adams, *Anal. Chem.* **43,** 1245 (1971).

20. D. E. Dorman, R E. Carhart, and J. D. Roberts, private communication (1972).

21. C. J. Carman, A. R. Tarpley, Jr., and J. H. Goldstein, *Macromolecules* **6,** 719 (1973).

22. D. M. Grant and B. V. Cheney, *J. Am. Chem. Soc.* **39,** 5315, 5319 (1967).

23. D. E. Dorman, E. P. Otocka, and F. A. Bovey, *Macromolecules* **5,** 574 (1972).

24. F. A. Bovey, F. C. Schilling, F. L. McCrackin, and H. L. Wagner, *Macromolecules* **9,** 76 (1976).

25. H. M. McConnell, *J. Chem. Phys.* **27,** 226 (1957).

26. J. A. Pople, *Proc. Roy. Soc.* **A239,** 541, 550 (1957).

27. J. A. Pople, W. G. Schneider and H. J. Bernstein, *High Resolution Nuclear Magnetic Resonance*, McGraw-Hill, New York, 1959, p. 179.

28. L. Pauling, *J. Chem. Phys.* **4,** 673 (1936).

29. J. A. Pople, *J. Chem. Phys.* **24,** 111 (1956).

30. J. S. Waugh and R. W. Fessenden, *J. Am. Chem. Soc.* **79,** 846 (1957).

31. J. S. Waugh and R. W. Fessenden, *J. Am. Chem. Soc.* **80,** 6697 (1958).

32. C. E. Johnson, Jr., and F. A. Bovey, *J. Chem. Phys.* **29,** 1012 (1958).

33. C. W. Haigh and R. B. Mallion, *Org. Magn. Reson.* **4,** 203 (1972).

34. C. W. Haigh, R. B. Mallion, and E. A. G. Armour, *Mol. Phys.* **18,** 751 (1970).

35. C. W. Haigh and R. B. Mallion, *Prog. NMR Spectrosc.,* **13,** 303 (1980).

36. L. W. Jenneskens, F. J. J. de Kanter, P. A. Kraakman, L. A. M. Turkenberg, W. E. Koolhaas, W. H. de Wolf, and F. Bickelhaupt, *J. Am. Chem. Soc.* **107,** 3716 (1985).

37. H. J. Bernstein, W. G. Schneider, and J. A. Pople, *Proc. Roy Soc.* **A236,** 515 (1956).

38. E. D. Becker and R. B. Bradley, *J. Chem. Phys.* **31,** 1413 (1959).

39. E. D. Becker, R. B. Bradley, and C. J. Watson, *J. Am. Chem. Soc.* **83,** 3743 (1961).

40. L. M. Jackman, F. Sondheimer, Y. Amiel, P. A. Ben-Ephraim, Y. Gaoni, R. Wolofsky, and A. A. Bothner-By, *J. Am. Chem. Soc.* **84,** 4307 (1962).

41. V. Boekelheide and J. B. Phillips, *Proc. Natl. Acad. Sci. U.S.A.* **51,** 551 (1964).

42. J. C. Calder and F. Sondheimer, *Chem. Comm.* **1966,** 904.

43. J. R. Lacher, J. W. Pollock, and J. D. Park, *J. Chem. Phys.* **20,** 1047 (1962).

44. D. J. Patel, M. E. H. Howden, and J. D. Roberts, *J. Am. Chem. Soc.* **85,** 3218 (1963).

45. J. R. Dyer, *Applications of Absorption Spectroscopy of Organic Compounds,* Prentice-Hall, Englewood Cliffs, New Jersey, 1965, pp. 80–81.

46. P. K. Freeman, M. F. Grostic, and F. A. Raymond, *J. Org. Chem.* **30,** 771 (1965).

47. F. A. L. Anet and A. J. R. Bourn, *J. Am. Chem. Soc.* **87,** 5350 (1965).

48. D. E. Dorman and F. A. Bovey, *J. Org. Chem.* **38,** 1719 (1979).

49. D. H. Williams, N. S. Bhacca, and C. Djerassi, *J. Am. Chem. Soc.* **86,** 2634 (1964).

50. L. M. Jackman, *Nuclear Magnetic Resonance Spectroscopy,* Pergamon, Oxford, 1959, pp. 112–113.

51. J. T. Arnold and M. E. Packard, *J. Chem. Phys.* **19,** 1608 (1951).

52. A. D. Cohen and C. Reid, *J. Chem. Phys.* **25,** 790 (1956).

53. E. D. Becker, U. Liddel, and J. N. Shoolery, *J. Mol. Spectry* **2,** 2 (1958).

54. M. Saunders and J. B. Hyne, *J. Chem. Phys.* **29,** 1319 (1958).

55. A. L. Porte, H. S. Gutowsky, and I. M. Hunsberger, *J. Am. Chem. Soc.* **82,** 5057 (1960).

56. K. Mislow and M. Raban, *Top. Stereochem.* **1,** 1 (1967).

57. J. J. Drysdale and W. D. Phillips, *J. Am. Chem. Soc.* **79,** 319 (1957).

58. P. M. Nair and J. D. Roberts, *J. Am. Chem. Soc.* **79,** 4565 (1957).

59. J. N. Shoolery and B. Crawford, Jr., *J. Mol. Spectrosc.* **1,** 270 (1957).

60. J. S. Waugh and F. A. Cotton, *J. Phys. Chem.* **65,** 562 (1962).

61. H. S. Gutowsky, *J. Chem. Phys.* **37,** 2196 (1962).

62. M. van Gorkom and G. E. Hall, *Quart. Rev. Chem. Soc.* **22,** 14 (1968).

63. W. B. Jennings, *Chem. Rev.* **76,** 307 (1975).

64. W. H. Pirkle, *J. Am. Chem. Soc.* **88,** 1837 (1966).

65. W. H. Pirkle and D. J. Hoover, *Top. Stereochem.* **13,** 263 (1982).

66. H. S. Finegold, *Proc. Chem. Soc.* **1960,** 283.

67. P. R. Shafer, D. R. Davis, M. Vogel, K. Nagarajan, and J. D. Roberts, *Proc. Natl. Acad. Sci. U.S.A.* **47,** 49 (1961).

68. J. G. Pritchard and P. C. Lauterbur, *J. Am. Chem. Soc.* **83,** 2105 (1961).

69. F. Kaplan and J. D. Roberts, *J. Am. Chem. Soc.* **83,** 4666 (1961).

70. G. C. Brumlik, R. K. Baumgarten, and A. I. Kosak, *Nature* **201,** 388 (1964).

71. A. E. Tonelli, F. C. Schilling, and F. A. Bovey, *J. Am. Chem. Soc.* **106,** 1157 (1984).

72. G. V. D. Tiers and F. A. Bovey, *J. Polym. Sci.* **A1,** 833 (1963).

73. F. A. Bovey, *High Resolution NMR of Macromolecules*, Academic Press, New York, 1972, Chapter III.

74. F. A. Bovey and L. W. Jelinski, *Chain Structure and Conformation of Macromolecules*, Academic Press, New York, 1982, pp. 40–47.

75. D. R. Eaton, A. D. Josey, W. D. Phillips, and R. E. Benson, *J. Chem. Phys.* **39,** 3513 (1963).

76. D. R. Eaton, A. D. Josey, W. D. Phillips, and R. E. Benson, *Disc. Faraday Soc.* **34,** 77 (1962).

77. E. Fermi, *Z. Phys.* **60,** 320 (1930).

78. F. W. Wehrli and T. Wirthlin, *Interpretation of Carbon-13 NMR Spectra*, Heyden, London, 1976, p. 94.

79. G. M. Whitesides and D. W. Lewis, *J. Am. Chem. Soc.* **92,** 6979 (1970).

80. H. L. Goering, J. N. Eikenberry, and G. S. Koermer, *J. Am. Chem. Soc.* **93,** 5913 (1971).

81. R. R. Fraser, M. A. Petit, and J. K. M. Saunders, *Chem. Comm.* **1971,** 1450.

82. T. Schaefer and W. G. Schneider, *Can. J. Chem.* **41,** 966 (1963).

83. G. A. Olah and G. D. Matescer, *J. Am. Chem. Soc.* **92,** 1430 (1970).

84. E. Breitmeir and G. Bauer, ^{13}C *NMR Spectroscopy*, Harwood Academic Publishers, Chur, Switzerland, 1984, pp. 44—45.

85. H. Spiesecke and W. G. Schneider, *J. Chem. Phys.* **35,** 731 (1961).

86. H. Spiesecke and W. G. Schneider, *Tetrahedron Lett.,* **1961** (14), 468.

87. G. L. Nelson, G. C. Levy, and J. D. Cargioli, *J. Am. Chem. Soc.* **94,** 3089 (1972).

88. W. F. Reynolds, I. R. Peat, M. H. Freedman, and J. R. Lyerla, *Can. J. Chem.* **51,** 1857 (1973).

89. G. V. D. Tiers, *J. Chem. Phys.* **29,** 963 (1958).

90. H. S. Gutowsky, *J. Chem. Phys.* **31,** 1683 (1959).

91. F. P. Hood and F. A. Bovey, unpublished observations.

92. K. L. Servis and R. L. Domenick, *J. Am. Chem. Soc.* **108,** 2211 (1986).

93. F. W. Wehrli and T. Wirthlin, *Interpretation of Carbon-13 NMR Spectra*, Heyden, (London) 1976, pp. 107—109.

94. G. V. D. Tiers, *J. Phys. Chem.* **64,** 373 (1960).

95. F. C. Schilling, unpublished observations.

CHAPTER 4

COUPLING OF NUCLEAR SPINS

4.1 INTRODUCTION

We have seen in Chapter 1 (Section 1.11) that magnetic nuclei in the same molecule may transmit information to each other concerning their spin states by slight polarizations of the spins or orbital motions of the valence electrons of the intervening bonds, and that, unlike direct dipole coupling, this coupling is independent of the applied magnetic field and is not averaged to zero by the tumbling of the molecular framework. There has been a great deal of theoretical work directed toward predicting and explaining nuclear coupling, but in this chapter we shall not describe this work in detail; rather we shall give the results of those studies that are particularly useful in interpreting the structures and conformations of organic molecules. The following general conclusions from theoretical and experimental studies are of particular significance:

a. The coupling of two nuclei via polarizations of the intervening electron *spins* depends upon the electron density at each nucleus. As only s atomic states have a finite electron density at the nuclei, it is to be expected that nuclear coupling will depend to some degree on the fraction of s character in the bonds. Coupling via electron spins is believed to be particularly important for proton–proton interactions because the electron densities about protons are well represented by $1s$ atomic orbitals. This type of coupling is also responsible for the hyperfine structure in electron spin resonance spectra. Its quantum mechanical form was first described by Fermi[1]; because this envisions a finite electron density at the nucleus, it is called the *Fermi contact term*.

b. Coupling of two nuclei may also occur via the *orbital motions* of the valence electrons. Each nuclear magnetic moment will induce currents in the molecular electron cloud and the resulting magnetic field will be felt by the other nuclei. This coupling is probably unimportant for proton–proton interactions, but will be significant for heavier atoms because of their greater electron density. Couplings to heavier nuclei are generally larger than to light nuclei.

c. The strength of the coupling via electron spins (and at least a part of that arising from electron orbital motions) is inversely proportional to the energies of the upper excited triplet states of the molecule. Some notion of the basis of this conclusion may be gained from the following simplified picture. Let us consider the $^1H^3H$ molecule, in which only two electrons transmit the coupling. (We might equally well consider $^1H_2$ itself, except that here the coupling is not observable, as we shall see.) The magnetic moments of the hydrogen and tritium nuclei are nearly equal (see Appendix A) and are so represented in Fig. 4.1; for both, $I = 1/2$. *If perfect pairing of the electron spins were rigorously required,* only state (a), with the nuclear spins likewise paired, would be possible, and there would be no splitting of the resonance line. The presence of unpaired nuclear states, such as (b) and (c), means that because of polarization by the nuclear spins perfect pairing of the electron spins cannot be maintained. This in turn means [as the inequality of the electron magnetic moments·in (b) and (c) is meant to imply] that coupling must require the contribution of excited triplet states having unpaired electron spins. The energies of these excited states are usually known only very approximately, if at all, and they are, therefore, customarily lumped together in an effective average excitation energy ΔE, just as in the calculation of shielding (Chapter 3, Section 3.3.)

d. Couplings between equivalent nuclei or groups of equivalent nuclei are not directly observable. In quantum mechanical terms, the observation

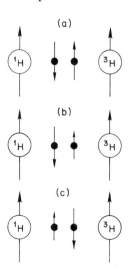

Fig. 4.1.   Schematic representation of nuclear and electron spins in the $^1H-^3H$ molecule.

of a splitting in the resonance of a molecule such as $^1H_2$, where the coupling is actually large, would involve transitions between symmetric and antisymmetric states, and these are strictly forbidden.

e. *All* contributions to the coupling of two nuclei (1 and 2) are proportional to the product of their magnetogyric ratios (p. 2), $\gamma_1\gamma_2$. This allows exact calculations of couplings to isotopic nuclei when the coupling to any isotope is known. For the HD molecule, a splitting of the proton resonance into a 1:1:1 triplet with a spacing of 43.5 Hz is observed. The deuterium resonance is a doublet with the same spacing. Multiplying this value by the ratio γ_{1H}/γ_{2H} (which equals 6.514), gives the $^1H - ^1H$ coupling as 283 Hz.

f. Couplings generally decrease rapidly as the number of intervening bonds increases and are usually unobservable if there are more than four or five. However, this is not always the case, and measurable couplings may be transmitted much further in systems of favorable geometry (see Section 4.3.5).

An important aspect of nuclear coupling with which this chapter will be primarily concerned is the calculation of *J* values and chemical shifts for systems of spins in which the couplings are comparable in magnitude to the chemical shift differences, both expressed in hertz. The spectra of such molecules often appear as complex systems of lines having spacings exhibiting no obvious regularity and intensities that deviate widely from binomial (Chapter 1, Section 1.11). From such spectra, one generally cannot obtain either *J* values or chemical shifts by simple measurements of line spacings. An example of such a spectrum is that of vinyl chloride (Fig. 4.16). An attempt to understand and solve this three-spin spectrum by "first-order" approximations will make this point clear. Spectral complexity increases rapidly with the number of coupled nuclei.

Spectra characterized by relatively large values of $J/\Delta\nu$ (see Section 4.2.1) are called "strong coupled." The analysis of such spectra (or indeed of any NMR spectrum) does not require the assumption of any particular physical model, i.e., we do not have to know what the molecule is, although we must know the number of spins involved. The task is to compute by quantum mechanical methods the energy levels and stationary-state wave functions of a system of coupled spins in a static magnetic field, and then to use perturbation methods and selection rules to give the probabilities of transitions occurring between the levels when a resonant rf field is applied. The line positions will be functions of the energy-level separations, and their relative intensities will correspond to the transition probabilities. If a suitable choice of parameters is made, the calculated spectrum will usually match the observed spectrum very closely. From the chemical shifts and couplings thus obtained, one may hope to deduce the structure of the molecule or, at least, to exclude incompatible structures. If the structure of the molecule is already known, chemical shifts and couplings may give clues to conformational or tautomeric preference, rates of isomerization or chemical exchange, and other important properties of the molecule.

Since there are many excellent sources to which the reader may turn for a description of these calculations, they will not be treated in detail here. We shall, instead, describe their results as applied to a number of spin systems of common occurrence in organic compounds. A small atlas of calculated spectra, sufficient to permit one to recognize some typical patterns, will be found in Appendix B. It would be nearly impossible to provide an atlas large enough to solve all problems. It is, therefore, important to be able to pick out certain lines and spacings that, even in a complex spectrum, will provide clues to key transitions and facilitate the full solution of the spectrum. For the latter, it is most convenient to employ a computer, which allows even those ignorant of quantum mechanical calculations to carry out rapid analyses of complex spectra that would require many hours of skilled hand computation. Computers associated with Fourier transform spectrometers (Chapter 2, Section 2.5) are quite sufficient for this task, at least for systems of up to six or seven spins, and many excellent soft-ware programs are available (e.g., from the Quantum Mechanical Exchange Library). It must also be recognized that the increasing use of superconducting magnets and the dominance of carbon-13 NMR has somewhat lessened the need to deal with strong-coupled spin systems. But proton NMR has never fallen out of fashion for the observation of biological molecules and has experienced a renaissance of more general application through 2D NMR (Chapter 6) and other advanced techniques.

4.2 DESCRIPTION AND ANALYSIS OF SPIN SYSTEMS

Nuclear spin systems are customarily designated by an alphabetical notation.[2] Nonequivalent nuclei of the same species (we shall soon consider the precise meaning of "nonequivalent") which have chemical shift differences comparable to their mutual couplings are designated by A, B, C, · · · . Another group of nuclei, separated from the first by large chemical shift differences or even of a different species, but among themselves by chemical shift differences comparable to the couplings, are designated X, Y, Z, $\cdots$. In some cases, still a third group with intermediate shifts may be present and would be denoted by K, L, M $\cdots$. Nuclei within a group are commonly (but not always) designated in alphabetical progression in order of increasing shielding. In these terms, the proton spectrum of the ethyl groups of ethyl orthoformate (Chapter 1, Fig. 1.16) would each be designated as an A_2X_3 system. If by some means its coupling to the ethyl protons could be demonstrated, the whole molecule would then be designated as an AM_6X_9 system. The ethyl group spectrum itself, however, could be treated as AM_2X_3 if the couplings between ethyl groups can be considered to be zero, as is in fact true. As pointed out in Chapter 1 (Section 1.11), $J/\Delta\nu$ for the ethyl group at 200 MHz is only about 0.01 and the multiplets exhibit binomial intensities. At 60 MHz (Chapter 7, Fig. 7.1), the intensities deviate noticeably from binomial values, although measurements of spacings give true J and $\Delta\nu$ values. Nevertheless, to reproduce the observed ethyl group spectrum accurately in terms of both spacings and intensities, one must treat it as an A_2B_3 system (Appendix B, Section VI).

4.2.1 Two-Spin Systems

When $J/\Delta\nu$ is very small for a two-spin system, as, for example, that represented by two nuclei of spin-1/2 and of different species, such as $^1H^{19}F$, $^1H^3H$, or $^{19}FCl_2C-C^1HCl_2$, the system is designated AX. In Fig. 4.2 the magnetic energy levels of such an AX system are shown; it is assumed that $\mu_A > \mu_X$. The resonant frequencies are ν_A and ν_X, ν_A exceeding ν_X. The four energy levels correspond to all possible combinations of alignments of the nuclear moments parallel and antiparallel to B_0. The highest energy state (1) is that in which both moments are antiparallel to B_0, and the lowest (4) is that in which both are parallel to B_0. In state 2 the moments are antiparallel to each other (i.e., paired) and the larger is antiparallel to B_0; in state 3 the moments are also antiparallel to each other and the smaller is antiparallel to B_0. Transitions of nucleus A will take place between states 1 and 3 and between states 2 and 4; transitions of nucleus X will take place between states 1 and 2 and between states 3 and 4. When $J = 0$ [(a) in Fig. 4.2], the two transitions of each nucleus are of the same energy, so only two lines, one for each nucleus, appear in the spectrum. When coupling is introduced [(b) and (c)], this is no longer true. It has been found that many spectra of systems of three or more spins can be satisfactorily analyzed only on the assumption that couplings may be both positive and negative. Since, as we shall see, only the *relative* signs of the couplings affect the appearance of the spectra, the question of sign cannot be important where only one coupling is involved, as here. Nevertheless, the

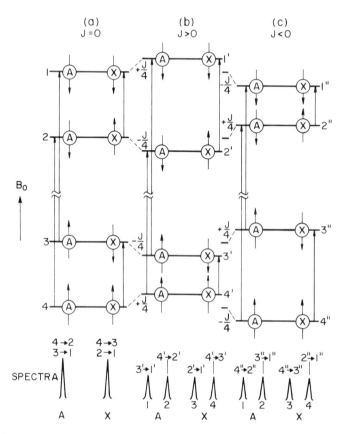

Fig. 4.2.   Energy levels and schematic spectra of the AX system.

arrangement of energy levels of even a two-spin system will be dependent
upon the sign of J. If $J > 0$ [(b) in Fig. 4.2], it can be shown that the
energy levels corresponding to parallel spins (1 and 4) will be raised by
$J/4$, and those corresponding to paired spins (2 and 3) will be decreased by
the same amount. When $J < 0$ [(c) in Fig. 4.2], the paired spins are
increased by $J/4$, and the unpaired are decreased. In both cases, there are
now two transitions of unequal magnitude for nucleus A and two for
nucleus X, and the difference in magnitude will be given by $4 \times J/4$ or J.
The singlet resonances of A and X are accordingly split by an amount J,
as shown in the schematic spectra on the bottom of Fig. 4.2. The reader
can easily verify that the spectra will be indistinguishable whether J is
positive or negative, although the peaks represent different transitions.

This last fact will become significant when we consider experimental means of determining relative signs of couplings in systems of three or more spins.

A positive coupling is to be expected, on theoretical grounds, between directly bonded nuclei, as in $^1H^3H$, HF, or $^{13}C-H$, but this is not always the case, as we shall see; for example, the sign of the $^{13}C-^{19}F$ coupling has been found to be opposite to that of $^{13}C-H$, and is therefore negative (see Chapter 9, Section 9.1.1).

As $J/\Delta\nu$ becomes larger, the energies of states 2 and 3 becomes closer together. In quantum mechanical terms, it becomes less realistic to assign each nucleus a specific quantum number, for there is now a mixing of states by the magnetic field. The transitions calculated for this system, assuming a positive J, are given in Table 4.1, where those of the AX system are given also for comparison. A simplification of form is brought about by defining a quantity C:

$$C = \frac{1}{2}[\Delta\nu^2 + J^2]^{1/2}$$

It is evident that when $\Delta\nu \gg J$ the AB transitions revert to those of the AX systems since C goes to $1/2\Delta\nu$. In Fig. 4.3, the AB energy levels are diagrammed for positive J; the extension to negative J is obvious. From Table 4.1 we can diagram the AB quartet as shown in Fig. 4.4. The

TABLE 4.1

Transitions of AB and AX Systems

(Line Positions are Referred to the Center of the Spectrum, $\frac{1}{2}(\nu_A + \nu_B)$)

Transition	Line	AX Line position	AX Relative intensity	AB Line position	AB Relative intensity
$3 \rightarrow 1$	1	$\frac{1}{2}(\Delta\nu) + \frac{1}{2}J$	1	$C + \frac{1}{2}J$	$C - \frac{1}{2}J$
$4 \rightarrow 2$	2	$\frac{1}{2}(\Delta\nu) - \frac{1}{2}J$	1	$C - \frac{1}{2}J$	$C + \frac{1}{2}J$
$2 \rightarrow 1$	3	$-\frac{1}{2}(\Delta\nu) + \frac{1}{2}J$	1	$-C + \frac{1}{2}J$	$C + \frac{1}{2}J$
$4 \rightarrow 3$	4	$-\frac{1}{2}(\Delta\nu) - \frac{1}{2}J$	1	$-C - \frac{1}{2}J$	$C - \frac{1}{2}J$

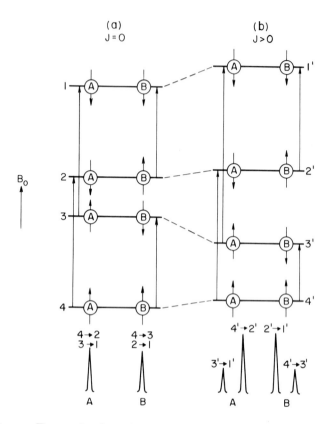

Fig. 4.3.   Energy levels and schematic spectra of the AB system.

following relationships are useful in solving AB quartets for $\Delta\nu$ and J and for constructing the spectrum given $\Delta\nu$ and J:

1. $J = \nu_1 - \nu_2 = \nu_3 - \nu_4$, the spacing of the doublets, as in the AX case.

2. $\Delta\nu = [(\nu_1 - \nu_4)(\nu_2 - \nu_3)]^{1/2}$, i.e., the geometric mean of the outer and inner line positions; this will approach the arithmetic mean for the AX case:

$$\Delta\nu_{AX} = 1/2[(\nu_1 - \nu_4) + (\nu_2 - \nu_3)]$$

3. Spacing of doublet centers:

$$(\Delta\nu^2 + J^2)^{1/2} = 2C$$

becoming of course $\Delta\nu$ when $\Delta\nu \gg J$.

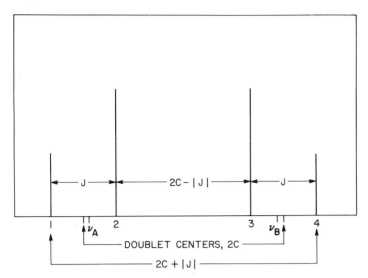

Fig. 4.4.   Schematic AB spectrum. [This is spectrum (c) of Fig. 4.5.]

4.   a. Spacing of center peaks:

$$\nu_2 - \nu_3 = (\Delta\nu^2 + J^2)^{1/2} - |J| = 2C - |J|$$

   b. Spacing of outer peaks:

$$\nu_1 - \nu_4 = (\Delta\nu^2 + J^2)^{1/2} + 2|J| = 2C + |J|$$

5.   Intensities:

$$\frac{i_2}{i_1} = \frac{i_3}{i_4} = \frac{\nu_1 - \nu_4}{\nu_2 - \nu_3}$$

i.e., the line intensities are inversely proportional to their spacings. The form of the spectrum is determined by $J/\Delta\nu$, and, as this ratio is increased without limit, the quartet approaches a singlet. The outer peaks are reduced to vanishingly small satellites, and the AB spectrum is reduced in effect to an A_2 spectrum. Calculated spectra in which $J/\Delta\nu$ is varied from 0.20 to 1.0 are shown in Fig. 4.5. As the satellites vanish, i.e., as $\Delta\nu$ becomes zero, we no longer can observe the coupling. Later we shall see, however, that this statement should not be taken too literally, for under some circumstances it is actually possible to measure the coupling between nuclei having the same chemical shift.

Although AB spectra are frequently observed as portions of more complex spectra, simple molecules giving only AB spectra are not particularly numerous. An example of such a molecule is 2-bromo-5-chlorothiophene, observed at 30.5 MHz (Fig. 4.6).[3]

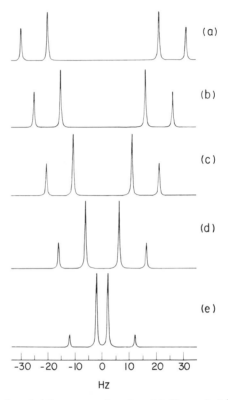

Fig. 4.5.   Calculated AB spectra for $J = 10$ Hz and $J/\Delta\nu$ equal to (a) 0.20, (b) 0.25, (c) 0.33, (d) 0.50, and (e) 1.00.

Analysis of this spectrum yields the parameters

$$\Delta\nu = 4.7 \text{ Hz } (0.15_4 \text{ ppm})$$

$$J = 3.9 \text{ Hz}$$

$$J/\Delta\nu = 0.83$$

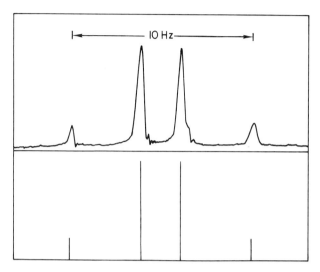

Fig. 4.6. The proton spectrum of 2-bromo-5-chlorothiophene, observed at 30.5 MHz. It is probable that proton 4 is the less shielding but the spectrum itself provides no proof of this point. [W. A. Anderson, *Phys. Rev.* **102**, 151 (1956).]

When $J/\Delta\nu$ is less than about 0.1, chemical shifts can be determined from the doublet centers within the usual experimental error, i.e., the spectrum may be treated as AX (Fig. 4.5).

4.2.2 Nuclear Equivalence and Nonequivalence

Let us consider more carefully the meaning of the terms "equivalence" and "nonequivalence." We recognize three types of nuclear equivalence. The lowest order of equivalence is *equivalence in chemical shift*. Nuclei that are equivalent in this sense may occupy quite different environments. Thus, methylacetylene, $CH_3C \equiv CH$, in chloroform solution shows a single resonance at 1.80 ppm from TMS for both methyl and acetylenic protons. In such cases, there is always the possibility that examination in a larger magnetic field or in a different solvent may remove the nonequivalence.

If a molecule has elements of symmetry, such as a twofold or higher axis or a mirror plane, magnetic nuclei of the same species that exchange their positions under the appropriate symmetry operation have *symmetry equivalence* and must have the same chemical shift. The six protons of benzene are equivalent in this sense. Likewise, the two protons and the two fluorine nuclei of 1,1-difluoroethylene form equivalent groups in this sense.

A group of spins A_n possesses *complete magnetic equivalence* if all n spins have the same chemical shift and are equally coupled to each of the m spins in every other magnetically equivalent spin group B_m. Thus the protons and fluorine nuclei of difluoromethane and of 1,1-difluoroallene,

$$\begin{array}{c} F \\ \diagdown \\ C = C = C \\ \diagup \\ F \end{array} \begin{array}{c} H \\ \diagup \\ \diagdown \\ H \end{array}$$

are magnetically equivalent, but those of 1,1-difluoroethylene,

$$\begin{array}{c} H_{(1)} \\ \diagdown \\ C = C \\ \diagup \\ H_{(1')} \end{array} \begin{array}{c} F_{(2)} \\ \diagup \\ \diagdown \\ F_{(2')} \end{array}$$

are not, since the *cis* couplings J_{12} and $J_{1'2'}$ are by no means equal to the *trans* couplings $J_{12'}$ and $J_{1'2}$. To make clear this distinction, molecules such as difluoromethane and 1,1-difluoroallene are designated A_2X_2, whereas 1,1-difluoroethylene is designated AA'XX'. Magnetic equivalence does not necessarily imply symmetry equivalence, but this is almost always the case, instances of fortuitous equivalence in both chemical shifts and couplings being very rare.

AA'BB' systems are of common occurrence. Benzene rings substituted by like groups in 1,2 positions (a) or by unlike groups in 1,4 positions (b) are frequently encountered examples:

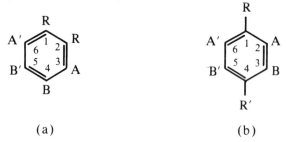

(a)   (b)

Their spectra are considered in Section 4.2.3.6.

Nuclei in nonrigid molecules or groups may achieve magnetic equivalence if group rotation is rapid but not if it is slow. (This point will be developed in Chapter 5, Section 5.4.) Methyl groups invariably constitute a set of three magnetically equivalent nuclei. Ethyl groups may usually be treated as A_2B_3 or A_2X_3, as in ethyl orthoformate (Chapter 1, Section 1.11), but in some cases must be treated as ABC_3 (or ABX_3) because of hindered rotation or because of the presence of a center of asymmetry in the molecule (see Chapter 3, Section 3.3.6).

4.2.3 Systems of Three and More Spins

4.2.3.1 AB_2 Systems

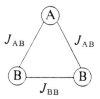

A system of two magnetically equivalent nuclei coupled to a third constitutes an AB_2 (or A_2B) system. Such spin systems are not really common, but occur often enough to deserve consideration. The most frequent examples are 1,2,3 (or 1,2,6) trisubstituted benzene rings, where R and R' may be like or unlike.

A further example is that of 2,6-lutidine.[4]

In an AB_2 system, there is one chemical shift difference $\Delta\nu$, two equal coupling constants J_{AB}, and a third, J_{BB}, not necessarily equal to J_{AB}. In the limit of the AX_2 case, one would of course expect a triplet for A and a doublet for X, a total of only 5 transitions. As $\Delta\nu$ decreases, the magnetic field begins to mix energy states appreciably, resulting in a splitting of levels that are degenerate when $J/\Delta\nu$ is small. Each line of the doublet splits into two, as does the center line of the triplet, making a total of eight lines. There is also a ninth transition, called a *combination line* because it corresponds to a simultaneous change of spin state of all three nuclei. This gives a weak line appearing beyond all the others on the B_2 side of the spectrum. It is usually difficult to see, but may be detectable if

$J/\Delta\nu$ is large. It is found that the coupling J_{BB} has no effect on the spectrum. Therefore, the AB_2 spectrum, like the AB spectrum, may be characterized by the single parameter $J_{AB}/\Delta\nu$. This should not be viewed as merely an obvious extension of previous conclusions concerning AB and A_2 spectra, for, as we shall see, in systems of four or more spins, couplings between equivalent nuclei *do* affect the spectrum.

In Table 4.2, line positions and relative intensities are given for the AB_2 system. It is convenient to define the following functions in order to simplify the notation:

$$C_+ = \tfrac{1}{2}(\Delta\nu^2 + \Delta\nu\cdot J + \tfrac{9}{4}J^2)^{1/2} \tag{4.1a}$$

$$C_- = \tfrac{1}{2}(\Delta\nu^2 - \Delta\nu\cdot J + \tfrac{9}{4}J^2)^{1/2} \tag{4.1b}$$

$$C_+\cos2\theta_+ = \tfrac{1}{2}\Delta\nu + \tfrac{1}{4}J \tag{4.2a}$$

TABLE 4.2

Transitions of an AB_2 System

Line	Origin	Line position[a]	Relative intensity[b]
1	A	$\tfrac{1}{2}(\nu_A + \nu_B) + \tfrac{3}{4}J + C_+$	$(\sqrt{2}\sin\theta_+ - \cos\theta_+)^2$
2	A	$\nu_B + C_+ + C_-$	$[\sqrt{2}\sin(\theta_+ - \theta_-) + \cos\theta_+\cos\theta_-]^2$
3	A	ν_A	1
4	A	$\tfrac{1}{2}(\nu_A + \nu_B) - \tfrac{3}{4}J + C_-$	$(\sqrt{2}\sin\theta_- + \cos\theta_-)^2$
5	B	$\nu_B + C_+ - C_-$	$[\sqrt{2}\cos(\theta_+ - \theta_-) + \cos\theta_+\sin\theta_-]^2$
6	B	$\tfrac{1}{2}(\nu_A + \nu_B) + \tfrac{3}{4}J - C_+$	$(\sqrt{2}\cos\theta_+ + \sin\theta_+)^2$
7	B	$\nu_B - C_+ + C_-$	$[\sqrt{2}\cos(\theta_+ - \theta_-) - \sin\theta_+\cos\theta_-]^2$
8	B	$\tfrac{1}{2}(\nu_A + \nu_B) - \tfrac{3}{4}J - C_-$	$(\sqrt{2}\cos\theta - \sin\theta_-)^2$
9	Comb.	$\nu_B - C_+ - C_-$	$[\sqrt{2}\sin(\theta_+ - \theta_-) + \sin\theta_+\sin\theta_-]^2$

[a] Since the AB_2 spectrum is not symmetrical, the line positions are referred to line 3 (i.e., ν_A) rather than $1/2(\nu_A + \nu_B)$, as in Table 4.1.

[b] Referred to line 3 as unity.

$$C_- \cos 2\theta_- = \tfrac{1}{2} \, \Delta\nu - \tfrac{1}{4} J \qquad\qquad (4.2b)$$

$$C_+ \sin 2\theta_+ = C_- \sin 2\theta_- = J/\sqrt{2} \qquad\qquad (4.3)$$

As in the AB system, no line can be associated unambiguously with an A or B nucleus, and the column labeled "origin" is intended to indicate only the customary, rather approximate association, which becomes rigorous in the limit of very large $\Delta\nu$. It is assumed that J_{AB} is positive. If it is negative, the appearance of the spectrum is unchanged but, as in the AB case, the detailed assignment of lines to particular transitions will be different.

In Appendix B, Section I, are shown a series of calculated AB_2 spectra with $\Delta\nu$ equal to 10 Hz and $J_{AB}/\Delta\nu$ varying from 0 to 2, i.e., J_{AB} varying from 0 to 20 Hz. For A_2B systems, the spectra will, of course, be mirror images of these. These features are of particular significance:

1. The chemical shift of nucleus A, ν_A, is always given by line 3. In strong-coupled systems, line 1 becomes very weak and might be overlooked. When $J/\Delta\nu$ exceeds 1, the spectra appear nearly symmetrical and might be mistaken for A_2B. One must be sure to count the lines correctly.

2. The chemical shift of nucleus B, ν_B, is given by the mean of lines 5 and 7. In many spectra, it may be difficult to resolve lines 5 and 6 (ν_B is also given by the mean of lines 2 and 9, but this is not very likely to be helpful).

The magnitude of J_{AB} is not given directly by any of the spectral spacings; but the separation of lines 1 and 6 gives $2C_+$ and that of lines 4 and 8 gives $2C_-$. From this information J_{AB} is obtained, knowing $\Delta\nu$. An equivalent calculation is

$$J_{AB} = \tfrac{1}{3}[(\nu_8 - \nu_6) + (\nu_4 - \nu_1)]$$

It will be seen from Appendix B, Section II, that as J_{AB} becomes infinitely large, the AB_2 spectrum is reduced to a triplet, the outer lines of which each have one-tenth the intensity of the center line.

Analysis of the spectrum of 2,6-lutidine at 40 MHz (Fig. 4.7) gives the following parameters.[5]

$$\Delta\nu = 21.9 \text{ Hz} \ (\nu_A - \nu_B = 0.55 \text{ ppm})$$

$$J = \pm 8.2 \text{ Hz}$$

$$J/\Delta\nu = 0.37$$

162   **4. Coupling of Nuclear Spins**

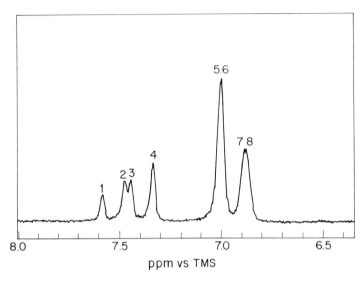

Fig. 4.7.   The proton spectrum of 2,6-lutidine observed at 60 MHz. [F. C. Schilling, private communication; see also H. J. Bernstein, J. A. Pople, and W. G. Schneider, *Can. J. Chem.* **35**, 65 (1957).]

4.2.3.2 AB$_3$ and AB$_n$ Systems

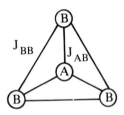

The AB$_3$ system, while not common, is encountered in molecules such as methyl mercaptan, methylacetylene, and methanol. (Methanol must be examined in solvents in which hydroxyl proton exchange is slow. The neat liquid commonly shows only two narrow singlets because of rapid exchange; this phenomenon is discussed in Chapter 5, Section 5.4.5.) In these molecules, a rapidly rotating methyl group forms a set of magnetically equivalent protons that are strongly coupled to a single proton. Whereas an AX$_3$ spectrum exhibits only 6 lines, i.e., an A quartet and an X doublet, an AB$_3$ spectrum may show as many as 16. Two of these are very weak combination lines, and only 14 are likely to be observed experimentally. Like AB and AB$_2$ spectra, AB$_3$ spectra and AB$_n$

spectra in general may be characterized by the single parameter $J_{AB}/\Delta\nu$. In Section II of Appendix B, a series of calculated spectra for $J_{AB}/\Delta\nu$ values of 0 to ∞ are shown. Unlike the AB_2 spectra, AB_3 spectra give J_{AB} directly as the spacings of lines 3 and 5 and of lines 8 and 13. Similar pairs of lines can be found in the spectra of all AB_n systems when n is odd. Chemical shifts in such spectra are not given directly by any line spacings or positions and must be obtained by calculation and spectral matching.

In Fig. 4.8 are shown the observed and calculated spectra for neat methyl mercaptan, as reported by Corio.[5] The calculated parameters that fit this spectrum are

$$\Delta\nu = 25.18 \pm 0.21 \text{ Hz}$$

$$J = \pm 7.42 \pm 0.17 \text{ Hz}$$

$$J/\Delta\nu = 0.29_4$$

AB_n spectra where $n > 3$ are uncommon, in general, and virtually nonexistent among organic compounds, so they will not be considered further here.

4.2.3.3 AMX Systems

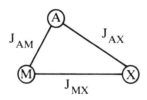

Three-spin systems in which all the nuclei are unlike must be characterized by three chemical shifts (or two chemical shift differences) and three coupling constants. The simplest case arises when all chemical shifts are large compared to the coupling constants. Such a spin system may be designated as AMX. The idealized spectrum will consist of 12 lines of equal intensity arranged in three quartets. A close approximation is the spectrum of trifluorochloroethylene, shown in Fig. 4.9. The first order analysis of this spectrum is shown schematically in the figure. The A signal may be considered to be first split to a doublet by J_{AM}; then each member of the doublet is split by J_{AX}. The M signal is similarly split by J_{MX} and J_{AM} and the X signal by J_{MX} and J_{AX}. (It is obvious that the order in which the splittings are taken is immaterial.) The results of measurements of multiplet splittings and spacings are

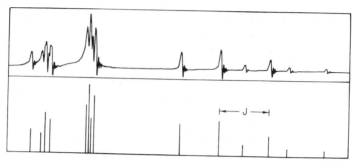

Fig. 4.8.   The proton spectrum of neat methyl mercaptan, CH_3SH, observed at 40 MHz. [P. L. Corio, *Chem. Rev.* **60**, 363 (1960).]

$$\nu_X - \nu_M = 1364.9 \text{ Hz} \qquad J_{AM} = 76 \text{ Hz}$$
$$J_{AX} = 56 \text{ Hz}$$
$$\nu_M - \nu_A = 983.6 \text{ Hz} \qquad J_{MX} = 116 \text{ Hz}$$

In the spectrum, the chemical shift scale is expressed in Φ units, which are ppm with respect to the solvent CCl_3F as zero (see ·Chapter 9, Section 9.1.1); each unit on the scale is equal to 56.4 Hz in this spectrum.

In analyzing such multiplets, the simplest procedure is to determine the larger of the two couplings by measuring the spacing of the first and third (or second and fourth) peaks and the smaller coupling by measuring the spacing of the first and second (or third and fourth) peaks.

4.2.3.4 ABX Systems

If the chemical shift difference between two of the three nuclei is comparable to their coupling to each other and both are coupled to a third nucleus well removed in chemical shift, the spectrum is designated as ABX. Such proton spectra are of common occurrence in organic molecules. They represent the simplest spin system in which the relative signs of the coupling constants affect the appearance of the spectrum; it is found that line intensities and positions depend on the relative signs of J_{AX} and J_{BX}, but are independent of the sign of J_{AB}. The ABX spectrum exhibits 14 lines; a fifteenth is customarily listed, but is of zero intensity. The AB portion of the spectrum consists of two pseudo-AB quartets, the total intensities of which are equal. These may be completely [Fig. 4.10(a)] or partially [Fig. 4-10(b)] overlapping, nonoverlapping

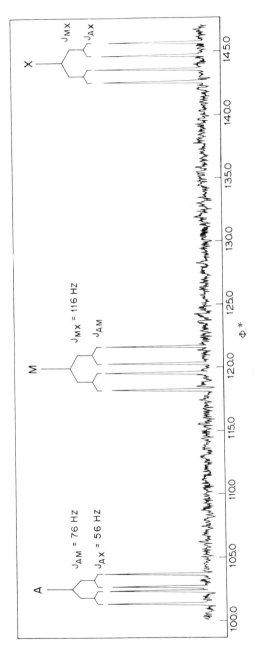

Fig. 4.9. The ^{19}F spectrum of trifluorochloroethylene, observed at 56.4 MHz as a 25% (vol./vol.) solution in CCl_3F. Chemical shifts are expressed on the Φ scale, CCl_3F being the reference zero (see text).

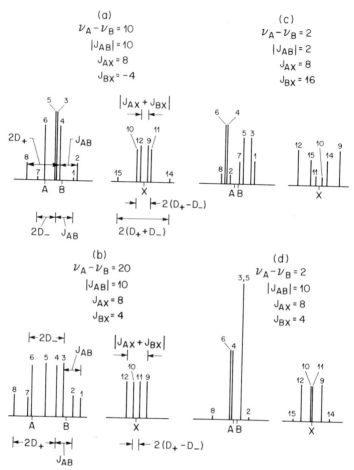

Fig. 4.10.   Calculated **ABX** spectra of four types: the "odd" and "even" AB subspectra (a) completely overlap, (b) partially overlap, and (c) do not overlap. Spectrum (d) is an example of a degenerate spectrum, having only 5 lines rather than the expected 8 lines in the AB part.

[Fig. 4.10(c)], or under certain circumstances, degenerate, i.e., having fewer than the expected eight lines [Fig. 4.10(d)].

The X portion of the spectrum consists of six lines, symmetrical about ν_X; two of these, lines 14 and 15, are combination lines in the sense that they decrease to vanishing intensity as $(\nu_A - \nu_B)$ increases. They are customarily very weak but become observable when $(\nu_A - \nu_B)$ is small, particularly if J_{AX} and J_{BX} have opposite signs. This portion of the spectrum is always symmetrical in true ABX spectra, but in spectra that are only approximately ABX (see Fig. 4.12), it will "lean" toward the AB part. The sum of the intensities of lines 14 and 11 equals that of line 9 (or 12), and the sum of lines 15 and 10 equals that of line 12 (or 9).

To express line positions and intensities, the positive quantities D_+ and D_- and the angles ϕ_+ and ϕ_- are defined

$$D_+ \cos 2\pi_+ = \tfrac{1}{2}(\nu_A - \nu_B) + \tfrac{1}{4}(J_{AX} - J_{BX}) \tag{4.4}$$

$$D_+ \sin 2\phi_+ = \tfrac{1}{2}J_{AB} \tag{4.5}$$

$$D_- \cos 2\phi_- = \tfrac{1}{2}(\nu_A - \nu_B) - \tfrac{1}{4}(J_{AX} - J_{BX}) \tag{4.6}$$

$$D_- \sin 2\phi_- = \tfrac{1}{2}J_{AB} \tag{4.7}$$

from which it follows that

$$D_+ = \tfrac{1}{2}\{[(\nu_A - \nu_B) + \tfrac{1}{2}(J_{AX} - J_{BX})]^2 + J_{AB}^2\}^{1/2} \tag{4.8a}$$

$$D_- = \tfrac{1}{2}\{[(\nu_A - \nu_B) - \tfrac{1}{2}(J_{AX} - J_{BX})]^2 + J_{AB}^2\}^{1/2} \tag{4.8b}$$

Line positions and intensities are listed in Table 4.3.

If J_{AX} and J_{BX} are small compared to $(\nu_A - \nu_B)$, the AB portion of the spectrum presents the appearance of an AB quartet, each line of which is split to a doublet. (See cases 1–4 in Section III of Appendix B.) It is tempting to take these doublet spacings as a measure of J_{AX} and J_{BX}. This is correct, however, only when $J_{AX} = J_{BX}$ and is otherwise only an approximation which becomes seriously in error as $(\nu_A - \nu_B)$ becomes small or $|(J_{AX} - J_{BX})|$ becomes large. No spacing in the spectrum gives J_{AX} or J_{BX} directly. However, if one picks out the two quartets, the separation of their centers gives $\tfrac{1}{2}|(J_{AX} + J_{BX})|$. D_+ and D_- are given by the appropriate spacings shown in Fig. 4.10(a,b); $|J_{AB}|$ is given by the outer spacings of the pseudo-AB quartets. This spacing occurs four times and is usually the most readily identifiable parameter.

The ABX spectrum contains a number of difficulties and ambiguities and contains less information than one might at first expect. Although we can obtain $|(J_{AX} + J_{BX})|$ and $|J_{AB}|$ directly from the spectrum, we cannot directly measure either the sign or magnitude of $(\nu_A - \nu_B)$ or

TABLE 4.3

Transitions of an ABX System

Line	Origin	Line position	Relative intensity
1	B	$\frac{1}{2}(\nu_A + \nu_B) - \frac{1}{4}(2|J_{AB}| + J_{AX} + J_{BX}) - D_-$	$1 - \sin 2\phi_-$
2	B	$\frac{1}{2}(\nu_A + \nu_B) - \frac{1}{4}(2|J_{AB}| - J_{AX} - J_{BX}) - D_+$	$1 - \sin 2\phi_+$
3	B	$\frac{1}{2}(\nu_A + \nu_B) - \frac{1}{4}(2|J_{AB}| - J_{AB} - J_{BX}) - D_-$	$1 + \sin 2\phi_-$
4	B	$\frac{1}{2}(\nu_A + \nu_B) + \frac{1}{2}(2|J_{AB}| + J_{AX} + J_{BX}) - D_+$	$1 + \sin 2\phi_+$
5	A	$\frac{1}{2}(\nu_A + \nu_B) - \frac{1}{4}(2|J_{AB}| + J_{AX} + J_{BX}) + D_-$	$1 + \sin 2\phi_-$
6	A	$\frac{1}{2}(\nu_A + \nu_B) - \frac{1}{4}(2|J_{AB}| - J_{AX} - J_{BX}) + D_+$	$1 + \sin 2\phi_+$
7	A	$\frac{1}{2}(\nu_A + \nu_B) - \frac{1}{4}(2|J_{AB}| - J_{AX} - J_{BX}) + D_-$	$1 - \sin 2\phi_-$
8	A	$\frac{1}{2}(\nu_A + \nu_B) + \frac{1}{4}(2|J_{AB}| + J_{AX} + J_{BX}) + D_+$	$1 - \sin 2\phi_+$
9	X	$\nu_X - \frac{1}{2}(J_{AX} + J_{BX})$	1
10	X	$\nu_X + D_+ - D_-$	$\cos^2(\phi_+ - \phi_-)$
11	X	$\nu_X - D_+ + D_-$	$\cos^2(\phi_+ - \phi_-)$
12	X	$\nu_X + \frac{1}{2}(J_{AX} + J_{BX})$	1
13	Comb.	$(\nu_A + \nu_B) - \nu_X$	0
14	Comb.	$\nu_X - D_+ - D_-$	$\sin^2(\phi_+ - \phi_-)$
15	Comb.	$\nu_X + D_+ + D_-$	$\sin^2(\phi_+ - \phi_-)$

$(J_{AX} - J_{BX})$. We can measure D_+ and D_- (see Fig. 4.10), but we do not know *a priori* which is which. The illustrative spectra in Fig. 4.10 contain assumptions as to these quantities; but when we are faced with real spectra where ν_A and ν_B are not known it is often a different matter.

Before proceeding to these real ambiguities, let us first dispose of one question concerning which many texts and discussions are misleading; this concerns the effect on the ABX spectrum of the relative signs of J_{AX} and J_{BX}. It is sometimes implied that their relative signs affect only the intensities and not the positions of the lines. This may appear to be the case within experimental error when $(\nu_A - \nu_B)$ is relatively large (e.g., cases 3 and 4 in Appendix B, Section III), but it can be seen from

Table 4.3 that it cannot in general be precisely true. Spectra 68 and 69 in Appendix B, Section III, make this clear [Fig. 4.10(a) is the same as 69]; a still more obvious example is afforded by comparing spectra 94 and 95 in the same appendix. Thus, this ambiguity is only apparent.

The troublesome consequences of the real ambiguities can be illustrated by attempting to solve spectrum (a) in Fig. 4.11 with no prior knowledge. We find by direct measurement that $|J_{AB} =| \; 10$ Hz and that $|(J_{AX} + J_{BX})| = 12$ Hz. Let us then analyze the subspectra represented by lines 1, 3, 5, and 7 and by lines 2, 4, 6, and 8 as though they were AB quartets (Section 4.2.1). We thus obtain *apparent* chemical shift differences for each quartet. This quantity we shall designate as δ_- for the 1, 3, 5, 7 quartet, for which it is found to be $|8.0|$ Hz; for the 2, 4, 6, 8 quartet we shall designate this quantity as δ_+ and find it equal to $|12.0|$ Hz.

It can be shown that

$$\delta_+ = |(\nu_A - \nu_B) + \tfrac{1}{2}(J_{AX} - J_{BX})| \tag{4.9a}$$

$$\delta_- = |(\nu_A - \nu_B) - \tfrac{1}{2}(J_{AX} - J_{BX})| \tag{4.9b}$$

However, we do not know the relative signs of δ_+ and δ_-.

From these data one finds two solutions that are entirely compatible with the AB part of the spectrum

$$(1) \qquad \nu_A - \nu_B = 10$$

$$|J_{AB}| = 10$$

$$J_{AX} = \pm 4$$

$$J_{BX} = \pm 8$$

these being the parameters in Fig. 4.11(a) and case 70, Appendix B, Section III. But there is another quite different solution that is equally valid for the AB part:

$$(2) \qquad \nu_A - \nu_B = 2$$

$$|J_{AB}| = 10$$

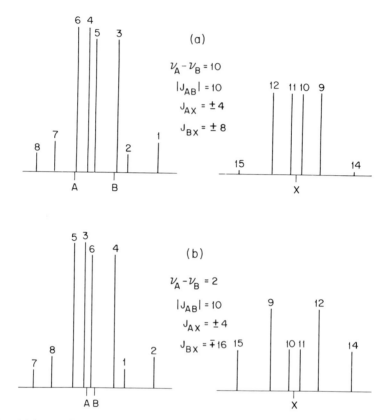

Fig. 4.11.   Calculated **ABX** spectra illustrating the fact that two quite
different sets of parameters can be found to fit the AB part,
which will, however, give very different line intensities in the
X spectrum. The line numbering shown corresponds to J_{AX}
and J_{BX} being both positive in (a) and to J_{AX} being positive
and J_{BX} negative in (b) If the other sign choice is made, the
"odd" and "even" AB subspectra are exchanged, the
ordering of the lines in each remaining the same.

$$J_{AX} = \pm 4$$

$$J_{BX} = \mp 16$$

The complete spectrum calculated for this second set of parameters is
shown in Fig. 4.11(b). We see that the AB part is identical with

Fig. 4.11(a). But we note that the X portions of the spectra are *not* identical, the lines being very different in relative intensity, and this furnishes a way out of our dilemma. Lines 14 and 15, scarcely observable in Fig. 4.11(a), have become much larger at the expense of lines 10 and 11 in Fig. 4.11(b).

To complete the analysis of the ABX spectra, we need to know $\frac{1}{2}(\nu_A + \nu_B)$. From Table 4.3, it is easily shown that this is given as the *mean* of the centers of the quartet subspectra. Knowing this quantity and $(\nu_A - \nu_B)$, it is evident that one can obtain ν_A and ν_B individually with respect to some reference line, most commonly internal tetramethylsilane (Chapter 2, Section 2.6.3).

The following relations among line spacings must always hold for all ABX spectra and (as we shall see in Section 4.2.3.5) for ABC spectra as well. These are helpful in avoiding errors in the assignments of transitions:

$$(3 - 1) = (4 - 2) = (7 - 5) = (8 - 6)$$

$$(2 - 1) = (4 - 3) = (11 - 9) = (12 - 10)$$

$$(10 - 11) = (8 - 4) - (5 - 1)$$

Fortunately, in many actual cases, the difficulties and uncertainties in ABX spectra are reduced or eliminated by knowledge of corresponding parameters in related structures. This can be illustrated by the spectrum of 2-chloro-3-aminopyridine (Fig. 4.12), which is reversed left to right in chemical shifts from the calculated spectra we have shown previously and perhaps ought properly to be called AYZ, but we shall retain the previous nomenclature. (It is also not strictly ABX, since the X proton is obviously by no means infinitely removed in chemical shift from the A and B protons.) As before, two completely different solutions will reproduce the AB part of the spectrum, but this can be resolved by comparisons in the X region. We also know that all couplings in such aromatic rings have the same sign, and furthermore, that the coupling between $H_{(4)}$ and $H_{(5)}$ is likely to be the largest and that between $H_{(4)}$ and $H_{(6)}$ the weakest; we also know that $H_{(6)}$ is likely to be the least shielded. We calculate these parameters:

$$\nu_X = 0.0 \text{ Hz } (7.58_7 \text{ ppm})$$

$$\nu_B = 49.0 \text{ Hz } (7.13_8 \text{ ppm})$$

$$\nu_A = 44.0 \text{ Hz } (7.08_7 \text{ ppm})$$

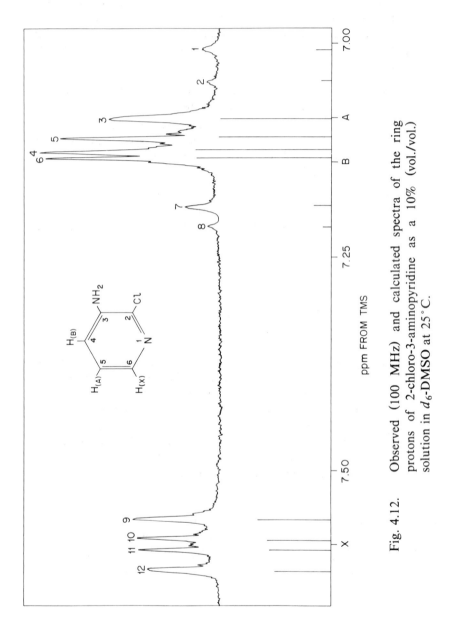

Fig. 4.12.   Observed (100 MHz) and calculated spectra of the ring protons of 2-chloro-3-aminopyridine as a 10% (vol./vol.) solution in d_6-DMSO at 25°C.

$$J_{AB} = 8.0 \text{ Hz}$$

$$J_{AX} = 4.2 \text{ Hz}$$

$$J_{BX} = 1.8 \text{ Hz}$$

The calculated spectrum (actually treated as ABC) is shown below the observed. Some intensity deviations are noticeable for certain lines; this is probably to be attributed to weak ^{14}N or amino proton couplings. The NH$_2$ resonance appears as a slightly broadened singlet at 5.50 ppm and is not included in the spectrum.

ABX systems can be deceiving in at least two ways. Figure 4.13 is case 59 in App. B, Section III, with a line width of 0.50 Hz provided by the computer. The X part would appear in practice as only a doublet because

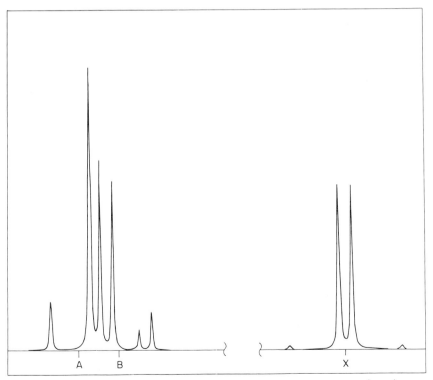

Fig. 4.13.   Calculated ABX spectrum for $\nu_A - \nu_B = 10$ Hz; $|J_{AB}| = 10$ Hz; $J_{AX} = 1$ Hz; $J_{BX} = -4$ Hz.

lines 9 and 11 and lines 10 and 12 cannot be resolved and the outer satellites are effectively zero. It might be concluded that either J_{AX} or J_{BX} is zero, which is not the case. In the AB part, only six lines appear for similar reasons. In Fig. 4.14 are shown two ABX spectra that are deceptive in another way. Spectrum (a) corresponds to a case where $(\nu_A - \nu_B)$ is small, and only nucleus B is coupled to X. Nevertheless, the X spectrum appears as a quartet (plus weak satellites), which would suggest two unequal nonzero X couplings. In spectrum (b), the couplings are unchanged, but $(\nu_A - \nu_B)$ is now zero, making this an AA'X system. The spectrum is now reduced to a doublet and triplet, with satellites to each that could easily escape detection. It closely resembles a simple A_2X spectrum, with J_{AX} equal to 2.0 Hz. Both spectra are instances of what has been termed "virtual coupling"[6]: nucleus X, although coupled to only one of the AB (or AA') nuclei, appears to be coupled to *both*. In case (b) the apparent coupling is half the actual value of 4.0 Hz. The X nucleus appears to "see" the combined spin states of A and A'. The spacing in the "triplet" is $\frac{1}{2}(J_{AX} + J_{BX})$, as we have already seen. Many instances of "virtual coupling" in ABX and more complex systems have been observed. In some cases, it is very easy to draw erroneous conclusions. An example is provided by the ^{19}F spectra of saturated fluorocarbon chains (see

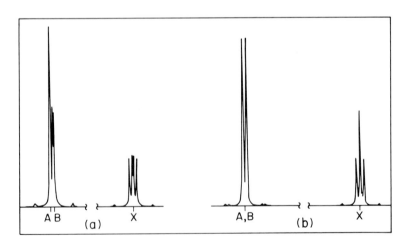

Fig. 4.14.   Calculated ABX spectra for

(a) $\nu_A - \nu_B = 2$ Hz   (b) $\nu_A - \nu_B = 0$ Hz
    $|J_{AB}| = 10$ Hz $|J_{AB}| = 10$ Hz
    $J_{AX} = 0$ Hz $J_{AX} = 0$ Hz
    $J_{BX} = 4$ Hz $J_{BX} = 4$ Hz

Chapter 9, Table 9.1, entries 1–4). In the spectrum of the perfluoroamine, the CF_2 resonance is a quartet from coupling to $F_{(3)}$, but is further split into equally spaced (5.1 Hz) septets from coupling to $F_{(2)}$ and $F_{(2')}$. This makes it appear that $J_{F_{(1)}F_{(2)}}$ and $J_{F_{(1)}F_{(2')}}$ are both equal to 5.1 Hz, although it is normally observed that vicinal $^{19}F-^{19}F$ couplings on saturated chains are nearly zero. The correct interpretation is that the septet spacing is actually equal to $1/2[J_{F_{(1)}F_{(2)}} + J_{F_{(1)}F_{(2')}}]$; $J_{F_{(1)}F_{(2)}}$ being *ca.* 0, $J_{F_{(1)}F_{(2')}}$ must be *ca.* 10.2 Hz, conclusions more in accord with other data from related but less ambiguous spectra. In this case, the "virtual" character of the spectrum is particularly pronounced because $F_{(2)}$ and $F_{(2')}$ are equivalent in chemical shift. It should be evident from this discussion that "virtual coupling" is not a special physical phenomenon but only a convenient term for well-understood spectral effects.

4.2.3.5 ABC Systems

When the X nucleus of the ABX system becomes close in chemical shift to the A and B nuclei so that $(\nu_A - \nu_B)$ and $(\nu_B - \nu_C)$ are of comparable magnitude, we have an ABC system. Such systems are of common occurrence; examples are vinyl groups (a), trisubstituted benzenes (b), disubstituted pyridines (c), substituted thiophenes (d), furans (e), and epoxides (f):

(a)   (b)

(c)   (d)

(e)   (f)

The ABC spectrum can contain up to 15 lines. Line 13 of the ABX spectrum (Table 4.3) is not of zero intensity in an ABC spectrum, although it is very weak and commonly unobservable. The combination lines 14 and 15 may be comparable to the fundamental lines in intensity. (In fact, there is no real distinction between them in strongly coupled systems, but the labels are convenient.) It is possible to determine the relative signs of all three couplings, not just two as in the ABX spectrum. The complete solution of the ABC spectrum is a considerable mathematical task. It is not possible to obtain simple algebraic expressions for the line positions and intensities, such as those given in Table 4.3. Approximate parameters can often be obtained by treating the spectrum as ABX. As we have already indicated, it is important to note that the same line spacing relationships which hold for the ABX spectrum hold for the ABC spectrum as well. From the parameters so obtained, a spectrum is calculated and compared to the observed. From the discrepancies, one can obtain a new and better set of parameters. This iterative procedure involves the use of a computer to achieve the desired convergence of calculated and observed spectra. One must be careful to identify the lines correctly, in particular not to confuse "fundamental" and "combination" lines. If line identification is uncertain, the results will be ambiguous. In some cases, spectra can be iteratively matched within experimental error by several different sets of parameters.

In Figs. 4.15, 4.16, and 4.17 some typical ABC spectra are shown. Their solutions are as follows. For styrene (observed at 60 MHz):

$$C_6H_5 \diagdown C = C \diagup H_{(B)}$$
$$H_{(A)} \diagup \qquad \diagdown H_{(C)}$$

$\nu_A = 0$ Hz (6.64 ppm)	$J_{AB} = 17.5$ Hz
$\nu_B = 59.4$ Hz (5.65 ppm)	$J_{AC} = 10.7$ Hz
$\nu_C = 89.4$ Hz (5.15)	$J_{BC} = 1.3$ Hz

For vinyl chloride (observed at 60 MHz; note that the AB part is degenerate, having only 6 observable lines):

$$Cl \diagdown C = C \diagup H_{(B)}$$
$$H_{(A)} \diagup \qquad \diagdown H_{(C)}$$

$\nu_A = 0$ Hz (6.26 ppm)	$J_{AB} = 14.8$ Hz
$\nu_B = 46.8$ Hz (5.48 ppm)	$J_{AC} = 7.4$ Hz
$\nu_C = 52.2$ Hz (5.39 ppm)	$J_{BC} = -1.3$ Hz

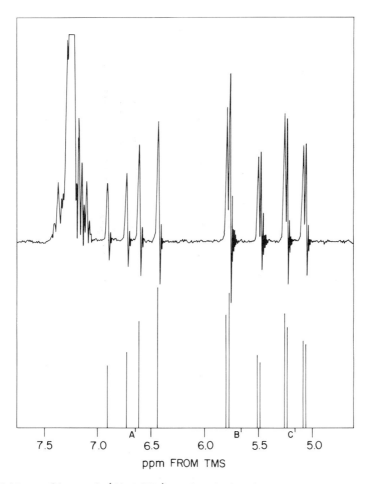

Fig. 4.15.   Observed (60 MHz) and calculated proton spectra of the vinyl group of styrene as a 25% (vol./vol.) solution in CCl_4 at 35°C.

For 1,2,4-trichlorobenzene (observed at 100 MHz):

$$\nu_A = 0 \quad \text{Hz } (7.41_3 \text{ ppm}) \qquad J_{AB} = 0.4 \text{ Hz}$$
$$\nu_B = 9.9 \quad \text{Hz } (7.31_4 \text{ ppm}) \qquad J_{AC} = 2.4 \text{ Hz}$$
$$\nu_C = 28.7 \text{ Hz } (7.12_6 \text{ ppm}) \qquad J_{BC} = 8.4 \text{ Hz}$$

In all cases, analysis is assisted by some prior knowledge of probable signs and magnitudes of couplings.

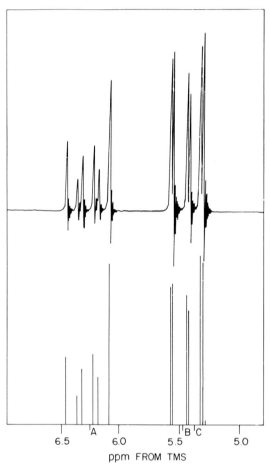

Fig. 4.16.   Observed (60 MHz) and calculated proton spectra of vinyl chloride as a 25% (vol./vol.) solution in CCl$_4$ at 35°.

In Appendix B (Section IV) some representative calculated ABC spectra are shown. The possible ranges and combinations of parameters are so large that a severe limitation has to be placed on the number of different magnitudes of the chemical shifts and coupling constants. This has been done by choosing ranges of parameters corresponding to the more common ABC proton spectra, particularly those of vinyl groups and trisubstituted benzene rings. In several instances, comparison may be made to ABX spectra with corresponding coupling constants.

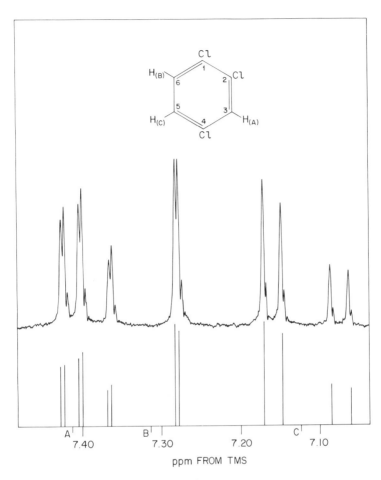

Fig. 4.17.   Observed (100 MHz) and calculated proton spectra of 1,2,4-trichlorobenzene as a 10% (vol./vol.) solution in CCl_4 at 25°C.

4.2.3.6 AA'XX' and AA'BB' Systems

Two pairs of spins, the members of each pair being equivalent in chemical shift but not magnetically equivalent (see Section 4.2.2) constitute an AA'XX' system.

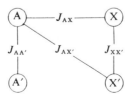

In systems of several spins, the total number of possible couplings may be quickly determined from the formula $\frac{1}{2}n(n-1)$, where n is the number of coupled nuclei. For this spin system, there are six couplings, but, from obvious symmetry considerations, only four different magnitudes. If $J_{AX} = J_{AX'}$, the system would be designated A_2X_2, as we have seen. Strict AA'XX' spectra are given by pairs of nuclei of different species:

$$
\begin{array}{ccc}
\text{F}\diagdown & & \diagup\text{H} \\
 & \text{C}=\text{C} & \\
\text{F}\diagup & & \diagdown\text{H}
\end{array}
\qquad
\begin{array}{ccc}
\text{F}\diagdown & & \diagup\text{F} \\
 & \text{C}=\text{C} & \\
\text{H}\diagup & & \diagdown\text{H}
\end{array}
$$

$$
\begin{array}{ccc}
\text{F}\diagdown & & \diagup\text{H} \\
 & \text{C}=\text{C} & \\
\text{H}\diagup & & \diagdown\text{F}
\end{array}
\qquad
\text{R—C}_6\text{H}_2\text{F}_2\text{—R (1,4-R, 2,3-H, 5,6-F ring)}
$$

Certain ^{19}F spectra, as, for example, that of

$$\text{C}_6\text{F}_3\text{Cl}_2 \text{ (ring: F, F, F, Cl, Cl)}$$

may approximate AA'XX' systems very closely because of the large values of $\nu_A - \nu_X$. These designations are convenient:

$$K = J_{AA'} + J_{XX'}$$

$$L = J_{AX} - J_{AX'}$$

$$M = J_{AA'} - J_{XX'}$$

$$N = J_{AX} + J_{AX'}$$

For expressing relative line intensities, we define these quantities:

$$\cos 2\theta_s : \sin 2\theta_s : 1 = K : L : (K^2 + L^2)^{1/2}$$

$$\cos 2\theta_a : \sin 2\theta_a : 1 = M : L : (M^2 + L^2)^{1/2}$$

The entire AA′XX′ spectrum contains 24 lines, but we shall consider only the A spectrum, as the X spectrum is identical in form. The 12 transitions of the A spectrum form a group centrosymmetric about ν_A, as shown in a typical spectrum (Fig. 4.18), in which the magnitudes of the couplings are indicated in the caption. Only 10 lines can actually be observed, as transitions 1 and 2 and transitions 3 and 4 are always degenerate. Line positions and relative intensities are given in Table 4.4. The line positions are given relative to ν_A. It can be seen from Table 4.4 that lines 1, 2 and 3, 4 together will always contain half the intensity of the A spectrum.

We observe further that

a. The separation of lines 1, 2 and 3, 4 equals $N = J_{AX} + J_{AX'}$.

b. The separation of lines 5 and 6 (and of 7 and 8) gives K.

c. The separation of lines 9 and 10 (and of 11 and 12) gives M.

d. Lines 5, 6, 7, and 8 form a quartet centered on ν_A.

e. Lines 9, 10, 11, and 12 form a quartet, also centered on ν_A.

There is some ambiguity in AA′XX′ spectra, for in the absence of prior information concerning the probable relative magnitudes and signs of $J_{AA'}$

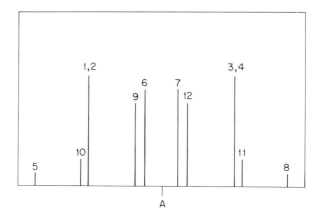

Fig. 4.18.   The A portion of a calculated AA′XX′ spectrum with the following parameters (the X portion is identical in form):

$$J_{AA'} = 9.0 \text{ Hz} \quad K = 12$$
$$J_{XX'} = 3.0 \text{ Hz} \quad M = 6$$
$$J_{AX} = 13.0 \text{ Hz} \quad N = 16$$
$$J_{AX'} = 3.0 \text{ Hz} \quad L = 10$$

and $J_{XX'}$ from related molecules, one does not know the relative signs of K and N and, therefore, cannot know how to assign the outer and inner lines of each quartet. (The inner lines, however, will always be more intense.) Thus, we can find K and M as a pair but cannot *a priori* know which is which, nor their signs.

    f. The spacings of line 5 and 7 (and of 6 and 8) give $(K^2 + L^2)^{1/2}$; the spacings of line 9 and 11 (and of 10 and 12) give $(M^2 + Li^2)^{1/2}$.

It can be seen that the relative signs of M and L do not affect the appearance of the spectrum and, therefore, cannot be determined. It is also found that the relative signs of K and N do not affect the appearance of the spectrum. [They do affect the appearance of the AA'BB' spectrum; (see below).] However, as with ABX systems, we often have prior knowledge of relative signs for many structural types, determined as described in Section 4.4.2, so these ambiguities can be resolved.

TABLE 4.4

Transitions of an AA'XX' System

Line	Line position[a]	Relative intensity
1	$\frac{1}{2}N$	1 ⎫
2	$\frac{1}{2}N$	1 ⎭ 2
3	$-\frac{1}{2}N$	1 ⎫
4	$-\frac{1}{2}N$	1 ⎭ 2
5	$\frac{1}{2}K + \frac{1}{2}(K^2 + L^2)^{1/2}$	$\sin^2 \theta_s$
6	$-\frac{1}{2}K + \frac{1}{2}(K^2 + L^2)^{1/2}$	$\cos^2 \theta_a$
7	$\frac{1}{2}K - \frac{1}{2}(K^2 + L^2)^{1/2}$	$\cos^2 \theta_s$
8	$-\frac{1}{2}K - \frac{1}{2}(K^2 + L^2)^{1/2}$	$\sin^2 \theta_s$
9	$\frac{1}{2}M + \frac{1}{2}(M^2 + L^2)^{1/2}$	$\sin^2 \theta_a$
10	$-\frac{1}{2}M + \frac{1}{2}(M^2 + L^2)^{1/2}$	$\cos^2 \theta_a$
11	$\frac{1}{2}M - \frac{1}{2}(M^2 + L^2)^{1/2}$	$\cos^2 \theta_a$
12	$-\frac{1}{2}M - \frac{1}{2}(M^2 + L^2)^{1/2}$	$\sin^2 \theta_a$

[a] Relative to ν_A.

Some calculated AA'XX' are included in Appendix B, Section V. The first spectrum in each of the eleven groups of AA'BB' spectra presented there is the A portion (or X portion; they are identical, as we have seen) of the AA'XX' spectrum with corresponding couplings. The effect of reducing the magnitude of $(\nu_A - \nu_B)$ in steps from ∞ (actually 10^4 Hz, large enough for our purposes) to 50 Hz, 40 Hz, 30 Hz, and 20 Hz can be seen.

When the chemical shift difference between the A and X nuclei is comparable in magnitude to the couplings, we have an AA'BB' spectrum. Many spin systems present spectra of this type. Some examples are o-disubstituted benzenes (a); p-disubstituted benzenes with unlike substituents (b); γ-substituted pyridines (c); thiophene (d); furan (e); 1,2-disubstituted ethanes with unlike substituents (f); and appropriately substituted cyclopropanes (g):

R_1

R_1

(a)

R_1

R_2

(b)

R

N

(c)

S

(d)

O

(e)

$R_1CH_2CH_2R_2$

(f)

R_1

A R_2 A'

B   B'

$H_{(B')}$

R_1 $H_{(B)}$ $H_{(A')}$

$H_{(A)}$ R_1

(g)

In treating such systems, we may again consider only the "A" portion of the spectrum as the complete spectrum is centrosymmetric about $1/2(\nu_A + \nu_B)$. However, the "A" and "B" portions of the spectrum are not themselves symmetric as in AA'XX' spectra. It is common in many discussions of this system to show only half the spectrum, but in Section V of Appendix B we present the complete spectra; this is particularly desirable when $(\nu_A - \nu_B)$ is small.

Transitions 1 and 2, and 3 and 4 of AA'XX' spectra (Table 4.4) are no longer degenerate in the AA'BB' spectrum, and so the total spectrum has 24 lines instead of only 20. These features can be seen by comparing the

AA'XX' spectrum in Section 5 of Appendix B to the corresponding AA'BB' spectrum. Some of the line positions and intensities of the AA'BB' spectrum can be expressed by explicit algebraic expressions and some cannot. In Table 4.5, all lines for the "A" part of the spectrum are tabulated, but the second column is blank for lines of the latter kind. It will be noted that none of the lines whose positions can be explicitly expressed are dependent on K; this often makes K hard to determine in AA'BB' spectra, particularly when it is large, as the appearance of the spectrum then tends to become independent of its magnitude and relative sign.

In addition to the parameters K, L, M, and N, which have the same definitions as for the AA'XX' spectrum with X and X' replaced by B and B', we must define the angles ϕ, ψ_+, and ψ_-:

$$\cos 2\phi : \sin 2\phi : 1 = (\nu_A - \nu_B) : N : [(\nu_A - \nu_B)^2 + N^2]^{1/2}$$

$$\cos 2\psi_+ : \sin 2\psi_+ : 1 = [(\nu_A - \nu_B) + M] : L : \{[(\nu_A - \nu_B) + M]^2 + L^2\}^{1/2}$$

$$\cos 2\psi_- : \sin 2\psi_- : 1 = [(\nu_A - \nu_B) - M] : L : \{[(\nu_A - \nu_B) - M]^2 + L^2\}^{1/2}$$

In addition, theory predicts four additional lines, but these are combination lines of such negligible intensity that they are not included in Table 4.5.

The separation of lines 1 and 3 (or the mirror-image lines in the "B" part of the spectrum) gives N, and from the sum of their spacings from the center of the spectrum, the chemical shifts ν_A and ν_B can then be obtained, using the expressions for lines 1 and 3 in Table 4.5. The splittings of lines 1 and 2 and of 3 and 4 are often not resolved, and they then appear as strong singlets and can be readily identified. As $(\nu_A - \nu_B)$ decreases, the spectral intensity tends to accumulate at the center (an effect obvious from inspection of Section V of Appendix B), and this may make the outer lines difficult to see.

Let us first consider a case of particular simplicity, where $J_{AB} = J_{AB'}$ and, therefore, $L = 0$. These conditions are sometimes approximated in actual spectra, although there is no known case where this is strictly true. Such a spin system can be properly designated as A_2B_2. It can be shown that the spectrum is now independent of the sign and magnitude of $J_{AA'}$ and $J_{BB'}$ (and, therefore, of the sign and magnitude of K and M), so that these may be set at zero. We thus have $K = M = L = 0$, and the appearance of the spectrum is governed only by the quantity

$$N/2(\nu_A - \nu_B) = J_{AB}/(\nu_A - \nu_B)$$

TABLE 4.5

Transitions of an AA'BB' System

Line	Line position[a]	Relative intensity
1	$\frac{1}{2}N + \frac{1}{2}[(\nu_A - \nu_B)^2 + N^2]^{1/2}$	$1 - \sin 2\phi$
2	—	—
3	$-\frac{1}{2}N + \frac{1}{2}[(\nu_A - \nu_B)^2 + N^2]^{1/2}$	$1 + \sin 2\phi$
4	—	—
5	—	—
6	—	—
7	—	—
8	—	—
9	$\frac{1}{2}\{[(\nu_A - \nu_B) + M]^2 + L^2\}^{1/2} + \frac{1}{2}(M^2 + L^2)^{1/2}$	$\sin^2(\theta_a - \psi_+)$
10	$\frac{1}{2}\{[(\nu_A - \nu_B) - M]^2 + L^2\}^{1/2} + \frac{1}{2}(M^2 + L^2)^{1/2}$	$\cos^2(\theta_a + \psi_-)$
11	$\frac{1}{2}\{[(\nu_A - \nu_B) + M]^2 + L^2\}^{1/2} - \frac{1}{2}(M^2 + L^2)^{1/2}$	$\cos^2(\theta_a - \psi_+)$
12	$\frac{1}{2}\{[(\nu_A - \nu_B) - M]^2 + L^2\}^{1/2} - \frac{1}{2}(M^2 + L^2)^{1/2}$	$\sin^2(\theta_a - \psi_-)$

[a] Relative to $\frac{1}{2}(\nu_A + \nu_B)$, i.e., center of spectrum.

just as for the AB_n systems already discussed (Section 4.3.2.3). A series of such spectra are given as group 1 in Section V of Appendix B. In the limit of $(\nu_A - \nu_B) = \infty$, each part of the spectrum is a 1:2:1 triplet, as would be expected. As $(\nu_A - \nu_B)$ decreases, the lines of the triplet become split, but since under these conditions lines 5, 8, 9, and 12 become forbidden and lines 10 and 11 become degenerate (Table 4.5), the half-spectra contain only 7 lines (again not counting two very weak combination lines) rather than 12. The position of the degenerate "A" lines 10 and 11 gives ν_A and that of the corresponding "B" lines gives ν_B.

In treating more complex AA'BB' spectra, it is appropriate and useful to analyze them in terms of relative magnitudes of K, M, L, and N. We shall use a different approach, similar to that used for ABC spectra. Eleven groups of spectra are calculated (Appendix B, Section V) corresponding to

(I) A_2B_2, as just discussed

(II) o-Disubstituted benzene rings [(a), p.   ]

(III) p-Disubstituted benzene rings [(b), p.   ]

(IV) $X-CH_2CH_2-Y$

It might seem that the protons in case IV form two fully equivalent groups and should give an A_2B_2 rather than an AA'BB' spectrum. But, as we shall see, there are actually, in general, two unequal vicinal couplings, J_{AB} and $J_{AB'}$. Four subgroups of IV are considered, corresponding to four ratios of the *trans/gauche* conformers. A fifth case, that in which the *trans* and both *gauche* conformers are equally populated, does actually correspond to an A_2B_2 case and is therefore included in I.

(V) Substituted cyclopropanes (g)

It should be emphasized that, like the previous spectra we have discussed, AA'BB' spectra contain a number of ambiguities. The AA'BB' spectrum is sensitive to the relative signs of K and N, but does not depend upon the relative signs of either M or L with respect to K and N. The result of this is that from the spectrum alone, without any knowledge of probable values of the couplings based on analogous molecules, we actually cannot tell which is $J_{AA'}$ and which is $J_{BB'}$, nor can we tell J_{AB} from $J_{AB'}$. But we shall see that, as before, in actual molecules most or all of these ambiguities can usually be resolved.

We shall now consider some experimental spectra and their solutions. The first is o-dichlorobenzene, which is of type II above. The experimental (100 MHz) and computed spectra are shown in Fig. 4.19. The solution of this spectrum yields these parameters:

$$\begin{aligned}
\nu_A - \nu_B &= 25.0 \text{ Hz} & J_{AA'} &= 0.3 \text{ Hz} & K &= 7.8 \\
\nu_A &= 7.22_8 \text{ ppm} & J_{BB'} &= 7.5 \text{ Hz} & M &= -7.2 \\
\nu_B &= 6.97_8 \text{ ppm} & J_{AB} &= 8.1_5 \text{ Hz} & N &= 9.7 \\
& & J_{AB'} &= 1.5_5 \text{ Hz} & L &= 6.6
\end{aligned}$$

A spectrum of type III is given by p-chlorobenzaldehyde (Fig. 4.20):

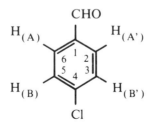

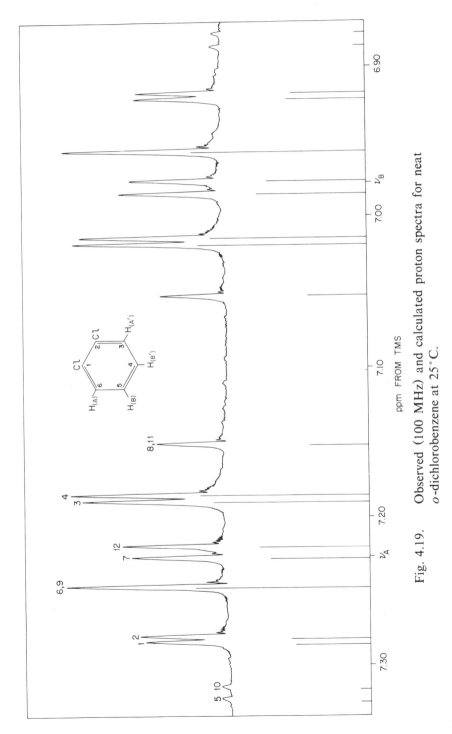

Fig. 4.19. Observed (100 MHz) and calculated proton spectra for neat
o-dichlorobenzene at 25°C.

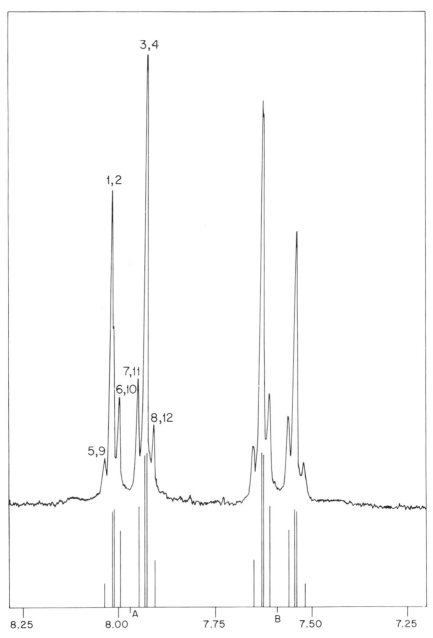

Fig. 4.20. Observed (100 MHz) and calculated proton spectra for *p*-chlorobenzaldehyde as 10% (wt./vol.) solution in trifluoroacetic acid at 25°C.

For the ring protons, these parameters are found (see also Ref. 7):

$\nu_A - \nu_B = 37.0$ Hz	$J_{AA'} = 2.1$ Hz	$K = 4.2$
$\nu_A = 7.97$ ppm	$J_{BB'} = 2.1$ Hz	$M = 0.0$
$\nu_B = 7.60$ ppm	$J_{AB} = 8.3$ Hz	$N = 8.7$
	$J_{AB'} = 0.4$ Hz	$L = 7.9$

It must be emphasized that the assignments of the A and B protons to low and high field cannot be made on the basis of the spectrum alone, but require additional information from molecular analogs. It can be seen that the B protons are slightly broadened compared to the A protons, probably owing to weak coupling to the aldehyde proton.

A spectrum of type IV is given by 1-bromo-2-chloroethane. In molecules such as this,

$$\underset{\text{(A)} \quad \text{(B)}}{ClCH_2CH_2Br}$$

which have conformational freedom, the vicinal couplings, J_{AB} and $J_{AB'}$, may vary substantially with the solvent because of varying conformer populations (Section 4.3.3). The observed and calculated spectra for a 10% solution in carbon tetrachloride are shown in Fig. 4.21. The spectrum contains 20 observable lines. The sum of the geminal couplings represents K and is in this case so large, probably of the order of -25, that, as described earlier, it becomes indeterminate.* We can, therefore, determine only $|M|$, the difference in the geminal couplings, and so cannot even tell *a priori* which is the larger. The parameters obtained in the authors' laboratory (see also Ref. 8) for a 10% (w./v.) solution of 1-bromo-2-chloroethane in carbon tetrachloride are shown here:

$J_{AA'}$	= indeterminate	K	= indeterminate
$J_{BB'}$	= indeterminate	M	= 1
J_{AB}	= 5.65 Hz	N	= 16.20
$J_{AB'}$	= 10.55 Hz	L	= -4.90

---

* Actually, by more sophisticated methods of spectral solution than the first-order perturbation approach upon which this chapter is based, it is found that the spectrum is not entirely insensitive to the relative signs and magnitudes of $J_{AA'}$ and $J_{BB'}$, and that they can be at least estimated, although of course one still cannot tell *a priori* which is which. The spectra is this chapter are actually generated by a computer program of this more sophisticated type (see preface), and from a knowledge that the protons on the carbon bearing the chlorine atom will be the less shielded (as implied in the structural formula above), one finds the best match of computed and experimental spectra when $J_{AA'} = ca. -12$ Hz, $J_{BB'} = ca. -13$ Hz assuming the vicinal couplings to be positive. The negative signs of these couplings are in accord with expectation (see Section 4.3.2).

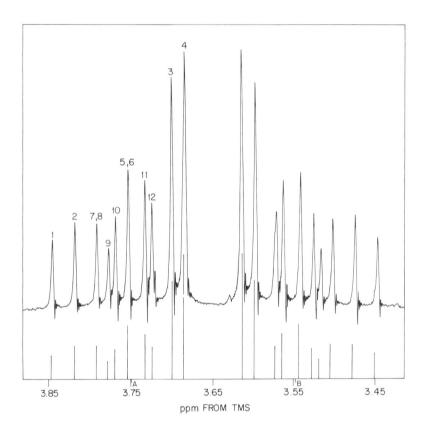

Fig. 4.21.   Observed (100 MHz) and calculated proton spectra for 1-
              bromo-2-chloroethane as 10% (vol./vol.) solution in CCl$_4$ at
              25°C.

Again, we must remind ourselves that from the spectrum alone we can tell
the relative signs of M and L, but we do not know the relative sign of N
with respect to L or M. N is here assumed to be positive because it is
independently known (Section 4.3.3) that J_{AB} and $J_{AB'}$ are positive in such
molecules.

A recent and useful discussion of the analysis of NMR spectra from a
more fundamental mathematical viewpoint than employed here is that of
Harris.[9]

4.3 DEPENDENCE OF J-COUPLING ON STRUCTURE AND GEOMETRY

4.3.1 Introduction

In this section we shall consider the dependence of scalar coupling on the structure and geometry — including conformation — of organic molecules. We shall deal mainly with proton–proton couplings, the magnitude of which will be assumed to be determined by analysis of the spectra in the manner described in Section 4.2. Proton–^{13}C and ^{13}C–^{13}C couplings will also be discussed (Section 4.3.6) (^{19}F, ^{14}N, ^{15}N, and ^{31}P are dealt with in Section 9.1 of Chapter 9). In Appendix C a tabulation of geminal and vicinal proton–proton couplings is given, while Appendix D contains a tabulation of ^{13}C couplings. These will be referred to in the subsequent discussion.

4.3.2 Geminal Proton–Proton Couplings

Coupling constants for nuclei separated by *n* chemical bonds are often designated as ^{n}J. The coupling between two protons bonded to the same atom — usually carbon — are accordingly represented as ^{2}J. Geminal couplings are found to vary over a large range and to be both positive and negative. Geminal couplings in sp^{3} hybridized groups are negative. (The determination of relative and absolute signs of coupling constants is discussed in Section 4.4.2.) Thus, for methane ^{2}J is −12.4 Hz (Appendix C, entry *1**). It is in the range −10 to −15 Hz for most sp^{3} hybridized groups (*2−17*) and increases algebraically, becoming less negative and then actually positive as the H−C−H angle increases. (Throughout this chapter we shall use the words *decrease* and *increase* in the algebraic sense.) In ethylene, which is sp^{2} hybridized, ^{2}J is +2.3 Hz (*103*); it is in range −4 to −8 Hz for cyclopropanes, (*21−23*), which are intermediate between sp^{2} and sp^{3} in hybridization.

The effects of adjacent substituents are very large and can often overwhelm geometrical effects. Both theory and observation indicate that the influence of substituents depends not only on their nature but also on their orientation. The presence of an electronegative substituent on the carbon bearing the geminal protons can greatly increase the coupling. A striking example is formaldehyde (*20*) for which the coupling is about +41 Hz. But if the electronegative substituent is one atom further removed, its effect is frequently to *decrease* the coupling. We have seen an example of this already in the negative geminal coupling of vinyl chloride (*125*) compared to positive values for styrene (*113*) and ethylene (*103*). In

---

*   The italic numbers are all entries in Appendix C.

CH_3CCl_3 (*2*) the coupling is somewhat more negative than in methane. In the rigid ring compound *53*, the geminal coupling is substantially more negative than in methane.

The influence of oxygen bonded directly to the geminal carbon can be large, as we have seen for formaldehyde, and is always in the direction of increasing ^2J. Methanol (*13*), diethyl sulfite (*16*), and diethyl acetal (*17*) appear to show this effect rather weakly; five-membered methylenedioxy groups (*39*) show it move strongly and dioxane (*51*) very strongly.

The presence of a π-electron system adjacent to the CH_2 decreases the coupling (cf. *9-12*, *14*, *15*, *37*, *38*, *52*). Molecular orbital theory[10–12] predicts that this effect will be greatest when the geminal internuclear axis is perpendicular to the nodal plane of the π system, as illustrated for a carbonyl group in (a), whereas if the internuclear axis is more nearly parallel to this plane (b) or if one of the protons is in the nodal plane (c), the effect will be small or even weak in the opposite direction.

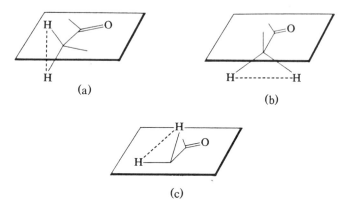

(a)   (b)

(c)

The methyl geminal couplings in acetic acid (*10*), acetone (*12*), toluene (*14*), and 4-picoline (*15*) appear to be consistent with a preference for the conformation shown as (d), where the cross-hatched bar represents the π system's nodal plane.

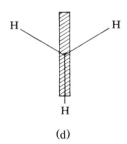

(d)

In compound *54*, one proton of each methylene group is probably in the nodal plane of the ring [conformation (c)] and ^2J is less negative than in toluene. The five-membered rings of *36−38* are necessarily in conformation (a). The large negative coupling in *52* indicates that it too is very probably planar as diketopiperazine itself is known to be in the crystal. The alternative boat conformation corresponds to (c). In nitriles (*8* and *9*), there are two nodal planes; conformations similar to (d) must predominate.

Couplings in the alkyl halides (*3−6*) are all less negative than in methane and do not conform to an electronegativity sequence or any other easy explanation.

We have seen that in cyclopropanes (*21−23*) the strained methylene groups have a bond hybridization intermediate between sp^2 and sp^3 and corresponding intermediate values of ^2J. In ethyleneimines and episulfides (*31−33*) the presence of directly bonded electronegative atoms increases ^2J further, as expected; this influence is even more evident in the epoxides (*24−30*). For cyclobutanes (*34−35*), it is difficult to disentangle substituent effects from geometrical effects, particularly as the ring is not planar (see Section 4.3.3).

In substituted ethylenes, we have already seen that increasing electronegativity of substituents decreases ^2J. This conclusion is amply confirmed by compounds *103−126*. As shown under entry *102*, ^2J may vary over an algebraic range of over 10 Hz, from quite strongly positive for metal substituents (*104, 105*) to negative values for CCl_3, OCH_3, and halogens. The oxime couplings (*18* and *19*) probably reflect the same influences. According to molecular orbital theory, this is the predicted effect of withdrawal of electrons from the antibonding orbitals of the methylene group.

4.3.3 Vicinal Proton−Proton Couplings

Like geminal couplings, vicinal couplings, ^3J, i.e., between protons on adjacent atoms (commonly adjacent carbon atoms), are markedly affected by substituents on the bonded carbons. They do not, however, vary over so large a range as geminal couplings and are rarely — probably never — negative. The variation with geometry is of particular importance to organic chemists and will be illustrated by two examples.

In 3,3,4,4,5,5-hexadeuterocyclohexanol, the replacement of protons by deuterium is an effective means of spectral simplification. Because of the smaller value of γ for deuterium (Chapter 1 and Appendix A), its coupling to a proton is smaller than the corresponding proton−proton coupling by a factor of $\gamma_{2_H}/\gamma_{1_H}$ ($=1/6.54$) and it, of course, does not appear in the proton spectrum. The strongly preferred conformation (by a factor of *ca.* 10) of the molecule is that in which the hydroxyl group is equatorial:

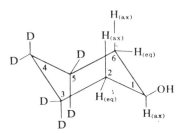

In Fig. 4.22 the spectrum obtained by Anet is shown.[13] The C–2 and equivalent C–6 protons appear as the AB part of an ABX spectrum, the C–1 proton being X. Cross-ring couplings are zero. The coupling of the C–1 proton, which is axial, to the axial protons on C–2 and C–6 is 10.2 Hz, whereas its coupling to the equatorial protons on these carbons is only 4.1 Hz. There is thus a strong dependence of these three-bond couplings on the dihedral angles between the C–H bonds. The angles are about 180° for J_{axax} and 60° for J_{axeq}.

A similar relationship of couplings is found for cis and trans protons on olefinic double bonds, as we have already seen in Section 4.2.3. Thus, in the spectrum of vinyl chloride (Fig. 4.16) $J_{cis} = 7.4$ Hz and $J_{trans} = 14.8$ Hz, and in that of styrene (Fig. 4.15) $J_{cis} = 10.7$ Hz and $J_{trans} = 17.5$ Hz. It is clear that the approximate relationship $J_{trans}/J_{cis} \simeq 2$ can be useful in assigning the structures of geometrical isomers.

Such observations emphasize the fact that J couplings are transmitted through bonding electrons and not through space like direct dipolar couplings, for, if the latter were the case, one would expect that cis or gauche protons, being closer together, would be more strongly coupled than trans. Karplus[14–16] has given a valence-bond interpretation of these results and concludes that the vicinal coupling of protons on adjacent sp^3 carbon atoms should depend on ϕ, the dihedral angle, as follows:

$$J = \begin{cases} 8.5 \quad \cos^2 \phi - 0.28 \text{ (Hz)} \\ \quad\quad 0° \leqslant \phi \leqslant 90° \\ 9.5 \quad \cos^2 \phi - 0.28 \text{ (Hz)} \\ \quad\quad 90° \leqslant \phi \leqslant 180° \end{cases} \qquad (4.10)$$

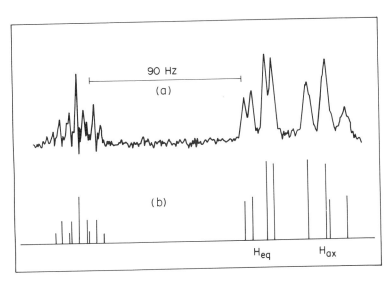

Fig. 4.22.   The observed (a) and calculated (b) spectra of 3,3,4,4,5,5-
hexadeuterocyclohexanol in D_2O, observed at 60 MHz.
[F. A. L. Anet, _J. Am. Chem. Soc._ **84**, 1053 (1962).] Data
for the calculated spectrum are

$$\nu_{2ax} - \nu_{2eq}(= \nu_{6ax} - \nu_{6eq}) = 38.0 \text{ Hz}$$

$$J_{1-2ax}(=J_{1-6ax}) = 10.17 \text{ Hz}$$

$$J_{1-2eq}(=J_{1-6eq}) = 4.08 \text{ Hz}$$

$$J_{2gem}(=J_{6gem}) = -12.2 \text{ Hz}$$

A plot of these functions (Fig. 4.23) shows that J should be 1.8 Hz when
$\phi = 60°$ (as for J_{axeq} and J_{eqeq} in undistorted cyclohexane rings or J_{gauche}
in staggered ethanes) and 9.2 Hz when $\phi = 180°$ (J_{axax} or J_{trans}). It
becomes slightly negative near 90°. Karplus's theory predicts that for
olefins the sigma-bond contributions should be 11.9 Hz to J_{trans} and 6.1 Hz
to J_{cis}, with +1.5 Hz as an additional contribution to each from the π
electrons. The predicted values are thus 13.4 Hz for J_{trans} and 7.6 Hz for
J_{cis}. Observed couplings are generally larger than these for both ethanes
and ethylenes and are strongly influenced by substituent effects as well as
by geometry, but the Karplus equations predict general trends of vicinal
couplings satisfactorily.

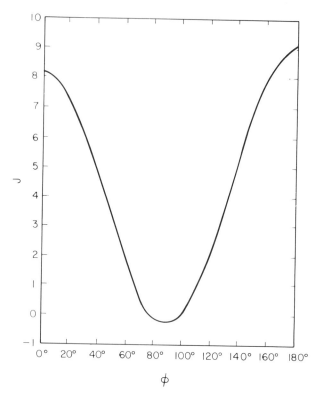

Fig. 4.23.   The Karplus function describing the magnitude of the vicinal proton–proton J coupling as a function of dihedral angle ϕ in the $H-C-C-H$ bond system.

In Appendix C, Section II, the ethyl compounds *56* through *75* illustrate the effects of substituents on the vicinal coupling of methyl groups. Here, the couplings are an average over three identical conformations; for each pair of vicinal protons, the coupling is an average of one trans and two gauche couplings:

gauche   *trans*   *gauche*

$$J_{\text{obs}} = (J_{\text{t}} + 2J_{\text{g}})/3 \qquad\qquad (4.11)$$

For ethane (56), a coupling of 8.0 Hz has been found, markedly greater than the value of *ca.* 4.3 Hz predicted by Eqs. (4.10) and (4.11). Electropositive atoms (74, 75) increase this appreciably, but nearly all other substituents decrease it. The dependence on electronegativity can be expressed approximately by

$$J_{CH_3-CH_2} = 7.9 - 0.4 \, \Delta E_x \qquad (4.12)$$

where ΔE_x is the difference in electronegativities between hydrogen and the substituent. The effect of oxygen (compounds 59, 60) is appreciable, particularly when a positive charge is present (57). In CH_3CH groups, ^3J is somewhat smaller [compare ethane (56) with propane compound 64], but shows similar trends, the effect of fluorine being particularly marked (86).

In compounds 89−101, the rotational isomers in general differ in energy, and the couplings are an average over unequal rotamer populations. We must, therefore, consider the means by which observed couplings can yield values that can be meaningfully intercompared and also compared to methyl couplings. For compounds of the type X_2CHCHY_2, there will be a trans conformer and two equienergy mirror-image gauche conformers; these are rapidly interconverting:

trans   *gauche*

These conformers will contribute to the averaged observed coupling in proportion to their mole fractions:

$$J_{obs} = X_t J_t + X_g J_g \qquad (4.13)$$

If the energy difference between conformers, $E_{gauche} - E_{trans}$, is expressed as ΔE, the dependence of J_{obs} on temperature will be given by

$$J_{obs} = \frac{J_t + 2J_g e^{-\Delta E/RT}}{1 + 2e^{-\Delta E/RT}} \qquad (4.14)$$

For such molecules, J_t and J_g cannot be obtained from a single measurement, but may be obtained from measurements at several temperatures plotted according to Eq. (4.14). One may also vary the rotamer ratio by varying the solvent and applying Eq. (4.13). At least two solvents in which the rotamer ratio differs substantially must be employed, and it must be known in each solvent; for this, infrared spectroscopy has been employed.[17,18]

In compounds such as *93* and *94*, which are X_2CHCHX_2, the protons are equivalent and ^{13}C side band spectra must be employed to obtain ^3J. This is described in Section 4.3.6.2.

Compounds of the type X_2CHCH_2X (or X_2CHCH_2Y) such as *91* and *92* present a similar problem; there is a symmetrical rotamer (I) and two equienergy mirror-image rotamers (II):

(I)   (II)

This gives an AB_2 (or AX_2) spectrum (Section 4.2.3.1) that is characterized by only a single coupling constant:

$$J_{AB} = X_I J_{Ig} + \tfrac{1}{2} X_{II}(J_t + J_{IIg}) \qquad (4.15)$$

If we assume that $J_{Ig} = J_{IIg} = J_g$, we may also write

$$J_{AB} = \frac{J_t e^{-\Delta E/RT} + J_g(1 + e^{-\Delta E/RT})}{1 + 2e^{-\Delta E/RT}} \qquad (4.16)$$

Compounds of the type XCH_2CH_2Y (or $XCH_2\,^{13}CH_2X$) represented by *89* and *90* give in general AA' BB' spectra. From XCH_2CH_2Y two geminal and two vicinal couplings may be extracted. The conformers are

I (*trans*)   II (*gauche*)

Assuming, as before, that all J_g are equal and also that all J_t are equal, we find

$$J_{AB} = X_I J_g + \tfrac{1}{2} X_{II}(J_g + J_t) \qquad (4.17)$$

$$J'_{AB} = X_I J_t + X_{II} J_g \qquad (4.18)$$

In the special case that the conformers are of equal enthalpy, i.e., that the *gauche—trans* ratio is 2, we find that $J_{AB} = J'_{AB} = \tfrac{1}{3}(J_t + 2J_g)$. The spectrum is then A_2B_2. Comparison with Eq. (4.11) shows that in this case J_{AB} may be compared directly with vicinal methyl couplings. In the general case, from Eqs. (4.18) and (4.17), J_g and J_t may be calculated directly without the need for solvent or temperature variation, but one must still have one independent measure of the *gauche—trans* ratio, usually from infrared.

For all of the compounds discussed, if J_t and J_g can be obtained from the observed ^3J, the quantity $\tfrac{1}{3}(J_t + 2J_g)$ can then be compared directly with methyl couplings. This quantity is shown in brackets for compounds *89*, *90*, *91*, *93*, and *95*. The effect of fluorine substitution in reducing ^3J shown by *95* is consistent with that shown by *86*. Compound *94* seems to indicate a marked effect of multiple bromine substitution, which would appear an extension of the results for *66* and *85*. Likewise, the observed ^3J for *93* seems to extend the trend shown by *63* and *84*; in this case, however, J_t and J_g are known, and the calculated value of $[^3J]$, 8.1, is about that for ethane. It seems very probable that the small values of J_{obs} for *93* and *94* indicate a strong preference for the gauche form:

This is surprising, for one would expect the trans form to be preferred, as minimizing bromine—bromine steric interactions.

For compounds of the type represented by *99*, i.e., with an asymmetric carbon atom, the methylene protons have different chemical shifts and will appear as the AB part of an ABX or ABC spectrum. Equations (4.13) and (4.14) can be applied to the coupling of each to the α proton, X. Molecules of this sort and the related, but more complex, situations presented by compounds *100* and *101*, which have two asymmetric centers, have been discussed in Chapter 3, Section 3.3.6, and will be considered further in Chapter 7. Compounds *96* and *97* also have two asymmetric centers, but can be treated essentially like *93* and *94*.

Couplings in cyclopropane rings are illustrated by compounds *138−144*. It is consistently observed that cis couplings are larger than trans, the J_{cis}/J_{trans} ratio being *ca.* 1.4−1.8. By comparison to the vicinal couplings in cyclopropane itself (*139*), it is apparent that the presence of substituents other than CH_3 groups reduces the vicinal couplings to the protons on the same carbon and increases those across the ring. Oxygen in a three-membered ring reduces all couplings significantly (*145−149*), as do sulfur (*150, 151*) and nitrogen (*152*) to a lesser degree.

Both cis and trans couplings in cyclobutane rings (*153−156*) show a strong dependence on the nature of the substituents. Rationalization is complicated by the fact that the rings are not planar, but are rapidly interconverting between two equivalent (mirror-image) [*154*, Fig. 4.24(a)] or nonequivalent [*155*, Fig. 4.24(b)] boat conformers. For the equivalent conformers (a):

$$J_{trans} = \tfrac{1}{2}(J_{axax} + J_{eqeq}) \tag{4.19a}$$

$$J_{cis} = J_{axeq} \text{ and } J'_{axeq} \tag{4.19b}$$

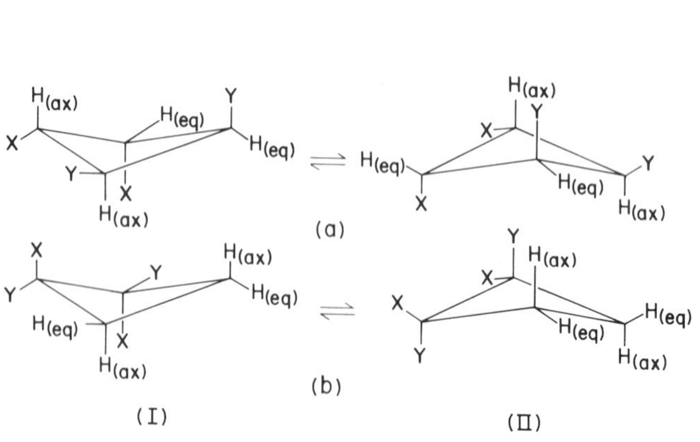

Fig. 4.24.   Inversion of substituted cyclobutanes.

For the nonequivalent conformers (b):

$$J_{\text{trans}} = X_I J_{I-\text{axax}} + X_{II} J_{II-\text{eqeq}} \qquad (4.20a)$$

$$J'_{\text{trans}} = X_I J_{I-\text{eqeq}} + X_{II} J_{II-\text{axax}} \qquad (4.20b)$$

$$J_{\text{cis}} = X_I J_{I-\text{axeq}} + X_{II} J_{II-\text{axeq}} \qquad (4.20c)$$

The observation of two different cis couplings for *154* makes it clear that here one cannot assume that geometrically similar couplings such as J_{axeq} and J_{axeq}' are equal, as is generally considered permissible for open-chain compounds.

As in four-membered rings, the couplings in five—membered rings (*157-159*) appear to be subject to both substituent and geometrical effects. The latter are particularly obvious for L-proline and L-alloproline (*158* and *159*), in which substituents are identical, but corresponding couplings vary widely.

Couplings in six-membered rings have already been discussed. These have received far more study than those in all other saturated cyclic compounds and have proved of great utility in structural determination, particularly of the steroids (*173* and *174*) and furanose sugar acetates (*175* and *176*). The few examples given are meant only to be illustrative of the large body of available data. The couplings in cyclohexane itself have been determined in an approximate manner from the low temperature spectrum of 1,1,2,2,4,4,5,5-d_8-cyclohexane (see Chapter 5, Section 5.4.3). From these (*161*) a [J] of 6.5 Hz is calculated, somewhat less than that for propane (*64*). In compounds *165* and *166* the bulky *tert*-butyl group ensures exclusive preference for the conformations shown. The effect of electronegative substituents such as OH or OAc is small, but oxygen in the ring (*175*) can reduce axial—axial couplings substantially.

From simple geometrical considerations alone, one would expect J_{axeq} and J_{eqeq} to be equal. Data on equatorial—equatorial couplings are extremely scarce, but Booth[19] has obtained evidence that when an electronegative substituent is axial, J_{axeq} is smaller than when it is equatorial (see entries *162* and *163*). Related orientation effects or perhaps ring distortion may account for the unusual couplings in *178*.

A word should be said concerning observed values of vicinal couplings in cyclohexanes (or cyclohexane-like compounds) in which two conformers are in rapid equilibrium, as, for example,

(I)   (II)

The proton on the carbon bearing the substituent X, designated $H_{(1)}$, will experience two observable couplings to the protons of a neighboring methylene group:

$$J_{12} = X_I J_{I-1ax2ax} + X_{II} J_{II-1eq2eq} \qquad (4.21a)$$

$$J'_{12} = X_I J_{I-1ax2eq} + X_{II} J_{II-1eq2ax} \qquad (4.21b)$$

This will be true even when the conformers are mirror images (as for compound *169*, for example) or identical, in which case

$$J_{12} = \tfrac{1}{2}(J_{I-1ax2ax} + J_{II-1eq2eq}) \qquad (4.22a)$$

$$J'_{12} = \tfrac{1}{2}(J_{I-1ax2eq} + J_{II-1eeq2ax}) \qquad (4.22b)$$

For 1,4-disubstituted six-membered rings of the type

there will be equienergy chair conformers. If these are in rapid equilibration, the CH_2CH_2 groups will give AA'BB' spectra, assuming negligible cross-ring couplings:

The spectrum will be like that of a 1,2-disubstituted ethane, XCH_2CH_2Y, as discussed earlier, with only the gauche conformers present. Therefore, from Eqs. (4.17) and (4.18),

$$J_{AB} = \tfrac{1}{2}(J_{axax} + J_{eqeq}) \tag{4.23a}$$

$$J'_{AB} = J_{axeq} \tag{4.23b}$$

Vicinal couplings across carbon–carbon double bonds are presented in entries *102–136* for open-chain olefinic compounds and in entries *187–192* for cyclic olefinic compounds. Geminal couplings are also tabulated; these have been discussed in Section 4.3.2. Entry *102* indicates that J_{trans} may vary over a twofold range and J_{cis} over a fivefold range, depending upon substitution at the double bonded carbons. The data for monosubstituted ethylenes (*104–126* and *134–136*) are arranged in approximate order of increasing substituent electronegativity. For the cis and trans couplings, there is a marked decrease with increasing electronegatively, as shown in Fig. 4.25. According to molecular orbital theory, these trends reflect a withdrawal of electrons from the antisymmetric bonding orbitals. Cis couplings in cyclic olefins show a marked dependence on ring size (compounds *187–192*), increasing with increasing ring size because of altered bond angles and hybridization.

4.3.4 Proton–Proton Couplings in Aromatic Rings

Proton–proton couplings in benzene rings do not vary over a wide range, as is indicated under entry *193* in Appendix C. The couplings in benzene itself cannot of course be determined directly but can be approximated by analysis of the ^{13}C side bands (Section 4.3.6.2), from the liquid crystal spectrum (Section 4.4.2), and by interpolation in trends of substituent influences.[20] These last values are shown under entry *194*. They are all of the same sign, which the liquid crystal spectrum shows to be positive. The effects of substituents are not large, but are troublesome from the theoretical point of view, being generally the opposite of those observed in substituted ethylenes. Nitro groups have a particularly marked effect. The correlation of J_{12} ($= J_{45}$) with the Pauling electronegativity E_x of the atom directly bonded to the ring is shown in Fig. 4.26. The vicinal coupling constants J_{23} ($= J_{34}$) were found to be almost unaffected by the nature of the substituent, so this effect falls off rapidly with distance. The *meta* coupling J_{15} also increases with electronegativity but J_{13} ($= J_{35}$) and J_{14} ($= J_{25}$) display, instead, *decreasing* trends.

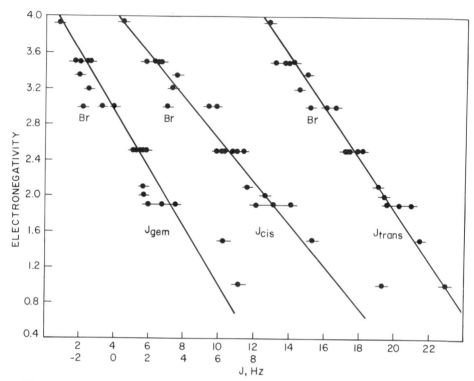

Fig. 4.25. J_{gem}, J_{cis}, and J_{trans} in vinyl groups $CH_2 = CHX$ plotted as a function of the electronegativity of X. Note that J_{cis} and J_{trans} are referred to the scale running from 0 to 24 Hz. {From T. Schaefer, *Can. J. Chem.* **40**, 1 (1962).]

Ranges of couplings for *ortho-*, *meta-*, and *para-*substituted benzene rings are shown under *195–197*. Some representative *meta* couplings in multiply substituted rings are shown under *198–201*.

Couplings in polynuclear aromatic hydrocarbons are similar to those in benzene rings, except that those between the β protons (B, B′) in naphthalene (*202*) and anthracene (*203*) are unexpectedly small; the AB coupling in pyrene (*204*) is of intermediate magnitude. Interring couplings are generally not large enough to cause more than peak broadening but in some cases can produce observable splitting.

Observed ranges of couplings in heterocyclic rings seem to be even smaller than in substituted benzenes. For pyridines, α,β-couplings are always markedly smaller than β,γ-couplings. Couplings for other heterocyclic rings are summarized in entries *205–218*.

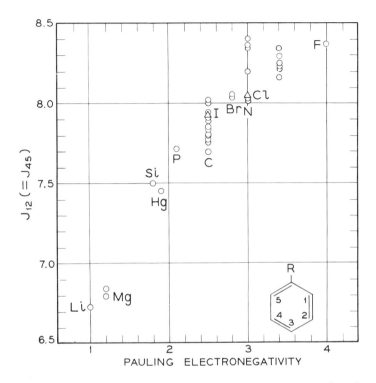

Fig. 4.26. J_{ortho} in monosubstituted benzenes as a function of the Pauling electronegativity of the atom directly bonded to the ring. [From S. Castellano and C. Sun, *J. Am. Chem. Soc.* **88**, 4741 (1966).]

4.3.5 Long-Range Proton—Proton Couplings

Proton—proton couplings through two and three bonds are almost always the largest. We have seen (Section 4.3.4) that appreciable proton couplings may be transmitted through four or five bonds in aromatic rings, where the π electrons undoubtedly play an important part. Couplings transmitted only by the σ framework usually decrease by at least an order of magnitude with each intervening bond. However, in certain rigid cyclic structures of favorable geometry, couplings through four bonds can be remarkably large. Thus, in *exo*-5-chlorobicyclo[2.1.1]hexane-*exo*-6-*tert*-butylcarboxamide (a), the coupling between the *endo* protons is 7 Hz, whereas the *exo—endo* coupling in *exo*-5-chlorobicyclo[2.2.1]hexane-*endo*-6-*tert*-butylcarboxamide (b) is zero.[21]

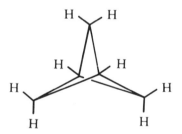

(a)   (b)

A number of cases of long-range coupling in strained bicyclic compounds have been collected by Barfield and Chakrabarti.[22] The most striking is bicyclo[1.1.1]pentane, in which the coupling between the bridgehead protons is found to be 18 Hz.[23] This is attributed to strong interaction between the backsides of the bridgehead CH bond orbitals.

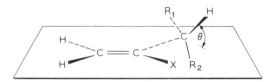

In nonaromatic molecules, long-range couplings may be substantially enhanced when unsaturated systems occur between the protons. In allylic systems, ^4J may reach a maximum of approximately 3 Hz, when θ, the angle between the plane containing the olefinic protons and the C—H bond of the allylic carbon atoms (Fig. 4.27) is 90°. Minima of zero occur when $\theta = 0°$ or 180°. For methyl groups ($R_1 = R_2 = H$), intermediate values

Fig. 4.27.   Definition of the angle θ in an allylic system.

are expected. Thus, in methyl methacrylate (*131* in Appendix C), the trans-like allylic coupling of $H_{(1)}$ to the α—CH_3 group is 1.0 Hz; the cis-like coupling is in this case larger, 2.0 Hz, rather than smaller as for the vicinal vinyl coupling. J_{12} is also *ca.* 2.0 Hz. The analysis of the spectrum is shown in Fig. 4.28; it provides an example of how multiplets of "first-order" appearance but nonbinomial intensities may result when two couplings have a simple ratio—in this case J_{12}/J_{13} ($\simeq J_{23}/J_{13}$) $\simeq 2.0$, giving a 1:3:4:4:3:1 sextuplet for $H_{(1)}$.

When R_1 or R_2 (Fig. 4.27) or both are other than hydrogen, changes in rotamer populations will affect the allylic coupling. Thus, in compound *134*, the four-bond couplings of the allyl proton $H_{(4)}$ to the vinyl protons $H_{(2)}$ and $H_{(3)}$ are of the order of 1 Hz with the cis-like coupling slightly the larger. When R_1 and R_2 are *tert*-butyl (*135*), the couplings become very small because the allyl proton is now preferentially in the plane of the vinyl group.

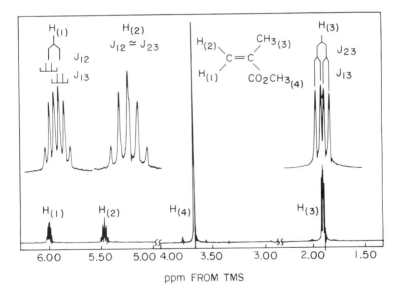

Fig. 4.28. Spectrum of methyl methacrylate, observed at 100 MHz, 20% (wt./vol.) solution in CCl$_4$. The assignments of the vinyl and allylic methyl protons are indicated; the ester methyl group appears as the narrow singlet at 6.33 τ. The multiplets are shown amplified and expanded (5×) above the full spectrum.

Other examples of allylic coupling are shown in Fig. 4.29, together with allenic and acetylenic couplings. (See the reviews of Sternhell[24] and Barfield and Chakrabarti.[22])

4.3.6 Carbon-13 Couplings

4.3.6.1 Carbon-13–Proton Couplings

Carbon–proton J couplings do not have the same vital significance in the determination of the structure and conformation of organic molecules

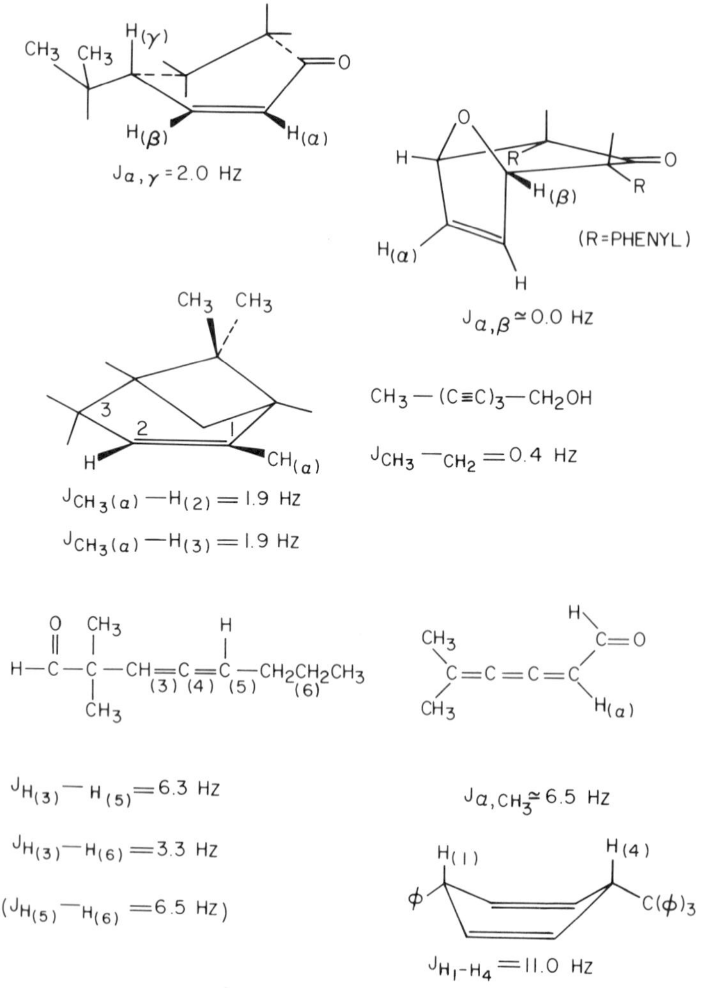
CH3 CH3
H(γ)
O
H(β)
H(α)
Ja,γ = 2.0 HZ
O
H
R
H(β)
R
O
H(α)
H
(R=PHENYL)
Jα,β ≃ 0.0 HZ
CH3 CH3
3
2
1
H
CH(α)
JCH3(α)—H(2) = 1.9 HZ
JCH3(α)—H(3) = 1.9 HZ
CH3—(C≡C)3—CH2OH
JCH3—CH2 = 0.4 HZ
O CH3 H
H—C—C—CH=C=C—CH2CH2CH3
(3) (4) (5) (6)
CH3
CH3
C=C=C=C
CH3
H
C=O
H(α)
JH(3)—H(5) = 6.3 HZ
JH(3)—H(6) = 3.3 HZ
(JH(5)—H(6) = 6.5 HZ)
Jα,CH3 ≃ 6.5 HZ
H(1)
H(4)
ϕ
C(ϕ)3
JH1-H4 = 11.0 HZ

Fig. 4.29. Some examples of long-range proton–proton couplings.

as we have seen to characterize proton—proton couplings but are nevertheless of substantial importance. For one thing, they give rise to multiplicity in carbon-13 spectra and on this account must be dealt with, usually by being eliminated through double resonance (see Section 4.4.3.2). The principal structural significance of direct, one-bond $^{13}C-^{1}H$ couplings is that they depend upon the fractional s character of the bonding carbon orbitals.[25-27] Examples of this behavior can be selected from Appendix D; typical examples are given in Table 4.6 in order of increasing fractional s character from 1/4 (sp^3) to 1/2 (sp). This relationship may be summarized as

$$^{1}J_{^{13}C-H} = 5\rho_{CH} \text{ (Hz)}, \qquad (4.24)$$

where ρ_{CH} is the percentage of s character in the bonding orbitals. Although a strong correlation undoubtedly exists, this simple expression has been somewhat eroded by more detailed investigations. It is probably incautious to use Eq. 4.24 as a quantitative measure of bond hybridization except perhaps for hydrocarbons. For example, it seems reassuring that $J_{^{13}C-H}$ in the methyl halides (Appendix D) shows little variation and is thus apparently independent of variations in bond polarity. But accumulation of halogens increases $J_{^{13}C-H}$ markedly, and it is further observed[28] that in methanes it increases with the inductive effect of the substituents as measured by the so—called Swain-Lupton factor.[29] Such substituent effects are of minor influence in olefinic and aromatic systems.

Carbon-13—proton couplings through two or more bonds are very much smaller than directly bonded couplings but do not necessarily fall off regularly with the number of bonds, ^3J being sometimes larger than ^2J. There is some dependence on s character for ^2J but no such trend is observable for ^3J. The latter, however, does exhibit a dependence on the dihedral angle between carbons separated by three bonds, paralleling the relationship for vicinal protons. The magnitude of the coupling is strongly dependent on the specific structure and the geometry of substituents, and is not a reliable measure of the dihedral angle (see Ref. 30).

4.3.6.2 ^{13}C Satellite Analysis; Observation of Couplings Between Structurally Equivalent Protons

We have seen (Section 4.2) that couplings between nuclei having the same chemical shift ordinarily do not give rise to splittings, but that in strongly coupled systems of four or more spins, — AA′BB′ systems, for example—couplings between structurally indistinguishable nuclei may have a marked effect on the observed transitions and can therefore be measured (Section 4.2.3.6).

TABLE 4.6

$^1J_{^{13}C-H}$ as a Function of s Character of
Bonding Orbitals

Compound	$^1J_{^{13}C-H}$ (Hz)
cyclohexane	123
CH_4	125
$(CH_3)_4C$	124
cyclopentane	128
$CH_3C \equiv CCH_3$	131
benzene	158
cyclobutane	136
cyclopropane	161
$CH_3C \equiv {}^{13}C - H$	251

There are many cases of interest, however, where all the nuclei are equivalent and the normal spectrum can provide no information concerning proton–proton couplings. Benzene, 1,2-dichloroethane, and 1,1,2,2-dibromoethane have already been seen as examples. Information can still be obtained for such compounds by taking advantage of the fact that direct $^{13}C-{}^1H$ couplings produce weak satellites in their proton spectra, straddling the much stronger ^{12}CH peak. The spacing of these satellites is equal to $J_{^{13}C-H}$, i.e., *ca.* 125–250 Hz. In a simple molecule such as methane or methyl chloride, the satellites are each singlets, but in more complex molecules they represent the spectrum of the molecule

$$X\diagdown_{^{12}CH}\text{———}_{^{13}CH}\diagup^X$$
$$X\diagup\qquad\qquad\diagdown^X$$

the number of molecules with two neighboring ^{13}C nuclei being only about 1 in 10^4. The effect of the ^{13}C nucleus on the chemical shift of the directly bonded proton is observable but very small (Chapter 3, Section 3.3.9), so that the spectrum is effectively AA'X, X representing the ^{13}C nucleus and not observed. In the proton spectrum, the $^{13}C-{}^1H$ coupling in effect plays

TABLE 4.7

Longer Range $^{13}C-^1H$ Couplings, Hz

Compound	$J_{^{13}C-H}$	$J_{^{13}C-C-H}$	$J_{^{13}C-C-C-H}$	$J_{^{13}C-O-C-H}$
A. Aliphatic Compounds				
$(CH_3)_2CH \cdot {}^{13}CH_2CH(CH_3)_2$			4.8	
$(CH_3)_2CH \cdot {}^{13}CHOH \cdot CH(CH_3)_2$	136		4.5	
$CH_3CH_2{}^{13}CD_2OH$		4	6.4	
$CH_{3(3)}CH_{2(2)}{}^{13}C(Cl)(CH_{3(1)})_2$		${}^{13}C-H_{(1)}$:3.9 ${}^{13}C-H_{(2)}$:3.7	${}^{13}C-H_{(3)}$ 5.7	
$CH_3{}^{13}CO_2CH_2CH_3$		6.0		3.1
$(CH_3CH_2)_2{}^{13}C=O$		5.7	4.7	
$CH_3CH_2{}^{13}CO_2CH_3$	-	6.5	5.3	4.0
${}^{13}CH_3C\equiv CH_{(1)}$	131.4	${}^{13}C-H_{(1)}$:50.8	3.6	
$CH_{3(2)}{}^{13}C\equiv CH_{(1)}$			${}^{13}C-H_{(2)}$:10.6	
$CH_{3(2)}C\equiv {}^{13}CH_{(1)}$	24		4.8	

(continues)

TABLE 4.7 (continued)

Compound	$J_{^{13}C-H}$	$J_{^{13}C-C-H}$	$J_{^{13}C-C-C-H}$	$J_{^{13}C-O-C-H}$ $J_{^{13}C-C-C-C-H}$
B. Aromatic Rings				
benzene	+158.8	+1.1	+7.6	−1.2
chlorobenzene	C$_2$–H$_2$ +164.8	C$_1$–H$_2$ −3.4	C$_1$–H$_3$ +11.1	C$_1$–H$_4$ −2.0
	C$_3$–H$_3$ +161.2	C$_2$–H$_3$ +1.4	C$_2$–H$_4$ +7.9	C$_2$–H$_5$ −1.2
	C$_4$–H$_4$ +161.3	C$_3$–H$_2$ +0.3	C$_2$–H$_6$ +5.0	C$_3$–H$_6$ −0.9
		C$_3$–H$_4$ +1.6	C$_3$–H$_5$ +8.2	
		C$_4$–H$_3$ +0.9	C$_4$–H$_2$ +7.4	
o-dichlorobenzene	C$_3$–H$_3$ +166.5	C$_2$–H$_3$ −3.5	C$_2$–H$_4$ +11.6	C$_2$–H$_5$ −1.9
	C$_4$–H$_4$ +164.0	C$_3$–H$_4$ +1.9	C$_2$–H$_6$ +7.9	C$_3$–H$_6$ −1.2
		C$_4$–H$_3$ +0.02	C$_3$–H$_5$ +8.4	
		C$_4$–H$_5$ +1.1	C$_4$–H$_6$ +8.6	
pyridine	C$_2$–H$_2$ +177.6	C$_2$–H$_3$ +3.1	C$_2$–H$_4$ +6.9	C$_2$–H$_5$ −0.9
	C$_3$–H$_3$ +163.0	C$_3$–H$_2$ +8.5	C$_2$–H$_6$ +11.2	C$_3$–H$_6$ −1.7
	C$_4$–H$_4$ +162.4	C$_3$–H$_4$ +0.8	C$_3$–H$_5$ +6.6	
		C$_4$–H$_3$ +0.7	C$_4$–H$_2$ +6.3	

(continues)

TABLE 4.7 (continued)

Compound	$J_{^{13}C-H}$	$J_{^{13}C-C-H}$	$J_{^{13}C-C-C-H}$
B. Aromatic Rings (cont.)			
Furan	C_2-H_2 +201.8	C_2-H_3 +11.0	C_2-H_4 +7.0
	C_3-H_3 +174.7	C_3-H_2 +13.8	C_2-H_5 +6.9
		C_3-H_4 +4.1	C_3-H_5 +6.0
Pyrrole			
	C_2-H_2 +182	C_2-H_3 7.6	C_2-H_4 7.6
	C_3-H_6 +170	C_3-H_2 7.8	C_2-H_5 7.6
		C_3-H_4 4.6	C_3-H_5 7.8
Thiophene			
	C_2-H_2 189	C_2-H_3 7.4	C_2-H_4 10.0
		C_3-H_2 4.7	C_2-H_5 5.2
	C_3H_3 168	C_3-H_4 5.9	C_3-H_5 9.5

the role of an artificial chemical shift difference between the ^{13}CH and ^{12}CH protons, much larger than the real one. In the 60 MHz proton spectra of 1,1,2,2-tetrabromoethane [Fig. 4.30(a)], the spacing in each satellite is 2.9 Hz, which represents the vicinal coupling already discussed (p. 199). The satellites of dioxane [Fig. 4.30(b)] present a somewhat more complex problem.[31] They can be treated, in effect, as an AA'BB' system in which the "chemical shift" difference corresponds to $J_{^{13}C-H}$ or 142.2 Hz. Analysis gives $J = 6.07$ Hz and $J' = 2.72$ Hz ($J_{gem} \simeq 10$ Hz), from which, by application of Eqs. (4.23a) and (4.23b), under the assumption that $J_{axeq} = J_{eqeq}$, it is found that $J_{axax} = 9.42$ Hz and J_{axeq} ($= J_{eqeq}$) = 2.72 Hz (Appendix C, entry *177*). At very low temperature, the process of ring inversion may be sufficiently retarded that the proton spectrum again becomes AA'BB' (without consideration of ^{13}C satellites) and may be analyzed for J_{axax}, J_{axeq}, J_{eqeq}, and J_{gem} (see Chapter 5, Section 5.4.3).

There are still other cases where all the protons of interest are attached to the same carbon, as in methane, formaldehyde, and methyl groups. The ^{13}C satellites of such groups are not informative and one must employ deuterium substitution.

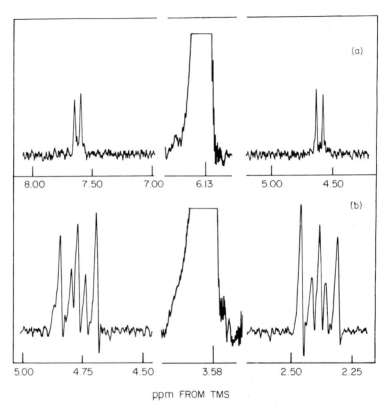

Fig. 4.30. ^{13}C satellites in the proton spectra of (a) 1,1,2,2-tetrabromoethane and (b) dioxane; observed as neat liquids at 60 MHz. Note breaks in the horizontal scale.

4.3.6.3 $^{13}C - ^{13}C$ Couplings

The observation of $^{13}C - ^{13}C$ couplings in natural abundance requires that the molecule have the ^{13}C isotope in two positions, which will be the case for approximately 1 molecule in 8300. Thus, sensitivity is a severe problem for all but the simplest molecules. In Fig. 4.31 is shown the 50.3 MHz proton-decoupled spectrum of carbon-2 of neat, unenriched tetrahydrofuran, in which the one-bond $^{13}C - ^{13}C$ coupling between C_2 and C_3 is 33.1 Hz, as given by the doublet straddling the major singlet resonance of the single $^{13}C_2$ molecules. The more shielded carbon−3 gives a similar spectrum with of course identical satellite spacing. It is evident that in complex spectra such doublet resonances would be difficult or impossible to observe. In Fig. 4.31 the doublet is not precisely centered on

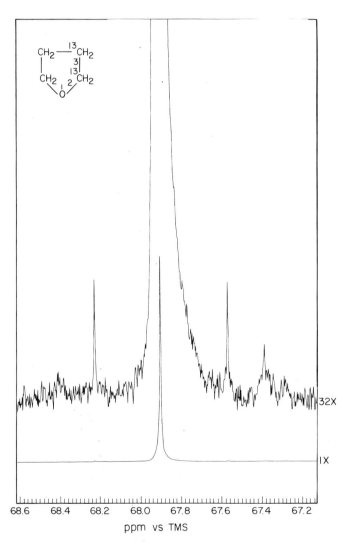

Fig. 4.31.   The 50.3 MHz proton-decoupled spectrum of carbon-2 in neat, unenriched tetrahydrofuran. The doublet straddling the large singlet arises from those molecules with carbon—13 nuclei in natural abundance in both the C-2 and C-3 (or C-4 and C-5) positions, with a one-bond $^{13}C-^{13}C$ coupling of 33.1 Hz. (F. C. Schilling, unpublished observations.)

the singlet but is displaced upfield by approximately 0.18 Hz or 0.0036 ppm owing to a nearest neighbor ^{13}C isotope effect (Chapter 3, Section 3.3.9).

By using special pulse sequences, the center carbon resonance may be suppressed, greatly facilitating such measurements. Such measurements have also been made by the difficult and expensive "brute force" procedure of single and even double carbon-13 labeling. Data obtained in this way have been tabulated by Hansen.[32]

One-bond carbon—carbon couplings, like one—bond carbon—proton couplings, are strongly dependent on the hybridization of the bonding orbitals.[33,34] Roberts *et al.* described the correlation by Eq. (4.25):

$$^1J_{^{13}C_a - ^{13}C_b} \text{ (Hz)} = 7.3 \frac{\rho_{C_a} \cdot \rho_{C_b}}{100} - 17 \qquad (4.25)$$

where ρ_{C_a} and ρ_{C_b} express the percent of s character of the bonding orbitals to carbon a and carbon b, respectively [see Eq. 4.24)]. A proportionality between $^1J_{^{13}C - ^{13}C}$ and $^1J_{^{13}C - ^1H}$ of course follows and this was found for a limited group of model compounds with methyl substituents to follow the relationship

$$^1J_{^{13}C_x - ^{13}CH_3} = 0.27 \cdot {}^1J_{^{13}C_x - ^1H} \qquad (4.26)$$

Two-bond ^{13}C—^{13}C couplings are generally very small—usually less than 5 Hz—and consequently difficult to measure. Like ^{13}C—^{1}H and ^{13}C—^{13}C couplings they also tend to increase with the s character of the bonding orbitals.

Three-bond ^{13}C—^{13}C couplings offer the possibility of dependence on the ^{13}C—C—C—^{13}C dihedral angle. This is indeed observed.[35-37] Again, the couplings are relatively small—0-5 Hz in aliphatic systems and up to *ca.* 14 Hz in aromatic rings.

4.4 SPIN DECOUPLING AND ASSOCIATED TECHNIQUES

4.4.1 Introduction

Double resonance or spin decoupling can often provide an effective means for simplification of complex spectra and for determining which nuclei are coupled in such spectra. For the second purpose, the use of the form of two-dimensional spectroscopy known as COSY is particularly

TABLE 4.8

Direct $^{13}C-^{13}C$ Couplings, $^{1}J_{^{13}C-^{13}C}$, in Selected Organic Compounds, Illustrating Dependence on s Character of Bonding Orbitals [a]

Entry	Compound	Bonding hybridization	$^{1}J_{^{13}C-^{13}C,Hz}$ (Hz)	
1	cyclopropane	$sp^5 - sp^5$		10
2	CH_3-CH_3	$sp^3 - sp^3$		34.6
3	(benzene ring)$C-CH_3$	$sp^3 - sp^2$		44.2
4	benzene	$sp^2 - sp^2$		57.0
5	$CH_2 = CH_2$	$sp^2 - sp^2$		67.6
6	$CH_3-C \equiv CH$	$sp^3 - sp$		67.4
7	CH_3CN	$sp^3 - sp$		56.5
8	$CH_2 = C = C(CH_3)_2$	$sp^2 - sp$		99.5
9	(benzene ring)$C-CN$	$sp^2 - sp$		80.0
10	$HC \equiv CH$	$sp - sp$		171.5
11	(piperidine ring, positions 2,3,4, NH)	$sp^3 - sp^3$	C_2-C_3	35.2
			C_3-C_4	33.0

[a] From F. J. Weigert and J. D. Roberts, J. Am. Chem. Soc. **94** 6021 (1972))

effective and is discussed in Chapter 6, Section 6.3.3. We shall first consider the basic decoupling experiment in relatively simple terms. Such experiments may employ CW methods as well as pulse techniques. We then discuss more advanced multipulse (but one-dimensional) methods, particularly those used for the identification of carbon-13 resonances and the "editing" of carbon-13 spectra.

Decoupling may be discussed under the two heads of *homonuclear* or *heteronuclear* depending on whether the two spins or two groups of spins are of the same or different species.

4.4.2 Homonuclear Decoupling

For spin decoupling (see Chapter 2, Section 2.3.4) to be effectively applied to spins of the same species, it is not necessary that the spectrum as a whole be of first order, but the nuclei that are being decoupled must be separated in chemical shift by at least 20–30 Hz. (Note that no such restriction applies to the COSY 2D method, in which the whole network of coupled spins—regardless of chemical shift differences—may be revealed by a single experiment.)

To take the simplest case, suppose that we have an AX system (two doublets, see Section 4.2.1) and that while observing A with the the usual weak rf field B_1, we simultaneously provide a second stronger rf field B_2 at the resonance frequency of X. We find that the A doublet collapses to a singlet, which appears at its midpoint. The field B_2 has caused the nucleus X to jump back and forth so rapidly between its possible spin states that it no longer perturbs the spin states of A. Double resonance is often discussed in terms of the saturation of X, but the phenomenon is really not so simple as such descriptions suggest. This is clearly shown by the fact that, at levels of B_2 below that required for the complete collapse of a multiplet, additional lines appear. Thus, on irradiation of X at low B_2 levels in an AX system, the spectrum of A may consist of as many as four lines in a frequency-sweep experiment or three lines in a field-sweep experiment (Chapter 2, Section 2.3). By reversing the experiment, we can also observe the collapse of the doublet of X upon irradiating A.

Decoupling was first proposed by Bloch.[38] Theoretical discussions are given by Bloom and Shoolery,[39] Baldeschwieler and Randall,[40] and Hoffman and Forsén.[41] These reviews present the fundamentals of double resonance very completely but deal with experimental methods—e.g., CW sweeping of the spectrum using audio sidebands—no longer in common practice.

The spectrum of acetaldehyde (25 vol. % in CCl_4), an AX_3 case, furnishes a simple illustration, as shown in Fig. 4.32. Irradiation of the

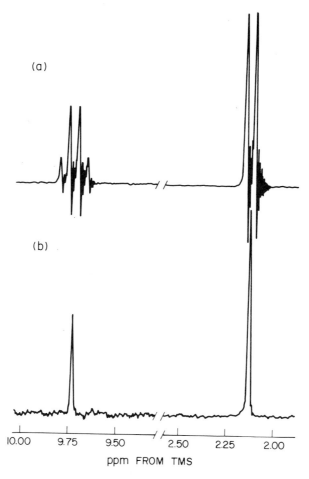

Fig. 4.32.   The normal (a) and double resonance (b) spectra of acetaldehyde, observed as a 25% (vol./vol.) solution in CCl_4 at 60 MHz.

aldehyde proton at the center of its quartet (J_{vic} = 2.8 Hz) collapses the CH_3 doublet to a singlet, while irradiation of the CH_3 doublet collapses the aldehyde quartet to a singlet. In this case, of course, the normal spectrum presents no interpretive problem. A more complex case is that of propylene oxide

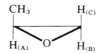

(Fig. 4.33), where the proton $H_{(A)}$, being coupled to $H_{(B)}$ and $H_{(C)}$ and the methyl group, appearing as a doublet at 1.26 ppm ($J_{CH_3-H_{(A)}} = 6.4$ Hz), gives a very complex multiplet at 2.83 ppm; this can be more clearly seen in the expanded spectrum (b). Upon irradiation at the center of the methyl doublet, the $H_{(A)}$ resonance collapses to a quartet [spectrum (c)]. The spectrum now becomes an ABC case, which can be treated with sufficient accuracy as AMX. We find

$$H_{(A)} = 2.83 \text{ ppm} \qquad J_{AB} = 5.4 \text{ Hz}$$
$$H_{(B)} = 2.59 \text{ ppm} \qquad J_{BC} = 5.4 \text{ Hz}$$
$$H_{(C)} = 2.28 \text{ ppm} \qquad J_{AC} = 2.6 \text{ Hz}$$

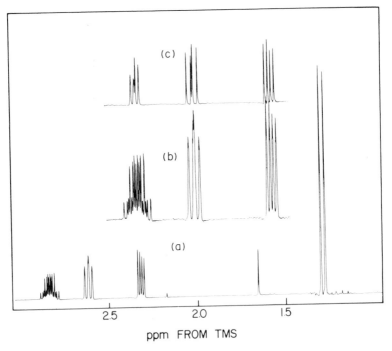

ppm FROM TMS

Fig. 4.33.   The double resonance proton spectrum of propylene oxide, observed at 200 MHz in 1% solution in $CDCl_3$ at 25°C. The doublet at 1.26 ppm corresponds to the methyl group. Protons A, B, and C appear in order of increasing shielding. (The peak at *ca.* 1.7 ppm is an impurity.) Spectrum (b) is an expansion of (a). In spectrum (c) the methyl protons are irradiated, reducing the multiplet of proton A to a quartet. See text. (P. A. Mirau, unpublished observations.)

Double resonance may be employed to determine the relative signs of coupling constants. This application is simplest when the spins are weakly coupled relative to the chemical shift differences. It allows one to identify transitions which share a common energy level. If particular transitions are irradiated with decoupling fields only strong enough to perturb and split the lines concerned, the experiment is sometimes termed spin "tickling."

Let us consider the AMX system shown in Fig. 4.34. We will assume that all three couplings give rise to observable splittings, and that they are all different in magnitude, so that the spectrum consists of three 1:1:1:1

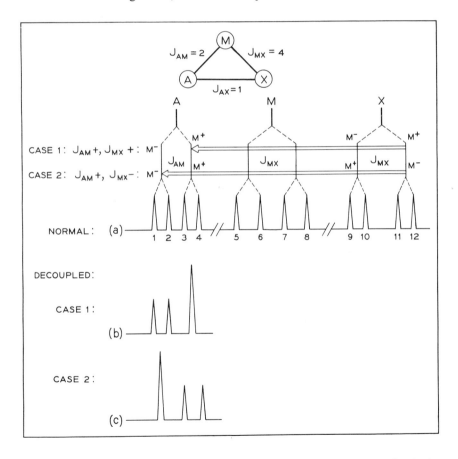

Fig. 4.34.   Schematic spectra illustrating the determination of relative signs of J couplings in an AMX spin system by double resonance.

quartets. As we have seen (Section 4.2.3.3) the relative signs of the coupling constants J_{AM}, J_{MX}, and J_{AX} cannot be learned from the normal spectrum, which is independent of them. The relative signs of J_{AM} and J_{MX} may, however, be determined by irradiating the X protons at a point midway between lines 11 and 12 and observing the behavior of the A multiplet. To understand this, let us first suppose that both J_{AM} and J_{MX} are positive (Case 1 in Fig. 4.34). If we then consider the transitions responsible for each of the lines in the A and X multiplets (see Fig. 4.2 and accompanying discussion), we find that the downfield doublets of both the A and X multiplets arise from transitions of just half the molecules, those in which the magnetic moment of M is directed against the polarizing field B_0. This state is designated as M^- (see Fig. 1.2). For Case 1, lines 1, 2, 9, and 10 arise from this state of M. The upfield doublets of A and X arise from the other half of the molecules having the moment of M directed with B_0, a state designated as M^+. When we irradiate between lines 11 and 12, we are causing decoupling only in these M^+ molecules. Accordingly, only the 3,4 doublet, arising also from M^+ molecules, will be collapsed to a singlet, as shown at (b); lines 1 and 2 will be unaffected. If we irradiate between lines 9 and 10, the 1,2 doublet will collapse instead.

Let us now suppose that J_{AM} is again positive, but that J_{MX} is now *negative* (Case 2). The upfield doublet of the X multiplet now arises from M^- molecules and, upon irradiation between lines 11 and 12, lines 1 and 2 will collapse [spectrum (c)]. It is readily shown that the behavior described for Case 1 will be the same if *both* J_{AM} and J_{MX} are negative, and that the results of Case 2 would be observed if the signs were reversed. Thus, only *relative* signs can be determined.

In principle the experiment might have been reversed, i.e., we could have adjusted B_2 to irradiate in the A region and observed X. The better separation of the doublets in X makes the first experiment the better choice. Similarly, one may determine the relative signs of J_{AM} and J_{AX} by appropriate irradiation of the X multiplet and observation of the M multiplet or vice versa. Only two such determinations are necessary and the choice will again depend on which multiplets exhibit optimum spacings.

An appropriate case, although not strictly AMX, is provided by 2,3-dibromopropionic acid,[42]

$$CH_2BrCHBrCOOH$$

in which the methylene protons are differentiated by the asymmetric center (Chapter 3, Section 3.3.6). (The carboxyl proton exchanges too rapidly to participate in the spin system.) To distinguish the geminal protons, one observes that the spectrum[43] (Fig. 4.35) exhibits couplings between $H_{(A)}$ and $H_{(B)}$ and between $H_{(A)}$ and $H_{(C)}$, which vary substantially with the solvent, whereas that between $H_{(B)}$ and $H_{(C)}$ does not. The latter are thus

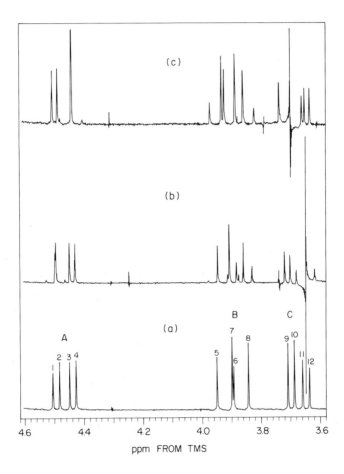

Fig. 4.35.   The 200 MHz proton spectra of 2,3-dibromopropionic acid observed as a 1% solution in CDCl$_3$ at 25°C; (a) normal spectrum; (b) spectrum upon irradiation between lines 11 and 12 of the C proton resonance. Note collapse of lines 1 and 2 of the A resonance; (c) spectrum upon irradiation between lines 9 and 10 of the C resonance. (P. A. Mirau, unpublished observations.)

the geminal protons. This is consistent also with the expectation that H_A, attached to a carbon bearing both a carboxyl group and a bromine atom, will be the least shielded. To distinguish between the geminal protons, it is necessary to assume that of the three rotational isomers, rotamer III

$H_{(C)}$ $H_{(A)}$ COOH $H_{(B)}$ Br Br

$H_{(B)}$ $H_{(A)}$ COOH Br $H_{(C)}$ Br

Br $H_{(A)}$ COOH $H_{(C)}$ $H_{(B)}$ Br

I   II   III

with the trans bromines is fairly strongly preferred. We have seen (Section 4.3.3) that trans proton couplings are greater than gauche and comparable to geminal in absolute magnitude, and it is therefore evident that proton B, with two large couplings, trans and geminal, must be designated as noted earlier. In the 200 MHz spectrum this corresponds to the resonance at 3.98_5 ppm, exhibiting couplings of 11.7 Hz and 10.3 Hz (absolute magnitudes) to $H_{(A)}$ and $H_{(C)}$, respectively.

If one irradiates between lines 11 and 12 in spectrum (a) (Fig. 4.35), it is observed that lines 1 and 2 collapse [spectrum (b)], whereas irradiation between lines 9 and 10 causes lines 3 and 4 to collapse [spectrum (c)]. Both observations demonstrate that J_{AB} and J_{BC} have opposite signs, which we have seen to be general for vicinal and geminal couplings. Parallel observations of the B multiplet while irradiating A demonstrate that J_{AB} and J_{AC} have like signs.[43] In principle, observation of B while irradiating C should provide the same information, but in fact one observes in Fig. 4.35 that these multiplets are too close in chemical shift and the results are complex.

We thus find that

$$H_{(A)} = 4.46_2 \text{ ppm} \qquad J_{AB} = \pm 11.7 \text{ Hz}$$
$$H_{(B)} = 3.89_5 \text{ ppm} \qquad J_{AC} = \pm 4.5 \text{ Hz}$$
$$H_{(C)} = 3.67_2 \text{ ppm} \qquad J_{BC} = \mp 11.3 \text{ Hz}$$

The determination of *absolute* signs of coupling constants is not an easy matter but fortunately is not of great practical significance. It is generally assumed that directly bonded $^{13}C - {}^{1}H$ couplings (Section 4.3.6.1) must be positive, and the signs of vicinal proton—proton couplings have been deduced from these by double resonance.[44] It can be shown that if the molecules are partially oriented by some means the direct nuclear dipole—dipole couplings will contribute to the spectrum, the form of which will then depend on the relative signs of the direct and electron-mediated couplings.[45-48] The absolute signs of the direct couplings are known if the direction of orientation of the molecules is known. The absolute signs of the indirect couplings may therefore be deduced. Buckingham and McLauchlan[49] deduced by electrostatic orientation of p-nitrotoluene that the *ortho* proton—proton couplings are positive.

A somewhat more conclusive result may be obtained by orientation in a liquid-crystal matrix. It has long been known that rodlike molecules such as p,p'-di-n-hexyloxyazoxybenzene

can form a nematic mesophase having properties intermediate between those of liquids and solids. In this liquid crystalline state the molecules exist in domains or clusters in which their long axes are parallel. Magnetic fields as low as 0.1 tesla are capable of orienting these clusters so that the long axes of the molecules are aligned with the magnetic field. Saupe and Englert[50] observed that small organic molecules may be dissolved in this matrix and then themselves become partially oriented. Benzene becomes oriented so that its hexagonal plane is preferentially oriented parallel to the field, but it rotates freely about its hexad axis, which itself may assume any orientation in the xy plane. Under these conditions, the benzene proton spectrum is neither the broad featureless band of the crystalline solid nor the narrow singlet of the liquid. Instead, one observes a complex centrosymmetric spectrum containing about 50 lines. It can be accurately simulated on the assumption that the direct dipole—dipole proton couplings, which must be negative, are of opposite sign to all three of the indirect couplings, which are therefore $J_{ortho} = +6.0$ Hz; $J_{meta} = +2.0$ Hz; and $J_{para} = +1.0$ Hz.

4.4.3 Heteronuclear Decoupling

4.4.3.1 ^{14}N Decoupling

In the normal spectrum of formamide (Fig. 4.36, top), the amide proton resonance is broad and featureless because it is actually two overlapping 1:1:1 triplets resulting from direct ^{14}N—^{1}H J coupling. But

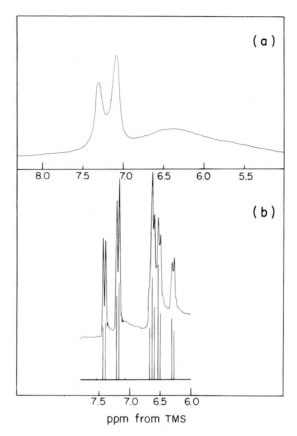

Fig. 4.36.   The normal (a) and double resonance (b) proton spectrum of formamide (60 MHz). In spectrum (b), the ^{14}N nucleus is irradiated at 4.3 MHz. (Courtesy of J. B. Stothers and T. G. Hill.)

the ^{14}N nucleus, having a spin of 1, possesses an electric quadrupole (Chapter 1, Section 1.9) that interacts with molecular electric field gradients in the tumbling molecule and causes transitions between the Zeeman levels. This causes the ^{14}N nucleus to decouple itself partially from the directly bonded protons, collapsing the 1:1:1 triplets. The formyl proton is principally a doublet resulting from the *trans* vicinal coupling to the amide proton $H_{(C)}$

but the weaker two-bond ^{14}N coupling produces a broadening here as well. By irradiation of the ^{14}N nucleus at 4.3 MHz (Fig. 4.36, bottom) this decoupling can be made complete and separate resonances observed for each amide proton. The proton spectrum becomes recognizable as ABC with the following parameters:

$$H_A = 7.30 \text{ ppm} \qquad J_{AB} = 2.1 \text{ Hz}$$
$$H_B = 6.65 \text{ ppm} \qquad J_{BC} = 2.1 \text{ Hz}$$
$$H_C = 6.44 \text{ ppm} \qquad J_{AC} = 13.3 \text{ Hz}$$

The direct $^{14}N-^{1}H$ couplings are *ca.* 50 Hz (see Chapter 9, Section 9.1.)

This collapse of multiplets arising from $^{14}N-^{1}H$ coupling is particularly marked in certain asymmetric molecules such as peptides, although observed not to be complete in all peptides.[51] Figure 4.37 shows the 100 MHz spectra of six molecules, including formamide, exhibiting varying degrees of collapse of CONH and NH_3^+ multiplets. In the relatively symmetric tetrahedral environment of the ammonium group of *n*-propylamine in trifluoroacetic acid [spectrum (a)], the $^{14}N-^{1}H$ triplet is observable although much broadened. In the other spectra in Fig. 4.37 the collapse of the $^{14}N-^{1}N$ triplets is more complete and other features become evident. The NH resonance of acetamide

in d_6-DMSO at 25°C [spectrum (d)] is a doublet corresponding to the chemical shifts of the two protons. As the temperature is increased to 81°C [spectrum (c)], these resonances broaden, in part because of the lessened contribution of the ^{14}N quadrupolar relaxation and increased participation of the 1:1:1 triplet. Roberts[52] observed that the broad singlet of formamide began to assume a triplet character above 50°C. For other compounds the collapse is more complete; most peptides evidence little or no ^{14}N quadrupolar broadening of the NH proton resonances at room temperature and give such well-resolved spectra that coupling of the NH protons to neighboring $C_\alpha H$ protons can be readily measured (Chapter 7, Section 7.3.1). This is observable in the spectrum of *N*-acetyl-*L*-alanyl-*L*-alanyl-*L*-alanyl methyl ester [spectrum (f)]. Quadrupolar relaxation is discussed further in Chapter 5, Section 5.2.4.

4.4.3.2 Broad-band $^{1}H-^{13}C$ Decoupling

The utility of carbon-13 NMR spectroscopy for the determination of molecular structure is largely dependent on the use of broad-band

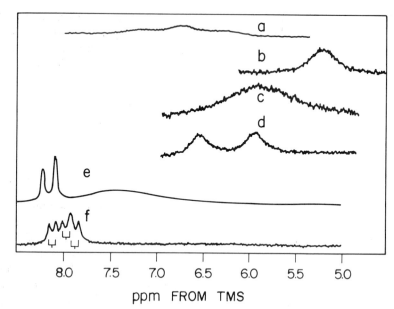

Fig. 4.37.   The 100 MHz proton spectra illustrating NH resonances in various structures: (a) $^+NH_3CH_2CH_2CH_3$ (in trifluoroacetic acid); (b) N-methylacetamide (in $CDCl_3$, 25°C); (c) acetamide (in d_6-DMSO, 81°C); (d) acetamide (in d_6-DMSO, 25°); (e) formamide (neat, 25°C); and (f) N-acetyl-L-alanyl-L-alanyl-L-alanyl methyl ester (in d_6-DMSO, 25°C). (A. I. Brewster, unpublished observations.)

proton—carbon decoupling methods. Reich *et al.*[53] reported great simplification of the carbon spectrum of cholesterol using a proton decoupling field modulated by random noise so as to cover the chemical shift range of the associated protons in the manner demonstrated by Ernst[54] for $^{19}F-\{^1H\}$ and $^1H-\{^{19}F\}$ spectroscopy. (In this notation, the nuclear species within the brackets is being irradiated and that outside the brackets is being observed.)

It is important to provide a proton decoupling field of sufficient strength to decouple all protons efficiently and to give spectra in which all the carbon resonances are of at least approximately equal intensity. Covering the chemical shift range of all the protons at superconducting frequencies may require supplying several watts of power to the decoupler coil in the probe (Chapter 2, Section 2.3.4). One must be able to remove the heat thus generated with a practicable flow of cooling gas. The problem can be particularly acute with biological samples in aqueous solvents owing to dieletric heating. There is therefore a need to provide effective decoupling with the least expenditure of power. Noise-modulated decoupling, standard in earlier spectrometers, is not particularly efficient in this regard because the irradiated nuclei are completely decoupled for only a small fraction of the random noise cycle time. This method has been superseded by more sophisticated schemes.

A widely used method is that of Grutzner and Santini,[55] in which the phase of the CW proton irradiation is periodically reversed. This is accomplished by square wave frequency modulation with a 50% duty cycle having a frequency of 20 to 100 Hz. With 10 watts of power, reasonably uniform decoupling can be attained over a bandwidth of approximately 1.3 kHz. Basus *et al.*[56] described a combination of square wave modulation with frequency modulation using a periodic linear ramp, the ramp being imposed at a frequency of at least 1.0 kHz; this scheme allows efficient decoupling at superconducting frequencies with as little as 2 watts of decoupling power.

Another decoupling scheme is the application of a sequence of pulses that rotates the components of ^{1}H$-^{13}$C proton doublets through large precession angles — e.g., 180° — repetitively.[57] Levitt and Freeman[58] showed that to be effective a sequence of such inversion pulses cannot be simply carried out repetitively about the same axis, for example, the $+x$ axis, since this involves some of the shortcomings of continuous wave decoupling, nor can it be carried out alternately about the $+x$ and $-x$ axes. Instead a 90°$(+x)$, 90°$(+x)$, 90°$(-x)$, 90°$(-x)$ cycle is preferred. This cycle may be repeated many times—up to 16 times in present-day instruments; this "supercycle" is known as MLEV-64. The 180° pulses may be replaced by "composite" pulses, which perform the inversion more exactly, i.e., return the magnetization more nearly to its starting point on each cycle. Such composite pulses may involve precession about the y axes as well. It can be shown that more perfect cycling results in greater decoupling efficiency.

A yet further refinement, useful for attaining optimum resolution at the highest superconducting frequencies, is attained by combining 90°, 180°, and 270° pulses along the $\pm x$ and $\pm y$ axes in varying combinations to produce the highest cycling precision.[57-60] The WALTZ-16 supercycle is of this kind and involves 36 pulses before repeating.

In Fig. 4.38 is shown the 200 MHz proton spectrum of ethylbenzene (a) and the 50.3 MHz proton coupled carbon-13 spectrum (b).[61] The aromatic portions of both spectra are omitted. In (b) the CH$_2$ triplet

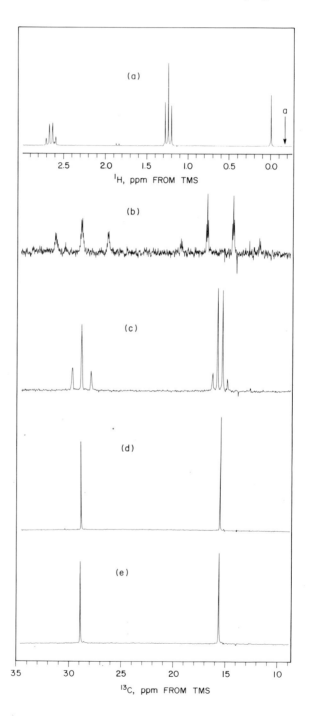

Fig. 4.38.   Spectra showing various modes of $^{13}C - \{^1H\}$ decoupling. (a) 200 MHz proton spectrum of ethylbenzene, observed at 25°C using a 1% solution in $CDCl_3$ (1 scan); (b) 50.3 MHz ^{13}C spectrum of the same solution as in (a), without proton decoupling (60 scans); (c) off-resonance ^{13}C spectrum as in (b) but with proton decoupling at the frequency indicated by the arrow in spectrum (a) (150 scans). Spectrum (d) shows the result of employing *noise modulated* proton decoupling; in (e) an essentially identical result is obtained using MLEV decoupling (see text) but with only 1/5 of the power. Both (d) and (e) represent 60 scans. (F. C. Schilling, unpublished observations.)

appears at 28.9 ppm and the CH_3 quartet at 15.6 ppm. It is to be noted that each peak of the triplet is itself a quartet because of two-bond coupling to the CH_3 protons, and that each peak of the quartet is a triplet owing to corresponding coupling to the methylene protons. The increase in sensitivity on proton decoupling arises from the collapse of this multiplicity as well as that of the larger splitting. In spectrum (c), "off-resonance" proton decoupling has been applied at the frequency represented by the arrow in spectrum (a). This result will be discussed in Section 4.4.3.3. In spectrum (d) a noise-modulated proton decoupling field has been employed, while spectrum (e), which is essentially identical, was obtained using the MLEV-16 technique. The rf power requirement for (e) was approximately one-fifth of that required for (d).

The broad-band decoupling techniques described in this section may also be used for decoupling protons from other nuclei such as ^{14}N, ^{15}N, and ^{19}F.

4.4.3.3 Off-Resonance Decoupling

It is often convenient to be able to determine the number of protons attached to particular carbons without having to deal with a confusing set of overlapping multiplets. This is particularly common for ^{13}C spectra. It can be shown[62,63] that if one applies the proton decoupling field at a frequency ν_2 removed from the proton resonance frequency of a particular multiplet by 100 to 500 Hz, the ^{13}C multiplet will be preserved but will exhibit a reduced spacing given by

$$J_R = {}^1J_{C-H} \cdot \frac{2\pi}{\gamma_H B_2} (\nu_H - \nu_2).$$
(4.27)

Here, $^1J_{C-H}$ is the direct $^{13}C-^1H$ coupling, B_2 the strength of the decoupling field, and ν_H the resonance of the proton or protons attached to the observed carbon. [This expression is an approximation, most nearly correct when $2\pi/\gamma_H B_2$ is substantially smaller than $(\nu_H - \nu_2)$.] We see then that the carbon multiplet spacing will be smaller the closer the proton irradiation frequency is to that of the particular proton concerned, being of course zero when they coincide and the multiplet collapses to a singlet. This relationship is illustrated in Fig. 4.38(c). Here, ν_2 [arrow in spectrum (a)] is 280 Hz upfield from the center of the CH_3 triplet [spectrum (a)], which in turn is 280 Hz upfield from the center of the CH_2 quartet. From Eq. (4.27) one expects that the multiplet spacing should be twice as great for the triplet as for the quartet, and this in fact is observed.

It is noteworthy, and is implied also by Eq. (4.27), that the two-bond and other smaller couplings are no longer manifested in the off-resonance spectrum. This gives a substantial increase in the signal-to-noise ratio even though the large splittings remain. One may also observe a marked deviation of the multiplets from binomial intensities, particularly obvious in the quartet, which is closer to complete collapse.

This technique may be used to identify protons with particular carbons and *vice versa*. Such a correlation may be more elegantly accomplished by 2D methods (Chapter 6, Section 6.3.1), although usually with more consumption of instrument time. The primary purpose of the off—resonance method—to identify carbon types, whether methyl, methylene, methine, or quaternary—may also be accomplished more effectively be spectral editing techniques to be described in Section 4.4.4.

4.4.4 Spectral Editing Techniques and Related Phenomena

4.4.4.1 The Spin Echo Method

The spectral editing techniques to be described enable one to select and identify specific nuclei—most commonly ^{13}C—on the basis of multiplicity arising from proton coupling. Among the pulse methods used for this purpose, the *spin echo* phenomenon has an important place. It was first proposed and named by Hahn.[64] It is illustrated in Fig. 4.39. At time zero (a), a 90° pulse is applied along the positive x' axis of the rotating frame, causing the magnetization M to precess into the positive y' axis (Chapter 1, Section. 1.4; Chapter 2, Section 2.3.3). The magnetization is the vector sum of individual spin vectors arising from nuclei in different parts of the sample and therefore experiencing slightly different values of the magnetic field B_0, which is never perfectly homogeneous. The individual vectors or *isochromats* will begin to fan out since some will be precessing slightly faster and some slightly slower than the rf field

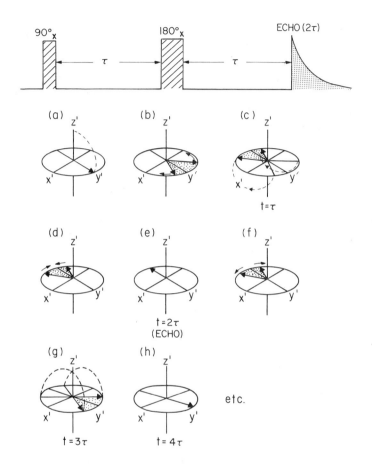

Fig. 4.39.   The spin echo experiment. (a) A 90° pulse along the x' axis causes the magnetization M to precess into the $+y'$ axis. The magnetization vectors corresponding to different parts of the sample fan out (b). After a time τ, a 180° pulse along the x' axis rotates the vectors into the $-y'$ axis (c). They continue to precess in the same sense as before (d) but now precess together and "refocus," i.e., all attain the same absorption phase signal (negative) at time 2τ (e). They then again disperse (f). ((g) and (h) refer to further 180° pulses, as in the Carr-Purcell method; Chap. 5, Sec. 5.3. (In the above and subsequent pulse representations the pulse durations are greatly exaggerated in comparison to the intervals between them.)

frequency ν_0; this is shown at (b). After time τ—usually a few milliseconds—we apply a 180° pulse. This rotates the spin vectors into the $-y'$ axis as shown at (c). Since they continue to precess in the same sense, they will now be rotating together rather than fanning out (d). After a second interval τ they will "refocus" as shown at (e). In this context, refocussing means that all vectors simultaneously attain the same negative (because along the $-y'$ axis) emission phase signal form corresponding to an inverted spin population. Beyond the signal buildup at 2τ, the vectors again dephase and become virtually unobservable. It should be remarked that refocussing of the vectors to an apparently "single" vector at (e) does not imply any narrowing of the observed resonance over that which would be normally observed. Upon detecting the signal, the varied isochromats contained within this apparently single vector will again become evident as signal broadening just as would have been the case if detection had followed the 90_x° pulse. The signal at 2τ is reduced somewhat in intensity, however, because the effects of the true spin—spin relaxation processes are not refocussed in the spin echo experiment.

In Chapter 5 (Sec. 5.3), modifications of the Hahn echo sequence which are more suitable for practical use are described.

Differences in chemical shift will also cause a fanning out of the vectors corresponding to each of the chemically shifted nuclei. The rate of precession of each nucleus in the rotating frame corresponds to the difference between its resonant frequency and ν_0, the carrier frequency. In Fig. 4.40 a pair of chemically shifted nuclei are shown. Immediately following the 90_x° pulse (a) their signals are in phase and are represented as a single magnetization vector. In (b) they are beginning to precess apart, nucleus A being further from the carrier than nucleus B. At time τ (c), a 180_x° pulse is applied. The vectors continue to rotate clockwise and refocus to build up a detectable negative signal at 2τ (d). Upon acquisition at this point the chemically shifted resonances of A and B will appear with correct phase. If detection had taken place at times shorter or longer than 2τ the resonance would appear with frequency-dependent phase errors. Thus, "refocussing" does not mean that chemical shift information is eliminated but only that correct phases are restored.

The spin echo experiment may also be carried out using 180_y° pulses. As shown in Fig. 4.40, (e) and (f), the vectors again fan together to form an echo, which, being now along the $+y'$ axis, has a positive sign.

Scalar coupling complicates the spin echo experiment in a way that is very significant for multipulse experiments and that depends on the manner in which the pulses are applied to the coupled spins. Let us consider a

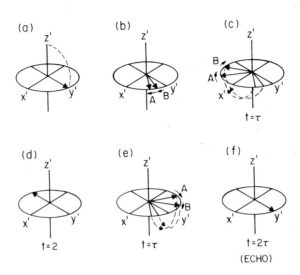

Fig. 4.40.   The spin echo experiment applied to a pair of chemically shifted nuclei, A and B. In (a) a 90° pulse is applied along the x' axis, turning the macroscopic magnetization into the $x'y'$ plane along the y' axis. In (b) the vectors corresponding to the A and B spins fan out, A being further from the carrier frequency and therefore precessing faster. In (c) a 180° pulse is applied along the x' axis, rotating the vectors as shown. They continue to precess in the same direction at the same relative rates, refocussing to a negative echo at (d). Alternatively, the 180° echo may be applied along the y' axis (e). Again the vectors precess together, now forming a *positive* echo at (f).

system of two coupled spins, AX. We first suppose these to be two different nuclear species for example, ^{1}H and ^{13}C, so that we may apply a 180_y° pulse to X without affecting A. In Fig. 4.41 the vector diagrams for spin X are shown. In (a) a 90_x° pulse has been applied to nucleus X, i.e., ^{13}C. The transverse magnetization is slightly off resonance and will precess away from the y' axis, splitting into two components α and β (b) because of scalar coupling to nucleus A. For a saturated ^{13}C$-^1$H system, this coupling is of the order of 125 Hz, as we have seen (Section 4.3.6.1). One-half of the doublet, corresponding to spin state α, is further from the carrier frequency ν_0 than the other half, corresponding to spin state β, and therefore precesses faster. Upon application of the 180_y° pulse at time τ, the vectors assume mirror image positions (c); as they are still moving in

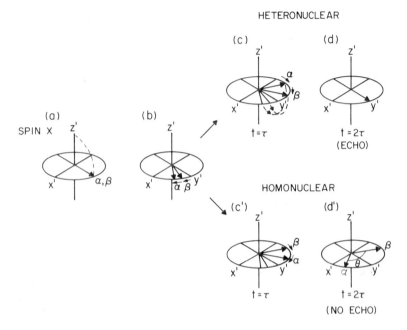

Fig. 4.41.   The spin echo experiment applied to a pair of coupled spins,
AX. We show the magnetization vectors for spin X. We
first show the behavior if A and X are different nuclear
species, e.g., ^{1}H and ^{13}C, so that a 180°_x pulse may be
applied to X without affecting A. At (a) a 90°_x pulse has
turned nucleus X (^{13}C) into the $+y'$ axis. The magnetization
is slightly off resonance and precesses away from the $+y'$
axis, splitting into two components α and β because of scalar
coupling to a single bonded A nucleus (^{1}H). The moment
corresponding to the α spin state precesses faster than the β
component, being further from the carrier. Upon application
of the 180°_y pulse to spin X at time τ (b), the vectors assume
mirror image positions (c). As they are still moving in the
same direction and relative velocities they refocus to form the
echo at (d).

If we now suppose that the AX system is homonuclear,
e.g., two protons, the 180°_y pulse affects both spins and also
exchanges the labels of the A spin components, as shown at
(c'). The vectors now precess apart (d') rather than refocus.
This will also be true of a ^{13}C—^{1}H system if 180° pulses are
simultaneously applied to each nucleus.

the same direction and with the same relative velocities they refocus to form the echo at (d).

If, however, the AX system is homonuclear (e.g., two protons) the $180_y°$ pulse affects both spins equally; this has the effect of *exchanging* the labels of the spin states, as shown at (c'). Their vectors will now diverge, as shown at (d'), rather than refocus as in the previous example, since β now cannot catch up with α.

The behavior of the vectors is shown in more detail in Fig. 4.42, which (in addition to the singlet of a quaternary carbon) shows the precession of the components of a doublet, a triplet, and a quartet, e.g., the ^{13}C resonances of a CH, CH_2, or CH_3 group, as a function of the choice of τ in units of $1/J$. In this case, it is assumed that $180_y°$ pulses are applied to the ^{13}C and ^1H *simultaneously*, so that the spin system in effect behaves as if it were homonuclear. The positions of the vectors are shown at echo time 2τ, i.e., at time τ following the 180° pulses. τ varies from 0 to $1/J$, where it attains its greatest development, either positive or negative. The form of the acquired multiplet is also shown for each choice of τ. If one decouples the protons at echo time, as shown in Fig. 4.43, it is evident (see the caption to Fig. 4.43) that one has a method for telling quaternary or CH_2 carbons on the one hand from CH or CH_2 carbons on the other, but without distinguishing those in each group. We shall describe more sophisticated methods in subsequent sections.

4.4.4.2 Cross-polarization and Sensitivity Enhancement

Another important phenomenon occurring during spectral editing pulse sequences—and having also broader significance—is *cross-polarization* and the accompanying enhancement of the observing sensitivity of dilute nuclei. (Cross-polarization in solid state spectra is accomplished differently and is discussed in Chapter 8, Section 8.5.) The proton is ubiquitous and has the largest magnetogyric ratio and greatest sensitivity of any nucleus (except tritium, ^3H). One may enhance the sensitivity of other nuclei by arranging for population transfers between appropriate energy levels in heteronuclear spin systems involving protons.

Such population transfers may be considered in terms of an AX system, as shown in Fig. 4.44 [same as diagram (b) in Fig. 4.2], which represents a proton (A) and carbon-13 nucleus (X) with a positive coupling. We have seen [Chapter 1, Eq. (1.7)] that for adjacent levels in such a spin—1/2 system the relative populations of spin i in the lower level are expressed by

$$\frac{W_{i\,(lower)}}{W_{i\,(upper)}} = 1 + \frac{2\mu_i B_0}{kT} = 1 + \frac{\hbar\gamma_i B_0}{kT} \qquad (4.28)$$

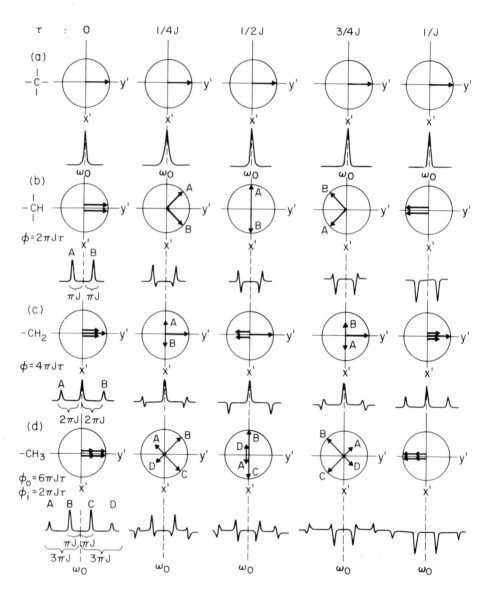

Fig. 4.42.   Behavior of vectors for ^{13}C spin systems with (a) no bonded
proton and (b) one, (c) two, and (d) three bonded protons
following 180_{y}° pulses at time τ experienced by both ^{13}C and
^{1}H nuclei.

for a Boltzmann distribution. We may therefore express the excess in the lower state, in units of $\hbar B_0/kT$, as γ_A and γ_X for protons and carbon-13 nuclei, respectively. The equilibrium populations of the levels are given by columns 2 and 3 in Table 4.9 , level 4 being the most populated and level 1 the least. We see that the relative excess of spin A in levels 3 and 4 over levels 1 and 2 is γ_A and that the excess of spin X in levels 2 and 4 over levels 1 and 3, respectively, is γ_X. We now irradiate the $4 \rightarrow 2$ transition with a selective pulse of sufficient duration to cause an inversion of the populations of these energy levels, i.e., a $180°$ rotation of the magnetic vector corresponding to this resonance. (An approximate measure of the width of such a selective pulse is that it should equal the reciprocal of the portion of the spectrum to be affected; thus, if the pulse is to be limited to a 1 Hz spectral width it should have a duration of *ca*. 1 *s*.) This would normally correspond to the irradiation of one of the ^{13}C satellites of the principal proton resonance. The effect of this pulse is to produce the population distribution shown in columns 4 and 5 in Table 4.9 and schematically illustrated in Fig. 4.44. We have now overpopulated level 3

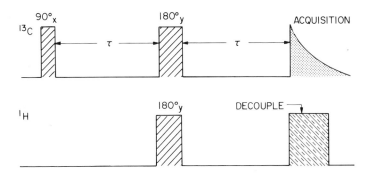

Fig. 4.43.   Pulse sequence for the determination of protons attached to ^{13}C, the latter being observed. The proton decoupled intensities acquired at time $\tau = 1/J$ are:

Carbon resonance	Decoupled intensity	I at $\tau = 1/J$
$-C-$, singlet	I	$+1$
$-CH$, doublet	$-I_o \cos(2\pi\tau J)$	-1
$-CH_2-$, triplet	$\frac{1}{2}I_o([1+\cos(4\pi\tau J)])$	$+1$
$-CH_3$, quartet	$\frac{1}{4}I_o[3\cos\pi\tau J + \cos 3\pi\tau J]$	-1

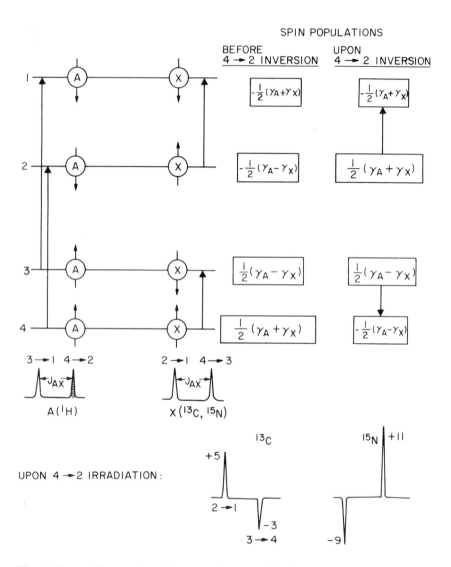

Fig. 4.44.   Energy level diagram for an AX spin system, showing effect of irradiating the $4 \rightarrow 2$ transition with an inverting pulse. At the left are shown the transitions for the normal system. If A is ^{1}H and X is ^{13}C, $\gamma_A = 4\gamma_X$. On the right are shown the populations of the energy levels before and after the $4 \rightarrow 2$ pulse. For the ^{1}H–^{13}C system the enhancements of the transitions are as shown below on the left. The result if X is ^{15}N ($\gamma_A = 10\gamma_X$) is shown at the lower right.

TABLE 4.9

Cross-polarization in an AX System

Level	Equilibrium population		Population after irradiation of 4 → 2 transition	
	Spin A	Spin X	Spin A	Spin X
1	$-\frac{1}{2}\gamma_A$	$-\frac{1}{2}\gamma_x$	$-\frac{1}{2}\gamma_A$	$-\frac{1}{2}\gamma_x$
2	$-\frac{1}{2}\gamma_A$	$+\frac{1}{2}\gamma_x$	$+\frac{1}{2}\gamma_A$	$+\frac{1}{2}\gamma_x$
3	$+\frac{1}{2}\gamma_A$	$-\frac{1}{2}\gamma_x$	$+\frac{1}{2}\gamma_A$	$-\frac{1}{2}\gamma_x$
4	$+\frac{1}{2}\gamma_A$	$+\frac{1}{2}\gamma_x$	$-\frac{1}{2}\gamma_A$	$+\frac{1}{2}\gamma_x$

with respect to level 4 and increased the population difference between level 1 and level 2. The result is an enhancement of the $2 \rightarrow 1$ ^{13}C transition by a factor of $\gamma_A + \gamma_X$ or 5 and the appearance of an inverted resonance corresponding to the $3 \rightarrow 4$ transition enhanced by a factor of $-(\gamma_A - \gamma_X)$, i.e. -3, over the normal transition. The corresponding observation for an $^{15}N-^1H$ ($\gamma_{1H}/\gamma_{15N} \simeq -10$) is also shown in Fig. 4.44.

For the multiplets of A_nX systems, this triangle replaces the Pascal triangle (p. __) in expressing peak intensities:

singlet   1

doublet -3 5

triplet -7 2   9

quartet -11 -9 15   13

The technique of selective population transfer (sometimes abbreviated as SPI) is useful but has the disadvantage that only one resonance at a time can be perturbed in the proton spectrum.

4.4.4.3 The INEPT Sequence
This pulse sequence is an acronym for "Insensitive Nuclei Enhanced by Polarization Transfer." We discuss it here primarily as an effective means for spectral editing on the basis of multiplicity selection.[65] It incorporates the phenomena described in Sections 4.4.4.1 and 4.4.4.2. The pulse

sequences for the simplest version are shown in Fig. 4.45. The insensitive nucleus X is coupled to the proton A. The effect of the pulses is shown in the vector diagrams of Fig. 4.46. In (a) the components of the proton doublet are shown prior to the 90°_x pulse. After the pulse (b), the components α and β precess at different rates, as we have seen. After a time τ equal to $1/4J$, their separation is $\pi/2$ ($90°$), as shown at (c). At this time, 180°_x pulses are applied simultaneously to both A and X [(d), (e)]; the effect of the pulse on the proton (A) is to rotate the vectors to the mirror image position (d), while the pulse on X simultaneously switches the labels of the components, as shown at (e). They continue to precess apart, reaching at time 2τ the π ($180°$) separation shown at (f). At this time, 90°_y pulses on α and β produce the opposed proton vectors at (g). Just prior to this, the X (^{13}C) vectors are similarly situated because the 180°_x proton pulse has also produced an inverted population across one of carbon transitions; this corresponds to the β vector at (h). The 90°_x pulse on x places the vectors as at (i), corresponding to the enhanced positive and negative doublet components for the insensitive nucleus, as represented in Fig. 4.44.

To eliminate signal strength contributions from the natural, unenhanced polarization of X, the second $90°$ proton pulse may be alternately applied along the $+y$ and $-y$ axes together with inversion of the receiver reference phase. Under these conditions, the ^{13}CH doublet has equal $+1$ and -1 components; for the ^{13}CH$_2$ triplet: $+1, 0, -1$; for ^{13}CH$_3$: $+1, +1, -1, -1$. Upon proton decoupling, all multiplets are thus reduced to zero intensity.

The INEPT sequence may be enhanced for spectral editing by adding a variable delay τ_2 and refocussing pulses for both nuclei, as shown in Fig. 4.47. Proton decoupling now gives singlets for appropriate values of

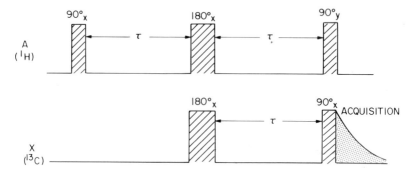

Fig. 4.45.   The pulse sequence for the simplest version of INEPT, shown here for a coupled heteronuclear AX system.

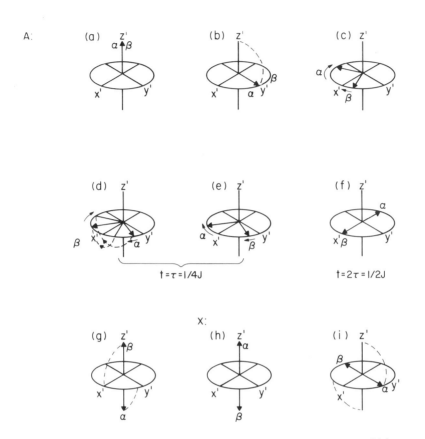

Fig. 4.46.   Behavior of the A nucleus magnetization vector [(a) through (g)] and the X nucleus magnetization vector [(h), (i)] during the simplest version of the INEPT sequence applied to an AX system (A = ^{1}H; X = ^{13}C or ^{15}N). See text for a fuller description.

τ_2. (Quaternary carbons remain null.) This is commonly referred to as the refocussed INEPT sequence. It facilitates the discrimination of carbon types. The value of τ_2 depends on the magnitude of the proton coupling J_{CH} and the nature of the X multiplet. The intensity of the decoupled signal as a function of the quantity $\theta = 2\pi J_{CH}\tau_2$ (expressed in radians) is shown in Fig. 4.48. The upper scale shows the position on the curves corresponding to values of τ_2 equal to $1/6J$, $1/4J$, and $3/8J$. [For $J = 125$ Hz, corresponding approximately to ^{13}C$-^1$H direct couplings in aliphatic structures (Section 4.3.6.1), these intervals are 1.33 ms, 2.0 ms, and 3.0 ms, respectively.] It will be observed that at a τ_2 of $1/6J$, all

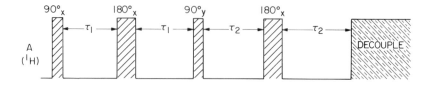

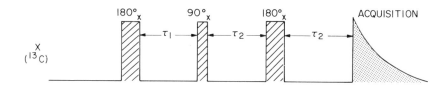

Fig. 4.47.   The refocussed decoupled INEPT sequence. The addition of
              180° pulses in both the A (^{1}H) and X (e.g. ^{13}C) channels
              refocusses the proton and ^{13}C chemical shifts.   Proton
              decoupling reduces all X multiplets to singlets (with
              enhancement), τ_2 being varied according to the multiplet to
              be detected (see text).

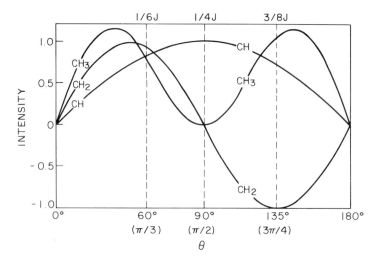

Fig. 4.48.   The dependence of the intensity of the signal of the
              insensitive or X nucleus, commonly ^{13}C, on $\theta = 2\pi J\tau_2$ in the
              refocussed decoupled INEPT experiment.

intensities nearly converge, although not at maxima; at a τ_2 of $1/4J$ the CH doublet reaches a maximum while the CH_2 triplet and quartet are of zero intensity; at a τ_2 of $3/8J$, the doublet and quartet will be positive singlets while the triplet vanishes. In no case does a quaternary carbon appear.

The application of the refocussed INEPT sequence is illustrated in Fig. 4.49 for isooctane:

$$CH_3 \diagdown \atop CH_3 \diagup \quad CH-CH_2-\underset{\underset{CH_3}{|}}{\overset{\overset{CH_3}{|}}{C_{IV}}}-CH_3$$

Spectrum (a) is the normal decoupled spectrum; the quaternary carbon resonance is reduced in intensity because of its long T_1 (see Chapter 5, Section 5.2.1). In (b), a τ_2 of $J/4$ is employed: the CH_3 and CH_2 resonances go to zero (Fig. 4.48) and only CH appears. In (c), τ_2 is $3/8J$: both methyl resonances and CH appear as in (a) but the CH_2 resonance is inverted. In (d), τ_2 is $1/6J$; the spectrum is similar to (a) except that the resonance of the quaternary carbon is absent. These experiments are sufficient to identify all the carbon types; in fact (b) and (c) would be enough.

4.4.4.4 The DEPT Sequence

The INEPT sequence has certain drawbacks. It is not tolerant with regard to the choice of J coupling, showing very noticeable intruding minor resonances if this is inaccurate or varies between multiplets. If employed in the coupled mode, it shows highly distorted multiplet intensities, as we have seen. These disadvantages are largely overcome in a development of INEPT termed DEPT for "Distortionless Enhancement by Polarization Transfer".[66] It is much less sensitive to the assumed J coupling and can give multiplets with correct signs and more nearly binomial intensities. It is, however, somewhat more complex to apply, since it involves spectral differences. The pulse sequence is shown in Fig. 4.50. First, a 90° proton pulse is applied (using the decoupler channel). After a delay $\tau = 1/2J$, during which the proton doublet components, assuming for the moment an AX system, acquire a 180° phase angle, a $180_x°$ pulse is applied to the protons. Simultaneously a $90_x°$ pulse is applied to the carbon spins. After an additional interval of $1/2J$, a $180_x°$ ^{13}C pulse is applied and at the same time a pulse of variable flip angle θ is applied to the protons. It is the value chosen for θ rather than the delay interval τ, which accomplishes

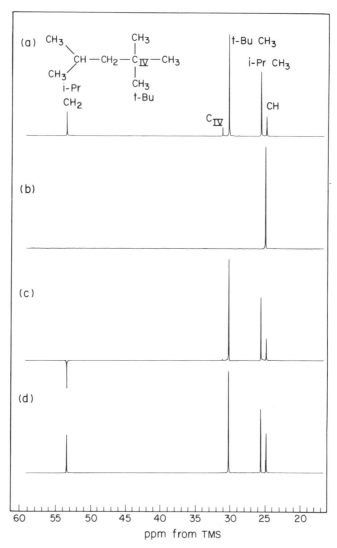

Fig. 4.49.   The refocussed decoupled INEPT experiment applied to
isooctane (2,2,4-tri-methylpentane), 40% (vol./vol.) solution
in C_6D_6 at 23°C. Carbon resonances observed at 50.3 MHz.
Spectrum (a) is the normal decoupled carbon−13 spectrum.
Because of relatively short pulse intervals (5.0 s) the C_{IV} and
CH resonances are reduced in intensity. In (b) a τ_2 of $1/4J$
is employed: the CH_3 and CH_2 resonances go to zero and
only CH appears. In (c), τ_2 is $3/8J$: both methyl resonances
and CH appear as in (a), but the CH_2 resonance is inverted.

In (d), τ_2 is $1/6J$; the spectrum is similar to (a) except that the resonance of the quaternary carbon is absent.

selectivity of multiplets. For doublets θ is 90° (see Table 4.10). After a third interval of $1/2J$ the signal is acquired, preferably with proton decoupling for spectral editing. The two $180_x^°$ pulses serve to refocus the chemical shifts in the manner already described (Section 4.4.4.1). In order to achieve transfer of magnetization from the protons to the ^{13}C spins, the θ pulse must be phase-shifted relative to the first 90° pulse.

The dependence of the decoupled carbon signal enhancement on θ is the same as shown for the decoupled INEPT experiment in Fig. 4.48. It can be shown that the following relationships apply. When θ is 45°, the CH_2 resonance is at its maximum but CH is nearly as great and CH_3 slightly greater. At 90°, CH is maximal while CH_2 and CH_3 are in principle zero. At 135°, CH_3 is near its maximum, while CH is still substantial and CH_2 is negative. Separate CH, CH_2, and CH_3 spectra may be obtained by appropriate addition and subtraction of spectra obtained with these three values of θ. The combining factors are indicated from theoretical considerations[66] but in practice some empirical adjustment is necessary. In Figs. 4.51 and 4.52 the results for isooctane are shown. They should be compared with the INEPT results in Fig. 4.49. By readily programmable manipulations, detailed in the caption of Fig. 4.52, clean spectra may be obtained for each type of carbon. For routine identification purposes, it is of course not always necessary that such complete separation of each type of carbon be achieved.

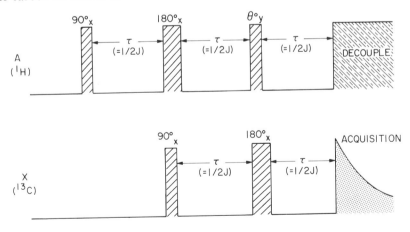

Fig. 4.50.   The DEPT pulse sequence, shown for an AX system, e.g. $^1H-^{13}C$ (see text).

248

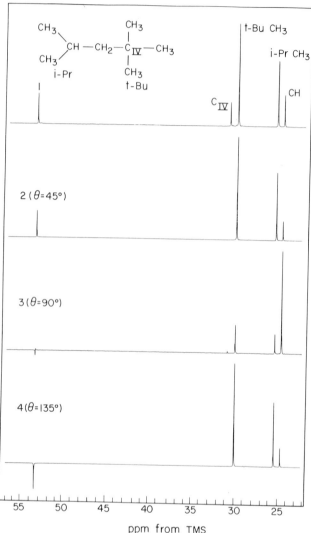

Fig. 4.51. The 50.3 MHz carbon−13 spectra illustrating DEPT sequence applied to isooctane (2,2,4−trimethylpentane), 40% (vol./vol.) solution in C_6D_6 at 23°C. Spectrum 1 is a normal proton-decoupled spectrum. Spectrum 2 corresponds to a pulse angle θ of 45° and eliminates all non−protonated carbon resonances with partial suppression of CH. Spectrum 3 corresponds to $\theta = 90°$ and to a partial suppression of all resonances except CH. Spectrum 4, with $\theta = 135°$, shows inversion of CH_2, suppression of the quarternary carbon resonance, and partial suppression of CH. (F. C. Schilling, unpublished results.)

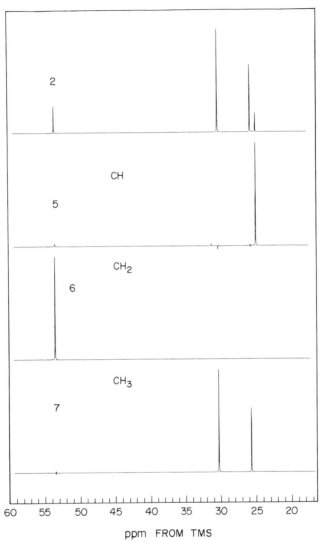

ppm FROM TMS

Fig. 4.52.   50.3 MHz carbon—13 spectra illustrating DEPT sequence
applied to isooctane; same conditions as Fig. 4.51.
Spectrum 2 is the same as spectrum 2 in Fig. 4.51.
Spectrum 5 corresponds to (spectrum 3) — 0.25
(spectrum 4) and shows only CH. Spectrum 6 corresponds
to (spectrum 2) — (spectrum 4) and shows only CH_2.
Spectrum 7 corresponds to (spectrum 2) + (spectrum 4) —
0.375 (spectrum 3), and shows only CH_3 (F. C. Schilling,
unpublished results.)

TABLE 4.10

Enhancements For ^{13}C in the Decoupled DEPT Experiment

Carbon type	Enhancement factor	Maximum intensity at θ value in fourth column	θ
CH	$(\gamma_H/\gamma_C) \cdot \sin\theta$	γ_H/γ_C	90°
CH$_2$	$(\gamma_H/\gamma_C) \cdot \sin 2\theta$	γ_H/γ_C	45°
CH$_3$	$0.75\,(\gamma_H/\gamma_C) \cdot (\sin\theta + \sin 3\theta)$	$1.15\,(\gamma_H/\gamma_C)$	135°

REFERENCES

1. E. Fermi, *Z. Phys.* **60**, 320 (1930).

2. J. A. Pople, W. G. Scheider, and H. J. Bernstein, *High Resolution Nuclear Magnetic Resonance*, McGraw-Hill, New York, 1959, p. 117.

3. W. A. Anderson, *Phys. Rev.* **102**, 151 (1956).

4. H. J. Bernstein, J. A. Pople, and W. G. Schneider, *Can. J. Chem.* **35**, 65 (1957).

5. P. L. Corio, *Chem. Rev.* **60**, 363 (1960).

6. J. I. Musher and E. J. Corey, *Tetrahedron* **18**, 791 (1962).

7. J. Martin and B. P. Dailey, *J. Chem. Phys.* **37**, 2594 (1962).

8. R. J. Abraham, L. Cavalli, and K. G. R. Pachler, *Mol. Phys.* **11**, 471 (1966).

9. R. K. Harris, Chap. 2 in "Nuclear Magnetic Resonance Spectroscopy," Pitman, London, 1983.

10. M. Barfield and D. M. Grant, *J. Am. Chem. Soc.* **83**, 4726 (1961).

11. M. Barfield and D. M. Grant, *J. Chem. Phys.* **36**, 2054 (1962).

12. M. Barfield and D. M. Grant, *J. Am. Chem. Soc.* **85**, 1901 (1963).

13. F. A. L. Anet, *J. Am. Chem. Soc.* **84**, 1053 (1962).

14. M. Karplus, *J. Chem. Phys.* **30**, 11 (1959).

15. M. Karplus, *J. Am. Chem. Soc.* **85**, 2870 (1963).

16. M. Karplus, *J. Chem. Phys.* **33**, 1842 (1960).

17. H. S. Gutowsky, G. G. Belford, and P. E. McMahon, *J. Chem. Phys.* **36**, 3353 (1962).

18. N. Sheppard and J. J. Turner, *Proc. Roy. Soc.* **A252**, 506 (1959).

19. H. Booth, *Tetrahedron Lett.* **7**, 411 (1965).

20. S. Castellano and C. Sun, *J. Am. Chem. Soc.* **88**, 4741 (1966).

21. J. Meinwald and A. Lewis, *J. Am. Chem. Soc.* **83**, 2769 (1961).

22. M. Barfield and B. Chakrabarti, *Chem. Rev.* **69**, 757 (1969).

23. K. B. Wiberg, G. Lampman, R. Ciula, D. S. Connor, P. Schertler, and J. Lavanish, *Tetrahedron* **21**, 2749 (1965).

24. S. Sternhell *Rev. Pure Appl. Chem.* **14**, 15 (1964).

25. J. N. Shoolery, *J. Chem. Phys.* **31**, 768 (1959).

26. N. Muller, *J. Chem. Phys.* **36**, 359 (1962).

27. M. D. Newton, J. M. Schulman, and M. M. Manus, *J. Am. Chem. Soc.* **96**, 17 (1974).

28. N. H. Werstiuk, R. Tailefer, R. A. Bell, and B. A. Sayer, *Can. J. Chem.* **51**, 3010 (1973).

29. E. M. Schulman, K. A. Christensen, D. M. Grant, and C. Walling, *J. Org. Chem.* **39**, 2686 (1974).

30. F. W. Wehrli and T. Wirthlin, *Interpretation of Carbon-13 NMR Spectra*, p. 48 *et seq.* Heyden, London (1976).

31. N. Sheppard and J. J. Turner, *Proc. Roy. Soc.* **A252**, 506 (1959).

32. P. E. Hansen, *Org. Magn. Res.* **11**, 215 (1978).

33. R. E. Carhart and J. D. Roberts, *Org. Magn. Res.* **3**, 139 (1971).

34. F. H. Weigert and J. D. Roberts, *J. Am. Chem. Soc.* **94**, 6021 (1972).

35. M. Barfield and M. Karplus, *J. Am. Chem. Soc.* **91**, 1 (1969).

36. M. Barfield, I. Burfitt, and D. Doddrell, *J. Am. Chem. Soc.* **97**, 2631 (1975).

37. J. L. Marshall and D. E. Miller, *J. Am. Chem. Soc.* **95**, 8305 (1973).

38. F. Bloch, *Phys. Rev.* **93**, 944 (1954).

39. A. L. Bloom and J. N. Shoolery, *Phys. Rev.* **97**, 1261 (1955).

40. J. D. Baldeschwieler and E. W. Randall, *Chem. Rev.* **63**, 81 (1963).

41. R. A. Hoffman and S. Forsén, *Prog. Nucl. Magn. Reson.* **1**, 15 (1966).

42. P. A. Mirau, unpublished results.

43. R. Freeman, K. A. McLauchlan, J. I. Musher, and K. G. R. Pachler, *Mol. Phys.* **5**, 321 (1962).

44. P. C. Lauterbur and R. J. Kurland, *J. Am. Chem. Soc.* **84**, 3405 (1962).

45. E. L. Hahn, *Phys. Rev.* **93**, 944 (1954).

46. M. W. P. Strandberg, *Phys. Rev.* **127**, 1162 (1962).

47. A. D. Buckingham and E. G. Lovering, *Trans. Faraday Soc.* **58**, 2077 (1962).

48. A. D. Buckingham and J. A. Pople, *Trans. Faraday Soc.* **59**, 2421 (1963).

49. A. D. Buckingham and K. A. McLauchlan, *Proc. Chem. Soc.* **1963**, 144.

50. A. Saupe and G. Englert, *Z. Naturforsch.* **19a**, 161, 172 (1964).

51. G. V. D. Tiers and F. A. Bovey, *J. Phys. Chem.* **63**, 302 (1959).

52. J. D. Roberts, *J. Am. Chem. Soc.* **78**, 4495 (1956).

53. H. J. Reich, M. Jautelat, M. T. Meese, F. J. Weigert, and J. D. Roberts, *J. Am. Chem. Soc.* **91**, 7445 (1969).

54. R. Ernst, *J. Chem. Phys.* **45**, 3845 (1966).

55. J. B. Grutzner and A. E. Santini, *J. Magn. Reson.* **19**, 178 (1975).

56. V. J. Basus, P. D. Ellis, H. D. W. Hill, and J. S. Waugh, *J. Magn. Reson.* **34**, 19 (1979).

57. A. J. Shaka, J. Keeler, and R. Freeman, *J. Magn. Reson.* **53**, 313 (1983).

58. M. H. Levitt and R. Freeman, *J. Magn. Reson.* **43**, 502 (1981).

59. J. S. Waugh, *J. Magn. Reson.* **49**, 517 (1982).

60. J. S. Waugh, *J. Magn. Reson.* **50**, 30 (1982).

61. F. C. Schilling, unpublished observations.

62. A. L. Bloom and J. N. Shoolery, *Phys. Rev.* **97**, 1261 (1955).

63. R. A. Forsén and S. Forsén, *Prog. Nucl. Mag. Reson.* **1**, 15 (1966).

64. E. L. Hahn, *Phys. Rev.* **80**, 580 (1950).

65. G. A. Morris and R. Freeman, *J. Am. Chem. Soc.* **101**, 760 (1979).

66. D. M. Doddrell, D. T. Pegg, and M. R. Bendall, *J. Magn. Reson.* **48**, 323 (1982).

CHAPTER 5

NUCLEAR RELAXATION AND CHEMICAL RATE PROCESSES

5.1 INTRODUCTION

We have discussed nuclear relaxation in an introductory manner in Chapter 1. We now consider in more detail the methods of measuring it, the causes and mechanisms which underlie it, and how its study may lead to important knowledge of molecular structure and motion and of chemical rate processes. Particular emphasis will be placed on the relaxation of carbon—13 nuclei because — for good reasons — they have received particular attention, with accumulation of extensive data and interpretation. However, some other nuclei, particularly the proton, must also claim our attention. Some significant aspects of nuclear relaxation are best observed by two-dimensional methods; these matters are postponed to Chapter 6.

5.2 SPIN LATTICE RELAXATION

We have seen that the most important and general mechanism for spin lattice relaxation is *dipole—dipole* relaxation. We must, however, consider other relaxation pathways that may contribute strongly or dominate under particular circumstances: *spin—rotation, chemical shift anisotropy, nuclear electric quadrupolar coupling*, and *scalar coupling*.

5.2.1 Dipole—Dipole Relaxation

5.2.1.1 Transition Probabilities

We consider a system of two spin-½ nuclei I and S which differ in chemical shift but do not experience scalar coupling. We show such a system in Fig. 5.1, which is similar to Fig. 4.2(a). Our attention now centers on the transition probabilities W_o, W_1 and W_2. W_o is the probability of a simultaneous inversion of both I and S, often termed a "zero quantum" transition or spin flip-flop; this is the basis of the phenomenon of spin diffusion, particularly important for protons when

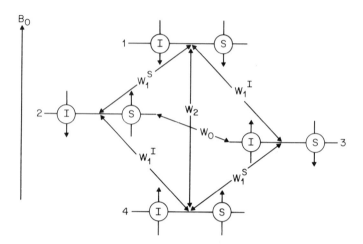

Fig. 5.1.   Energy levels of an AX spin system. Both nuclei have a spin
              I of $1/2$ and scalar coupling of 0.

molecular motion is relatively slow, as in solids (Chapter 8). W_1^I is the
probability of a one-quantum transition that inverts spin I, while W_1^S is the
corresponding probability for spin S. W_2 is the probability of simultaneous
transitions of both spins in the same direction. Only the transitions
represented by W_1^I and W_1^S are ordinarily directly observable as spectral
lines.

Let us assume that we are observing spin I, the Z magnetization of
which is M_I^Z. The Z magnetization of spin S is M_S^Z. The rate of change
of the Z magnetization of spin I is given by

$$\frac{d\, M_I^Z}{dt} = -\rho(M_I^Z - M_I^o) - \sigma(M_S^Z - M_S^o) \tag{5.1}$$

as shown by Solomon.[1] Here, M_I^o and M_S^o are the equilibrium
magnetizations and

$$\rho_{IS} = W_o + 2W_1^I + W_2 \tag{5.2}$$

$$\sigma_{IS} = W_2 - W_o \tag{5.3}$$

where the transition probabilities are those shown in Fig. 5.1. It may be
shown[2] that these probabilities are related to the motional frequencies as
follows:

$$W_0 = \frac{1}{20} K^2 J_0(\omega_S - \omega_I) \qquad (5.4)$$

$$W_1^I = \frac{3}{40} K^2 J_1(\omega_I) \qquad (5.5)$$

$$W_1^S = \frac{3}{20} K^2 J_1(\omega_S) \qquad (5.6)$$

$$W_2 = \frac{3}{10} K^2 J_2(\omega_S + \omega_I) \qquad (5.7)$$

where ω_S and ω_I are the respective resonant frequencies of spins S and I, and K is given

$$K = \frac{\gamma_I \gamma_S \hbar}{r_{IS}^3} \qquad (5.8)$$

(Chapter 1, Sections 1.1, 1.5). Here, r_{IS} is the internuclear distance between I and S. The spectral density functions have the form

$$J_n(\omega) = \frac{2\tau_c}{1 + \omega^2 \tau_c^2} \qquad (5.9)$$

This spectral density function corresponds to the simple motional model discussed in Chapter 1, Section 1.5, i.e., a sphere rotating by small random diffusive steps in a viscous continuum. We have seen that the correlation time τ_c describes the time required for the molecule to lose memory of its previous state of motion, and that in practice, it may be regarded as the time required to rotate approximately one radian.

5.2.1.2 Selective and Non—selective Relaxation

We consider a homonuclear system for which $\omega_s \simeq \omega_I = \omega$ but for which the chemical shift difference is still experimentally observable. Protons are the commonest case. We may imagine experiments in which we excite individual spins in a *selective* manner (Fig. 5.2). In the present case, we suppose that we are inverting the magnetization only of spin I, leaving spin S undisturbed. We then find that at time zero

$$M_I^Z - M_I^o = -2M_I^o \qquad (5.10)$$

and $\qquad M_S^Z - M_S^o = 0 \qquad (5.11)$

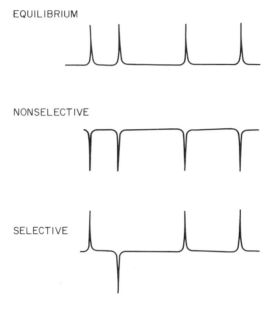

Fig. 5.2.   Resonance signals of a system of spins. Top: at equilibrium;
              middle: all spins inverted; relaxation back to equilibrium
              from this state is *nonselective*; bottom: only one spin inverted;
              relaxation in this case is *selective*.

and therefore from Eqs. (5.1) and (5.4) through (5.7) we find

$$\frac{d\,M_I^Z}{dt} = \frac{K^2}{20}(M_I^Z - M_I^o)[J_0(0) + 3J_1(\omega) + 6J_2(2\omega)] \qquad (5.12)$$

and therefore since

$$\frac{d\,M_I^Z}{dt} = \frac{1}{T_1(I)}(M_I^Z - M_I^o) \qquad (5.13)$$

$$\frac{1}{T_1} = \frac{K^2}{20}[J_0(0) + 3J_1(\omega) + 6J_2(2\omega)] \qquad (5.14)$$

We note that under these circumstance the spin flip–flop process (Chap. 1,
Section 1.6) participates. Under extreme narrowing conditions

$$\frac{1}{T_1(I)} = \frac{\gamma^4 \hbar^2}{r_{IS}^6} \cdot \tau_c \qquad (5.15)$$

For *nonselective* excitation (Fig. 5.2), we have again

$$M_I^Z - M_I^o = -2M_I^o \qquad (5.16)$$

but also that:

$$M_S^Z - M_S^o = -2M_S^o \qquad (5.17)$$

For a homonuclear system, M_I^o and M_S^o are very nearly equal and therefore we may write

$$\frac{d M_i^Z}{dt} = -(\rho + \sigma)(M_I^Z - M_I^o) \qquad (5.18)$$

$$= \frac{K^2}{20}(M_I^Z - M_I^o)[3J_1(\omega) + 12J_2(2\omega)] \qquad (5.19)$$

or

$$\frac{1}{T_1} = \frac{K^2}{20}[3J_1(\omega) + 12J_2(2\omega)] \qquad (5.20)$$

The flip–flop process does not contribute since both spins — or *all* spins in a multi–spin system — are pointing in the same direction. In the extreme narrowing limit

$$\frac{1}{T_1(I)} = \frac{3}{2} \cdot \frac{\gamma^4 \hbar^2}{r_{IS}^6} \cdot \tau_c \qquad (5.21)$$

as we have seen in Chapter 1, Section 1.5. The classical derivation of Bloembergen *et al.*[3] leads to a coefficient of the second spectral density term half that given in Eq. (5.20), and therefore to a coefficient of 9/10 in the extreme narrowing case. It appears that this is not correct.

We see that non–selective and selective relaxation rates [Eqs. (5.14) and (5.20)] depend on the internuclear distance r_{IS} in the same way but depend on molecular motion in different ways. This has important consequences for the observation of molecular motion — particularly of larger molecules — through proton relaxation.

We find for *cross relaxation* — simultaneous transitions of both spins — that

$$\frac{dM_I^Z}{dt} = \frac{K^2}{20}(M_I^Z - M_I^o)[6J_2(2\omega) - J_0(0)] \qquad (5.22)$$

or
$$\frac{1}{T_1(I)} = \frac{\gamma^4\hbar^2}{20r_{IS}^6}[6J_2(2\omega) - J_0(0)] \qquad (5.23)$$

Cross relaxation dominates proton relaxation under conditions of relatively slow motion. In solids, the zero−quantum term dominates and is the mechanism of *spin diffusion*.

5.2.1.3 Carbon−13 Relaxation

The measurement of the relaxation of carbon−13 nuclei in natural abundance (1.1%) has received extensive application for the study of molecular dynamics. There are several reasons for this:

a. Carbon−13 resonances are generally well resolved in chemical shift so that separate T_1 measurements are feasible.

b. Carbon−13 nuclei are spatially isolated so that mutual dipole−dipole influence is negligible. Only nearby protons are effective for relaxation and for other than quaternary carbons only directly bonded protons need be considered, thus simplifying evaluation of the r_{IS} term in the relaxation equation.

c. Carbon atoms are generally not at the periphery of the molecule and so only intramolecular influences need be considered.

For reasons we have already discussed (Chapter 4, Section 4.4.3.2), it is customary to decouple all protons by broad-band irradiation when observing carbon−13 spectra. This is the practice also in T_1 measurements and in this case has the further advantage of simplifying the relaxation equation. Let us suppose that in Eq. (5.2) spin I is ^{13}C and spin S is a proton. The proton irradiation causes M_S^Z in Eq. (5.1) to become zero, and it can be shown[4] that under these conditions, in effect

$$\frac{1}{T_{1(C)}} = -\rho_{CH} \qquad (5.24)$$

$$= \frac{K^2}{20}\left[J_0(\omega_H - \omega_C) + 3J_1(\omega_C) + 6J_2(\omega_H + \omega_C)\right] \qquad (5.25)$$

where J_n is of the form shown in Eq. (5.9) and $K = \gamma_C\gamma_H\hbar/r_{CH}^3$ [See Eq. (5.10)]. A plot of Eq. (5.25) is shown as a function of τ_c in Fig. 5.3 for two values of the magnetic field corresponding to carbon resonant frequencies $\nu_0 (= \omega_0/2\pi)$ of 50.3 MHz and 125 MHz. As one expects, they are of the same form as in Fig. 1.9 for $^1H-^1H$ relaxation. Figure 5.3 shows the relaxation of a carbon−13 bonded directly to a single proton with r_{CH} taken as 1.09 Å.

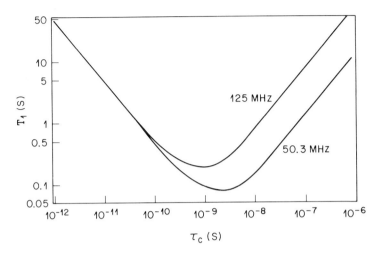

Fig. 5.3.   Spin lattice relaxation time T_1 for a carbon-13 nucleus with one directly bonded proton as a function of the correlation time τ_c at 50.3 MHz and at 125 MHz. The proton is irradiated to remove scalar coupling [Eq. (5.25)].

Under extreme narrowing conditions (where $\tau_c \ll \omega_c^{-1}$) for N_H directly bonded protons

$$\frac{1}{T_{1(C)}} = \frac{N_H \, \gamma_H^2 \, \gamma_C^2 \, \hbar^2}{r_{CH}^6} \cdot \tau_c \qquad (5.26)$$

$$= 2.1 \times 10^{10} \cdot \tau_c \cdot N_H \qquad (5.27)$$

or

$$\tau_c = \frac{4.7 \times 10^{-11}}{N_H \cdot T_1} \qquad (5.28)$$

Equation (5.28) provides a ready estimate of the correlation times for small molecules in mobile solvents, but it is in general best to employ the full equation.

When the proton is irradiated there is an enhancement in the value of the carbon-13 magnetization. This is discussed in Section 5.2.8.

5.2.2 Spin Rotation

A molecule in a fluid medium is a rotating charge system and therefore generates a magnetic moment, apart from the nuclear moments.

Molecular collisions cause changes in the rate of rotation and in the direction of the axis (or axes) of rotation, and therefore cause this magnetic field to fluctuate. It has been shown[5,6] that the rate of relaxation of a magnetic nucleus at or near the center of gravity of an isotropically tumbling molecule is given by

$$\frac{1}{T_{1\,(SR)}} = \frac{2\,IkT}{3\hbar^2} \cdot C^2 \cdot \tau_{SR} \tag{5.29}$$

Here, I is the moment of inertia of the molecule, C is the averaged value of the spin-rotation coupling tensor (which is very weak) and τ_{SR} is the spin rotation correlation time. There are few measurements of τ_{SR} but it appears to be quite different from τ_c. It *increases* with temperature and is longest in the vapor. It expresses the average time that the molecule remains in a given angular momentum state and is reciprocally related to τ_c. It is principally for this reason that $T_{1\,(SR)}$ becomes shorter in liquids as the temperature is increased. Measurements on liquids in the presence of the vapor phase may be misleading because molecules in the liquid enter the vapor space, relax, and return to the liquid.

Because of the opposing effects of the dipole–dipole and spin–rotation contributions, some liquids show a weak maximum in the value of T_1 as a function of temperature. In Fig. 5.4 the temperature dependence of the carbon–13 T_1 in $^{13}CH_3I$ is shown.[7] The experimental T_1 is the lowest curve. The uppermost curve (without points) is the dipole–dipole contribution calculated from other data. The third (middle) curve is the spin-rotation effect obtained as the difference between the other two.

The spin rotation mechanism is generally less significant for protons than for carbon–13. We shall discuss its relative importance for carbon–13 relaxation further in Section 5.2.8.

5.2.3 Chemical Shift Anisotropy

We have seen (Chapter 3, Section 3.3.1; see also Chapter 8, Section 8.1) that the chemical shift of an observed nucleus is a directional quantity and depends on the orientation of the molecule with respect to the magnetic field. In the liquid state the rapid motion of the molecule averages the shielding to the isotropic value that we normally see for solution spectra, but on the time scale of molecular motion the nucleus experiences fluctuations in the local magnetic field arising from this source. This gives rise to the following relationship in the fast motion limit:

$$\frac{1}{T_{1\,(CSA)}} = \frac{2}{15}\,\gamma^2 B_o^2 (\sigma_{\parallel} - \sigma_{\perp})^2 \tau_c \tag{5.30}$$

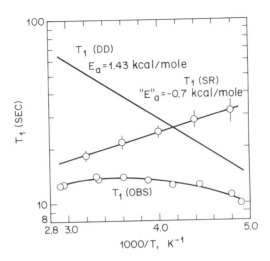

Fig. 5.4.   Carbon—13 relaxation times in methyl iodide-^{13}C. T_1 (DD) is calculated from other data and subtracted (as reciprocal) from $1/T_1$ (experimental) to yield T_1 (SR), the spin-rotation contribution. [From K. T. Gillen, M. Schwartz, and J. H. Noggle, *Mol. Phys.* **20**, 899 (1970)].

where $\sigma_{\parallel}$ and $\sigma_{\perp}$ are as defined in Chapter 3, Section 3.3.1, for an anisotropic group with axial symmetry. The quadratic dependence on the field strength provides a means for evaluating the contributions to spin lattice relaxation from this cause. It may be expected to be of considerable significance for large unsaturated and aromatic carbons at superconducting frequencies (see Section 5.2.8).

5.2.4 Quadrupolar Relaxation

We have seen (Chapter 1, Section 1.9) that a particularly potent source of relaxation is electric rather than magnetic in origin. Every nucleus is an electric monopole, and although no nucleus may be an electric dipole, all nuclei with spin greater than 1/2 (i.e., 1, 3/2, 5/2,) are electric quadrupoles. A quadrupole does not interact with a homogeneous electric field, i.e., where the potential gradient $\partial V/\partial z$ is a constant, but it does interact with an electric field when a gradient $\partial^2 V/\partial z^2$ is present. The electric quadrupole moment of the nucleus is expressed in terms of eQ, where e is the unit atomic charge (the charge on a proton) and Q has the dimensions of distance squared. Q is negative or positive depending on whether the distribution of nuclear charge has the form of an oblate or prolate ellipsoid (see Chapter 1 and Fig. 1.15). In Table 5.1 the quantity Q is expressed in terms of the unit $10^{-28}\,m^2$ (or "barn") for some common quadrupolar nuclei. These values are selected from Appendix A.

The perturbation of the Zeeman energy levels of the quadrupolar nucleus by interaction with molecular electric field gradients is an important phenomenon in the solid state and is discussed in Chapter 8 (Section 8.8). Here we assume that molecular motion is sufficiently rapid — as in the liquid state — to erase that interaction. The contribution to spin lattice and spin–spin relaxation remains substantial and is given in the limit of extreme motional narrowing ($\omega_o \tau_c \ll 1$) by

$$\frac{1}{T_{1\,(Q)}} = \frac{1}{T_{2\,(Q)}} = \frac{3\pi^2}{10} \cdot \frac{(2I+3)}{I^2(2I-1)} \left[1 + \frac{\eta^2}{3}\right] \left[\frac{e^2 qQ}{h}\right]^2 \cdot \tau_c \quad (5.31)$$

The quantity $3\pi^2/10[(2I+3)/I^2(2I-1)]$ has these values: for $I = 1$: 14.8044; for $I = 3/2$: 3.9478; for $I = 5/2$: 0.9475; for $I = 7/2$: 0.4028. The electric field gradient is expressed by q. The symmetry parameter η expresses the departure of the electric field gradient from cylindrical symmetry at the observed nucleus. For cylindrical symmetry — which is common — η is zero. The quantity $e^2 qQ/h$ is called the *quadrupole coupling constant*. The relaxation is governed primarily by its magnitude.

TABLE 5.1
Quadrupole Moments of Some Common Nuclei

Nucleus	Spin I	Quadrupole moment Q (in 10^{-28} m²)
^{2}H	1	0.0028
^{6}Li	1	−0.0008
^{7}Li	3/2	−0.04
^{10}B	3	0.085
^{11}B	3/2	0.041
^{14}N	1	0.01
^{17}O	5/2	−0.026
^{79}Br	3/2	0.37
^{81}Br	3/2	0.31
^{35}Cl	3/2	−0.10
^{37}Cl	3/2	−0.079
^{127}I	5/2	−0.79

In fluid systems τ_c is of the order of 10^{-11} to 10^{-12}s. For ^{14}N in an environment of high local symmetry — as in NH_4^+ or $N^+(CH_3)_4$ — the quadrupole coupling constant is weak and the resonance lines are less than 1 Hz in width. In most organic ^{14}N compounds, however, this quantity is of the order of megahertz and linewidths may be tens to hundreds of hertz. It is primarily for this reason that the observation of ^{15}N ($I = 1/2$) has gained favor despite a natural abundance of only 0.37% (Chapter 1, Table 1.1). ^{17}O also suffers from the same disadvantage and from low abundance (0.037%) but no spin−1/2 isotope is available. The quadrupole coupling constants of the chlorine isotopes in organic compounds are so large — of the order of 10^2 MHz — that it is not practical to observe them. An advantageous aspect of this property is the fact that they also effectively decouple themselves from other nuclei and do not usually perturb the ^{1}H or ^{13}C spectra of the compounds in which they are present.

The quadrupole moment of deuterium is small enough (Table 5.1) to make it feasible to observe this isotope in liquid systems. In D_2O, for example, its linewidth is approximately 1 Hz. As with other quadrupolar nuclei, the quadrupolar relaxation dominates over all other mechanisms and therefore simplifies the theoretical interpretation of molecular dynamics. This becomes particularly important in the solid state (Chapter 8, Section 8.8).

5.2.5 Scalar Relaxation

The quadrupolar relaxation of a nucleus with spin $I \geqslant 1$ may be manifested not only in the broadening of its resonance but also in that of a bonded spin−1/2 nucleus which experiences scalar coupling to the quadrupolar nucleus. A commonly observed example is that of ^{14}N with bonded protons, as in amines, amides, and peptides. We have seen in Chapter 4 (Section 4.4.3.1) that in most ^{14}N-containing molecules the 1:1:1 triplet resonance of ^{14}N-bonded protons with *ca.* 50 Hz spacing is collapsed in varying degrees by the rapid transitions of the ^{14}N nucleus between its spin states, expressed by Eq. 5.31. The same effect is of course also observed for the ^{14}N multiplet. This phenomenon is less marked for the $^{14}NH_3$ protons in $^+NH_3CH_2CH_2CH_3$ cation, which still show a broadened 50 Hz triplet. But for amides and peptides the triplet can no longer be observed. The collapse is more complete at lower temperatures where τ_c is longer and $1/T_{1(Q)}$ greater. In most peptides — particularly if chiral — triplet collapse is complete[8] but three−bond vicinal coupling of the ^{14}N−H protons is unaffected.

The multiplet collapse in these cases is the same as would be expected if proton exchange were occurring. If it actually arises from chemical exchange it is termed *scalar relaxation of the first kind* (see

Section 5.4.5). If it occurs because of quadrupole relaxation without exchange it is called *scalar relaxation of the second kind.*

5.2.6 Relaxation by Paramagnetic Species

We have seen (Chapter 1, Section 1.5) that paramagnetic ions can cause marked decreases in both T_1 and T_2 because of the very large fluctuating magnetic fields generated by unpaired electron spins, of the order of 10^3 times the largest nuclear moments. The general dipole–dipole theory of Bloembergen, Purcell, and Pound[3] applied to the spin lattice relaxation of water protons in solutions of paramagnetic ions gives[9]

$$\frac{1}{T_{1(P)}} = \frac{4\pi^2\gamma^2\eta N_p\mu_{eff}^2}{kT} \tag{5.32}$$

where γ is the magnetogyric ratio of the proton, η is the solution viscosity, N_p is the number of paramagnetic ions per milliliter, and μ_{eff} is the effective magnetic moment of the paramagnetic species. If each electronic moment maintains its orientation during the course of the diffusion together of the ion and the water molecule, i.e., if the spin lattice relaxation time T_{1e} is longer than the diffusional correlation time t, then μ_{eff} will be equal to the magnetic moment found by static susceptibility measurements. If, however, $T_{1e} < t$, then T_{1e} governs the relaxation and μ_{eff} will be smaller than the true value. Some ions give values of μ_{eff} calculated from Eq. (5.32) that are close to the static susceptibility values. It is found that Cu^{+2}, Fe^{3+}, Mn^{2+}, and Cr^{3+}, having relatively long T_{1e} values, have values of μ_{eff} nearly equal to the static values, while Co^{2+}, Fe^{2+}, Ni^{2+}, and most of the rare earths are much less effective in reducing T_1 of the solvent.

In discussing the chemical shift effects of paramagnetic metals (Chapter 3, Section 3.3.7), we have seen that in addition to the dipole component there may also be a "contact" component arising from the presence of unpaired electron spin at the observed nucleus — the same interaction that gives rise to hyperfine splittings in electron spin resonance spectra. There is also a contact contribution to $1/T_1$ of the solvent protons, but it has been shown by Bloembergen[10] that for the usual values of T_{1e} this effect is small, although reduced values of T_2 may be observed.

It has long been known that molecular oxygen is an effective paramagnetic relaxing agent, and should be removed by sweeping with nitrogen or inert gases when measuring spin lattice relaxation values.

The use of paramagnetic complexes in eliminating nuclear Overhauser enhancements is briefly discussed in Section 5.2.8.

5.2.7 Measurement of T_1

5.2.7.1 CW Methods

The basic requirement in the measurement of T_1 is that we perturb the spin system in a reproducible manner and measure its rate of return to equilibrium. This may be done in a number of ways. In earlier days, before the universal adoption of pulse Fourier transform spectroscopy, a preferred method was to sweep through resonance by varying the magnetic field at a rate such that $dB_o/dt \ll |\gamma| B_o^2$, a condition termed *adiabatic fast passage*. This results in an inversion of the z magnetization and is equivalent to a 180° pulse. The return to equilibrium is monitored by making sampling sweeps in the reverse direction at varied intervals. We discuss the pulse equivalent of this experiment in Section 5.2.7.2.

A simpler approach is to saturate the spin system (Chapter 1, Section 1.8) by the momentary application of a strong rf field B_1 and to follow the regrowth of the magnetization with a rapid-sweep recorder. One may even omit the rf saturation by beginning to monitor the magnetization immediately after inserting the sample into the magnetic field. Such methods are at best only approximate and require that T_1 be at least 5−10 s.

5.2.7.2 Inversion−Recovery

The pulse sequence and vector diagrams for this experiment are shown in Fig. 5.5. This is the standard method of T_1 measurement. Software for carrying it out automatically is provided with every modern pulse spectrometer except the simplest. The z magnetization M^z (a) is inverted by a 180° pulse (b). The magnetization begins to grow back through zero toward its equilibrium value M^o, as shown at (c). To sample this regrowth, the magnetization vector is turned back along the y' axis (d) at time τ, where it is observed and recorded (after transformation) as M_t^z. Following the acquisition of the signal at time τ a delay τ_D is allowed. This interval should be long enough compared to T_1 to permit attainment of the equilibrium value before the application of the next 180° pulse; the appropriate interval may be judged from the value of the quantity $(1-e^{-\tau/T_1})$, which is 0.993 if $\tau = 5T_1$, 0.982 if $\tau_D = 4T_1$, and 0.950 if $\tau = 3T_1$. This sequence is repeated for 6 to 8 values of τ, based on an estimated value of T_1. The longest value of τ must be such as to provide a value of the equilibrium magnetization, M^o. This may be judged by the same criterion as for τ_D.

The recovery of the z magnetization obeys Eq. (1.14) in Chapter 1 [or, in macroscopic terms, Eq. (1.44) with B_1 of negligible magnitude]:

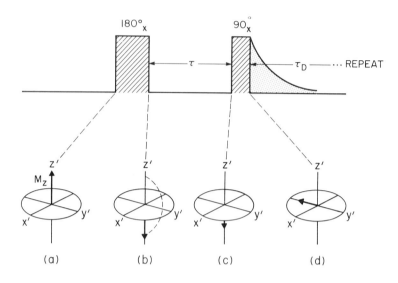

Fig. 5.5.   The inversion-recovery pulse sequence for the measurement
of T_1. (Pulse widths are greatly exaggerated in comparison
to pulse intervals.)

$$\frac{dM^z}{dt} = - \frac{M^o - M^z}{T_1} \tag{5.33}$$

Replacing M^z by M_t^z and integrating we have

$$M^o - M_t^z = Ae^{-\tau/T_1} \tag{5.34}$$

The constant A depends on the initial conditions. For an inversion-
recovery experiment M_t^z at $t = 0$ equals $-M^o$ and so $A = 2M^o$. To
determine T_1, one thus plots M_t^z versus t according to this relationship:

$$\ln(M^o - M_t^z) = \ln 2 + \ln M^o - \tau/T_1 \tag{5.35}$$

T_1 is given by the reciprocal of the slope of this plot.

   In Fig. 5.6 is shown the stacked plot resulting from an
inversion–recovery measurement of the T_1 values of the six distinguishable
carbons of ethylbenzene (10% solution in CDCl$_3$ at 30°C), observed at
50.3 MHz:

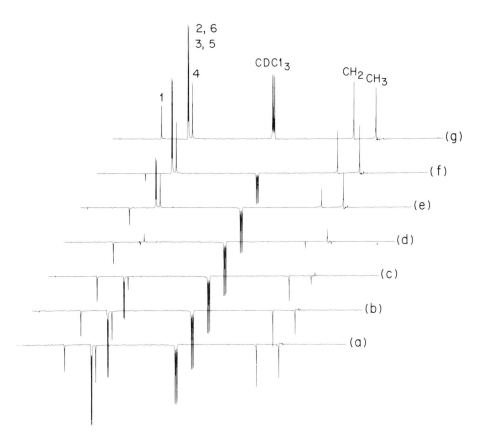

Fig. 5.6.   Stacked plot of 50.3 MHz carbon—13 spectra from an inversion-recovery experiment for the measurement of T_1 values of 10% (vol./vol.) ethylbenzene in $CDCl_3$ at 23°C. The 90° sampling pulses were applied at these times (τ) after inversion: (a) 1.0 s, (b) 2.0 s, (c) 4.0 s, (d) 8.0 s, (e) 16.0 s, (f) 32.0 s, and (g) 300 s. The resonances are identified on spectrum (g). (P. A. Mirau, private communication.)

The acquisition of each spectrum required 20 accumulations. The delay times τ are indicated in the caption. In Fig. 5.7 plots of the C_1, CH_2, and CH_3 magnetization according to Eq. (5.35) are shown. From these and similar plots of the other carbon signals, the NT_1 values [see Eqs. (5.27) and (5.28)] are calculated. The T_1 values are thus in effect normalized to

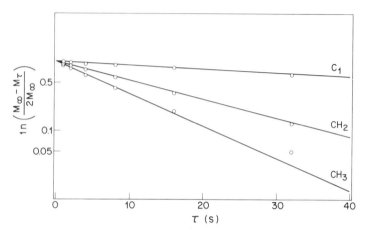

Fig. 5.7.   Semilog plot of the carbon–13 magnetization of C_1, CH_2, and CH_3 of ethylbenzene (10% in $CDCl_3$, 23°C) corresponding to spectral intensities of Fig. 5.6 and Eq. (5.35). The derived T_1 values are given in Table 5.2, item 3.

the values expected for carbons with a single attached proton. For the non–protonated carbon C_1, T_1 is much longer than for the others, as expected.

T_1 values can also be estimated from null times. As M_t^z passes through zero, Eq. (5.35) becomes

$$T_1 = \frac{\tau_{(null)}}{\ln 2} = \frac{\tau_{(null)}}{0.693} \qquad (5.36)$$

or

$$T_1 = \frac{8.0}{0.693} \cong 11.5 \text{ s}$$

for C_2 and C_3 of ethylbenzene (Fig. 5.6). The value obtained from the semilog plot is 13.5 s, the null estimate being only approximate. We also observe that the carbon of the $CDCl_3$ solvent exhibits a T_1 of at least 80 s estimated from the null, whereas protiochloroform shows a value of 32 s (Table 5.2). This reflects the fact that deuterium is only about 1/40th as effective as the proton in dipole relaxation; however, that the $CDCl_3$ T_1 is as short as it is shows that other factors contribute, probably mainly spin rotation. We shall discuss these matters further in Section 5.2.9.

Freeman and Hill[11] have proposed an extended pulse sequence in which the free induction decays from the M_∞ pulses are added and those from the shorter t values are subtracted in the computer memory. This gives an accumulated spectrum which after transformation shows large signals at small t values decaying to zero at "infinite" time. This procedure minimizes effects of instrumental drift over long accumulations. However, it is not actually widely needed or used with modern instruments.

5.2.7.3 Progressive Saturation

The inversion-recovery method can require many hours of spectrum accumulation because of the long delays needed to achieve spin equilibrium. The method of *progressive saturation*, although less accurate, is faster. It employs a series of 90_x° pulses to achieve a dynamic steady state between rf absorption and spin lattice relaxation. The magnitude of the steady state signal (early signals are discarded) depends on the interval t between pulses according to

$$M^o - M_t^z = M^o e^{-t/T_1} \qquad (5.37)$$

and therefore a plot of $\ln(M^o - M_t^z)$ versus t yields T_1 as the reciprocal of the slope in much the same manner as for the inversion-recovery method.

It is important for the successful application of this method that no y' magnetization remain prior to each pulse, since this would be rotated to the $-z'$ axis and invalidate Eq. (5.37). One must not, in other words, allow this to be a spin echo experiment (Chapter 4, Section 4.4.4.1; this chapter, Section 5.3) but rather confine it to longitudinal relaxation. This may be assured by various means that we shall not detail here. The method is well adapted to carbon−13 relaxation measurement since in this case broad−band proton irradiation ensures rapid ^{13}C spin−spin relaxation (Chapter 4, Section 4.4.3.2).

5.2.8 The Nuclear Overhauser Effect

5.2.8.1 Introduction

This important effect has many ramifications. As in the treatment of carbon−13 relaxation (Section 5.2.1.3), let us take spin I as ^{13}C and spin S as ^{1}H. We irradiate the protons so that spin S is saturated and $M_s^t = 0$. Then under steady state conditions we can show from Eqs. (5.1) and (5.2)

$$\frac{M_I(S \text{ irrad})}{M_I^o} = 1 + \frac{\sigma}{\rho} \cdot \frac{M_S^o}{M_I^o} \qquad (5.38)$$

or, since equilibrium spin populations and magnetizations are proportional to magnetogyric ratios, we may write

$$\frac{M_I(S \text{ irrad})}{M_I^o} = 1 + \frac{\sigma}{\rho} \cdot \frac{\gamma_S}{\gamma_I}$$

For carbon−13 enhancement under proton irradiation

$$\frac{M_C(^1\text{H irrad})}{M_C^o} = 1 + \left[\frac{6J_2(\omega_H + \omega_C) - J_0(\omega_H - \omega_C)}{J_0(\omega_H - \omega_C) + 3J_1\omega_C + 6J_2(\omega_H + \omega_C)} \right] \cdot \frac{\gamma_H}{\gamma_C} \quad (5.39)$$

$$= 1 + \eta_{C-H} \qquad\qquad\qquad (5.40)$$

where η is termed the $^{13}C-\{^1H\}$ NOE enhancement factor. The ratio γ_H/γ_C equals 3.977. Under extreme motional narrowing, Eq. (5.33) becomes

$$\frac{M_C(^1\text{H irrad})}{M_C^o} = 1 + 0.5 \cdot \frac{\gamma_H}{\gamma_C} \qquad (5.41a)$$

$$= 2.988 \qquad\qquad (5.41b)$$

or $\qquad\qquad\qquad\qquad \eta = 1.988 \qquad\qquad\qquad (5.41c)$

The nuclear Overhauser effect depends upon dipole−dipole relaxation. Other competing relaxation pathways detract from it, and since such pathways are common, especially in smaller molecules, the carbon−proton NOE is often less than maximal even under fast−motion conditions. This may afford insight into these other mechanisms of relaxation since

$$\frac{1}{T_1(\text{other})} = \frac{1}{T_1(\text{obs})} \left[1 - \frac{\eta(\text{obs})}{\eta_o} \right] \qquad (5.42)$$

where η_o is the maximal NOE, assuming fast motion. The NOE will also be less than maximal if the motional narrowing limit does not apply. In Fig. 5.8(a) are shown plots of Eq. (5.39) for the $^{13}C-\{^1H\}$ Overhauser enhancement at carbon−13 frequencies of 50.3 MHz and 125 MHz. Corresponding plots of T_1 for a carbon with a single bonded proton ($^{13}C-^1H$ bond, length = 1.09 Å) are shown in Fig. 5.3, as already discussed (Section 5.2.1.3). The inflection points of the NOE occur at the minima of the T_1 plots.

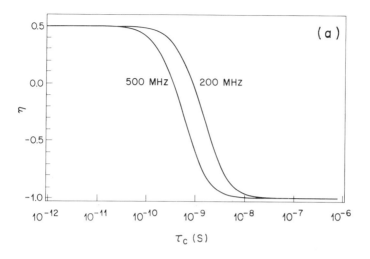

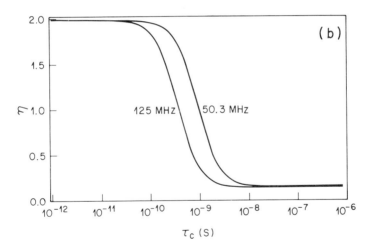

Fig. 5.8. (a) Nuclear Overhauser enhancement factor η for a proton, produced by irradiation of a neighboring proton(s), as a function of the correlation time τ_c, calculated for proton frequencies of 200 MHz and 500 MHz [Eq. (5.43)]. (b) Nuclear Overhauser enhancement factor η for a carbon–13 nucleus, produced by irradiation of neighboring proton(s), as a function of the correlation time τ_c, calculated for carbon–13 frequencies of 50.3 MHz and 125 MHz [Eq. (5.39)].

In quantitative carbon−13 NMR measurements, where η may vary between carbons, it is sometimes desirable to quench all NOE's by means of a paramagnetic reagent such as chromium acetylacetonate. On this and on quantitative carbon−13 NMR generally the reader is referred to a discussion by Shoolery.[12]

The *proton−proton* Overhauser effect was the first to be used for making resonance assignments (see Chap. 3, Ref. 47) and has much potential for dynamic and structural measurements. Since the magnetogyric ratio term is now unity, Eq. (5.39) may be rewritten as

$$\frac{M_H(\text{irrad})}{M_H^o} = 1 + \left[\frac{6J_2(2\omega_H) - J_0(0)}{J_0(0) + 3J_1(\omega_H) + 6J_2(2\omega_H)}\right] \qquad (5.43)$$

$$= 1 + \eta_{HH} \qquad (5.44)$$

It must of course be assumed that the protons to be observed and to be irradiated differ sufficiently in chemical shift, as in double resonance experiments (Chapter 4, Section 4.4.2), to make the experiment feasible. A plot of Eq. (5.40) is shown in Fig. 5.8(b) for magnetic fields corresponding to a proton ν_o ($= \omega_o/2\pi$) of 200 MHz and 500 MHz. The maximum enhancement η is only 0.5, and it is further notable that η passes through zero when $\tau_c\omega_o \simeq 1$ and becomes negative with an asymptote of -1.

The proton−proton NOE can be very effectively employed in the 2D mode, and further discussion will be postponed until Chapter 6.

5.2.8.2 Experimental Measurement

The simplest experimental procedure for the measurement of $^{13}C - \{^1H\}$ nuclear Overhauser enhancements might appear to be to obtain the integrated carbon−13 spectra with and without broad-band proton irradiation (Chapter 4, Section 4.4.3.2). In practice, this is found to be undersirable because of the overlapping multiplets of the coupled spectrum, which make it difficult to measure enhancement ratios for individual carbon resonances. A better method is to gate the proton decoupler on only during data acquisition. The multiplets collapse instantaneously, but the NOE requires a time of the order of T_1 to build up. Figure 5.9(a) shows the pulse sequence employed.

If it is desired to retain both the coupling and the NOE, this may be done by inverse gating of the proton decoupler, i.e., gating it on *except*

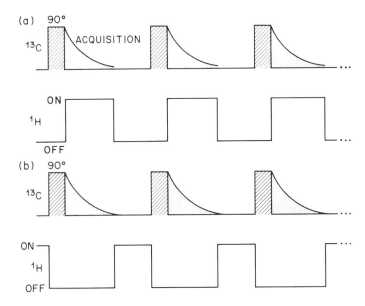

Fig. 5.9.   Gated decoupling pulse sequences for observation of carbon−13: (a) Sequence for eliminating the NOE and also collapsing the multiplicity arising from proton coupling; (b) inverse gated decoupling sequence for retaining both the NOE and the multiplicity (see text). (Note that carbon pulse widths are greatly exaggerated in comparison to pulse intervals and decoupler "ON" times.)

during data acquisition. In this way carbon−proton couplings may be measured with greater effective sensitivity [Fig. 5.9(b)].

5.2.9 Relative Significance of Relaxation Mechanisms

Of the various spin lattice relaxation mechanisms we have discussed, the principal contributors (in diamagnetic solutions) to carbon−13 relaxation − the most widely observed nucleus for this purpose − are the dipole−dipole (DD) and spin rotation (SR) mechanisms. The SR interaction becomes prominent in small, symmetrical molecules. The DD mechanism generally dominates in larger molecules and is the exclusive interaction in macromolecules (Chapter 7). Chemical shift anisotropy (CSA), which is usually negligible when measurements are made with low-field instruments, as in earlier work, becomes prominent in large unsaturated molecules observed at superconducting frequencies. In solutions containing paramagnetic substances, this interaction almost always dominates.

The contributions to the observed T_1 combine in reciprocal fashion i.e., the overall rate is (as might be expected) the sum of the rates attributable to individual mechanisms:

$$\frac{1}{T_{1\,(obs)}} = \frac{1}{T_{1\,(DD)}} + \frac{1}{T_{1\,(SR)}} + \frac{1}{T_{1\,(CSA)}} + \frac{1}{T_{1\,(S)}} + \frac{1}{T_{1\,(Q)}} \quad (5.45)$$

The last two terms, $1/T_{1\,(S)}$ and $1/T_{1\,(Q)}$, represent the scalar and quadrupolar mechanisms, which as we have seen occur only under special circumstances and usually affect specific protons (carbons less commonly) in the molecule.

The DD contribution in small to moderately large molecules decreases with increasing temperature. The SR contribution has the opposite dependence. The CSA contribution increases with the square of the magnetic field. These characteristics enable one to sort out their relative contributions in a given molecule. Thus, if $T_{1\,(other)}$ represents relaxation by other than dipolar mechanisms [see Eq. (5.38)], and B_o' and B_o'' represent two different instrumental field strengths (preferably with a ratio of at least 2), we may write [see Eq. (5.26)]

$$\frac{1}{T_{1\,(other,\,B_o')}} = \frac{1}{T_{1\,(SR)}} + \frac{2}{15}\,\gamma_c^2(B_o')^2(\sigma_\| - \sigma_\perp)^2\tau_c \quad (5.46a)$$

$$\frac{1}{T_{1\,(other),\,B_o''}} = \frac{1}{T_{1\,(SR)}} + \frac{2}{15}\,\gamma_c 2(B_o'')^2(\sigma_\| - \sigma_\perp)^2\tau_c \quad (5.46b)$$

From the difference between these equations [the $1/T_{1\,(SR)}$ term drops out], one may obtain both the CSA contribution and the chemical shift anisotropy.

Relaxation measurements can be used to learn several things of interest concerning organic molecules. For example, even casual inspection of carbon−13 spectra collected with relatively short pulse intervals may reveal assignments of quaternary or carbonyl carbons because their T_1 values are usually severalfold longer than those of carbons with bonded protons. [However, spectral editing methods such as INEPT and DEPT (Chapter 4, Section 4.4.4) provide a more complete and elegant way of sorting out carbon types.] One may also detect rapid internal motions, most commonly of methyl groups. Carbon-13 relaxation measurements have provided valuable insights into the modes of motion of molecules in the liquid state. Proton−proton relaxation measurements can provide important structural information, particularly when carried out in the 2D mode; this is treated in Chapter 6, Section 6.3.2. Finally, both proton and carbon—13 relaxation measurements give very significant information concerning the modes and rates of motion of macromolecules both in the solid and solution states. This is discussed in Chapter 7, Section 7.3.

In Table 5.2 carbon—13 T_1 and NOE values for a selection of organic molecules are shown. Most of the data are taken from the reviews of Lyerla and Levy,[13] Wright et al.[14] and Levy et al.[15] which should be consulted for additional information and more extensive discussion. The T_1 and η values are taken at temperatures in the 25—35°C range (occasionally higher), sometimes neat and sometimes in solution. They cannot be strictly compared between molecules but are intended rather to be instructive when compared within the same molecule. Values of $T_{1\ (DD)}$ are given in some cases where the factor η has been measured (in the extreme narrowing limit); these are the values of T_1 that would have been observed if dipole—dipole relaxation were the only mechanism. They are calculated from the relationship

$$T_{1\ (DD)} = T_{1\ (obs)} \cdot \frac{\eta_o\ (=1.98)}{\eta\ (obs)} \tag{5.47}$$

which is readily derived from Eqs. (5.42) and (5.45).

For benzene (entry 1), DD is the principal contribution but the less than full NOE shows that there is another pathway; because of symmetry and rapid motion, it is likely that this is SR. The same is also the case for the ring carbons of toluene (entry 2). Another influence may also be seen here, in that the relaxation of C_2 and C_3 is less efficient than that of C_4. This is to be attributed to faster rotation about the C_1-C_4 axis of the molecule than about perpendicular axes. Rotation about the C_1-C_4 axis does not affect the C_4-H vector but does cause tumbling of the C_2-H and C_3-H vectors. The motion of these latter vectors is faster on average and so has a smaller component at the carbon—13 resonant frequency. The C_2 and C_3 relaxations are therefore slower. The quaternary carbon C_1 and the methyl carbon have a marked contribution from SR; for the methyl carbon, this is mainly to be attributed to rapid internal rotation as the torsional barrier is nearly zero. For ethylbenzene (entry 3) the same conclusions apply except that the SR contributions may be relatively somewhat smaller.

For phenol (entry 4) the higher viscosity, arising from hydrogen bonding, and resulting slower motion, leads to a complete dominance of the DD mechanism. Upon dilution, the T_1 values become longer. Molecular

TABLE 5.2 (continued)

Spin Lattice Relaxation Mechanisms in Organic Molecules

Number	Compound	Carbon	T_1 (obs), s	η	T_1 (DD), s	Principal non DD mechanism	Reference
1	benzene		29	1.6	35	SR	a
2		1	89	0.56		SR	a
		2	24	1.6	30		
		3	24	1.7	28		
		4	17	1.6	30		
		CH_3	16	0.61		SR, rapid internal rotation	

278

TABLE 5.2 (continued)

Spin Lattice Relaxation Mechanisms in Organic Molecules

Number	Compound	Carbon	$T_{1(obs)}$, s	η	$T_{1(DD)}$, s	Principal non DD mechanism	Reference
3	CH$_2$CH$_3$ (benzene ring, positions 1,2,3,4)	1	79.3			SR	b
		2	13.5				
		3	13.5				
		4	10.8				
		CH$_2$	15.8				
		CH$_3$	9.4			SR, rapid internal rotation	

279

TABLE 5.2 (continued)

Spin Lattice Relaxation Mechanisms in Organic Molecules

Number	Compound	Carbon	$T_{1\,(obs)}$, s	η	$T_{1\,(DD)}$, s	Principal non DD mechanism	Reference
4	OH (neat) 	1	18.4	*ca.* 2		all DD	a
		2	2.8	2.0		all DD	
		3	2.8	2.0		all DD	
		4	1.9	2.0		all DD	

TABLE 5.2 (continued)

Spin Lattice Relaxation Mechanisms in Organic Molecules

Number	Compound	Carbon	T_1 (obs), s	η	T_1 (DD), s	Principal non DD mechanism	Reference
5		1	61	0.9		SR	a
		2	5.9	2.0		all DD	
		3	5.9	2.0		all DD	
		4	3.2	2.0		all DD	
6		1					c
		2	5.4			probably all DD	
		3	5.3			probably all DD	

281

TABLE 5.2 (continued)

Spin Lattice Relaxation Mechanisms in Organic Molecules

Number	Compound	Carbon	$T_{1\,(obs)}$, s	η	$T_{1\,(DD)}$, s	Principal non DD mechanism	Reference
6 (cont'd.)		4	1.1			DD dominates	
		α	25 MHz:82 63 MHz:15			SR	
		β	25 MHz:136 63 MHz:30			CSA	
7	cyclobutane		36	1.4	70	SR	d
8	cyclooctane		10.3	2.0		all DD	d

TABLE 5.2 (continued)

Spin Lattice Relaxation Mechanisms in Organic Molecules

Number	Compound	Carbon	$T_{1\,(obs)}$, s	η	$T_{1\,(DD)}$, s	Principal non DD mechanism	Reference
9	$(CH_3CH_2CH_2CH_2CH_2)_2$ (neat, 39°C)	1	7.1			SR	e
		2	7.4			probably all DD	
		3	6.3			probably all DD	
		4	5.4			probably all DD	
		5	5.1			probably all DD	

TABLE 5.2 (continued)

Spin Lattice Relaxation Mechanisms in Organic Molecules

Number	Compound	Carbon	$T_{1\,(obs)}$, s	η	$T_{1\,(DD)}$, s	Principal non_DD mechanism	Reference
10	$CH_3(CH_2)_8CH_2OH$ (neat, 42 °C)	1	0.65			probably all DD	f
		2	0.77			all carbons	
		3	0.77				
		4–6	*ca.* 0.8				
		7	1.1				

TABLE 5.2 (continued)

Spin Lattice Relaxation Mechanisms in Organic Molecules

Number	Compound	Carbon	T_1 (obs), s	η	T_1 (DD), s	Principal non DD mechanism	Reference
10 (cont'd.)		8	1.6				
		9	2.2				
		10	3.1				
11	$CHCl_3$		32			SR	g
12	$CHBr_3$		1.6	0		probably all S	g

T_1 (obs) (38°C)

η

285

TABLE 5.2 (continued)

Number	Carbon	T_1 (obs.) 38°C		η		CSA	Ref.
		22.6 MHz	67.9 MHz	22.6 MHz	67.9 MHz		
13	1,2,4,7,11* 12,15,16	0.39 ± .06	0.37 ± .03	1.9	1.9		h
	3,6,8,9,17*	0.70 ± .16	0.70 ± 0.10	1.9	1.9		
	10,13*	4.5 ± 0.2	4.2 ± 0.10	1.9	1.9		
	5	5.6	3.2	1.6	0.8	5.3	

REFERENCES

a. G. C. Levy, J. D. Cargioli, and F. A. L. Anet, *J. Am. Chem. Soc.* **94**, 699 (1972).
b. P. A. Mirau, private communication (1986).
c. G. C. Levy, J. D. Cargioli, and F. A. L. Anet, *J. Am. Chem. Soc.* **95**, 1527 (1973).

TABLE 5.2 (continued)

d. S. Berger, F. R. Kreissl, and J. D. Roberts, *J. Am. Chem. Soc.* **96**, 4348 (1974).

e. J. R. Lyerla, Jr., H. M. McIntyre, and D. A. Torchia, *Macromolecules* **7**, 11 (1974).

f. D. Doddrell and A. Allerhand, *J. Am. Chem. Soc.* **93**, 1558 (1971).

g. T. C. Farrar, S. J. Druck, R. R. Shoup, and E. D. Becker, *J. Am. Chem. Soc.* **94**, 699 (1972).

h. G. C. Levy and U. Edlund, *J. Am. Chem. Soc.* **97**, 5031 (1975).

* Limits are range for all carbons in the group.

rotation exhibits similar anisotropy to that shown by compounds 1−3. For biphenyl (entry 5) preference for rotation about the long axis is more marked while for the diphenyldiyne (entry 6) it becomes very strong. For the β-carbon, CSA is believed to be the major contributor as the ring protons are too far away (recall the inverse sixth power distance dependence) to make a significant DD contribution.

The behaviors of cyclobutane (entry 7) and of cyclooctane (entry 8) provide a significant comparison. The small size and rapid internal motion ("butterfly" ring inversion; see Chapter 4, Section 4.3.3) of cyclobutane permit a significant SR pathway, whereas the larger size and slower internal motion (Section 5.4.2) of cyclooctane lead to complete DD dominance.

The measurements of n-decane (entry 9) are part of a larger study of neat hydrocarbons, including also C_7, C_{12}, C_{15}, C_{18}, C_{20}, and 2-methyl-C_{19}.[16] A complete theory of the motion of such segmented molecules is not available. It will be seen that (ignoring a possible SR contribution to the methyl carbon relaxation) by applying Eq. (5.28) to these T_1 values, the correlation time increases from 2.2 ps (picoseconds) at the ends of the chain to 4.6 ps in the middle of the chain. As one proceeds up the series it is found that this trend becomes more marked; for the C_{20} hydrocarbon the correlation times of the center carbons are sevenfold greater than for the methyl carbons. One may best understand this behavior by dividing the complex motion of the chain into (a) overall tumbling, $1/\tau_o$, and (b) local segmental motion, $1/\tau_s$. For a particular carbon, carbon-i

$$\frac{1}{\tau_i(\text{obs})} = \frac{1}{\tau_{o-i}} + \frac{1}{\tau_{s-i}} \tag{5.48}$$

The motion represented by $1/\tau_{o-i}$ may be regarded as the same for all carbons in the chain regardless of position, as if the molecule were stiff and inflexible. The local motion is brought about by rotation about carbon−carbon bonds and is influenced by the mass and mobility of the groups attached to those bonds, that is, by proximity to the chain end. This division of motion is of course to some degree artificial; one may readily picture the overall tumbling as due to biased local segmental motion. Nevertheless, it has a certain reality in that for macromolecular chains of more than about 100−150 bonds the observed correlation time becomes independent of chain length (chain ends being ignored), overall motion becomes a negligible contributor to T_1 (see Chapter 7, Section 7.4), and all observable motion is segmental.

The more rapid motion near the chain end seems intuitively reasonable, since internal bond rotation is retarded by the necessity for some cooperation between adjacent bonds to avoid sweeping large chain segments through the solution. This effect becomes more marked with increasing chain length, as we have just seen.

The motion behavior of decanol-1 (entry 10) shows a significant contrast to that of *n*-decane. The motions of all carbons are slower because of the higher viscosity of the neat liquid, but those near the hydroxyl are strikingly retarded by the hydrogen bonding of the latter, which has a marked anchoring effect. A charged amino group has a similar effect but that of an uncharged amino group is negligible.[17]

Chloroform (entry 11) appears to be a case in which spin rotation is the principal contribution in addition to DD, whereas in bromoform (entry 12) scalar relaxation, arising from the large $^{13}C-^{79}Br$ and $^{13}C-^{81}Br$ J couplings (>100 Hz) and nuclear quadrupolar couplings (Table 5.1), dominates strongly. Substantial scalar relaxation of carbon−13 is rare and almost always arises from directly bonded bromine.

In the 3-chlorocholestene (entry 13) carbon-5 is in the right setting for CSA relaxation: no directly bonded proton, nearby unsaturation, and a large molecule. A threefold increase in field decreases both T_1 and η substantially but has no effect on the other carbons.

5.3 SPIN−SPIN RELAXATION

We have described the basic concept of spin−spin or transverse relaxation in Chapter 1 (Section 1.3) and have seen that T_2 is by definition an inverse measure of the linewidth

$$T_2 = \frac{1}{\pi \delta \nu} \tag{1.26}$$

Such line broadening in solids and viscous liquids is primarily caused by dipole−dipole interactions of the same kind we have seen to cause spin lattice or longitudinal relaxation (Section 5.2.1). Another contribution to line broadening is inhomogeneity, δB_o, in the laboratory magnetic field. This may be minimized by careful shimming of the field, discussed in Chapter 2, Section 2.2.5. It is usually included in the experimental values of the spin−spin relaxation time, commonly termed T_2^*:

$$\frac{1}{T_2^*} = \frac{1}{T_2} + \frac{\gamma \delta B_o}{2} \tag{5.49}$$

We have seen (Chapter 4, Section 4.4.4.1) that the "true" T_2 may be measured by the method of Hahn spin echos. In this method we apply first a 90_x° pulse, followed after a time interval τ (short compared to T_1) by a 180_x° pulse, which causes the diverging spin isochromats to refocus and produce an echo after a further interval τ (Fig. 4.39). The echo consists in fact of a free induction buildup to a maximum, followed by a free induction decay. It may be Fourier transformed (see Chapter 6, Section 6.2) or observed as such.

In the original Hahn experiment, it is necessary to repeat the $90_x^\circ - \tau - 180_x^\circ$ sequence several times with varied values of τ and then plot the decline of the echo magnitude as a function of τ to obtain τ_2. This is somewhat troublesome, and it may also be shown that the echo will decay not only as a function of "true" T_2 but also as a function of diffusion of the nuclei between portions of the sample experiencing slightly different magnetic fields. The effect of diffusion is more serious as τ is extended. Carr and Purcell[18] showed that the Hahn experiment may be modified to increase its convenience and at the same time greatly reduce the effect of diffusion. This method employs the sequence

$$90_x^\circ - \tau - 180_x^\circ - 2\tau - 180_x^\circ - 2\tau - 180_x^\circ - 2\tau ...$$

where the 180_x° pulses (after the first) are applied at 3τ, 5τ, 7τ, etc., giving rise to a train of echos at 2τ, 4τ, 6τ, 8τ ..., as may be seen from Fig. 4.39 (g), (h), ... The effect of diffusion is proportional to τ rather than to the accumulated interval and so may be made as small as desired by making τ short. It may also be seen from Fig. 4.39 that the echos will alternate in sign.

A yet further improvement in the Hahn echo sequence has been devised by Meiboom and Gill.[19] This is similar to the Carr–Purcell sequence except that the $180°$ pulses are applied along the *positive y' axis*. We have seen [Fig. 4.40 (e) and (f)] that this results in positive echos. It may be further shown (see Martin *et al.*[20] for a good discussion) that there is also an automatic correction for imperfections in the effective flip angle of the 180_{+y}° pulses. The echos obtained for a sample of $CH_3^{13}COOH$ (60% enriched at the carboxyl carbon) by the Carr–Purcell method and by the Meiboom–Gill modification are compared in Fig. 5.10.[20] It will be seen that in the latter the echos persist longer, giving a more correct and accurate measure of T_2.

An aspect of spin–spin relaxation of great interest is that of *chemical rate processes*. In such processes, the line broadening does not arise from

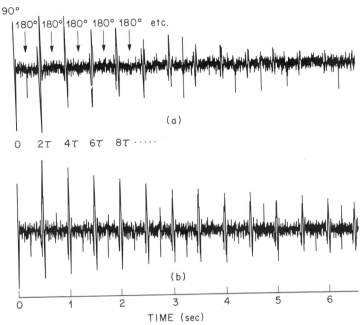

Fig. 5.10.   (a) Carr—Purcell echoes for 60% enriched $CH_3^{13}COOH$.
The signs of the echoes alternate. The 90° pulse is applied
at the start and the 180° pulses at τ, 3τ, 5τ, etc., as
indicated. (b) Echoes from the Meiboom—Gill modification
of the Carr—Purcell experiment; all are of the same sign and
persist longer than in (a). [From T. C. Farrar and E. D.
Becker, *Pulse and Fourier Transform NMR*, Academic
Press, New York, 1971, p. 26.]

dipole—dipole interactions, which are not normally affected by chemical
reactions, but from other causes that we now discuss.

5.4 RATE PROCESSES

5.4.1 Lifetimes and Linewidths

Molecules often exist in two (or more) coexisting states or subspecies
that are in rapid equilibration with each other. These may be different
conformations, different ionized forms, different positions of protonation,
complexed forms, etc. In the ultraviolet or infrared spectrum of such a
molecule, we may expect to see absorption peaks corresponding to each
state. If δt is the lifetime of each form (we shall suppose for the present

that these exist in equal amounts, and therefore, that δt is the same for both), then from the uncertainty principle, one may expect a broadening of the absorption peaks given by

$$\delta \nu \simeq \frac{1}{2\pi \cdot \delta t} \qquad\qquad (5.50)$$

At the high frequencies associated with optical spectra ($10^{13}-10^{15}$ s^{-1}), this broadening cannot be appreciable compared to the resolving power of optical spectrometers unless δt is very small, $10^{-12}-10^{-13}$ s or less. Conformations, complexes, and ionized states of organic molecules usually have lifetimes considerably longer than this. However, in NMR spectroscopy the observing frequency is one-millionth as great, spectral lines may be less than 1 Hz in width, and those corresponding to interchanging subspecies may be only a hertz apart. Very marked broadening, sufficient to merge the lines for each subspecies, may occur even when lifetimes are as long as $10^{-1}-10^{-3}$ s. If their lifetimes are much shorter than this, the two forms will give a single narrow peak and be entirely indistinguishable. These lifetimes ($10^{-1}-10^{-5}$ s) correspond to a range of rates of reaction and isomerization that is of great importance in chemistry. Furthermore, it is possible, as we shall see, to employ NMR for the measurement of rates that would be difficult or impossible to measure in any other way, particularly *virtual* reactions, where the reactant and the product are identical.

5.4.2 Modified Bloch Equations

Equation (5.50) provides only an approximate estimate of the line broadening produced by chemical rate processes. For a more detailed description of the spectral behavior, we must turn again to the Bloch equations (Chapter 1, Section 1.7).

Let us suppose that we are observing a particular magnetic nucleus involved in a rapid, reversible exchange between two different molecular sites or environments, A and B. We assume that in each of these sites this nucleus would, if the rate of exchange were sufficiently slowed, appear as a single resonance line; i.e., spin coupling to other nuclei will be taken as zero. Gutowsky, McCall, and Slichter[21] gave the first theoretical treatment of the line shape under these conditions. McConnell[22] used a somewhat simpler and more direct approach, which we shall employ in the following discussion.

Let us first consider that there is no exchange of magnetic nuclei between sites A and B, and that the spectrum therefore consists of two lines separated by a chemical shift difference of $2\pi(\nu_A - \nu_B)$ Hz or

$(\omega_A - \omega_B)$ rad/s. There are two resonant Larmor frequencies, ω_A and ω_B. In the rotating frame, the in-phase and out-of-phase components of magnetization in the x,y plane (p.   ) will be the sums of the contributions of the nuclei in the two sites (employing the notation of Section 1.7):

$$u = u_A + u_B \qquad (5.51)$$

$$v = v_A + v_B \qquad (5.52)$$

Likewise the z component will be given by

$$M_z = M_z^A + M_z^B \qquad (5.53)$$

We now permit exchange of the observed nucleus between the sites A and B. The residence times in each site are designated τ_A and τ_B; the rates of transfer from each site are given by first-order rate constants that are the reciprocals of these quantities:

$$A \underset{1/\tau_B}{\overset{1/\tau_A}{\rightleftarrows}} B$$

We note that if the fractional populations of the A and B sites are designated p_A and p_B, then clearly,

$$\frac{\tau_A}{\tau_B} = \frac{p_A}{p_B} = \frac{k_B}{k_A} \qquad (5.54)$$

where k_B is the first-order rate constant for the transfer from site B to site A, and k_A is the corresponding quantity for site A. From this relationship it follows that

$$p_A = \frac{\tau_A}{\tau_A + \tau_B} \qquad (5.55)$$

$$p_B = \frac{\tau_B}{\tau_A + \tau_B} \qquad (5.56)$$

It is assumed that the exchanges of nuclei between sites A and B take place in jumps that are very short in relation to the relaxation times of the nuclei when present in these sites, i.e., that no change in magnetization takes place during the actual transfer. Let T_{1A} and T_{1B} be the spin lattice relaxation times of the nucleus in sites A and B, respectively, and let T_{2A} and T_{2B} be the corresponding spin–spin relaxation times. The *total* spin lattice relaxation rates of the nucleus in sites A and B are given by

$$\frac{1}{T'_{1A}} = \frac{1}{T_{1A}} + \frac{1}{\tau_A} \tag{5.57}$$

$$\frac{1}{T'_{1B}} = \frac{1}{T_{1B}} + \frac{1}{\tau_B} \tag{5.58}$$

Corresponding equations describe the spin–spin relaxation rates in each site:

$$\frac{1}{T'_{2A}} = \frac{1}{T_{2A}} + \frac{1}{\tau_A} \tag{5.59}$$

$$\frac{1}{T'_{2B}} = \frac{1}{T_{2B}} + \frac{1}{\tau_B} \tag{5.60}$$

These equations express the fact that the transfer of the observed nucleus from A to B is, in effect, a contribution to its relaxation in site A, proportional to its rate of transfer; correspondingly, of course, its transfer from B contributes $1/\tau_B$ to its relaxation at B.

Using these definitions, we may then rewrite Eqs. (1.43)–(1.45) as modified for two exchanging sites by McConnell[22]:

$$\frac{du_A}{dt} + \frac{u_A}{T'_{2A}} - \frac{u_B}{\tau_B} + (\omega_A - \omega)v_A = 0 \tag{5.61}$$

$$\frac{du_B}{dt} + \frac{u_B}{T'_{2B}} - \frac{u_A}{\tau_A} - (\omega_B - \omega)v_B = 0 \tag{5.62}$$

$$\frac{dv_A}{dt} + \frac{v_A}{T'_{2A}} - \frac{v_B}{\tau_B} + (\omega_A - \omega)u_A + \gamma_A B_{1A} M_z^A = 0 \tag{5.63}$$

$$\frac{dv_B}{dt} + \frac{v_B}{T'_{2B}} - \frac{v_A}{\tau_A} - (\omega_B - \omega)u_B + \gamma_B B_{1B} M_z^B = 0 \tag{5.64}$$

$$\frac{dM_z^A}{dt} + \frac{M_z^A}{T'_{1A}} - \frac{M_o^A}{T_{1A}} - \frac{M_z^B}{\tau_B} - \gamma_1 B_1 v_A = 0 \tag{5.65}$$

$$\frac{dM_z^B}{dt} + \frac{M_z^B}{T_{1B}'} - \frac{M_o^B}{T_{1B}} - \frac{M_z^A}{\tau_A} - \gamma_1 B_1 v_B = 0 \qquad (5.66)$$

The term u/T_2 in Eq. (1.43) has been replaced by two terms in Eq. (5.61): u_A/T_{2A}' represents the rate of decrease of u_A due to transfer of spins, i.e., of magnetization, from A to B; u_B/τ_B is the rate of increase of u_A due to transfer of spins from B to A. The corresponding significance of terms in B is evident, as are the analogous equations, Eqs. (5.63) and (5.64) for the out-of-phase magnetization.

The v mode signal line shape will be given by a rather complicated expression.[23] We shall consider four special cases that are useful and can be more simply set down.

a. Slow Exchange. As we have said, if exchange of the observed nucleus between sites A and B is very slow, the spectrum will consist of two lines. The criterion for this condition is that τ_A and τ_B should be large compared to $(\omega_A - \omega_B)^{-1}$. Under these circumstances, it is found that for the signal of the A site

$$v_A = -M_o \frac{\gamma B_1 T_{2A}' p_A}{1 + T_{2A}'^2(\omega_A - \omega)} \qquad (5.67)$$

Comparison to Eqs. (1.47) and (1.25) shows that the width of the A signal will be (in hertz)

$$\pi \delta v_A = \frac{1}{T_{2A}} + \frac{1}{\tau_A} \qquad (5.68a)$$

keeping in mind the definition of T_{2A}' [Eq. (5.57)]. Similarly for site B

$$\pi \delta v_B = \frac{1}{T_{2B}} + \frac{1}{\tau_B} \qquad (5.68b)$$

Exchange thus results in an excess broadening, $(\pi \tau_i)^{-1}$, of the signal at each site over that observed in the absence of exchange. From Eq. (5.54), we see that if the signals are unequal in intensity the weaker one will be the more broadened. Equations (5.68a) and (5.68b) can be used to estimate rates if $1/T_{2A}$ and $1/T_{2B}$ are either known or negligibly small and if the signals are not so greatly broadened as to approach overlapping. The larger $(\omega_A - \omega_B)$, the greater the rate that can be measured in this way with reasonable accuracy.

It should be noted that if the signals are well separated τ_A and τ_B can be obtained without any knowledge of $(\omega_A - \omega_B)$. Indeed, if an observed peak is broadened by exchange with other sites, the rate of exchange with respect to the observed site can be obtained *without any knowledge of the number or nature of the other sites.*

b. Very Fast Exchange. Fast exchange for this purpose is defined as a rate such that τ_A and τ_B are small compared to $(\omega_A - \omega_B)^{-1}$. If the rate is such that the two signals are completely collapsed to a single resonance (which is not narrowed further no matter how rapid the exchange), we find

$$v = -M_o \frac{\gamma B_1 T_2'}{1 + T_2'^2 (p_A\omega_A + p_B\omega_B - \omega)^2} \tag{5.69}$$

This defines a signal centered at an averaged frequency:

$$v_o = p_A v_A + p_B v_B \tag{5.70}$$

characterized by an averaged spin–spin relaxation time:

$$\frac{1}{T_2'} = \frac{p_A}{T_{2A}} + \frac{p_B}{T_{2B}} \tag{5.71}$$

The linewidth is given by

$$\delta v = \frac{1}{\pi T_2'} \tag{5.72}$$

If the signal is not completely collapsed, i.e., if the process is slow enough to contribute to its width but is still well beyond the rate corresponding to separated signals, this approximation is useful:

$$\frac{1}{T_2''} = \frac{1}{T_2'} + p_A^2 p_B^2 (\omega_A - \omega_B)^2 (\tau_A + \tau_B)^2 \tag{5.73}$$

The second term on the right corresponds to the excess broadening due to slow exchange. For equal populations of sites A and B, we have $p_A = p_B = 1/2$ and $\tau_A = \tau_B = 2\tau$ and the excess broadening becomes (in hertz)

$$4\tau\pi(v_A - v_B)^2 \tag{5.74}$$

i.e., inversely proportional to the exchange rate.

c. *Equal Populations of A and B Sites.* Under the circumstances where $p_A = p_B$, a more exact solution than Eq. (5.74) is feasible. The line shape can be expressed in convenient form over the entire range of rates from two separated peaks, through "coalescence" into a broadened resonance and collapse to a narrow singlet. We shall, however, further assume that $1/T_{2A}$ and $1/T_{2B}$ can be neglected in comparison to separation of the peaks under conditions of slow exchange. We then find

$$v = -\frac{M_o}{4} \frac{\gamma B_1 \tau (v_A - v_B)^2}{[\frac{1}{2}(v_A + v_B) - v]^2 + 4\pi^2 \tau^2 (v_A - v)^2 (v_B - v)^2} \quad (5.75)$$

The meaning of τ under these conditions must be emphasized:

$$\frac{1}{\tau} = k = \frac{2}{\tau_A} = \frac{2}{\tau_B} = 2k_A = 2k_B \quad (5.76a)$$

Here, k is the rate constant for the reaction in *both* directions. Since the rates are the same in both directions, we may say that

$$k_1 = \frac{1}{2\tau} \quad (5.76b)$$

where k_1 is the pseudo-first order or "one-way" rate constant, i.e., what one normally wishes to determine. In Fig. 5.11, Eq. (5.75) is plotted for a number of values of the quantity $\tau(v_A - v_B)$. When this quantity is large, the spectrum appears as two peaks; when small, it appears as one. When it is equal to $\sqrt{2}/2\pi$, the peaks just merge to give the flat-topped resonance shown at (e). The single peak then narrows as $\tau(v_A - v_B)$ continues to decrease.

Figure 5.12 shows a plot of linewidth as a function of $\tau(v_A - v_B)$. This has general applicability to any system that can be described by Eq. (5.75).

d. *Unequal Populations of A and B Sites.*

$$v = (P + QR)/(4\pi^2 P^2 + R^2) \quad (5.77)$$

where

$$P = \tau\{-[\frac{1}{2}(v_A - v_B) - v]^2 + \frac{1}{4}(v_A - v_B)^2]$$

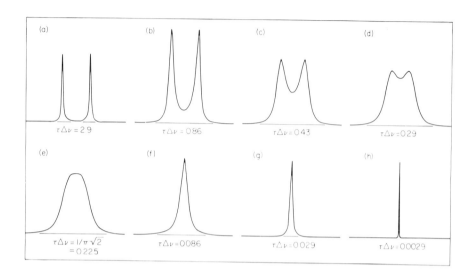

Fig. 5.11.   Line shapes calculated from Eq. (5.75) for varying values of
              $\tau(\nu_A - \nu_B) = \tau\Delta\nu$. (The seemingly rather arbitrary values of
              $\tau\Delta\nu$ arise from the fact that these curves were calculated
              specifically for the inversion of d_{11}-cyclohexane, where
              $\Delta\nu = ca.$ 29 Hz at 60 MHz.)

$$Q = \tau[\tfrac{1}{2}(\nu_A + \nu_B) - \nu - \tfrac{1}{2}(p_A - p_B)(\nu_A - \nu_B)^2]$$

$$R = \tau(\nu_A + \nu_B) - \nu + \tfrac{1}{2}(p_A - p_B)(\nu_A - \nu_B)$$

and

$$\tau = \tau_A\tau_B/(\tau_A + \tau_B)$$

Nearly all earlier work on chemical kinetics by NMR centered on the
proton, with occasional observation of the ^{19}F nucleus where appropriate.
An extensive review to about 1973 is to be found in *Dynamic Nuclear
Magnetic Resonance Spectroscopy*, L. M. Jackman and F. A. Cotton, Eds.,
Academic Press, New York, 1975. More recently, carbon−13 has been
employed. Because of its much greater chemical shift range, faster
reactions can be measured.[24]

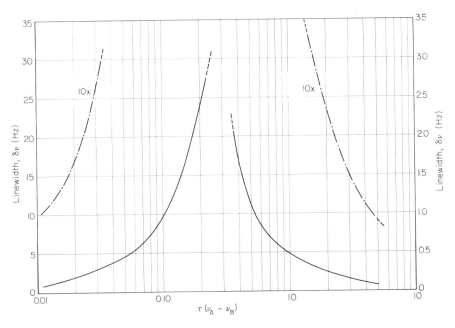

Fig. 5.12.   Line width as a function of $\tau(\nu_A - \nu_B)$ for equal populations of A and B sites. The solid curve on the right represents the line width for either member of the doublet, i.e., below coalescence. The solid curve on the left represents the line width above coalescence. The left-hand scale is to be used for both solid curves. The dot-dashed curves are 10× expansions of the solid curves, useful for relatively narrow lines, and are to be read on the right-hand scale.

5.4.3 Ring Inversions

Among the many types of rate processes amenable to NMR study, the inversion of cyclic molecules has attracted particular attention and been much studied and reviewed.[25] We consider first the inversion of *cyclohexane*. The measurement of this rate depends on the fact that the equatorial and axial protons have considerably different chemical shifts (Chapter 3, Section. 3.3.4.6), the equatorial protons appearing at lower field. (This is an example of a process that cannot be studied by carbon−13 NMR, although other ring inversions can be.) Upon inversion of the ring, they exchange environments.

At normal temperatures, this process is very rapid and all the protons appear as a single narrow peak. At temperatures low enough so that the residence time in each chair conformation becomes larger than the reciprocal of the frequency difference $\Delta \nu$ between $H_{(eq)}$ and $H_{(ax)}$, one may expect peaks for each to appear. This is indeed the case, but the low-temperature proton spectrum of cyclohexane itself is highly complex,[26] being actually that of an A_6B_6 spectrum (loosely so described, as the protons in each group are not magnetically equivalent). Spectral complexity can be avoided by replacing all but one of the protons by deuterium. The deuterium proton couplings can then be removed by irradiating the deuterium nuclei.[27-29] The spectrum of d_{11}-cyclohexane at $-89°C$, with deuterium irradiation, consists of two sharp peaks, as shown in Fig. 5.13.[29] As the temperature is raised, the peaks become broader. At $-75°C$, the widths of the peaks and the application of Eq. (5.68) indicates that τ is of the order of 0.04 s. At $-60°C$, the peaks just merge to a single line; at this temperature, therefore, $\tau = (\pi \Delta \nu \sqrt{2})^{-1}$ or *ca.* 0.008 s. As we continue to raise the temperature the single line continues to narrow until at room temperature we see the very narrow line (*ca.* 0.3 Hz, determined principally by field inhomogeneity) characteristic of small molecules.

In the Eyring formulation, the residence time τ may be expressed [see Eq. (5.76b)] as

$$\frac{1}{2\tau} = k_1 = \kappa \, \frac{kT}{h} \, e^{\Delta S_{cc}^{\neq}/R} \cdot e^{-\Delta H_{cc}^{\neq}/RT} \tag{5.78}$$

$\Delta S_{cc}^{\neq}$ and $\Delta H_{cc}^{\neq}$ are the activation enthalpy and energy for the overall chair-to-chair reaction path. From this we have

$$\Delta G_{cc}^{\neq} = 2.303 \, RT (10.319 + \log T - \log k_1) \tag{5.79a}$$

$$\Delta H_{cc}^{\neq} = - R \left[\frac{\partial \ln k_1}{\partial (1/T)} \right]_p - RT \tag{5.79b}$$

$$\Delta S_{cc}^{\neq} = (\Delta H^{\pm} - \Delta G^{\neq})/T = -(\partial \, \Delta G^{\neq}/\partial T)_p \tag{5.79c}$$

TABLE 5.3

Peak Positions, Line Widths, Peak-to-Valley Ratios, and Kinetic Data Versus Temperature for d_{11}-Cyclohexane[a]

Temperature (°C)	Peak position (ppm from TMS)[b]			ω (Hz)[c]	Peak-to-valley ratio	k_1 (s⁻¹)[d] (= 1/2τ)	$\Delta G^{\neq}_{cc}$ (kcal/mole)
	Equatorial	Averaged	Axial				
−47.0		1.37		5.3 ± 0.1		24_2	10.6_0
−48.7		1.37		5.2 ± 0.1		25_0	10.5_0
−52.0		1.37		6.1 ± 0.1		21_5	10.4_2
−53.2		1.36		8.1 ± 0.2		16_3	10.4_8
−57.0		1.36		15.3 ± 0.3		97	10.5_0
				16.1 ± 0.3		94	10.5_2
				26.2 ± 0.4		69	10.5_0
−60.3		1.35					
−61.3	1.48		1.25		1.08 ± 0.03	54	10.5_5
					1.31 ± 0.01	45	10.5_3
−63.2	1.51		1.21		1.22 ± 0.04	$47._5$	10.4_5
−64.5	1.54		1.21		1.70 ± 0.07	$37._3$	10.5_5
−67.3	1.56		1.13	12.0 ± 0.2	2.57 ± 0.10	$30._7$	10.4_8
						$29._4$	10.5_0
−67.8	1.57		1.13	7.9 ± 0.2	3.89 ± 0.10	$23._6$	10.5_3
						$23._0$	10.5_4
−75.0	1.60		1.12	4.1 ± 0.1	11.0 ± 0.2	(12.9)	10.3_9)
					11.0 ± 0.2	(13.4)	(10.3_8)

[a] 50 vol. % in CS₂; 2% tetramethylsilane as reference.

[b] Error ± 0.01.

[c] Corrected for reference peak width.

[d] Chair-to-chair.

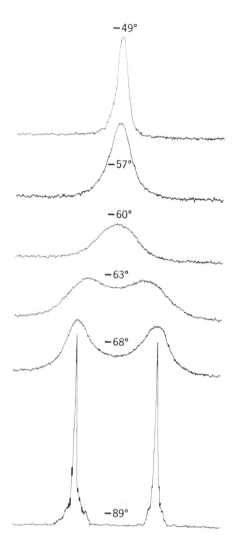

Fig. 5.13.   The spectrum of the lone proton of d_{11}-cyclohexane (50% vol./vol. in CS_2) as a function of temperature (°C) observed at 60 MHz with deuterium irradiation at 9.4 MHz.

In Table 5.3 are given the peak positions, widths, and "peak-to-valley" ratios observed for the C_6HD_{11} system as a function of temperature. These are not necessarily the most accurate data of this sort, but are given in full as being fairly typical. The peak width, which is the primary experimental observation, has been corrected for instrumental broadening by subtracting the observed linewidth of the tetramethylsilane reference peak, which, of course, is not subject to kinetic broadening. This procedure obviates the need to use the more complex version of Eq. (5.75) involving T_2' and, though not rigorous, is adequate when $1/T_2'$ does not represent a major fraction of the linewidth. From the variation of k_1 with temperature, $\Delta H_{cc}^{\neq}$ is obtained using Eq. (5.79b) and is found to be equal to 10.5 ± 0.5 kcal; a plot of the data is shown in Fig. 5.14. $\Delta G_{cc}^{\neq}$ is obtained at each temperature from Eq. (5.79a), k_1 being obtained from the line shape and Fig. 5.12. $\Delta S_{cc}^{\neq}$ is obtained from Eq. (5.79c), and it can be seen from the last column that it is nearly zero, as there is no observable dependence of $\Delta G_{cc}^{\neq}$ on T.

The inversion of many six-membered rings, including cyclohexane, presents some special features which, while not peculiar to NMR measurements as such, are appropriate to consider here, in order that $\Delta G_{cc}^{\neq}$

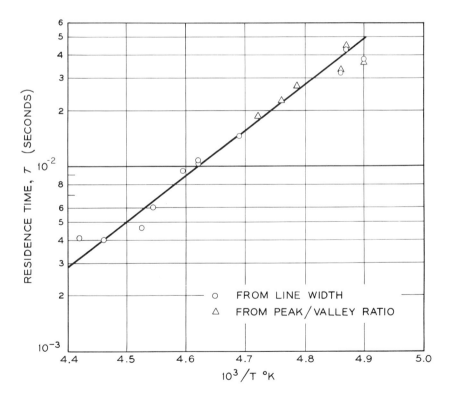

Fig. 5.14.   Arrhenius plot of $\tau(=1/2k_1)$ versus $1/T$ for the lone proton of d_{11}-cyclohexane.

and $\Delta S_{cc}^{\neq}$ may be fully interpreted. In the calculation of these quantities, it is supposed that only one barrier is present, and that the transmission coefficient κ is unity, i.e., that no molecule moving along the reaction coordinate is ever reversed in its path and returned to the starting form. It is recognized, however, that this picture is oversimplified and that the most likely path for chair-to-chair isomerization is through a boat form as a metastable intermediate. This boat form is flexible, in contrast to the rigid chair form, and therefore has a higher entropy than the chair ground state, but because it involves a complete or partial eclipsing of the $C-H$ bonds it has a torsional energy (plus additional van der Waals interactions) of about 5 kcal referred to the chair form. Potential energy calculations[30] indicate that the most probable conformation of the transition state is the "twist-boat" form shown in Fig. 5.15 rather than the usual boat form. The existence of this metastable state means that there are, in effect, two barriers that an inverting molecule must surmount in passing from chair to chair form, the boat form representing the valley between them. We may assume that a molecule in the boat form has an equal chance of passing over either barrier, the interconversion of boat forms being much more rapid than the boat-to-chair isomerization. The rate of chair-to-boat isomerization (in a given direction), k_{cb}, must equal $2k_{cc}$, the observed chair-to-chair rate, since half the boat-form molecules return to the ground state over the same barrier they have just surmounted. It follows that

$$k_{cb} = 2k_{cc} = \frac{2kT}{h} e^{\Delta S_{cc}^{\neq}/R} \cdot e^{-\Delta H_{cc}^{\neq}/RT} \tag{5.80}$$

κ being assumed to be unity for the path from chair to boat over each barrier. We thus have

$$\Delta S_{cb}^{\neq} = \Delta S_{cc}^{\neq} + R \ln 2 \tag{5.81a}$$

$$\Delta G_{cb}^{\neq} = \Delta G_{cc}^{\neq} - RT \ln 2 \tag{5.81b}$$

$\Delta H_{cc}^{\neq}$ is not affected by these statistical considerations and so $\Delta H_{cc}^{\neq} = \Delta H_{cb}^{\neq}$. We thus have

$$\Delta G_{cb}^{\neq}(206.0^\circ K) = 10.2 \text{ kcal}$$

$$\Delta H_{cb}^{\neq} = 10.5 \pm 0.5 \text{ kcal}$$

$$\Delta S_{cb}^{\neq} = 1.4 \pm 1.0 \text{ cal-mole}^{-1}\text{K}^{-1}$$

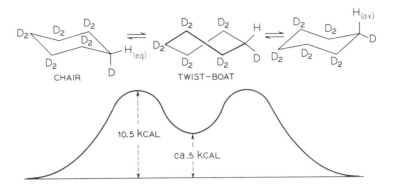

Fig. 5.15.   Inversion barriers for cyclohexane.

In this system, the conformers are not identical, but are of equal energy, as indicated by the equal intensity of the peaks. There is no appreciable isotope effect. In monosubstituted cyclohexanes, however, the conformer having the equatorial substituent has a substantially lower energy, and we therefore deal with exchanging populations which are unequal:

This complicates the interpretation, but by employing the appropriate line—shape function (see Eq. 5.77), rates for the inversion in each direction can be measured.

Four-membered rings undergo a butterfly inversion (see Chapter 4, Section 4.3.3) that is too fast to measure by NMR. The rates of pseudorotation of cyclopentanes are likewise not accessible. Inversion of rings larger than cyclohexane is generally faster than that of cyclohexane itself.[31] Thus, cycloheptane appears to be too fast to measure, while cyclooctane-d_{14} exhibits a $\Delta G^{\neq}$ of 8.1 kcal; and C_9, C_{10}, and C_{12} cycloparaffins have $\Delta G^{\neq}$ values of 6—7 kcal, about as low as can be reliably estimated by NMR. Stiffening elements such as double bonds, carbonyl groups, or substituents — including fluorines — retard inversion. The inversion of cycloheptatriene (tropilidene) can be measured by observing the partial collapse of the multiplet corresponding to the methylene group. The free energy of activation $\Delta G^{\neq}$ is only 6.1 kcal, again at the lower limit of barriers measurable in this way.

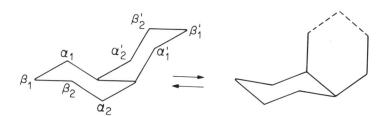

The first example of the use of carbon−13 for the measurement of ring inversion was the study of *cis*-decalin by Dalling *et al.*[32] It furnishes a dramatic example of a case where proton NMR is entirely ineffectual because the proton spectrum hardly changes with temperature. At temperatures above about 40°C, the cyclohexane rings in *cis*-decalin are inverting so rapidly that there are in effect only two types of CH_2 groups on the NMR time scale, which may be designated as α and β:

These resonances are shown in Fig. 5.16(a), observed at 50°C.[33] At −80°C [Fig. 5.16(b)] both the α- and β-carbon resonances are split because the α_1 and α_2 carbons, and likewise the β_1 and β_2 carbons are not equivalent when the molecule is inverting at a rate slow compared to their chemical shift differences. As represented in the preceding structural formula (left), C_{α_1} experiences gauche interactions with both $C_{\alpha'_1}$ and $C_{\beta'_2}$ of the other ring, whereas C_{α_2} experiences trans interactions with $C_{\alpha'_2}$ and $C_{\beta'_1}$ (intraring interactions are all gauche for all carbons and are ignored). The bridge-head carbons see the same environment and their resonance at 36.0 ppm is not affected by the "freezing" of the conformation. The inversion is characterized by an activation enthalpy $\Delta H^{\neq} = 12.3 \pm 0.11$ kcal mol^{-1} and an activation entropy $\Delta S^{\neq} = 0.15 \pm 0.44$ cal mol^{-1}−K^{-1}.[34] These values were obtained by combining line shape analysis with saturation transfer (see Section 5.4.7.1).

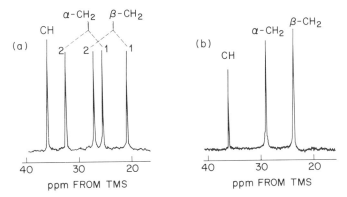

Fig. 5.16.   Carbon−13 spectra of *cis*-decalin in a mixture of toluene and perdeuterotoluene at (a) 50°C and (b) −80°C. The bridgehead CH signal at 36 ppm is unaffected by temperature. [From B. E. Mann, *J. Mag. Res.* **21**, 18 (1976); *Prog. Nucl. Magn. Reson. Spectrosc.* **11**, 79 (1977).]

5.4.4 Rotation about "Single" Bonds

At room temperature, the spectrum of dimethylformamide shows a separate peak for each methyl group (Chapter 3, Section 3.3.4.5). The amide group is planar, and the partial double−bond character of the C−N bond causes exchange of the methyl groups to be much slower than might be expected:

$$O{=\!\!\!\!\!=}C-N\begin{smallmatrix}CH_{3(a)}\\CH_{3(b)}\end{smallmatrix} \rightleftharpoons O{=\!\!\!\!\!=}C-N\begin{smallmatrix}CH_{3(b)}\\CH_{3(a)}\end{smallmatrix}$$

We have seen (Chapter 3, Section 3.3.4.5) that the methyl group cis to the carbonyl is the more shielded. As we raise the temperature, the sharp peaks observed at room temperature begin to broaden (Fig. 5.17) and at 118°C they coalesce. For the neat liquid, we find from these data and application of Eq. (5.79a) a value for $\Delta G^{\neq}$ of 20.2 kcal mol^{-1} at 118°C. Nearly the same value is found at the other temperatures, indicating that $\Delta H^{\neq}$ is *ca.* 20 kcal. and $\Delta S^{\neq}$ is nearly zero.

Restricted rotation is also observed in:

$$O{=\!\!\!\!\!=}N-N\begin{smallmatrix}CH_3\\CH_3\end{smallmatrix} \qquad\qquad R{-}O{\diagdown}N{\diagup}O \rightleftharpoons R{-}O{\diagdown}N{=\!\!\!\!\!=}O$$

N,N'-Dimethylnitrosamines   Alkyl nitrites

$$\underset{H}{\overset{R}{>}}C=N\diagdown^{OH} \rightleftharpoons \underset{H}{\overset{R}{>}}C=N\diagdown_{OH}$$

Oximes

In the alkyl nitrite and the oxime isomerizations, the populations of the two conformers are not necessarily equal. Jackman[35] reviewed the extensive proton literature on kinetics of amides and related compounds; an earlier review is that of Stewart and Siddall.[36] Carbon−13 NMR may also be employed and sometimes is advantageous. For example, the rate of rotation about the peptide bond in acetyl-L-proline

cis (c)   trans (t)

and glycyl-L-proline (in D_2O solution)

cis (c)   trans (t)

has been measured by Cheng and Bovey[37] employing carbon−13 resonances: for acetyl-L-proline, the acetyl carbonyl carbon (the other cis and trans resonances do not coalesce below $100°C$), and for glycyl-L-proline, the glycyl C_α carbon. For such compounds the proton resonances

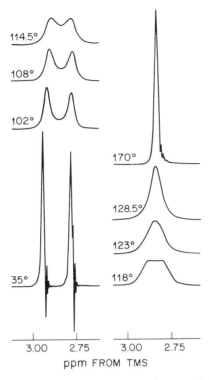

Fig. 5.17.   The methyl proton spectrum of N, N'-dimethylformamide
(neat) as a function of temperature (°C), observed at
60 MHz.

are too complex because of proton–proton coupling, but of course ^{13}C-
proton multiplicity may be removed by proton irradiation. From the
relative magnitudes of the cis and trans carbon resonances, it is found that

$$\ln K° = 496/K - 1.17 \tag{5.82}$$

where $K° = $ [trans]/[cis]; thus at 300 K (27 °C) $K° = 1.61$ and at 373 K
(100 °C) $K° = 1.17$. The trans conformer is favored by enthalpy but
disfavored by entropy. From line shape analysis it is found

	$\Delta H_{t \to c}^{\neq}(kcal\ mol^{-1})$	$\Delta S_{t \to c}^{\neq}(cal\ mol^{-1}\ K^{-1})$
Acetyl-L-proline	$19.5 \pm 0.1_0$	0.0
Glycyl-L-proline	$19.9 \pm 0.0_5$	0.0

These activation parameters are not significantly different from those of N,N-dimethylformamide. Indeed, all amides and peptides show very nearly the same activation barrier. Measurements by carbon–13 NMR have been reviewed by Mann.[38]

Benzaldehyde is believed to be planar and the $C_1 - C_\alpha$ bond may be expected to have considerable double–bond character because of resonance:

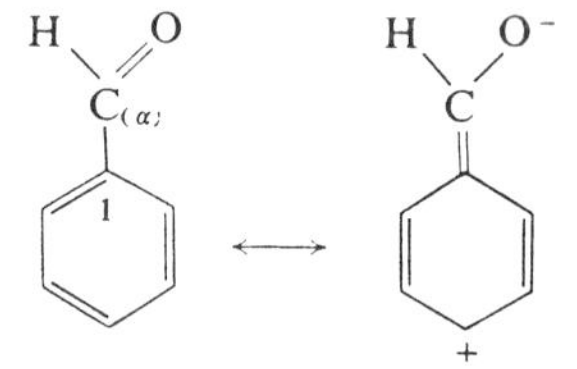

Anet and Ahmad[39] have observed that in 4-N,N'-dimethylamino-3,5-d_2-benzaldehyde, the ring protons give a single line at room temperature (the deuterium nuclei being irradiated

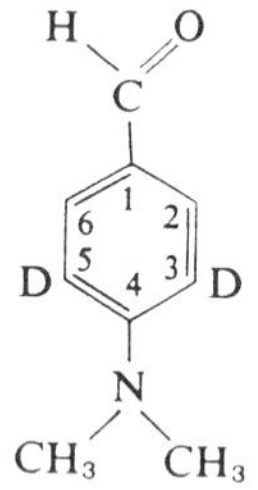

to eliminate their coupling), but below $-85°$ appear as an AB quartet ($\Delta\nu_{AB} = 8.4$ Hz at 60 MHz; $J = 2.5$ Hz). Rotation of the aldehyde group is now slow, and, being coplanar with the ring, it causes the 2 and 6 protons to be nonequivalent; $\Delta G^{\neq}$ for this process is 10.8 kcal.

Analogous behavior is shown by benzaldehyde itself, but only below about $-123°C$ as the barrier is lower (7.9 kcal). Carbon–13 NMR is more favorable for this measurement because the *ortho* carbons are very sensitive to the orientation of the carbonyl group; it is found that $\Delta H^{\neq} = 8.3 \pm 0.2$ kcal mol^{-1} and $\Delta S^{\neq} = 3.6 \pm 1.2$ cal mol^{-1}-K^{-1}.[40,41]

In the compounds so far discussed, the rotation occurs about bonds with some degree of double bond character. For $sp^3 - sp^3$ bonds in substituted ethanes, rotational barriers are usually only about 3 kcal mol^{-1} and spectra of the conformers cannot be seen at the lowest temperatures that can be attained in the NMR spectrometer, about $-150°$C. Information can in principle be obtained even for such compounds if the conformer populations change with temperature, that is, if they differ appreciably in energy, conformer. But one must be sure that temperature affects only the populations and not the chemical shifts *per se* for each conformer. As we shall see, this cannot be guaranteed, particularly for ^{19}F nuclei.

Multiple substitution by fluorine and larger halogen atoms may raise the barrier to the 8—10 kcal level, which permits "freezing out" the individual conformer spectra. A simple example is provided by difluorotetrachloroethane,[42] which gives a single peak at room temperature; at low temperatures (below *ca.* $-75°$C), this splits into two peaks corresponding to the trans and two equienergy mirror-image gauche conformers.

The two fluorine nuclei are identical within each conformer, and so no splitting arising from coupling can be seen. The larger peak at 67.98 Φ^* (see Chapter 9, Section 9.1.1) is presumed to represent the statistically favored gauche conformer. The barrier is approximately 9.6 kcal. From Fig. 5.18 it is evident that the ^{19}F chemical shifts in each conformer change markedly with temperature; the conditions indicated in the previous paragraph clearly do not hold, and an attempt to obtain the populations and NMR parameters of the conformers from the temperature dependence of Φ^* above coalescence could not succeed. Indeed, the ^{19}F chemical shift at 30°C falls *outside* those of the individual conformers and is clearly far from the weighted average one might ordinarily assume. It is very possible that proton chemical shifts may show similar (but smaller) deviations.

This and a number of more complex cases are discussed by Newmark and Sederholm,[43] who developed a computer program, based on the Bloch—McConnell equations and Eyring rate theory, which permits calculation of spectra of substituted ethanes for varying assumed values for the heights of the rotational barriers. Barriers are determined by the choice that best matches the observed spectra as a function of temperature. The reader should also consult S. Sternhell in *Dynamic Nuclear Magnetic*

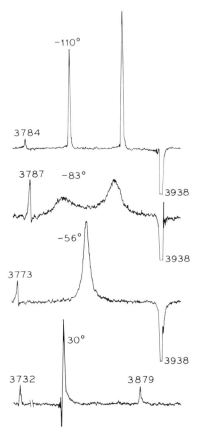

Fig. 5.18. ^{19}F spectra of $CFCl_2 \cdot CFCl_2$ in CCl_3F as a function of temperature (°C), observed at 56.4 MHz. At 30°, the ^{19}F spectrum is a single peak, the position of which is given by the CCl_3F audio side bands on each side, expressed in hertz upfield from the principal CCl_3F resonance (not shown). At −110° two sharp peaks are observed corresponding to trans and guache conformers.

Resonance Spectroscopy, L. M. Jackman and F. A. Cotton, Eds., Academic Press, New York, 1975, pp. 164−170.

5.4.5 Proton Exchange

Proton exchange may be measured by NMR in three ways:

a. By use of D_2O or other solvents containing labile deuterons, exchange with substrate protons may be measured analytically by observing the disappearance of the substrate proton peaks. This method does not actually exploit the rate-dependent characteristics inherent in NMR spectra but is useful for relative slow exchange processes ($k < ca.$ $10^{-2}s^{-1}$).

b. By observation of the broadening and possible coalescence of substrate and solvent peaks. This requires little elaboration.

c. By observation of the collapse of spin multiplets.

So far, we have dealt with systems in which spin coupling can be neglected in order to avoid interpretive complications. Actually, however, the collapse of spin multiplets can be employed to measure rates of exchange of magnetic nuclei even between equivalent sites. For example, the resonance of the hydroxyl proton, $H_{(1)}$, of pure isopropyl alcohol, $(CH_3)_2CH_{(2)}OH_{(1)}$, is a doublet because of coupling to $H_{(2)}$. But, upon addition of a trace of acidic catalyst, the peaks broaden, and, if the quantity of catalyst is large enough, the doublet may collapse to a singlet. This spectrum can be described by an equation analogous to Eq. (5.75) but with $(\nu_A - \nu_B)$ replaced by J_{12}. The reason for this behavior is, of course, that the hydroxyl protons undergo intermolecular exchange, and, in the presence of a sufficient quantity of catalyst, their residence time on any one molecule is so short compared to $1/J_{12}$ that the coupling to $H_{(2)}$ is in effect "turned off." Following the departure of a hydroxyl proton, the chances are nearly even that the next proton to arrive at that site will be in the other spin state. When this occurs rapidly, no appreciable coupling is possible.

Similarly, the hydroxyl proton of pure ethyl alcohol is a 1:2:1 triplet that collapses in the presence of acidic impurities, as we have already seen (Chapter 3, Section 3.3.5). Arnold[44] has calculated the line shape for this case. Results are shown in Fig. 5.19 for three values of τ.

The exchange of ammonium protons in aqueous solutions of methylammonium ion, $CH_3NH_3^+$, has been measured by Grunwald, Lowenstein, and Meiboom[45] by observing the collapse of the proton quartet of the methyl group, arising from coupling to the three NH_3^+ protons. They calculated the methyl line shape for several values of τJ and allowed for a finite spin—spin relaxation time in the absence of exchange. In Fig. 5.20 some results, covering a 64-fold range of values of τ, each at two values of $4\pi/T_2J$ are shown. These results are, of course, general for the collapse of any binomial quartet by an analogous exchange process.

5.4.6 Bond Isomerization

5.4.6.1 Cyclooctatetraene

Cyclooctatetraene has alternate single and double bonds in a tub-shaped nonaromatic structure. It can undergo isomerization by two distinct pathways, (a) bond shift and (b) ring inversion:

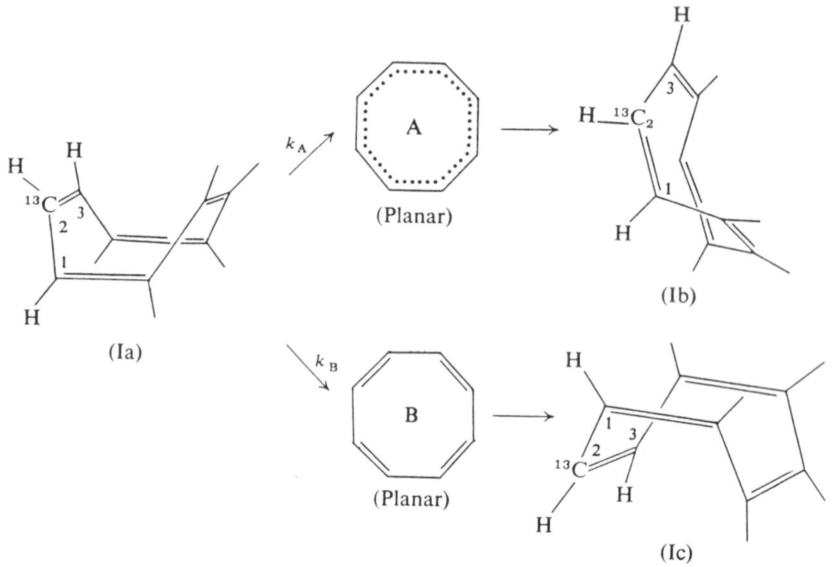

Let us assume that the bonds shift by the upper path, i.e., via the planar form A. If k_A were small, a given proton, for example, that on $C_{(2)}$, would experience a fairly large cis olefinic coupling (ca. 12 Hz) to $H_{(3)}$ and a weak coupling to $H_{(1)}$, the dihedral angle between the $C_{(2)}-H_{(2)}$ and $C_{(1)}-H_{(1)}$ bonds being about 90°. These couplings cannot be directly measured because all the protons are equivalent under all conditions. As we have seen, however, there is a natural abundance of about 1.1% of ^{13}C in carbon compounds. In a C_8 compound, about 8% of the molecules will have one ^{13}C atom. (The chances of having two ^{13}C atoms in the same molecule are too small to concern us.) We shall suppose this ^{13}C to be at position 2; this is, of course, permissible since all carbon positions are equivalent. The ^{13}C satellite peaks (Chapter 4, Section 4.3.6.1) will each appear as a doublet with spacing J_{23}, each member of which is split by J_{12}. Anet[46] has observed that at $-55°$ each ^{13}C satellite of cyclooctatetraene does indeed appear as a doublet of doublets, with $J_{23} = 11.8$ Hz and J_{12} ca. 2 Hz. At room temperature, this is replaced by a more complex

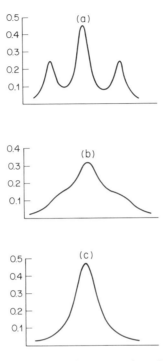

Fig. 5.19.   Collapse of a 1:2:1 triplet as a function of τJ; (a) $\tau J = 1.00$; (b) $\tau J = 0.250$; (c) $\tau J = 0.0625$. [From J. T. Arnold, *Phys. Rev.* **102**, 136 (1956).]

spectrum that can be fully described only as an 8-spin system. The partial collapse of the doublet permits an estimate of about 13.7 kcal for the $\Delta G^{\neq}$ of the process corresponding to k_A.

Cyclooctatetraene may also undergo inversion via the planar form B. This process has no effect on the couplings J_{12} and J_{13} and, therefore, cannot be observed for cyclooctatetraene itself. It can be studied by attaching a group such as

$$CH_3\diagdown \underset{/}{\overset{}{C}}\diagup^{CH_3}_{\diagdown OH}$$

to the ring. If the ring were static or only undergoing process A, the methyl groups would be nonequivalent (Chapter 3, Section 3.3.7) and would show two signals. These would collapse to a singlet when the ring

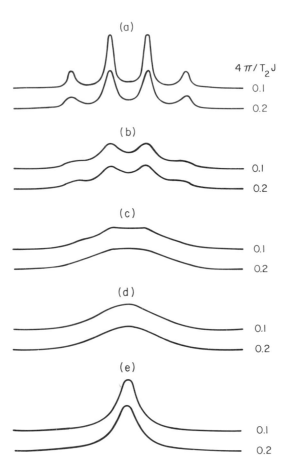

Fig. 5.20.   Collapse of a 1:3:3:1 quartet as a function of τJ, each at two
different values of $4\pi/T_2J$, where T_2 is the spin–spin
relaxation time; $2\pi\tau J$ is: (a) 8, (b) 2, (c) 1, (d) 0.5, and (e)
0.125. [From E. Grunwald, A. Lowenstein, and S. Meiboom,
J. Chem. Phys. **27**, 630, 641 (1957).]

inverts but would not be affected by process A. In this way Anet[47] found
that process B is considerably slower than bond exchange, $\Delta G^{\neq}$ being
about 2 kcal greater than for A.

The isomerization of cyclooctatetraene and its metal complexes has
been extensively studied by both proton and carbon-13 NMR and the
subject has been reviewed by Anet[48] and Mann.[49]

5.4.6.2 Bullvalene

A fascinating problem in bond shift and structure fluctuation is provided by bullvalene, a compound proposed by Doering and Roth[50] and synthesized by Schröder.[51] Bullvalene has the structure of a trivinylcyclopropane with two different kinds of aliphatic protons and two different kinds of olefinic protons.

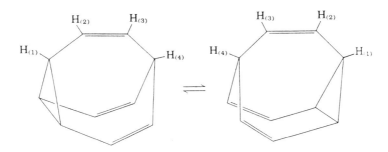

At room temperature, the proton spectrum shows only a single peak, which is about 60 Hz in width at 60 MHz; see Fig. 5.21.[52] This peak narrows rapidly as the temperature is raised, and is only 3 Hz in width at 83°C (not shown in Fig. 5.21). When the temperature is lowered below about 10°C, the single peak becomes resolved into two peaks. At -20°C, these peaks are quite narrow, that at lower field (5.6 ppm) corresponding to $H_{(2)}$ and $H_{(3)}$, and the other (2.2 ppm) to $H_{(1)}$ and $H_{(4)}$; the two types of protons appearing in each peak are very similar in chemical shift, and couplings between the olefinic and aliphatic protons are weak.

It is evident that bullvalence has a fluctuating structure and undergoes rapid degenerate Cope rearrangements between two equivalent forms. In any single rearrangement, one $H_{(2)}$ proton is converted to an $H_{(3)}$ proton and vice versa; and the $H_{(4)}$ and remaining $H_{(1)}$ are interconverted. Saunders[52] has devised a computer program to describe this complex system and, by matching its output to the observed spectra, has found that this rearrangement proceeds with a normal frequency factor, i.e., $\Delta S^{\neq}$ is *ca.* zero, and with an activation energy of only 11.8 kcal. This is remarkably small for a reaction involving the fission of carbon—carbon double bonds.

5.4.7 Other Methods of NMR Rate Measurement

The analysis of line shapes is direct and appealing and has been the dominant method for the study of rate processes. However, there are other NMR methods. We shall not treat these in detail but the following

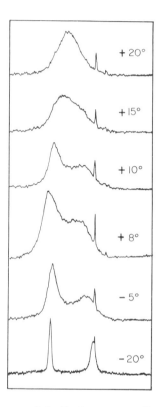

Fig. 5.21.   Proton spectra of bullvalene as a function of temperature
                (°C). The sharp peak appearing at the right arises from an
                impurity. (Spectra reproduced through the courtesy of
                M. Saunders.)

deserve mention. They can often enable rates to be measured that are too
slow for line shape analysis.

5.4.7.1 Saturation Transfer

If we have a system in which a magnetic nucleus is reversibly
transferred between nonequivalent sites A and B, and if we strongly
irradiate signal B so as to disturb the spin populations at this site, we may
expect this disturbance to be felt likewise at site A, provided that the spin
lattice relaxation time at site A is not negligibly short compared to the
time of transfer between A and B. We expect that upon irradiation of B,
signal A will decay to a new equilibrium value and will grow back to its
original intensity when irradiation of B ceases. It has been shown by

Forsén and Hoffman[53,54] that the decay of the magnetization of A is given by

$$M_A^Z(t) = M_A^o \left[\frac{\tau_{1A}}{\tau_A} e^{-t/\tau_{1A}} + \frac{\tau_{1A}}{T_{1A}} \right] \qquad (5.83)$$

where M_A^o is the normal equilibrium magnetization at A (see Section 5.2.1.1), T_{1A} is the spin—lattice relaxation time at A, and

$$\frac{1}{\tau_{1A}} -= \frac{1}{\tau_A} + \frac{1}{T_{1A}} \qquad (5.84)$$

τ_A being the residence time at site A. From Eq. (5.79) it is found that the new equilibrium magnetization at A under irradiation of B is given by

$$M_A^{o'} = M_A^o \left[\frac{\tau_{1A}}{T_{1A}} \right] = M_A^o \left[1 - \frac{\tau_{1A}}{\tau_A} \right] \qquad (5.85)$$

From the measurement of $M_A^{o'}$ and M_A^o, τ_{1A}/T_{1A} is obtained; by plotting $\ln[M_A^Z(t) - M_A^o]$ versus t, one obtains τ_{1A} as the slope of the resulting straight line [Eq. (5.83)], and therefore, both τ_{1A} and T_{1A} are obtained from a single experiment. By reversing the experiment, one may obtain these quantities for site B. No complication is introduced by unequal populations of A and B.

Forsén and Hoffman measured the acid-catalyzed rate of exchange of hydroxyl protons between salicylaldehyde (A) and 2-hydroxyacetophenone (B) in a mixture having a mole ratio of 5.65:1. In Fig. 5.22, the top spectrum is that of salicylaldehyde and the bottom is that of 2-hydroxyacetophenone. Downward arrows denote the time of application of the second rf field to the hydroxyl peak of the other compound, and upward arrows, denote the time when it is turned off. The intensity of B, the minor component, is more greatly affected than that of A, as would be expected. From these spectra, it was found that $\tau_A = 11.6$ s and $\tau_B = 2.05$ s.

From Eq. (5.83), it is evident that this method is limited to systems where the exchange rate is comparable to the reciprocal of the spin lattice relaxation time at either site. Since T_1 is usually 5—20 s for most organic molecules, it can be applied to relatively slow reactions not accessible to other methods. If the molecules are large, T_1 may be much shorter, and then it can be used for faster reactions.

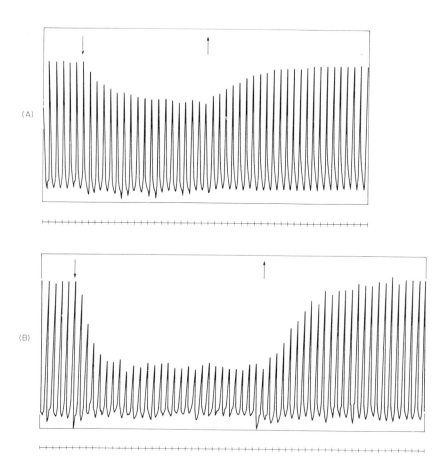

Fig. 5.22.   The decay and recovery of NMR signals of the OH protons
of (A) salicylaldehyde and (B) 2-hydroxyacetophenone in a
5.65:1 (molar) mixture. Downward arrows denote the time
of application of the saturating rf field, and upward arrows
denote the time when it is turned off. In each case, we are
observing the OH peak of one compound while irradiating
the OH line of the other. [From S. Forsén and
R. A. Hoffman, *J. Chem. Phys.* **39**, 2892 (1963).]

5.4.7.2 Spin Echo

Another way of observing chemical exchange is by measurement of T_2
by *spin echo*. We have already discussed the spin echo phenomenon in
Chapter 4, Section 4.4.4.1 (see also Section 5.3). In a system in which
exchange between nonequivalent sites is occurring, each such exchange

contributes to the irreversible dephasing of the precessing nuclear moments since it carries a nucleus from an environment characterized by one precessional frequency to another environment with a different precessional frequency. This idea is implicit in the development of the rate-modified Bloch equations (Section 5.4.2). The result is a shortening of T_2. If in the train of pulses of the Carr—Purcell method the pulse separation 2τ is very short compared to the exchange lifetime, this dephasing is negligible; but if 2τ is long compared to the lifetime, many exchanges will take place between pulses, and the echo intensity will be minimal. If echo decay is measured as a function of the pulse rate, one can measure lifetimes over a wide range of exchange rates.[55,56] The method is particularly useful for extending the range of kinetic measurements on both sides of the coalescence condition, since line shape analysis becomes inaccurate at both rate extremes. However, it has proved in practice to be in general less accurate overall than the line shape method and is also limited to cases where all observed nuclei are equivalent in the fast-exchange limit.

5.4.7.3 Two-Dimensional Methods

There are certain advantages in making rate measurements in complex systems by means of two-dimensional NMR methods. This will be discussed in Chapter 6.

REFERENCES

1. I. Solomon, *Phys. Rev.* **99**, 559 (1959).

2. K. F. Kuhlmann, D. M. Grant, and R. K. Harris, *J. Chem. Phys.* **52**, 3439 (1970).

3. N. Bloembergen, E. M. Purcell, and R. V. Pound, *Phys. Rev.* **73**, 679 (1948).

4. J. R. Lyerla and G. C. Levy in *Topics in Carbon—13 NMR Spectroscopy*, Vol. 1, G. C. Levy, Ed., Academic Press, New York, 1974, pp. 84—95.

5. P. S. Hubbard, *Phys. Rev.* **131**, 1155 (1963).

6. D. K. Green and J. G. Powles, *Proc. Phys. Soc.*, London **85**, 87 (1965).

7. K. T. Gillen, M. Schwartz, and J. H. Noggle, *Mol. Phys.* **20**, 899 (1970).

8. G. V. D. Tiers and F. A. Bovey, *J. Phys. Chem.* **63**, 302 (1959).

9. J. A. Pople, W. G. Schneider, and H. J. Bernstein, *High Resolution Nuclear Magnetic Resonance*, McGraw-Hill, New York, 1959, p. 207.

10. N. Bloembergen, *J. Chem. Phys.* **27**, 572 (1957).

11. R. Freeman and H. D. W. Hill, *J. Chem. Phys.* **54**, 3367 (1971).

12. J. N. Shoolery, *Prog. Nucl. Mag. Res. Spectrosc.* **11**, 79 (1977).

13. J. R. Lyerla, Jr., and G. C. Levy in *Topics in Carbon−13 NMR Spectroscopy*, Vol. 1, G. C. Levy, Ed., Wiley (Interscience), New York, (1974) p. 81.

14. D. A. Wright, D. E. Axelson, and G. C. Levy in *Topics in Carbon−13 NMR Spectroscopy*, Vol. 3, G. C. Levy, Ed., Wiley (Interscience), New York, 1979, p. 103.

15. G. C. Levy, R. L. Lichter, and G. L. Nelson in *Carbon−13 Nuclear Magnetic Resonance Spectroscopy*, 2nd ed., Wiley (Interscience), New York, 1980, Chapter 8.

16. J. R. Lyerla, Jr., H. M. McIntyre, and D. A. Torchia, *Macromolecules* **7**, 11 (1974).

17. G. C. Levy, R. A. Komoroski, and J. A. Halstead, *J. Am. Chem. Soc.* **96**, 5456 (1974).

18. H. Y. Carr and E. M. Purcell, *Phys. Rev.* **94** 630 (1954).

19. S. Meiboom and D. Gill, *Rev. Sci. Instrum.* **29** 688 (1958).

20. M. L. Martin, G. J. Martin, and J. J. Delpuech, *Practical NMR Spectroscopy*, Heyden, London; Chapter 7, Section 5.2.2.

21. H. S. Gutowsky, D. W. McCall, and C. P. Slichter, *J. Chem. Phys.* **21**, 279 (1953).

22. H. M. McConnell, *J. Chem. Phys.* **28**, 430 (1958).

23. H. S. Gutowsky and C. H. Holm, *J. Chem. Phys.* **25**, 1228 (1956).

24. B. E. Mann, *Prog. Nucl. Mag. Res. Spectrosc.* **11**, 95 (1977).

25. F. A. L. Anet and R. Anet in *Dynamic Nuclear Magnetic Resonance Spectrocopy*, L. M. Jackman and F. A. Cotton, Eds., Academic Press, New York, 1975.

26. F. R. Jensen, D. S. Noyce, C. H. Sederholm, and A. J. Berlin, *J. Am. Chem. Soc.* **82**, 1256 (1960); **84**, 386 (1962).

27. F. A. L. Anet, M. Ahmad, and L. D. Hall, *Proc. Chem. Soc.* **145** (1964).

28. F. A. Anet and M. A. Haq, *J. Am. Chem. Soc.* **86**, 956 (1964).

29. F. A. Bovey, F. P. Hood, E. W. Anderson, and R. L. Kornegay, *Proc. Chem. Soc.* **146** (1964); *J. Chem. Phys.* **41**, 2041 (1964).

30. F. A. L. Anet and R. Anet in *Dynamic Nuclear Magnetic Resonance Spectroscopy*, L. M. Jackman and F. A. Cotton, Eds., Academic Press, New York, 1975, pp. 554–574.

31. F. A. L. Anet and R. Anet in *Dynamic Nuclear Magnetic Resonance Spectroscopy*, L. M. Jackman and F. A. Cotton, Eds., Academic Press, New York, 1975, pp. 592–613.

32. D. K. Dalling, D. M. Grant, and L. F. Johnson, *J. Am. Chem. Soc.* **93**, 4297 (1971).

33. B. E. Mann, *Prog. Nucl. Mag. Res. Spectrosc.* **11**, 79 (1977).

34. B. E. Mann, *J. Mag. Res.* **21**, 18 (1976).

35. L. M. Jackman in *Dynamic Nuclear Magnetic Resonance Spectroscopy*, L. M. Jackman and F. A. Cotton, Eds., Academic Press, New York, 1975, pp. 203-250.

36. W. E. Stewart and T. H. Siddall III, *Chem. Rev.* **70**, 517 (1970).

37. H. N. Cheng and F. A. Bovey, *Biopolymers* **16**, 1465 (1977).

38. B. E. Mann, *J. Mag. Res.* **21**, 18 (1976).

39. F. A. L. Anet and M. Ahmad, *J. Am. Chem. Soc.* **86**, 119 (1964).

40. T. Drakenberg, R. Jost, and J. Sommer, *J. Chem. Soc. Chem. Comm.*, **1974**, 1011.

41. L. Lunazzi, D. Macciantelli, and C. A. Boicelli, *Tetrahedron Lett.*, **1975**, 1205.

42. F. A. Bovey and F. P. Hood, unpublished observations.

43. R. A. Newmark and C. H. Sederholm, *J. Chem. Phys.* **43**, 602 (1965).

44. J. T. Arnold, *Phys. Rev.* **102**, 136 (1956).

45. E. Grunwald, A. Lowenstein, and S. Meiboom, *J. Chem. Phys.* **27**, 630, 641 (1957).

46. F. A. L. Anet, *J. Am. Chem. Soc.* **84**, 671 (1962).

47. F. A. L. Anet, A. J. R. Bourn, and Y. S. Lin, *J. Am. Chem. Soc.* **86**, 3576 (1964).

48. F. A. L. Anet in *Dynamic Nuclear Magnetic Resonance Spectroscopy*, L. M. Jackman and F. A. Cotton, Eds., Academic Press, New York, 1975, pp. 598–604.

49. B. E. Mann, *Prog. Nucl. Mag. Res. Spectros.* **11**, 95 (1977).

50. W. von E. Doering and W. R. Roth, *Angew. Chem.* **75**, 27 (1963); *Tetrahedron* **19**, 715 (1963).

51. G. Schröder, *Angew. Chem.* **75**, 722 (1963).

52. M. Saunders, *Tetrahedron Lett.*, **1963**, 1699.

53. S. Forsén and R. A. Hoffman, *Acta. Chem. Scand.* **17**, 1787 (1963).

54. S. Forsén and R. A. Hoffman, *J. Chem. Phys.* **39**, 2892 (1963); **40**, 1189 (1964).

55. Z. Luz and S. Meiboom, *J. Chem. Phys.* **39**, 366 (1964).

56. A. Allerhand and H. S. Gutowsky, *J. Chem. Phys.* **41**, 2115 (1964).

CHAPTER 6

TWO-DIMENSIONAL NUCLEAR MAGNETIC RESONANCE SPECTROSCOPY

P. A. Mirau and F. A. Bovey

6.1 INTRODUCTION

We have seen in Chapter 4 (Section 4.4.4) that if we allow a period for the nuclear spin magnetization vectors to precess in the xy plane and for interactions between them to evolve, instead of immediately transforming them following the pulse in the usual way, important information concerning the spin system may be obtained. Such an experiment may thus be divided into three time periods:

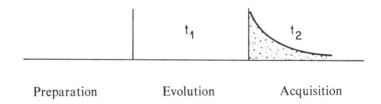

| Preparation | Evolution | Acquisition |

During the preparation period, the spins are allowed to come to equilibrium, i.e., the populations of the Zeeman levels are allowed to equilibrate with their surroundings. This interval allows the establishment of reproducible starting conditions for the remainder of the experiment. During the evolution period, t_1, the x, y, and z components evolve under all the forces acting on the nuclei, including the interactions between them. We have seen in Chapter 4 several examples involving scalar interactions, but other interactions may be employed including (of particular interest) direct through-space dipole–dipole interactions (Chapter 1, Section 1.6; Chapter 5, Section 5.2.1, Section 5.2.8).

A particularly powerful class of experiments involves the introduction of a second frequency dimension into the spectrum. This requires the corresponding introduction of a second time dimension into the experiment in addition to the acquisition time t_2, which is, of course, common to all pulse NMR measurements. This second time dependence is supplied by systematically incrementing the evolution time t_1, as illustrated in Fig. 6.1. Following each interval t_1, a second $90°$ pulse is applied after which exchange of magnetization between spins may occur. The free induction decay (FID) is then acquired in t_2 and transformed. In this schematic representation, the development of a two-dimensional or 2D spectrum is shown in general terms. The pulse sequence is actually appropriate for the production of a *chemical shift correlated* or COSY spectrum, in which the correlating influence is the J coupling between spins. In a typical example,

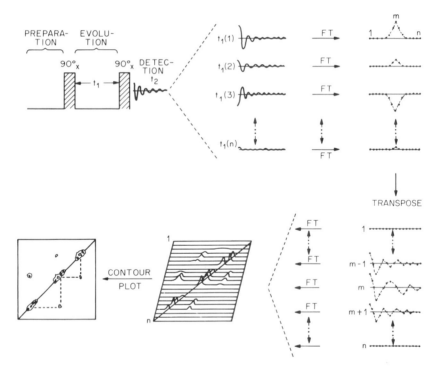

Fig. 6.1.   Schematic representation of the formation of a 2D spectrum. The pulse sequence shown is actually appropriate for the development of a *chemical shift correlated* or COSY spectrum. The free induction decays and their Fourier transforms are appropriate for a single resonance, while the final spectra show two sets of J-coupled spins. See text.

we might employ 1K or 1024 such t_1 increments increasing from 0.5 to 500 ms. The free induction decay is different for each increment because the interacting spins modulate each other's response. Each FID is detected in t_2 and transformed. We thus generate a series of 1024 matrix *rows*, each consisting of 1024 points (for a square matrix) representing the frequency-domain spectrum for a particular value of t_1. The *columns* of the matrix contain information about how the free induction decays were modulated as a function of t_1. We now perform a *transpose* operation, which consists of looking vertically down the columns at 1024 successive points and constructing 1024 new free induction decays from the information thus obtained. (For simplicity, the spectrum is represented as a single resonance at this stage, although in reality it will always be more complex.) We then perform a second Fourier transform on these new FIDs, obtaining a square two-dimensional data matrix, which is actually a surface in three-dimensional space and may be represented either as a stacked plot or as a contour plot, as shown. The stacked plot conveys relative peak intensities more vividly than the contour plot. However, it is very time-consuming to record and does not clearly show complex relationships. The contour plot is much preferred for most purposes.

Those nuclei that did not exchange magnetization during the second 90° pulse have the same frequencies during t_1 and t_2—designated F_1 and F_2, respectively—and and give rise to the normal spectrum along the diagonal, corresponding at each point to $F_1 = F_2$. Those nuclei that exchange magnetization owing to scalar coupling have a final frequency differing from the initial. This gives rise to off-diagonal contours or *cross peaks* connecting the coupled nuclei, as shown in Fig. 6.1. We thus obtain in a single experiment a picture of all the J-coupled connectivities in the molecule. It may be imagined that this technique is of great importance in assigning the resonances of complex molecules. It will be described further in Section 6.3.3.

In the following discussion, we will describe the most useful forms of 2D NMR, using as illustrations molecules that for the most part are not of highly complex structure, since we believe these relatively simple systems are more effective in showing the principles involved.

Two-dimensional NMR was first proposed in 1971 by J. Jeener,[1] but this proposal was not published. The basic experiment was later described by Aue *et al.*[2] in terms of density matrix theory. Even before this, in 1974, Ernst described two-dimensional spectra of trichloroethane and a mixture of dioxane and tetrachloroethane obtained by Jeener's technique, but this work also was not published.[3] In 1975, Ernst[4] reviewed the various forms which two-dimensional NMR might take and Hester *et al.*[5] and Kumar *et al.*[6,7] described 2D experiments on solids. The field was properly launched the next year, when several laboratories reported a variety of experiments. The early history has been reviewed by Freeman and Morris.[8] General reviews of the technique have been provided by Bax,[9] Benn and Günther, [10] Bax and Lerner[11] and Bax.[12]

6.2 *J*-RESOLVED SPECTROSCOPY

We have seen (Chapter 4, Section 4.4.4.1) that the echo train in a spin echo experiment is modulated by the *J* coupling when a system of coupled spins is observed, and that this allows us to edit ^{13}C spectra in terms of the number of protons bonded to each carbon. As an introduction to the discussion of *J*-resolved 2D spectroscopy, we consider the *J*-modulated spin echo experiment in further detail. In Fig. 6.2 is shown a series of proton spectra of 1,1,2-trichloroethane, $CH_2Cl \cdot CHCl_2$, in which $J_{AX} = 6.0$ Hz. The spectra were obtained by Fourier transformation of the second half of each of the echos in a Carr—Purcell echo train (Chap. 5, Sec. 5.3) in

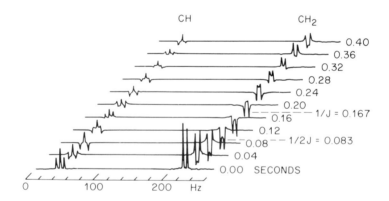

Fig. 6.2.   100 MHz proton spectra of 1,1,2-trichloroethane, $CH_2Cl \cdot CHCl_2$, for which $^3J = 6.0$ Hz. The spectra represent Fourier transformations of the echos in a Carr—Purcell sequence in which $2\tau = 0.020$ s. The time scale to the right represents the duration of the sequence. (R. Freeman and H. D. W. Hill in *Dynamic Nuclear Magnetic Resonance Spectroscopy*, L. M. Jackman and F. A. Cotton, Eds., Academic Press, New York, 1975, p. 137.)

which $2\tau = 0.020\,\text{s}$.[13] It will be observed that the triplet of the A proton and the doublet of the X protons pass through the phases shown in Fig. 4.42 as a function of time, expressed in units of $1/J$; the triplet center peak is unaffected and the triplet outer peaks pass through their phases twice as fast as those of the doublet. This J dependence is troublesome with respect to the measurement of T_2, but can be very valuable for the observation of J. Even very weak couplings, too small to appear as observable splittings in the 1D spectrum, can be measured if the system is allowed to evolve long enough. In Fig. 6.3 is shown the result of a Carr—Purcell experiment entirely analogous to that shown in Fig. 6.2 but performed on 2,4,5-trichloronitrobenzene[13]

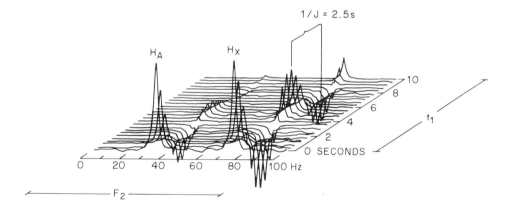

Fig. 6.3.   100 MHz proton spectra of 2,4,5-trichloronitrobenzene. The spectra represent Fourier transformations of the echos in a Carr—Purcell sequence in which the echo spacing $2\tau = 0.4$ s. The time scale to the right represents the duration of the 25 echo sequence. (R. Freeman and H. D. W. Hill in *Dynamic Nuclear Magnetic Resonance Spectroscopy*, L. M. Jackman and F. A. Cotton, Eds., Academic Press, New York, 1975, p. 138.)

in which the cross-ring *para* coupling is only 0.4 Hz. This is too small to be resolved in the 1D spectrum, but causes an amplitude modulation of both the H_A and H_X resonances with a peak-to-valley interval of 2.5 s, corresponding to $J = 0.4$ Hz. In both spectra, the time scale may be regarded as corresponding to t_1. A second Fourier transform of the "interferogram" represented by the intensity oscillation along this scale in Fig. 6.3 will yield a doublet with 0.4 Hz spacing. Along this dimension of the stacked plot, the chemical shift difference between H_A and H_X does not appear. We have thus achieved at least a partial separation of chemical shift and J coupling. Similarly, a Fourier transform of the phase modulation data represented by the time dimension of Fig. 6.2 will yield a doublet and triplet that are both centered at the same frequency (taken as $\nu = 0$). But the other dimension, analogous to F_2, still contains *both* chemical shift and J coupling information. We must therefore consider further the means of generating a truly J-resolved homonuclear 2D spectrum.

The pulse sequence employed is shown in Fig. 6.4; it is very similar to that shown in Fig. 4.39. The behavior of the vectors may be explained in a manner similar to that given in the caption of Fig. 4.41 for a homonuclear system of two coupled spins, except that in the present case the 180° pulse is applied along the x' axis rather than the y' axis. It turns out that in the 2D spectrum the resonances will have the coordinates shown in Table 6.1. As shown schematically in Fig. 6.5(a), for a doublet and triplet, this means that (assuming the scales of the F_2 and J axes are the same) the component peaks of each multiplet will lie on a line making a 45° angle with the F_2 axis. If the spectra are rotated counterclockwise by 45° about their centers, which may be done by a computer routine, then projections on the F_2 axis contain only chemical shift information [Fig. 6.5(b)] and a complete separation of J coupling and chemical shift has been achieved.

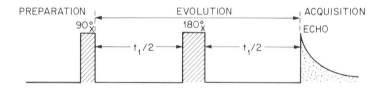

Fig. 6.4.   Pulse sequence for generation of a homonuclear J-resolved 2D spectrum. For a doublet the behavior of the vectors is essentially given by Fig. 4.41, except that here the 180° pulse is along the x' rather than the y' axis.

TABLE 6.1

Coordinates of 2D *J*-resolved Spectrum

Multiplicity	F_2 axis	J_{axis} (zero at center)
doublet (nucleus A)	$\nu_A \pm \dfrac{J}{2}$	$\pm \dfrac{J}{2}$
triplet (nucleus B)	outer $\quad \nu_B \pm J$ center $\quad \nu_B$	$\pm J$ 0
quartet (nucleus C)	outer $\quad \nu_C \pm \dfrac{3J}{2}$	$\pm \dfrac{3J}{2}$
	inner $\quad \nu_C \pm \dfrac{J}{2}$	$\pm \dfrac{J}{2}$

In many cases, the *J*-axis scale is so much expanded over that of the chemical shift axis—tenfold, for example—that the effect of such a rotation is minimal.

In Fig. 6.6 is shown the 200 MHz, *J*-resolved, 2D proton spectrum of 2,3-dibromopropionic acid, an AMX case.[14] The 1D spectrum appears in Fig. 4.35(a) and is discussed in Chapter 4, Section 4.4.2. Slices through each chemical shift along the *J* axis give the multiplets, all centered at $\nu = 0$, shown in Fig. 6.7. Another useful feature is the projection on the F_2 axis that yields the homonuclear decoupled spectrum.

Heteronuclear *J*-resolved spectroscopy requires somewhat more elaborate pulse schemes. The most generally applicable scheme, termed the *gated decoupler method*[15,16] employs the pulse sequence shown in Fig. 6.8, which is essentially the same as that shown in Fig. 4.43 except that the proton decoupler is gated on during the first half of the evolution time, as well as during acquisition, and there is no 180° proton pulse. The method provides magnetic inhomogeneity refocusing and nuclear Overhauser enhancement of signals of the rare nucleus (usually ^{13}C). In addition, of course, the evolution time t_1 is incremented. The behavior of the ^{13}C vectors is illustrated for the components of a carbon doublet, *i.e.*, a CH group, in the caption to Fig. 6.8. The 90_x° pulse turns both

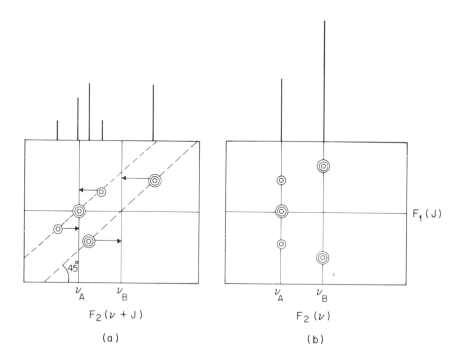

Fig. 6.5.   Schematic homonuclear J-resolved 2D spectra. Spectrum
             (a) exhibits a triplet at ν_A and a doublet at ν_B. It does not
             represent a real system, since the couplings have different
             magnitudes. In (a) the F_2 axis contains information
             concerning both J couplings and chemical shifts, ν. After a
             counterclockwise rotation of 45° (b), the F_2 axis contains
             only chemical shift information. In both (a) and (b),
             projections on the F_1 axis give J information only.

components of the doublet into the y' axis [(a) in Fig. 6.8]. The proton
decoupler being on, these components do not precess apart but remain as a
single vector, designated I. Upon application of the $180_y°$ pulse (b), vector
I undergoes phase reversal to position II. The decoupler is now turned off
and the component vectors begin to precess apart, developing a phase
difference of $\pm Jt_1/2$ at (c). During acquisition, the components collapse to
a single vector again. As a result of the J coupling, one observes in effect
an amplitude modulation of this vector as a function of t_1. Because the
divergence of the vectors occurs during only one half of the evolution time,
the apparent J coupling in the resulting 2D spectrum has only half the

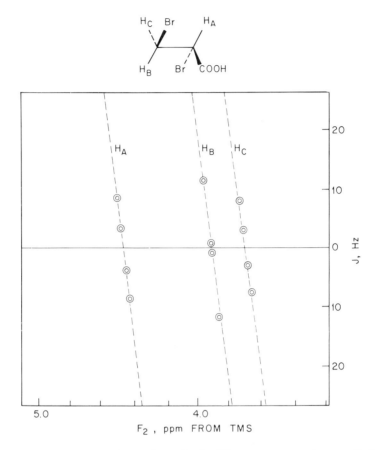

Fig. 6.6. The 200 MHz *J*-resolved 2D proton spectrum of 2,3-dibromopropionic acid (1% in $CDCl_3$, 25°C); for 1D spectrum see Fig. 4.35. The *J* couplings are (Chapter 4, Section 4.4.2): $J_{AB} = \pm 11.7$ Hz; $J_{AC} = \pm 4.5$ Hz; $J_{BC} = \mp 11.3$ Hz. (P. A. Mirau, private communication.)

true value. This is easily allowed for, however, and in fact the necessary doubling factor may be included in the 2D data processing software.

In Fig. 6.9 the 50.3 MHz *J*-resolved ^{13}C spectrum of 2,2,4-trimethylpentane (isoöctane) is shown.[14] Along the F_2 axis is shown also the 1D spectrum, which is the same as spectrum (a) in Fig. 4.49. The resonances appear elongated because of the *ca.* fivefold expansion of the *J* axis. The spectrum represents 256 t_1 values incremented from 2 to 512 ms, giving a sweep width in the F_1 dimension of 500 Hz.

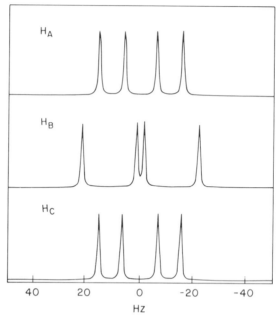

Fig. 6.7.   Vertical slices through each of the chemical shift positions
(measured on the central $J = 0$ axis) in Fig. 6.6.

The form of the multiplets can be readily seen by even casual
inspection of Fig. 6.9. This is usually possible even for much more
complex spectra and constitutes the principal value of the method. For
accurate measurement of the magnitudes of the coupling—in this instance,
all very nearly equal—vertical slices through the chemical shift positions
are again useful (Fig. 6.10).

It is evident that the identification of carbon types can be more
elegantly carried out by this method than by the INEPT and DEPT
editing procedures discussed in Chapter 4 (Section 4.4.4). In addition, the
2D method is not as sensitive to the setting of the empirical parameters.
However, it must be pointed out that 2D experiments are inherently more
time-consuming and lower in sensitivity than 1D methods. When one must
identify carbon types that occur in small concentrations, as for example in
the "defects" associated with inverted monomer units in poly(propylene
oxide),[17] the 1D procedures may be more practical.

A variant of this method is the so-called proton-flip experiment,[18,19]
which provides a selective 180° proton pulse to invert only certain chosen
^{13}C satellites in the proton spectrum.

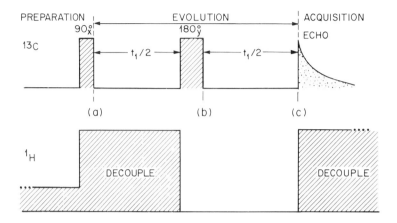

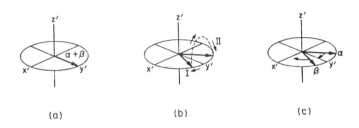

Fig. 6.8.   Pulse sequence for the gated decoupler method for heteronuclear *J*-resolved 2D spectrum generation. In (a) the 90° ^{13}C pulse turns both vectors of the ^{13}C doublet into the y' axis. Both components α and β remain together in (b) because of the proton decoupling but undergo a phase inversion under the influence of the 180_y° carbon pulse. Following this, the decoupler is gated off and the doublet vectors are free to evolve and develop a phase difference $\pm\pi J t_1/2$. During acquisition, the components again precess at the same rate (c). As t_1 is incremented, one observes an amplitude modulation of the precessing singlet vector brought about by the *J* coupling.

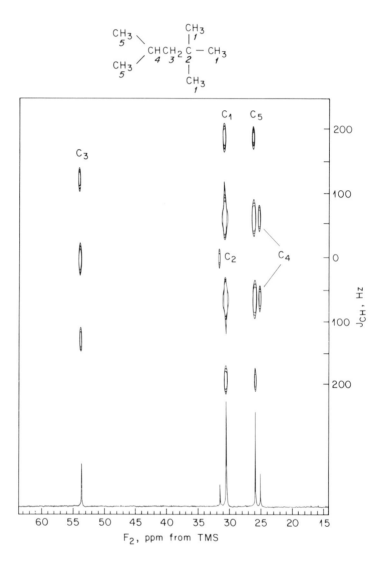

Fig. 6.9.   The 50.3 MHz *J*-resolved carbon-13 spectrum of isooctane
              (2,2,4-trimethylpentane) observed on a 40% (vol./vol.)
              solution in C_6D_6 at 23°C; see Chapter 4, Fig. 4.51.
              (P. A. Mirau, private communication.)

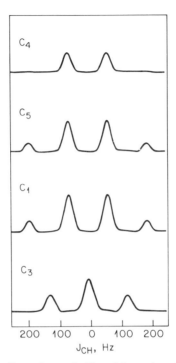

Fig. 6.10. Vertical slices through Fig. 6.9 at the chemical shift positions indicated.

6.3 CORRELATED SPECTROSCOPY

We now discuss more fully the principal forms of correlated spectroscopy which were introduced at the beginning of this chapter. Let us first consider the correlation of chemical shifts of different nuclear species, a frequently very useful aid in spectral interpretation of one of the nuclei when the resonance assignments of the other are known.

6.3.1 Heteronuclear Correlated Spectroscopy

In this experiment, different species of nuclei—most commonly ^1H and ^{13}C—are related through scalar coupling. As in the one-dimensional INEPT and DEPT procedures (Chapter 4, Section 4.4.4), the pulse sequence used involves polarization transfer and, therefore, enhancement of the signals of the less sensitive nucleus occurs. The basic experiment has been described by Ernst, Freeman, and coworkers.[20–23] Although relatively sophisticated, it can still be described meaningfully in terms of vector diagrams.

The pulse sequence is given in Fig. 6.11, and the behavior of the ^{1}H and ^{13}C magnetization vectors is diagrammed in Fig. 6.12. Here, we observe a carbon-13 (X) doublet arising from the presence of one bonded proton (A). The sequence could equally well apply for any other rare spin-1/2 species X, such as, for example, ^{15}N. During the incremented evolution period t_1, the ^{1}H spins are labelled by their precessional frequencies. The degree of precession of ^{1}H is actually measured by observing ^{13}C, the transfer of information between them occurring in the mixing period. After the initial 90_x° proton pulse places the proton magnetization in the xy plane (a, b), the α and β components of the doublet fan out by an angle ϕ determined by the magnitude of the CH coupling. The doublet components precess by an average angle θ determined by the chemical shift of the proton (c). The 180_x° carbon pulse (c, d), applied after a time $t_1/2$, exchanges the α and β ^{1}H spin states in the manner we have already seen (Chapter 4, Section 4.4.4.1). Their relative precessional frequencies are also exchanged with the result that they are refocussed and form an echo after the end of the evolution period (e). This is in effect equivalent to decoupling the ^{13}C from the ^{1}H during the whole period t_1. After a time $\Delta_1 = 1/2J$ following the refocussing, the phase angle θ, which contains the chemical shift information, equals $180°$

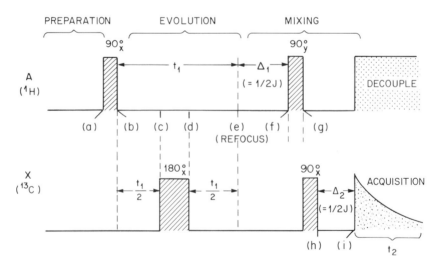

Fig. 6.11.   Pulse sequence for the generation of a heterocorrelated 2D spectrum. See text and caption to Fig. 6.12. (Pulse widths are of the order of μs and are greatly exaggerated in comparison to pulse intervals).

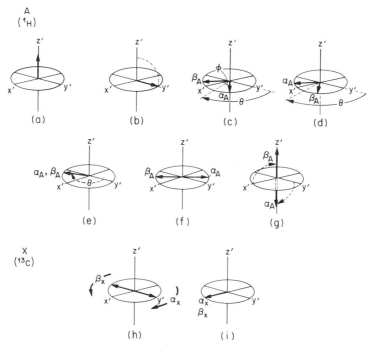

Fig. 6.12.   Vector diagrams for proton and carbon-13 magnetizations accompanying the heterocorrelated 2D NMR pulse sequence shown in Fig. 6.11 and described in the text. The letter designations correspond to the times so designated in Fig. 6.11.

(f). The resulting antiparallel ^{1}H vectors (f) are transformed into $\pm z$ vectors by a 90_y° pulse (f, g). The magnitude of these z components, which is determined by the phase angle θ, now contains the proton chemical shift information, since the 90_y° proton pulse only rotates the x-components of the vectors. If $\theta = \pi/2, 3\pi/2, 5\pi/2....$, the ^{1}H magnetization at (g) will be a maximum, whereas if $\theta = \pi, 2\pi, 3\pi...$, the ^{1}H polarization at (g) will be zero. This is most important because, as we have seen in Chapter 4 (Section 4.4.4.2), proton polarization leads to simultaneous polarization of the X or ^{13}C magnetization, and this in turn can be transformed into $\pm y'$ magnetization by a 90_x° ^{13}C pulse (h). These vectors will refocus in an interval $\Delta_2 = 1/2J$ and can be observed as a singlet (i) upon irradiation of the protons. In this way, the proton chemical shift modulates the carbon-13 spectrum. It is the latter which we are actually observing.

The choice of $1/2J$ for Δ_2 is correct only for a CH doublet. For a CH_2 triplet and CH_3 quartet, the relationships shown in Figs. 4.42 and 4.48 apply. A compromise value of $\Delta_2 = 1/3J$ is customarily employed as being satisfactory when CH, CH_2, and CH_3 are all present in the compound under observation.

A heterocorrelated $^1H-^{13}C$ spectrum of cytidine (10% solution in DMSO-d_6) is shown in Fig. 6.13.[14] Also included for clarity are the 1D proton and carbon-13 spectra. These, of course, must be separately

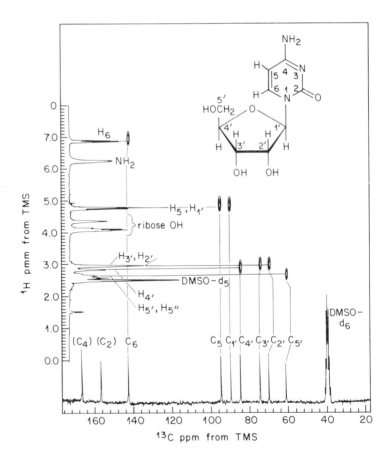

Fig. 6.13.   A heterocorrelated $^1H-^{13}C$ 2D spectrum of cytidine (10% solution in DMSO-d_6). The 50.3 MHz 1D ^{13}C and 200 MHz ^1H spectra are also shown along the vertical (^1H) and horizontal (^{13}C) axes, respectively.

generated and would ordinarily not be shown. The seven elongated resonances in the 2D plane constitute the actual heterocorrelated spectrum. They connect C_6, C_5, $C_{1'}$, $C_{4'}$, $C_{3'}$, $C_{2'}$, and $C_{5'}$ (in order of increasing shielding) with the corresponding attached protons; $H_{3'}$, and $H_{2'}$ overlap while H_5 and $H_{1'}$ nearly overlap. Here, C_4 and C_2 show no correlating peaks because they have no directly attached protons and the longer range $^1H-^{13}C$ J couplings are much too weak (Chapter 4, Section 4.3.6.1). For the same reason, the NH_2 and ribose OH proton resonances likewise show no correlating peaks. One might expect a 2D resonance connecting the DMSO$-d_4$ proton resonance (1% of the DMSO$-d_6$) with the DMSO$-d_6$ ^{13}C resonance (40.6 ppm) since both resonances are strong. However, it must be realized that only one DMSO molecule in 10^4 has a $^{13}CHD_2$ group, and this is too weak to show up.

6.3.2 Nuclear Overhauser (NOESY) 2D Spectroscopy

The format of the data in 2D nuclear Overhauser effect spectroscopy (NOESY) resembles that for the COSY experiment, which we have introduced in Section 6.1 and will discuss further in Section 6.3.3. In both, chemical shifts of the same nuclear species are correlated. In the COSY spectrum, the correlating influence is the through-bond J coupling, while in the NOESY spectrum it is the exchange of z magnetization resulting from through-space dipole—dipole interactions. An equivalent effect is also produced by chemical exchange, when this can occur. In the one-dimensional NOE experiment (Chapter 5, Section 5.2.8), a selective pulse is applied to one of the spins and the change in intensity of the resonances of other spins is observed. Because of the inverse sixth-power dependence of the dipole—dipole interaction, the NOE is a powerful measure of structure. In the 2D experiment, the spins are all excited at the same time with nonselective pulses but are labelled by their precessional frequencies. This *frequency labelling* is equivalent to selectively exciting each of the spins and separately measuring each pairwise interaction. In addition to the inverse sixth-power dependence, which it shares with the 1D NOE experiment, the NOESY experiment has the major advantage that all the pair-wise interactions can be observed in a single experiment.

In Fig. 6.14 are shown the three-pulse NOESY sequence and a series of vector diagrams to show how a two-spin system might behave under this sequence. The first 90_x° pulse, following the preparation period, converts the z magnetization of the two chemically shifted spins A and B (which we may think of as protons in the subsequent discussion) into observable magnetization at (b). We suppose for simplicity that spin A coincides in Larmor frequency with the carrier ν_0 (see Chapter 1, Section 1.4) and so does not precess in the rotating frame during the evolution time t_1, while

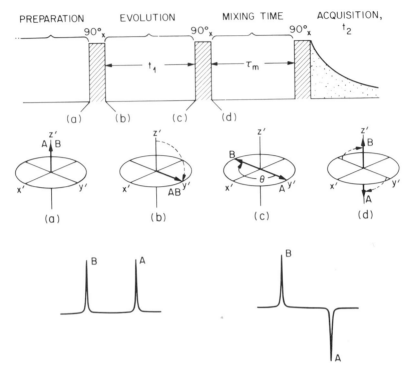

Fig. 6.14.   The three-pulse sequence for the generation of a nuclear
              Overhauser correlated or NOESY spectrum. Also shown are
              the magnetization vectors for a system of two chemically
              shifted spins A and B of the same species, the evolution time
              t_1 being chosen to be equal to $1/2\delta_{AB}$, where δ_{AB} is the
              chemical shift difference between A and B.

spin B precesses. At (c) we show the vectors after a particular choice of t_1
equal to $1/2\delta_{AB}$, where δ_{AB} is the chemical shift difference in hertz between
A and B. During this interval, a $180°$ phase angle θ will accumulate
between the vectors. The second $90°_x$ pulse at (c) produces the $\pm z$
polarization of the spins shown at (d), the optimum condition for exchange
of magnetization. Other choices of t_1 will result in less-than-optimum z
polarization; for example, if $t_1 = 1/4\delta_{AB}$, the B vector will be along x' and
the second $90°_x$ pulse will not affect it. Since t_1 is systematically
incremented over a series of several hundred values (typically 512)
covering the chemical shifts of all the protons in a multiproton system,
each proton will experience its optimum t_1 value. The net effect is a
frequency labelling of each spin.

Any process that leads to an exchange of magnetization during the mixing time, τ_m can be measured by this method. The mixing time is chosen to be of the order of the spin lattice relaxation time. The effect of interactions during the mixing time is that spins labeled by their frequencies in t_1 may precess at a different frequency at the end of the mixing time. This gives rise to cross peaks in the 2D spectrum.

The third 90_x° pulse converts z magnetization to observable $\pm y$ magnetization in the usual manner. There are four frequencies that may be detected during t_2 for a two-spin system:

1. Spins labeled with the precessional frequency of a spin A and detected at the A frequency, *i.e.*, A spins that have not experienced any exchange of magnetization.

2. Spins labelled as spin A but undergoing magnetization exchange to become B spins during the mixing time.

3. Spins labelled and detected as spin B, *i.e.*, B spins that have not experienced any exchange of magnetization.

4. Spins labelled as spin B but undergoing magnetization exchange to become A spins during the mixing time.

The spins of type 1 and 3 that have been labelled and detected at the same frequency appear on the diagonal, while spins of type 2 and 4 that have exchanged magnetization appear as off-diagonal or cross peaks because they precessed at different frequencies during t_1 and t_2. The intensities of the peaks depend on the probability of magnetization exchange and the length of the mixing time τ_m. When τ_m is very short compared to the time constant for magnetization exchange, the off-diagonal peaks will be very weak compared to the diagonal peaks. Another extreme is represented by choosing τ_m to be long enough to allow the z magnetization to come to equilibrium during τ_m. In this case, the frequency labelled magnetizations A and B have a 50% probability of being converted during τ_m, and the intensities of the diagonal and cross peaks will be equal. If τ_m greatly exceeds the relaxation times of the system, the spin system will come to thermal equilibrium with the lattice (Chapter 1, Section 1.5; Chapter 5, Section 5.2.1) and no peaks will be detected, either on or off the diagonal, because the spin system will have lost all memory of the frequency labelling. By measuring the peak intensities as a function of mixing time, *i.e.*, by carrying out several NOESY experiments at varied τ_m values, one may determine the rates of magnetization transfer between spins, and from these one may calculate internuclear distances and correlation times.

We shall illustrate the use of the NOESY experiment by two examples. The first may be regarded as a "true" NOE case in the sense that the proton spins observed communicate only by dipole—dipole interaction. In the second case, the experiment is carried out in the same way but the exchange of the spin A and spin B magnetizations comes about by isomerization of the molecule between two forms.

We first consider a polymer that is also discussed in Chapter 7 (Section 7.3.1), poly-γ-benzyl L-glutamate (PBLG). This polypeptide may exist in either α—helical or random coil forms depending on the solvent and temperature. We are concerned here with the α—helical form, shown in Fig. 6.15. The structure of the crystalline α—helical form is well known from X-ray diffraction. The α-helix also exists in solution (demonstrated principally by its optical properties) and is assumed to have a conformation very similar to that of the solid state. This may be tested through measurement of interproton distances by NOESY.

In Fig. 6.16 is shown the 500 MHz proton spectrum of a PBLG 20-mer in α—helical form.[14] The proton resonances which principally concern us are identified: NH, αH, and βH. The protons of the β— and γ—CH_2 groups differ from each other in chemical shift within each group (see Chapter 3, Section 3.3.6) but the difference is not resolved in Fig. 6.16. The side chains of PBLG are large enough that the helical 20-mer is nearly as broad as it is long and may be treated as tumbling isotropically in solution.

In Fig. 6.17 shows the 500 MHz proton NOESY spectrum of PBLG under the same conditions as Fig. 6.16, obtained with a 76 ms mixing time.

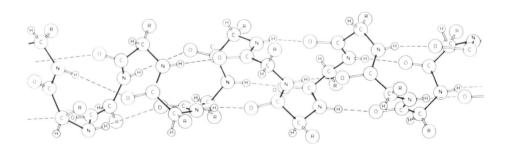

Fig. 6.15.   A fragment of α—helix (14 amino acid residues). For poly-γ-benzyl L-glutamate (PBLG); the side-chain R = $-CH_2CH_2COOCH_2 \cdot C_6H_5$.

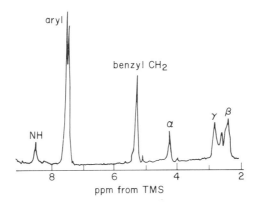

Fig. 6.16.   The 500 MHz 1D proton spectrum of a poly-γ-benzyl L-
              glutamate (PBLG) 20-mer in α—helical form in 95:5
              chloroform:trifluoroacetic acid at 20°C [P. A. Mirau and
              F. A. Bovey, *J. Am. Chem. Soc.* **108**, 5130 (1986).]

The diagonal resonances are readily identified by comparison to the 1D
spectrum.   Cross peaks are observed between the main chain protons:
NH—αH, NH—βH, and αH—βH (we observe two symmetrical sets of
cross peaks above and below the diagonal; we will employ the latter).
These may be identified by constructing vertical lines from the diagonal
resonances, as shown.   (There is also a very weak cross peak between NH
and γH protons.)   The NH—αH cross peak is expected from the chemical
structure of PBLG, but the interproton distance, and therefore the cross
peak intensity, depends on the rotation angle of the NH—αH bond.   The
NH and βH protons are not bonded to adjacent atoms, so the strong cross
peaks between them must be the reflection of a particular conformation
which is strongly preferred in solution.   (This conclusion is strengthened
when we recall the inverse sixth—power dependence of the proton—proton
interaction.)   From the very weak NH—γH cross peaks, we can set limits
on the proton separation since cross peaks are not expected for protons
separated by more than 4Å.

In Fig. 6.18 are shown the intensities of the NH—αH, NH—βH, and
αH—βH cross peaks measured at seven values of the mixing time τ_m.
From the slopes of these plots, one may obtain the interproton distances
r_{HH} from (see Chapter 5, Section 5.2.1.2)

$$\frac{1}{T_1(c)} = \frac{\gamma^4 \hbar^2}{20 r_{HH}^6} [6J_2(2\omega) - J_o(0)] \qquad (6.1)$$

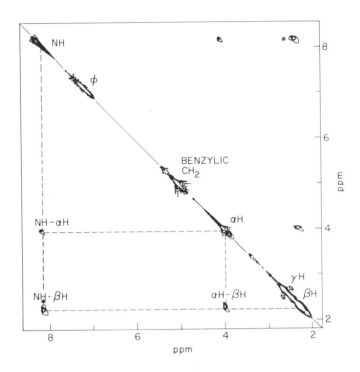

Fig. 6.17.   The 500 MHz 2D NOESY proton spectrum of a PBLG 20-mer under the same conditions as the 1D spectrum in Fig. 6.16. The mixing time τ_m was 76 ms. A total of 256 1K spectra were acquired with a t_1 increment of 0.2 ms, giving a sweep width of 5 kHz in both dimensions. [P. A. Mirau and F. A. Bovey, *J. Am. Chem. Soc.* **108**, 5130 (1986).]

where

$$J_n(n\omega) = \frac{2\tau_c}{1 + (n\omega)^2\tau_c^2} \tag{6.2}$$

The rate of cross relaxation $1/T_1(c)$ may be obtained from the cross peak growth. We require also the correlation time τ_c. This may be obtained from the rate of diagonal peak decrease, which parallels the cross peak increase, combined with the nonselective relaxation rates measured by the inversion-recovery method (Chapter 5, Section 5.2.7.2); this procedure is described in Ref. 24. For this system, τ_c is found to be 10^{-9} s. In

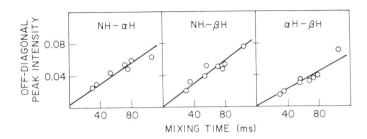

Fig. 6.18.   The rise of the cross peak intensities as a function of mixing time τ_m for the PBLG spectrum shown in Fig. 6.17. The intensities are normalized to the intensities of the corresponding diagonal resonances. [P. A. Mirau and F. A. Bovey, *J. Am. Chem. Soc.* **108**, 5310 (1986).]

Table 6.2 are shown the cross-relaxation rates, calculated interproton distances r_{HH} and values of r_{HH} expected for the standard α-helix from X-ray diffraction.

Thus, NOESY may be employed for quantitative structural determinations in solution. It has been broadly applied to semi-quantitative conformational studies on polypeptides, proteins, and polynucleotides.

TABLE 6.2

Cross-Relaxation Rates, Calculated Interproton Distances,

and X-ray Distances for α-Helical PBLG

Interaction	$1/T_1(c)\,(s^{-1})$	$r_{HH}(\text{Å})$ Calculated	Expected for α-helix
NH-αH	0.54	2.20	2.48
NH-βH	0.55	2.20	2.26
αH-βH	0.72	2.10	2.20

Figure 6.19 shows the 1D 200 MHz proton spectrum of glycylsarcosine, which exists as two conformational isomers

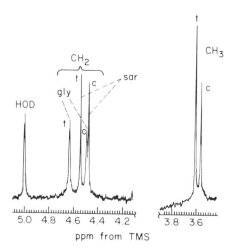

Because of the partial double bond character of the imide bond these conformers exchange relatively slowly at 65°C, with a time constant of the order of one second. Since this is long compared to the reciprocal of the chemical shift difference between the resonances of the cis and trans forms — 20 to 50 ms — the lines are sharp. At 65°C the trans/cis ratio is *ca.* 60/40.

Fig. 6.19.   The 200 MHz 1D proton spectrum of glycylsarcosine observed in D_2O solution at 30°C. (P. A. Mirau, private communication.)

Figure 6.20 shows the NOESY spectrum of glycylsarcosine under the same conditions as Fig. 6.19 and with a mixing time of 500 ms.[14] Since the cross peaks between the resonances of the cis and trans isomers are substantially weaker than the diagonal resonances, it is evident that the time constant for equilibration is longer than this, as already suggested. By measuring the growth of the cross peaks as a function of τ_m, the equilibration rate could be measured quantitatively.

6.3.3 Homonuclear *J*-Correlated (COSY) Spectroscopy

Chemical shift correlation via *J* coupling is a very useful technique for the interpretation of complex NMR spectra, particularly proton spectra, as it furnishes virtually automatic assignments of resonances. The simple pulse sequence employed has already been shown in Section 6.1 (Fig. 6.1) and is indicated also in Fig. 6.21 together with a variant spin echo version

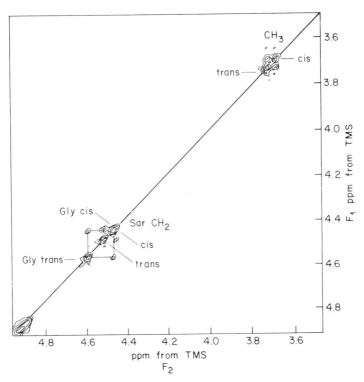

Fig. 6.20.   The 200 MHz 2D NOESY spectrum of glycylsarcosine observed under the same conditions as the 1D spectrum of Fig. 6.19 and a with a mixing time τ_m of 500 ms. (P. A. Mirau, private communication.)

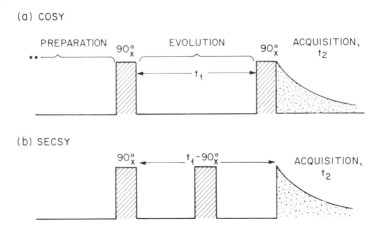

Fig. 6.21.   The pulse sequences for the generation of chemical shift
correlated (COSY) and spin echo correlated (SECSY) 2D
spectra.

which we shall discuss later. Unlike 2D J-resolved (Section 6.2) and 2D
nuclear Overhauser spectroscopy (Section 6.3.2), COSY spectroscopy
cannot be usefully represented by simple vector pictures. The reason lies in
the fact that the cross peaks are generated by the evolution of the coupled
spin system during the t_1 and t_2 periods. This evolution can be described
as a product of the vectors of the coupled spins but cannot readily be
graphically represented. Explanations can be provided by either the
density matrix[9] or product operator formalism,[25] but are beyond the scope
of this book. One may still picture the evolution time as causing frequency
labeling of the spins, as in the NOESY experiment.

In the COSY experiment, the spins precess under the influence of both
the chemical shift and the J coupling. It is the precession under the latter
influence during t_1 and t_2 that gives rise to the cross peaks. For this
reason, t_1 and t_2 must be of the order of $1/2J$. Cross peaks due to J
coupling are also created by the $90_x^\circ - t_1 - 90_x^\circ$ part of the NOESY pulse
sequence. These are considered as artifacts in NOESY spectra and can be
almost completely suppressed by cycling the pulse phases (Chapter 2,
Section 2.3.5).

In Fig. 6.22 are shown 400 MHz proton spectra of 9,11-
bisdehydrobenzo[18]annulene.[26,27]

Spectrum (a) is the COSY spectrum; 1D spectra are shown along each axis as an aid to interpretation. These would not ordinarily be included. As in NOESY spectra, correlated resonances may be identified by extending vertical and horizontal tie lines from each cross peak to the diagonal. In Fig. 6.22 only cross peaks arising from vicinal coupling (*ca.* 6−10 Hz) are detectable. Longer range couplings (< 1 Hz) are too weak to generate cross peaks under the conditions employed. (Spectrum (b) is the SECSY spectrum, discussed next.)

The spin echo version of COSY, called SECSY,[28] also involves two 90°_x pulses, but the second occurs in the middle of the evolution period to allow chemical shift refocusing as in a spin echo experiment (Chapter 4, Section 4.4.4.1; see also Figs. 6.4, 6.8, and 6.11). The information in the SECSY spectrum is the same as in the COSY spectrum but the format is different. As may be seen in Fig. 6.22(b), the 1D spectrum appears on a horizontal axis through the center of the 2D plane and cross peaks appear off the horizontal. Correlations are established by drawing a vertical tie line from the center horizontal line to the cross peak. A tie line is drawn at 45° from the cross peak and extended until it intersects another cross peak, and a vertical tie line constructed back to the horizontal. The two vertical lines then show the positions of the coupled spins. Such a plot is less elegant than the COSY spectrum but has the advantage that the sweep width in the t_1 (vertical) dimension can be much smaller than in the t_2 dimension, since it need only be as large as the maximum frequency difference between coupled spins. In Fig. 6.22(b) this is only a limited advantage because of the large chemical shift differences between the A and B and the D and E spins, but in general there is a saving in time because less digital resolution, *i.e.*, fewer t_1 values, is required.

The COSY experiment is an important tool for establishing resonance assignments and the chemical identification of new materials, but it is not universally applicable to all molecules of interest. One limitation is the

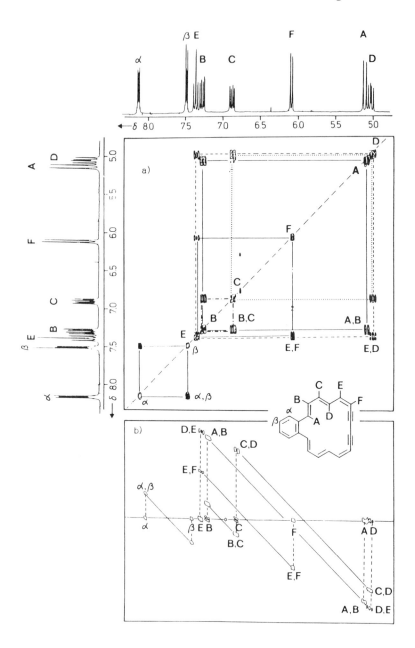

Fig. 6.22.   The 400 MHz proton spectra of 9,11-bisdehydro-
benzo[18]annulene. Spectrum (a) is the COSY spectrum
with the 1D spectra shown along each axis. The dimensions

of the 2D matrix and of the transform are both 256 × 1024; (b) is the SECSY spectrum, dimensions likewise 256 × 1024 (Refs. 25 and 26).

antiphase nature of the cross peaks. In an AX system, for example, the 1D spectrum consists of doublets at the chemical shift positions of the A and X spins. The cross peaks will be four-spot patterns with two of the peaks positive and two negative (antiphase). For small molecules with sharp lines, this presents no problems, as both the positive and negative peaks are resolved. However, when the linewidths are approximately as large as the coupling constants, the positive and negative lobes of the cross peaks can overlap and begin to cancel each other. Thus, for normal proton spectra, it is usually difficult to obtain COSY spectra for large, slowly tumbling molecules (*i.e.*, those with correlation times longer than 25 nsec). In addition, complex spin coupling patterns result in highly complex cross peaks that are even more likely to cancel as the linewidth becomes large.

Another limitation arises for spins with small coupling constants. As mentioned, the cross peaks are generated by the evolution of the coupled spin system during the t_1 and t_2 periods. To efficiently generate cross peaks, the length of these periods must be on the order of $1/J$. Since the t_1 and t_2 periods in a 2D experiment are typically on the order of 100–200 msec, cross peaks for spins with 1 Hz coupling constants are not very large. However, introduction of delays (on the order of $1/J$) before and after the second 90° pulse allows a sufficient time for the evolution of such weak coupling. This modification of COSY is known as "delayed" COSY[9] and should be used for the correlation of weakly coupled peaks. Introduction of the delay lowers the signal-to-noise ratio of the experiment because the spins can relax by spin–spin relaxation during the added delay.

The COSY experiment provides information about spins that are coupled directly to each other. Another type of COSY experiment involves *relayed coherence transfer*. The format of these spectra are similar to what we have seen, except that cross peaks are observed for both direct and indirect connectivities. For example, no cross peaks are observed between the protons labelled D and F in Fig. 6.22 because these protons are not appreciably coupled to each other. In a relayed COSY spectrum additional cross peaks would appear at AC, BD, CE, and DF because these protons are indirectly coupled to each other. Determination of whether connectivity is direct or indirect may be made by comparing the COSY and relayed COSY spectra.

6.4 MULTIPLE QUANTUM SPECTROSCOPY

The correlation of chemical shifts by *multiple quantum coherences* is another class of frequently used 2D experiments. As with the COSY spectrum, it is not possible to draw vector diagrams to show why these experiments work, although the principles can be understood using the density matrix or product operator formalism. The correlating influence in a multiple quantum experience is the *coherence* between J-coupled spins. Such coherences are not directly visible in our NMR spectrometer, where we essentially detect the magnetization along the y' axis in the rotating frame. The strategy in the study of multiple quantum coherences is to apply a series of pulses to create the coherence between the coupled spins. While we cannot directly detect such coherences, we can measure their effect on the phase and amplitude of the observable magnetization. Thus, multiple quantum coherences are indirectly detected.

The most useful applications of multiple quantum spectroscopy in 2D NMR are (1) the correlation of chemical shifts (as in COSY spectroscopy) and (2) multiple quantum filtering. In the correlation of chemical shifts by double quantum coherences, for example, a map is generated showing the connectivities between the coupled spin systems, as in COSY. However, these experiments differ from COSY in the format of the data and some of the features appearing in the COSY spectrum but not in the correlation by multiple quantum coherences. These features can be understood by considering the idea of multiple quantum filters. The idea is to reduce the amount of information in the spectrum by designing a pulse sequence which selects for certain orders of coherences. The maximum order of the coherence is determined by the number of coupled spins in the system. A single peak can have only a single quantum coherence. Two coupled spins can have a coherence order as high as two, and three coupled spins can have triple quantum coherence. Multiple quantum filtering takes advantage of the fact that coherences of different order behave differently from each other depending on the phases of the pulses applied to the system. It is possible to design a pulse sequence selective only for triple-quantum coherences (*i.e.*, a three quantum filter). In a 2D experiment with such a filter, correlation would only be observed for a spin system with three or more coupled spins. In many cases, such as proteins, this results in a simplification of the spectrum and aids in the unambiguous assignment of resonances.[29] The use of multiple quantum 2D NMR spectroscopy has been reviewed by Munowitz and Pines.[30]

REFERENCES

1. J. Jeener, Ampère International Summer School, Basko Polje, Yugoslavia, 1971.

2. W. P. Aue, E. Bartholdi, and R. R. Ernst, *J. Chem. Phys.* **B64**, 2229 (1976).

3. R. R. Ernst, *VIth Int. Conf. Magn. Reson. Biol. Syst.*, Kandersteg, Switzerland, 1974.

4. R. R. Ernst, *Chimia* **29**, 179 (1975).

5. R. K. Hester, J. L. Ackerman, B. L. Neff, and J. S. Waugh, *Phys. Rev. Lett.* **34**, 993 (1975).

6. A. Kumar, D. D. Welti, and R. R. Ernst, *J. Magn. Reson.* **18**, 69 (1975).

7. A. Kumar, D. D. Welti, and R. R. Ernst, *Naturwissenschaften* **62**, 34 (1975).

8. R. Freeman and G. A. Morris, *Bull. Magn. Reson,* **1** (1), 5 (1979).

9. A. Bax, *Two Dimensional Nuclear Magnetic Resonance in Liquids*, Delft University Press and D. Reidel Publishing Co., Dordrecht, Holland, 1982.

10. R. Benn and H. Günther, *Angew. Chem. Int. Ed. Engl.* **22**, 350 (1983). This article also reviews other multipulse methods, not necessarily two-dimensional.

11. A. Bax and L. Lerner, *Science* **232**, 960 (1986).

12. A. Bax, *Bull. Magn. Reson.* **7** (4), 167 (1985).

13. R. Freeman and H. D. W. Hill in *Dynamic Nuclear Magnetic Resonance Spectroscopy*, L. M. Jackman and F. A. Cotton, Eds., Academic Press, 1975, p. 131.

14. P. A. Mirau, private communication, 1986.

15. G. Bodenhausen, R. Freeman, and D. L. Turner, *J. Chem. Phys.* **65**, 839 (1976).

16. R. Freeman, S. P. Kempsell, and M. H. Levitt, *J. Magn. Reson.* **34**, 663 (1979).

17. F. C. Schilling and A. E. Tonelli, *Macromolecules* **19**, 1337 (1986).

18. G. Bodenhausen, R. Freeman, R. Niedermeyer, and D. L. Turner, *J. Magn. Reson.* **24**, 291 (1976).

19. A. Kumar, W. P. Aue, P. Bachmann, J. Karhan, L. Miller, and R. R. Ernst, *Proc. XIXth Congress Ampère*, Heidelberg, 1976.

20. A. A. Maudsley and R. R. Ernst, *Chem. Phys. Lett.* **50**, 368 (1977).

21. G. Bodenhausen and R. Freeman, *J. Magn. Reson.* **28**, 471 (1977).

22. A. A. Maudsley, L. Miller, and R. R. Ernst, *J. Magn. Reson.* **28** 463 (1977).

23. R. Freeman and G. A. Morris, *J. Chem. Soc. Chem. Comm.* **1978**, 684.

24. P. A. Mirau and F. A. Bovey, *J. Am. Chem. Soc.* **108**, 5130 (1986).

25. W. O. Sorenson, G. W. Ech, M. H. Levitt, G. Bodenhausen, and R. R. Ernst, *Prog. NMR Spectros.* **16**, 163 (1983).

26. N. Darby, T. M. Cresp, F. Sondheimer, *J. Org. Chem.* **42**, 1960 (1977).

27. H. Günther, M.-E. Günther, D. Mondeshka, H. Schmicker, F. Sondheimer, N. Darby, and T. M. Cresp, *Chem. Ber.* **112**, 71 (1979).

28. K. Nagayama, A. Kumar, K. Wüthrich, and R. Ernst, *J. Magn. Reson.* **40**, 321 (1980).

29. M. H. Levitt and R. R. Ernst, *J. Chem. Phys.* **83**, 3297 (1985).

30. M. Munowitz and A. Pines, *Science* **233**, 525 (1986).

CHAPTER 7

MACROMOLECULES

7.1 INTRODUCTION

NMR spectroscopy is a method of great power for the study of macromolecular substances. Early studies dealt with polymers in the *solid state* and the spectra were of the broad-band type (Chapter 8, Section 8.1). These were almost always proton spectra and their width and detailed shape, particularly when combined with measurements of spin lattice and spin—spin relaxation (Chapter 5, Section 5.2, Section 5.3), could be interpreted to provide information concerning molecular motion. The broadness of the peaks, however, concealed all structural information.

When one observes polymers in *solution*, the resonance lines are greatly narrowed by motional averaging effects, and one may obtain detailed information concerning the structure of the chains. Modern instrumental methods (Chapter 2) have made it possible to obtain dynamic information from polymer solutions — principally by carbon—13 NMR — and also to obtain high resolution structural information from solid polymers (Chapter 8). Therefore, the barrier between two once quite separate modes of polymer investigation — solid- and liquid-state NMR spectroscopy — has all but disappeared.

7.2 CHAIN MICROSTRUCTURE

Although structural information can be obtained for polymers in the solid state, the narrowest lines and greatest detail are obtained from polymer solutions, where segmental motion is usually rapid enough — except for large biopolymers in the native state — to provide spectra comparable to those of small molecules.

In general, the chemical shifts and *J* couplings of random coil macromolecules are closely similar to those of small molecules of analogous structure. The spectrum of ethyl orthoformate (see also Chapter 1, Section 1.11, and Fig. 1.14) is compared with that of poly(vinyl ethyl ether)

$$OCH_2CH_3$$
$$|$$
$$[\ CH_\alpha-CH_{2_\beta}\ ]_n$$

in Fig. 7.1. The polymer exhibits the same ethyl proton resonances and, in addition, those of the main chain β protons, appearing at 1.5 ppm, and of the α protons, appearing under the CH_2 quartet. The broadening of the polymer lines is in part due to the slower motions of the large molecules but mainly arises from structural complexities, which we will discuss in Sec. 7.2.1.

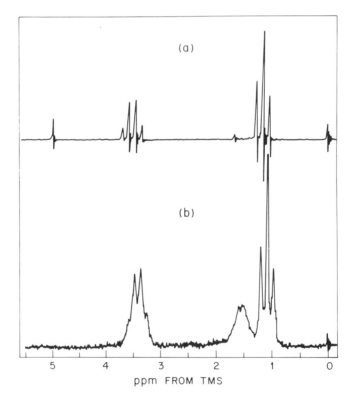

Fig. 7.1.   60 MHz proton spectra of (a) ethyl orthoformate $HC(OCH_2CH_3)_3$ and (b) poly(vinyl ethyl ether), both observed in 15% (wt./vol.) solution in carbon tetrachloride at *ca.* 25°C. (A. I. R. Brewster, private communication.)

In general, linewidth is only weakly dependent on molecular weight or the macroscopic viscosity of polymer solutions. Polymer solutions so viscous that they can barely be poured from the NMR tube may yet give well-resolved spectra. This is because the rate of local segmental motion of the chains is almost independent of molecular weight and viscosity.

7.2.1 Stereochemical Configuration

NMR is very informative concerning the stereochemical configuration of vinyl and diene polymer chains, revealing details seen by no other means.

7.2.1.1 Poly(methyl methacrylate)

In Fig. 7.2 are shown the 500 MHz proton spectra of poly(methyl methacrylate) prepared (a) with a free radical initiator and (b) with fluorenyllithium in toluene, an anionic initiator.[1] The marked differences between these spectra reflect the nature of the initiator. To interpret these spectra, we refer to the previous discussion of molecular asymmetry and chemical shift non-equivalence (Chapter 3, Section 3.3.6). We consider the chain in terms of sequences of two monomer units or *dyads*, analogous to the 2,4-disubstituted pentane models considered earlier. The dyad sequences are of two types. The *racemic* (*r*), dyad has a twofold symmetry axis; the two methylene protons consequently are in equivalent environments on a time average over the chain conformations. These protons therefore have the same chemical shift and appear as a singlet despite the strong geminal coupling between them. The isotactic or *meso* (*m*) dyad has a plane

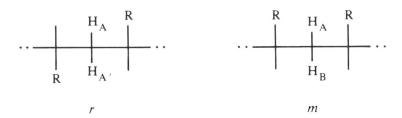

of symmetry but no twofold axis, and so the two protons are nonequivalent and have different chemical shifts. When there is no vicinal coupling to neighboring protons, as is the case in poly(methyl methacrylate), the syndiotactic sequences should exhibit a methylene singlet while the isotactic form should give two doublets, each with a spacing equal to the geminal coupling, *ca.* −15 Hz. We see in Fig. 7.2 that the methylene

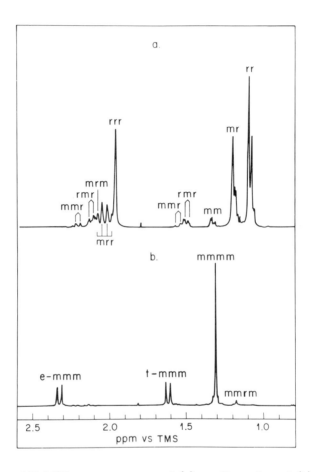

Fig. 7.2.   500 MHz proton spectra of (a) syndiotactic and (b) isotactic
                resonances poly(methyl methacrylate); the OCH₃ resonances
                at *ca.* 3.4 ppm are not shown. Observed as 10% solutions in
                chlorobenzene-d_5 at 100°C. [F. C. Schilling, F. A. Bovey,
                M. D. Bruch,   and   S. A. Kozlowski,   *Macromolecules* **18**,
                1418 (1985).]

spectrum of the anionic polymer (b) is almost exclusively a pair of
doublets; quantitative assessment shows that this polymer is 95% isotactic.
The methylene spectrum of the free radical polymer (a) is more complex,
but the principal resonance — at *ca.* 1.9 ppm — is a singlet, showing that
this polymer is predominantly syndiotactic but more irregular than (b).

This is generally the case for vinyl polymers prepared with free radical initiators.

Thus, proton NMR provides absolute stereochemical information concerning polymer chains without recourse to X-ray or other methods. Somewhat more detailed, but not absolute, information can be gained from the methyl proton resonances near 1.2 ppm. (The ester methyl resonance at *ca.* 3.6 ppm — not shown— is less sensitive to stereochemistry.) In both spectra we note three peaks — or, more correctly, groups of peaks — appearing in similar positions but with greatly different intensities. These correspond to the α—methyl groups in the center monomer unit of the three possible *triad* sequences: isotactic, syndiotactic, and heterotactic.[2] These

Isotactic, *mm*   Syndiotactic, *rr*

Heterotactic, *mr (or rm)*

may be more simply and appropriately designated by the *m* and *r* terminology, as indicated. Measurement of the relative intensities of the *mm*, *mr*, and *rr* α—methyl peaks, which appear from left to right in both spectra in this order, gives a valid statistical representation of the structure of each polymer.

From the triad data we may gain considerable insight into the mechanism of propagation. This is one of the principal uses of such information. Let us designate by P_m the probability that the polymer chain will add a monomer unit to give the same configuration as that of the last unit at its growing end, i.e., that an *m* dyad will be generated. We assume that P_m is independent of the stereochemical configuration of the growing chain. The generation of the chain is a Bernoulli trial process; it is like reaching into a large jar of balls marked m and r and withdrawing a ball at random. The proportion of m balls in the jar is P_m. Since two monomer additions are required to form a triad sequence, it is readily evident that the probabilities of their formation are

$$[mm] = P_m^2 \tag{7.1}$$

$$[mr] = 2P_m(1 - P_m) \tag{7.2}$$

$$[rr] = (1 - P_m)^2 \tag{7.3}$$

The heterotactic sequence must be given double weighting because both directions, mr and rm — observationally indistinguishable — must be counted. A plot of these relationships is shown in Fig. 7.3. It will be noted that the proportion of mr (heterotactic) units rises to a maximum when P_m is 0.5, corresponding to a strictly random or atactic configuration for which $[mm]:[mr]:[rr]$ will be 1:2:1. For any polymer, if Bernoullian, the $[mm]$, $[mr]$, and $[rr]$ sequence intensities will lie on a single vertical line in Fig. 7.3, corresponding to a single value of P_m. Spectrum (a) in Fig. 7.2 corresponds to these simple statistics, P_m being 0.25 ± 0.01. The polymer corresponding to spectrum (b) does not. The propagation statistics in this case can be interpreted to indicate that the probability of isotactic placement is dependent upon the stereochemical configuration of the growing chain and cannot properly be expressed by a single parameter such as P_m. Free radical and cationic propagations always give predominantly syndiotactic chains. Anionic initiators may also do so if strongly complexing ether solvents such as dioxane or glycol dimethyl ether are employed rather than hydrocarbon solvents as in polymer (b) (Fig. 7.2).

It is evident that in the spectra of Fig. 7.2 there is fine structure in both the methylene and methyl regions, which we have not discussed. In spectrum (a) this corresponds principally to residual resonances of the stereoirregular portions of the chains; in (b) such residual resonances are less conspicuous. These arise from sensitivity to longer stereochemical sequences than dyad and triad. In Table 7.1 are shown planar zigzag projections of such sequences, together with their frequency of occurrence as a function of P_m, assuming Bernoullian propagation. The tetrads — and all "even-ads" — refer to observations of β—methylene protons (or β carbons), while the "odd-ads" refer to substituents on the α carbons (or α carbons themselves). Resonances for tetrad sequences or "even-ads"

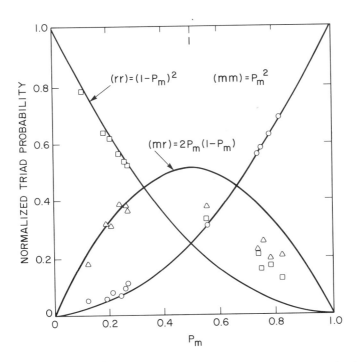

Fig. 7.3.   Probabilities of isotactic (mm), heterotactic (mr), and syndiotactic (rr) triads as a function of P_m, the probability of m placement. The points on the left-hand side represent polymers prepared with free radical initiators and those on the right-hand side polymers prepared with anionic initiators.

should appear as fine structure in the dyad spectra, while pentad sequences or higher "odd-ads" should appear as fine structure on the triad resonances. The assignments to longer sequences as indicated on the spectra are based on Bernoullian probabilities in spectrum (a); those in spectrum (b) are primarily based on (a). It may be noted that r-centered tetrads, e.g., mrr, do not necessarily appear as singlets if the sequence as a whole lacks a twofold axis.

The numbers of observationally distinguishable configurational sequences, or n-ads, designated $N(n)$, obey the following relationship:

n	2	3	4	5	6	7	8	9
$N(n)$	2	3	6	10	20	36	72	136

TABLE 7.1

α Substituent

	Designation	Projection	Bernoullian probability
Triad	Isotactic, mm (i)	*(projection)*	P_m^2
	Heterotactic, mr (h)	*(projection)*	$2P_m(1-P_m)$
	Syndiotactic, rr (s)	*(projection)*	$(1-P_m)^2$
Pentad	mmmm (isotactic)	*(projection)*	P_m^4
	mmmr	*(projection)*	$2P_m^3(1-P_m)$
	rmmr	*(projection)*	$P_m^2(1-P_m)_2$
	mmrm	*(projection)*	$2P_m^3(1-P_m)$
	mmrr	*(projection)*	$2P_m^2(1-P_m)^2$
	rmrm (heterotactic)	*(projection)*	$2P_m^2(1-P_m)^2$
	rmrr	*(projection)*	$2P_m(1-P_m)^3$
	mrrm	*(projection)*	$P_m^2(1-P_m)^2$
	rrrm	*(projection)*	$2P_m(1-P_m)^3$
	rrrr (syndiotactic)	*(projection)*	$(1-P_m)^4$

β -CH₂

	Designation	Projection	Bernoullian probability
Dyad	meso, m	*(projection)*	P_m
	racemic, r	*(projection)*	$(1-P_m)$
Tetrad	mmm	*(projection)*	P_m^3
	mmr	*(projection)*	$2P_m^2(1-P_m)$
	rmr	*(projection)*	$P_m(1-P_m)^2$
	mrm	*(projection)*	$P_m^2(1-P_m)$
	rrm	*(projection)*	$2P_m(1-P_m)^2$
	rrr	*(projection)*	$(1-P_m)^3$

or in general

$$N(n) = 2^{n-2} + 2^{m-1} \qquad (7.4)$$

where $m = n/2$ if n is even and $m = (n-1)/2$ if n is odd. Discrimination of these longer sequences in unlikely to be possible beyond $n = 6$ (hexads) or $n = 7$ (heptads). The observation of such sequences permits rather searching tests of polymerization mechanisms.

7.2.1.2 Polypropylene

Another example is provided by *polypropylene*,

$$\begin{array}{c} CH_3 \\ | \\ [\quad CH \quad - CH_2 \quad]_n \\ \quad \alpha \qquad \beta \end{array}$$

particularly instructive as it is one of the few vinyl polymers that can be prepared in both isotactic and syndiotactic forms with coordination catalysts. The proton spectra are relatively complex because of vicinal coupling between α and β protons and α and methyl protons as well as geminal methylene proton coupling in isotactic sequences. In Fig. 7.4 are shown 220 MHz proton spectra of isotactic (a) and syndiotactic (b) polypropylene.[3-6] The β protons of the syndiotactic polymer appear as a triplet at 1.03 ppm corresponding to a single chemical shift and J coupling to two neighboring α protons. In the isotactic polymer they appear as widely spaced multiplets at 1.27 ppm and 0.87 ppm, corresponding to *syn* and *anti* positions in the trans—trans conformation:

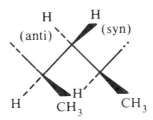

Analysis of these spectra yields the following values for the vicinal main-chain couplings (in both polymers the vicinal CH_3–H_α coupling is 6.5 Hz and the geminal methylene proton coupling is 13.5 Hz):

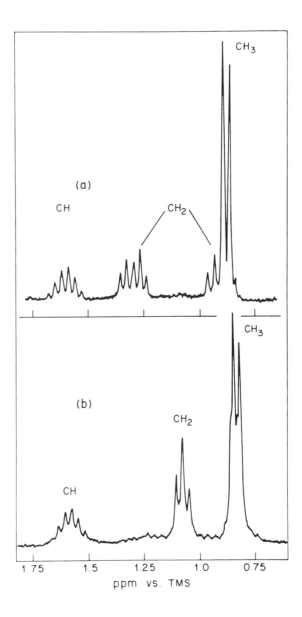

Fig. 7.4.   220 MHz proton spectra of (a) isotactic and (b) syndiotactic
               polypropylenes, observed in *o*-dichlorobenzene at 165°C.
               [R. C. Ferguson, *Trans. N.Y. Acad. Sci.* **29**, 495 (1967).]

Isotactic

$$J_{H_\alpha-H_{syn}} \quad : \quad 6.0 \text{ Hz}$$

$$J_{H_\alpha-H_{anti}} \quad : \quad 7.0 \text{ Hz}$$

Syndiotactic

$$J_{H_\alpha-CH_2} \quad : \quad 4.8, 8.3 \text{ Hz}$$

We have seen (Chapter 4, Section 4.3.3) that such couplings are strongly dependent upon the dihedral angle between vicinal protons. In polymers, a gauche arrangement has a proton–proton coupling constant of *ca.* 2–4 Hz and protons in a trans conformation show couplings of 8–13 Hz. A time-averaged value of J is observed due to rapid conformational equilibration, and from this the populations of gauche and trans conformers can be obtained. Without describing such calculations in detail, we may say that for polypropylene substantial proportions of both gauche and trans conformations are found for syndiotactic sequences, whereas isotactic chains show a strong preference for ··(*gt*) (*gt*) (*gt*)·· sequences, i.e., a threefold helical conformation.

The proton spectrum of atactic polypropylene (not shown) is virtually uninterpretable, being a complex of overlapping multiplets. In Fig. 7.5 are shown the 25 MHz ^{13}C spectra of isotactic (a), atactic (b), and syndiotactic (c) polypropylene, observed in 20% 1,2,4-trichlorobenzene solution.[7] In these spectra the multiplicity arising from ^{1}H–^{13}C J couplings has been removed by proton irradiation (see Chapter 4, Section 4.4.3.) The isotactic and syndiotactic polymers give very similar spectra but with readily observable chemical shift differences, especially for the methyl carbons. This sensitivity to stereochemical configuration is particularly clear in the spectrum of the atactic polymer, in which the methyl resonance is split into peaks corresponding to 9 of the 10 possible pentad sequences (Table 7.1). It is also noteworthy that the syndiotactic polymer is less stereoregular than the isotactic one and that the configurational statistics of both the atactic and syndiotactic polymers depart markedly from Bernoullian. This is generally the case for chains generated by Ziegler–Natta coordination catalysts.

Carbon–13 chemical shifts can be effectively rationalized through recognition of a γ–*gauche shielding effect*, which we have discussed in Chapter 3, Section 3.3.3. This treatment is based on the observation that when two carbons separated by three bonds are gauche to each other, they shield each other by *ca.* 5 ppm compared to the chemical shifts of the

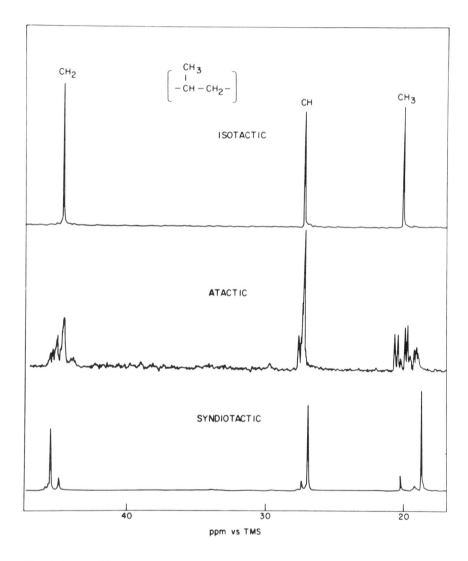

Fig. 7.5.   The 25-MHz carbon—13 spectra of three preparations of
polypropylene: (a)   isotactic,   (b)   atactic,   and   (c)
syndiotactic, observed as 20% (wt./vol.) solutions in 1,2,4-
trichlorobenzene at 140°C. [A. Zambelli, D. E. Dorman,
A. I. R. Brewster, and F. A. Bovey, *Macromolecules* **6**, 925
(1973).]

corresponding trans conformation.[8] In Fig. 7.6 (top) we see that in polypropylene the methyl group is gauche to the methine carbon (C_α) when the chain is trans or gauche — but not when it is gauche+. Thus, the methyl group experiences differing numbers of gauche interactions

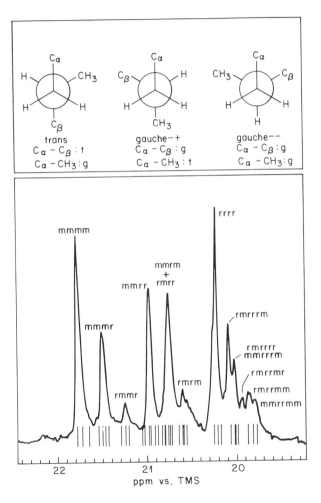

Fig. 7.6.   Top: conformations of a fragment of a polypropylene chain. Bottom: 90 MHz carbon−13 NMR spectrum of the methyl region of atactic polypropylene. The "stick" spectrum below the experimental spectrum shows the predicted chemical shifts of the 36 heptad sequences (probabilities assumed equal).

depending upon the local chain conformation. The methyl carbon chemical shift can be accurately predicted for all 36 heptad configurational sequences (Fig. 7.6, bottom) from the theoretically calculated populations of gauche and trans states of the main-chain bonds flanking the methine carbon in each sequence. The C_α and C_β chemical shifts may be calculated as a function of stereochemical configuration in the same way, and the treatment may be extended to other polymers.[8]

7.2.2 Geometrical Isomerism

Geometrical isomerism in diene polymers may be observed and measured by NMR. In addition to the cis and trans structures found in natural isoprene polymers, the formation of such polymers by chain propagation of diene monomers may proceed also by incorporation of the monomer through one double bond rather than by 1,4 addition. Thus from butadiene the following isomeric chains may be produced in pure form by appropriate choice of coordination catalysts:

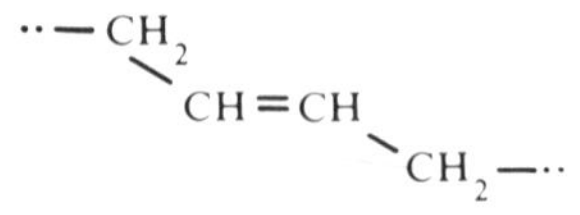

$$\cdots -CH_2 \diagdown \atop CH=CH \diagdown \atop CH_2-\cdots$$

trans - 1,4

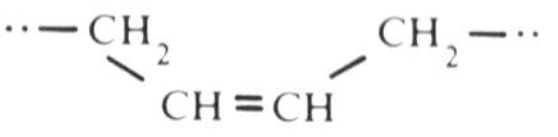

$$\cdots -CH_2 \diagdown \atop CH=CH \diagup CH_2-\cdots$$

cis - 1,4

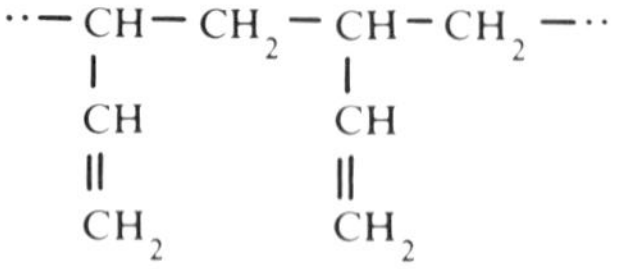

$$\cdots -CH-CH_2-CH-CH_2-\cdots$$
$$| \qquad\qquad |$$
$$CH \qquad CH$$
$$|| \qquad\qquad ||$$
$$CH_2 \qquad CH_2$$

Isotactic 1,2

$$CH_2$$
$$||$$
$$CH$$
$$|$$
$$\cdots -CH-CH_2-CH-CH_2-\cdots$$
$$|$$
$$CH$$
$$||$$
$$CH_2$$

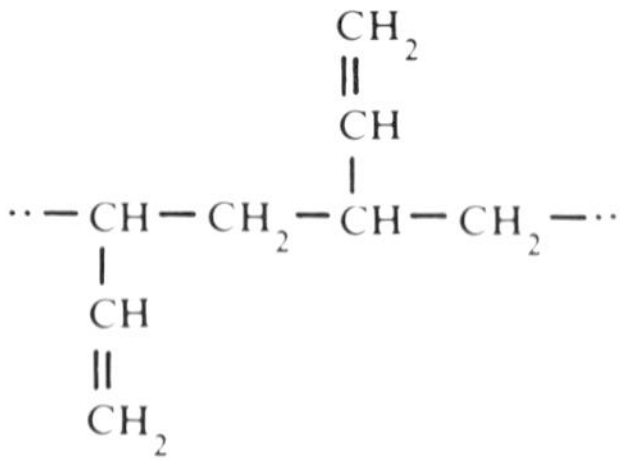

Syndiotactic 1,2

7.2.2.1 Polybutadiene

Polybutadiene formed by free radical propagation contains all the above structures. In Fig. 7.7 shows the carbon—13 spectra of (a) *cis*-1,4-polybutadiene and (b) *trans*-1,4-polybutadiene, observed in deuterochloroform solution.[9] The olefinic carbons are only moderately

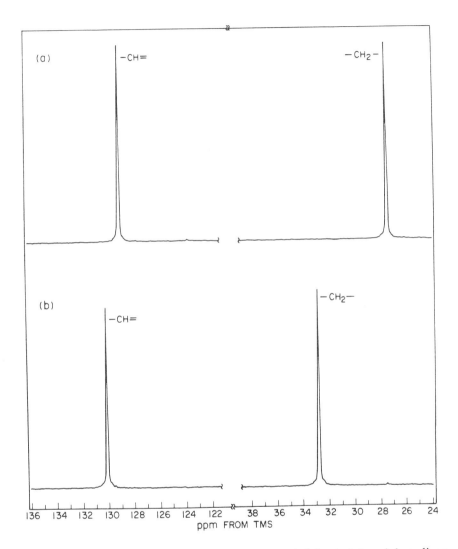

Fig. 7.7.   50.3 MHz carbon−13 spectra of (a) *cis*-1,4,-polybutadiene
and (b) *trans*-1,4-polybutadiene, observed in CDCl₃ at 40°C.
(F. C. Schilling, private communication.)

sensitive to geometrical isomerism but the methylene carbons are highly
sensitive, being substantially more shielded (*ca.* 8 ppm) in the cis polymer,
an effect no doubt closely related to the *gauche−γ* shielding effect
(Section 7.2.1.2).

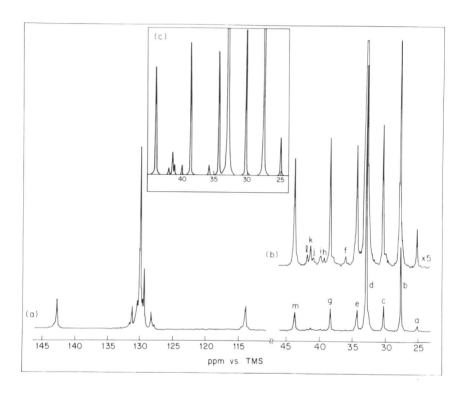

Fig. 7.8.   50.3 MHz carbon—13 spectrum of a free radical polybutadiene, observed as a 20% (wt./vol.) solution in CDCl$_3$ at 50°C. (L. W. Jelinski, private communication; see also F. A. Bovey, *Chain Structure and Conformation of Macromolecules*, Academic Press, New York, 1982, pp. 105—110); (a) olefinic carbon spectrum; (b) aliphatic carbon spectrum; (c) computer simulation of (b) based on a random distribution of cis, transp, and 1,2 units in the proportion 23:8:19.

Chains of mixed structure exhibit more complex spectra because of sequence effects. In Fig. 7.8 is shown the 50.3 MHz ^{13}C spectrum of a deuterochloroform solution of polybutadiene produced by free radical initiation.[10] At the left (a) is the region of olefinic carbon resonance, not fully analyzed. The olefinic carbon singlets of the pendant vinyl groups flank those of the 1,4 units. At the right (b) are the aliphatic carbon resonances — mainly 1,4 (and 1,2) methylene groups. This part of the spectrum is shown at two values of gain — 11X and 5X — to show the small resonances of the sequences containing 1,2 units. Major peak b corresponds to central methylenes in cis—cis units; the principal·peak d is that of the central 1,4-unit methylene group in trans—trans and trans—cis units, not discriminated. Peaks a, c, e, and m correspond to sequences involving 1,4 units and one 1,2 unit, while the very small resonances f through l represent sequences containing two 1,2 units. The overall composition of the polymer is 23% cis-1,4, 58% *trans*-1,4, and 19% 1,2. Spectrum (c) is a computer simulation of (b) based on the assumption of a random distribution of units in these proportions. The satisfactory fit shows that free radical propagation in butadiene polymerization is a Bernoullian process with regard to the generation of these isomeric structures.

7.2.2.2 Polyisoprene

Figure 7.9 shows the 50.3 MHz ^{13}C spectra of cis-1,4-polyisoprene and trans-1,4-polyisoprene, observed in solution.[9] Again, the olefinic carbons are relatively insensitive to isomerism at the double bond but the CH_3 and CH_2-1 carbons markedly shield each other in a cis arrangement compared to the trans. The CH_2-4 carbon, cis to a carbon atom in both isomers, is less affected.

Spectrum (a) is that of natural rubber or *hevea brasiliensis*. The biochemical pathway to natural rubber is an enzymatic process in which isoprene as such plays no part. The polymer is highly stereoregular, no trace of the trans structure being observable. Synthetic cis-1,4-polyisoprene is produced commercially using lithium alkyls or Ziegler—Natta catalysts. It contains 2—6% of trans units.

7.2.3 Regioregularity

Vinyl monomers may in principle propagate in either a head-to-tail (a) or a head-to-head: tail-to-tail mode (b):

$$
\begin{array}{c}
\quad\ \ R\qquad\ \ R\qquad\ \ R\qquad\ \ R \\
\quad\ \ | \qquad\ \ | \qquad\ \ | \qquad\ \ | \\
(a)\qquad \cdots-CH_2-CH-CH_2-CH-CH_2-CH-CH_2-CH-\cdots
\end{array}
$$

$$
\begin{array}{c}
\quad\ R\ \ R\qquad\qquad\ R\ \ R \\
\quad\ |\ \ |\qquad\qquad\ |\ \ | \\
(b)\qquad \cdots-CH_2-CH-CH-CH_2-CH_2-CH-CH-CH_2-\cdots
\end{array}
$$

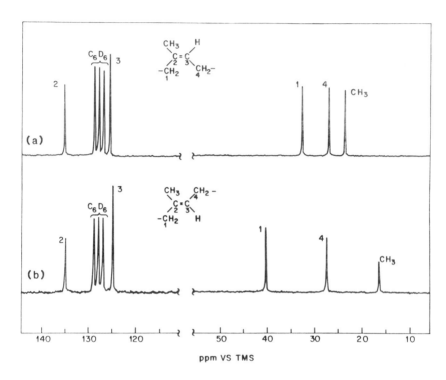

Fig. 7.9.   50.3 MHz carbon—13 spectrum of (a) cis- and (b) trans-
1,4-polyisoprene, observed in 10% (wt./vol.) solution in C_6D_6
at 60°C. (F. C. Schilling, private communication.)

In general, head-to-tail propagation is overwhelmingly preferred, but the
other direction of addition is sometimes of considerable importance.
Fluorine-substituted ethylenes are particularly subject to the generation of
inverted units, presumably because fluorine atoms are relatively
undemanding sterically. At the same time, the physical properties of their
polymers may depend strongly on the presence of such units. The presence
of ^{19}F offers an additional means for detailed study, as ^{19}F chemical shifts
are highly sensitive to structural variables.

7.2.3.1 Poly(vinyl fluoride)

Poly(vinyl fluoride) — a commercial plastic having high resistance to
weathering — has a substantial proportion of head-to-head units.[11,12] The
188 MHz ^{19}F spectrum of a commercial polymer is shown in Fig. 7.10.

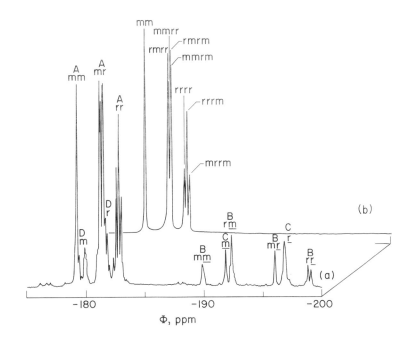

Fig. 7.10.   188 MHz ^{19}F spectra of (a) commercial poly(vinyl fluoride),
and (b) poly(vinyl fluoride) prepared without monomer
reversals; both spectra observed as 8% solutions in
N,Ndimethylformamide-d_7 at 130°C. [R. E. Cais and
J. M. Kometani in *NMR and Macromolecules, ACS Symp.
Ser., No. 247, J. C. Randall, ed., 1984, pp. 153−166.]*

The ^{19}F−^{1}H J coupling multiplicity has been removed by proton
irradiation as in ^{13}C spectroscopy (Chapter 4, Section 4.4.3.2). The
resonance assignments, designated by capital letters, are

$$\cdots - CH_2 - \underset{\longrightarrow}{\overset{A}{CHF}} - CH_2 - \underset{\longrightarrow}{\overset{A}{CHF}} - CH_2 - \underset{\longrightarrow}{\overset{B}{CHF}} - \underset{\longleftarrow}{\overset{C}{CHF}} - CH_2 - CH_2 - \underset{\longrightarrow}{\overset{D}{CHF}} - CH_2 - \underset{\longrightarrow}{\overset{A}{CHF}} - \cdots$$

The stereochemical assignments are also indicated in Fig. 7.10. The m
and r designations that are not underlined represent the usual relationships
between substituents in 1,3 positions (in planar zigzag projection):

The underlined designations represent the substituents in 1,2 positions (also in planar zigzag projection):

It may be further noted in connection with the B resonances that $r\underline{m}$ and $\underline{m}r$ are quite different structures and do not differ merely in direction as with rm and mr. The fraction of inverted units in *ca.* 11%.

Spectrum (b) is that of a poly(vinyl fluoride) prepared without inversions by a special chemical route. The upfield portion of spectrum (a), as well as the D resonances, are absent, thus clearly identifying them as defect peaks. This material has a melting point of 210°C, compared to the normal polymer's melting point of 190°C. Both polymers are nearly atactic.

Two-dimensional ^{19}F NMR spectroscopy has been employed to make the assignments to pentad configurational sequences indicated in Fig. 7.10.[13] The 188 MHz COSY spectrum (Chapter 6, Section 6.2.1) of the inversion-free polymer is shown in Fig. 7.11. It shows correlations established through off-diagonal cross peaks arising from four-bond J couplings between fluorines having central positions in pentads forming overlapping sequences in the same configurational hexad. Thus, for example, this ^{19}F–^{19}F coupling, although probably only about 7–8 Hz and giving rise to no spectral splittings, causes the $rmmr$ and $mmrm$ resonances to be correlated and to exhibit cross peaks because they occupy the $rmmrm$ hexad:

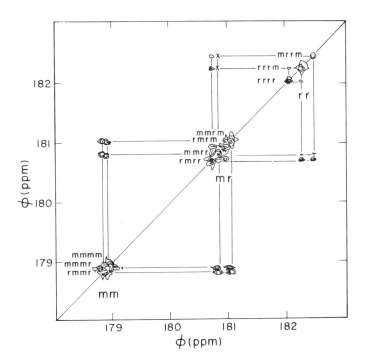

Spectral assignments can be made for all resonances from the network of connectivities thus established.

Fig. 7.11.   188 MHz two-dimensional ^{19}F COSY spectrum of defect-free poly(vinyl fluoride), same conditions as in Fig. 7.10. [M. D. Bruch, F. A. Bovey, and R. E. Cais, Macromolecules, 317, 2547 (1984).]

7.2.3.2 Poly(vinylidene fluoride)

Poly(vinylidene fluoride) is a material of great interest because of its piezoelectric and pyroelectric properties. NMR spectroscopy shows that it contains 3–6% of inverted units:[11,14–17]

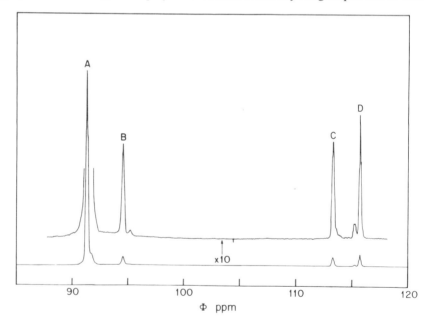

The 188 MHz ^{19}F spectrum,[18] shown in Fig. 7.12, is made somewhat simpler by the absence of asymmetric carbons. Assignments to sequences in terms of the above structure are indicated.

7.2.4 Copolymer Structure

Copolymers are broadly divided into three types, *random*, *block*, and *graft*. Block and graft copolymers contain relatively long sequences of one

Fig. 7.12.   188 MHz ^{19}F spectrum of commercial poly(vinylidene fluoride) observed in 11% (wt./vol.) solution in dimethylformamide-d_7 at 27°C. [R. E. Cais and J. M. Kometani, *Macromolecules* **15**, 849 (1982).]

monomer bonded to similar sequences of another. Although they are of major scientific and technological interest, their overall composition is usually known from their method of synthesis and they do not present microstructural problems essentially different from those of homopolymers. Our attention will be confined to the "random" type, in which two or more types of comonomer units are present in each chain. We shall discuss only copolymers of vinyl monomers. Copolyesters and copolyamides are also significant but their composition is also usually readily predictable from the ratio of comonomers employed.

The compositions of copolymers of vinyl and diene monomers are not in general the same as that of the monomer mixtures from which they are formed and cannot be deduced from the homopolymerization rates of the monomers involved. We shall not enter here into the details of the mechanism by which the monomers enter the copolymer chain, which is described in any textbook of polymer chemistry. From the comonomer sequences and their frequency of occurrence, observed by NMR, these mechanistic features can be worked out with a discrimination greatly exceeding that of more traditional methods.

For a random copolymerization of two monomers m_1 and m_2, dyad, triad, and tetrad sequences may be represented as follows, ignoring stereochemistry:

Dyads	m_1m_1	m_1m (or m_2m_1)	m_2m_2
Triads	$m_1m_1m_1$		$m_2m_2m_2$
	$m_1m_1m_2$ (or $m_2m_1m_1$)		$m_1m_2m_2$ (or $m_2m_2m_1$)
	$m_2m_1m_2$		$m_1m_2m_1$
Tetrads	$m_1m_1m_1m_1$	$m_1m_1m_2m_1$ ($m_1m_2m_1m_1$)	$m_2m_2m_2m_2$
	$m_1m_1m_1m_2$ ($m_2m_1m_1m_1$)	$m_1m_1m_2m_2$ ($m_2m_2m_1m_1$)	$m_2m_2m_2m_1$ ($m_1m_2m_2m_2$)
		$m_2m_1m_2m_1$ ($m_1m_2m_1m_2$)	
	$m_2m_1m_1m_2$	$m_2m_1m_2m_2$ ($m_2m_2m_1m_2$)	$m_1m_2m_2m_1$

The system vinylidene chloride (m_1):isobutylene (m_2) is appropriate to consider since the copolymer has no asymmetric carbons and no vicinal J coupling. The proton NMR spectrum thus conveys only compositional sequence information. In Fig. 7.13 60 MHz proton NMR spectra of the homopolymers (a) and (b) are shown.[19-21] The homopolymer of vinylidene chloride gives a single resonance for the methylene protons (a); the homopolymer of isobutylene (which can be prepared with cationic but not with free radical initiators) gives singlet resonances of 3:1 intensity for the methyl and methylene protons. The spectrum of a copolymer, prepared

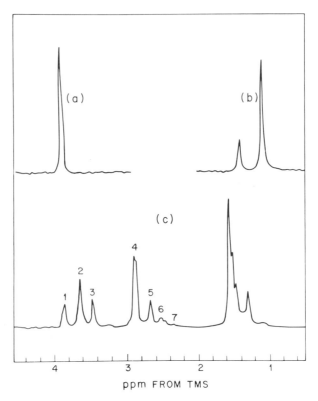

Fig. 7.13.   60 MHz proton spectra of (a) poly(vinylidene chloride); (b) polyisobutylene; (c) a vinylidene chloride (m_1):isobutylene (m_2) copolymer containing 70 mole % m_1. Peaks are identified with monomer tetrad sequences: (1) $m_1m_1m_1m_1$; (2) $m_1m_1m_1m_2$; (3) $m_2m_1m_1m_2$; (4) $m_1m_1m_2m_1$; (5) $m_2m_1m_2m_1$; (6) $m_1m_1m_2m_2$; (7) $m_2m_1m_2m_2$. (K. H. Hellwege, U. Johnsen, and K. Kolbe, *Kolloid-Z* **214**, 45 (1966).]

with a free radical initiator and containing 70 mole % vinylidene chloride, is shown in (c). The methylene resonances are grouped in three chemical shift ranges: m_1m_1, centered peaks at low field; peaks of methylene protons in m_1m_2, centered units

$$
\begin{array}{ccc}
\text{Cl} & & \text{CH}_3 \\
| & & | \\
\cdots - \text{C} - \text{CH}_2 - \text{C} - \cdots \\
| & & | \\
\text{Cl} & & \text{CH}_3
\end{array}
$$

near 3 ppm; and CH_2 and CH_3 resonances of m_2—centered sequences at high field. It is evident that tetrad sequences are involved (assignments given in the figure caption). If only dyad sequences were distinguished there would be only three methylene resonances, corresponding to m_1m_1, m_1m_2(or m_2m_1), and m_2m_2 sequences. (The upfield isobutylene peaks show considerable overlap and assignments here are less certain, but these resonances are not required for the analysis.) From measurements of the intensities of these resonances it can be shown that a growing chain radical ending in a vinylidene chloride unit is 3.3 times as likely to add another vinylidene chloride as to add isobutylene; while a growing chain ending in isobutylene is 20 times more likely to add vinylidene chloride.

A larger number of copolymer systems have been studied in this way, including those in which stereochemical and geometrical isomerism are present. The reactivity ratios thus obtained agree with those determined by more ponderous analytical methods, usually elemental analysis. The latter methods, of course, having nothing to say concerning structural details.

7.3 BIOPOLYMERS

We consider both polymers of biological origin — proteins and nucleic acids — and synthetic polymers of related structure. NMR is particularly effective in the study of such macromolecules. Even in solution, biopolymers commonly assume well-defined secondary structures such as helices, while synthetic polymers do not. The information furnished by NMR usually concerns this aspect of the chains rather than their covalent structure, which is generally known.

7.3.1 Amino Acids and Polypeptides

By NMR we can readily detect and measure the state of protonation of acidic and basic substances. We have seen this in Chapter 3, Section 3.3.2, where we observe in Fig. 3.4 that the protons of n—propylamine, particularly those of the methylene group to which the amino group is bonded, are strongly deshielded when the $-NH_2$ becomes $-NH_3^+$. In aqueous solution, protons (i.e., those which do not exchange with the solvent) become more shielded as the pH is raised, while ^{13}C nuclei generally become less shielded. (The fundamental origin of this behavior is unclear.) In Fig. 7.14, the 100 MHz spectra of the nonexchangeable protons of the amino acid *histidine* in acid, neutral, and basic D_2O solution are shown. (D_2O is used to avoid an excessively large solvent peak at *ca.* 5 ppm.) Smoothed plots of the chemical shifts of the α, β, C—2, and C—4

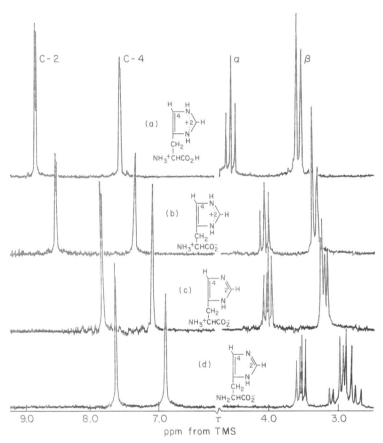

Fig. 7.14.   100 MHz spectra of histidine in D_2O as (a) dication
$(pD \sim 1)$;   (b)   zwitterion   with   imidazole   ring   charged
$(pD \sim 5)$;   (c)   zwitterion   with   imidazole   ring   uncharged
$(pD \sim 8)$;   (d)   anion $(pD \sim 12)$ (F. A. Bovey   and
A. I. R. Brewster, unpublished observations; see F. A. Bovey,
High Resolution NMR of Macromolecules, Academic Press,
1972, pp. 250–253.)

protons as a function of pD are shown in Fig. 7.15. The least shielded
proton and the one most affected by increasing the pD is that at C–2. It
reflects the state of ionization of the imidazole ring but not those of the
amino and carboxyl groups. The α–proton is not sensitive to the state of
ionization of the ring. The β– and C–4 proton plots reflect all four states
of ionization shown in Fig. 7.14 and exhibit three steps. The midpoints of

the steps correspond to the pK_a's of dissociation of the protonated species involved, C−2 and C−4 both reflecting that of the ring.

Histidine residues often play an important role in protein structure and function (Section 7.3.2) and may appear at or near the catalytic site in enzymes. For example, all four of the histidines in ribonuclease may be titrated separately, and the pK_a values thus obtained, particularly in the presence of inhibitors, provide valuable clues to enzyme function.[22] In the titration of proteins, of course, the protonation of amino and carboxyl groups will contribute significantly only if these groups occur on side chains.

The transition between the random coil and α−helical conformations of synthetic polypeptides has been intensively studied by NMR and has yielded results of interest.[22] In helix-supporting solvents such as chloroform, poly−γ−benzyl L−glutamate

$$\left[NH - \underset{\alpha}{CH} - \overset{\overset{\textstyle O}{\|}}{C} \right]_n$$

with side chain:
$$\underset{\alpha}{CH} - \underset{\beta}{CH_2} - \underset{\gamma}{CH_2} - \overset{\overset{\textstyle O}{\|}}{C} - O - CH_2 \cdot C_6H_5$$

gives a proton spectrum in which only the side-chain resonances are visible, and these are much broadened [Fig. 7.16, spectrum (a)].[23] (Further discussion of the α helix has been given in Chapter 6, Section 6.3.2.) Upon addition of 10% trifluoroacetic acid, the peaks narrow markedly and resonances appear for NH protons near 8 ppm and α protons near 4 ppm. Optical measurements show that the polymer is still in the α−helical form, so this transition represents the breaking up of associations between helices. As the trifluoroacetic acid content is increased, new resonances appear in the NH and α−CH resonance regions. These represent random coil segments of the polypeptide chains. They increase in intensity with increased trifluoroacetic acid content until at 20% [spectrum (f)] the polymer becomes all random coil, as indicated also by optical measurements. The behavior of the resonances at intermediate stages shows that they represent nonexchanging or slowly exchanging populations, which in turn suggests that the disrupting action of the trifluoroacetic acid — probably mainly a breakage of CO··NH hydrogen bonds within the helix — begins at the ends of the helix and works inward. It may be noted

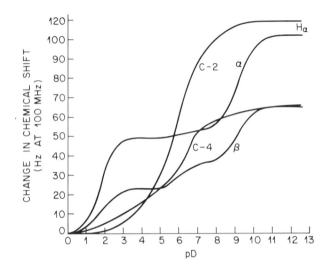

Fig. 7.15.   Chemical shifts (expressed in hertz at 100 MHz upfield from
             positions at lowest pH) of C–2, C–4, α– and β– protons of
             histidine versus *pD* (F. A. Bovey and A. I. R. Brewster,
             unpublished observations; see F. A. Bovey, *High Resolution*
             NMR of Macromolecules, Academic Press, New York, 1972,
             pp. 250–253.)

that when the helix–coil transition is made to take place all at once by a
temperature jump it is found to do so at microsecond rates.

In small polypeptides the splitting of the NH resonances gives a
measure of the NH–C$_\alpha$H *J* coupling and thus an indication of the chain
conformation. In longer, helix-forming sequences, the peaks are usually
too broad to permit the measurement of this coupling.

7.3.2 Proteins

Globular proteins in the native state are composed of polypeptide
chains, partly helical, which are rather tightly intertwined to form compact
spheres or ellipsoids. The complete structures of more than three hundred
such proteins are known from X-ray diffraction studies. There is little or
no free segmental motion in such structures and so the rotation of the
molecule as a whole determines the linewidth. If the molecular weight is
large, the correlation time (Chapter 1, Section 1.5, and Chapter 5,
Section 5.2.1) will be too long to permit truly high resolution spectra
except perhaps for certain side-chain protons. Thus, native ribonuclease
(molecular weight *ca.* 13,000) gives a moderately well-resolved spectrum in
aqueous solution, whereas that of bovine serum albumin (molecular weight

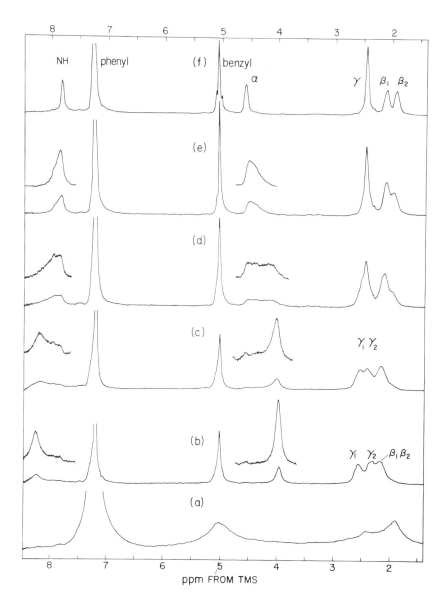

Fig. 7.16.   300 MHz spectra of poly-γ-benzyl *L*-glutamate of $\overline{DP}$ ~ 160 in CDCl₃ (5% wt./vol.) at 32°C. The volume percent of trifluoroacetic acid added is (a) 0%; (b) 10%; (c) 12%); (d) 13%; (e) 16%, (f) 20%. [F. A. Bovey and F. C. Schilling, *Macromol. Rev.* **9**, 1 (1975).]

ca. 65,000) is much more poorly resolved. Unfolding in 8 molar urea gives a spectrum composed of narrow lines since segmental motion is now possible, but of course most of the secondary structure is thus disrupted. The effect of molecular weight is illustrated by the 360 MHz proton spectrum of native carboxypeptidase inhibitor in D_2O solution (Fig. 7.17), which has a molecular weight of only *ca.* 4000. The resonances at low field (7−8 ppm) arise from aromatic and histidine ring protons (peptide NH, arginine NH, and histidine NH protons would appear at still lower field − 8.0 − 8.5 ppm − but have been exchanged for deuterium). The band of resonances at 4−5 ppm arises mainly from backbone α−CH protons and is followed by a wide variety of sidechain protons and, at highest field (0.4−1.5 ppm), by methyl protons of valine, leucine, and isoleucine. The latter show small splittings (*ca.* 6 Hz) arising from vicinal couplings.

Two-dimensional NMR (Chapter 6) has been extensively employed in the study of protein structure, both in confirming known X-ray structures and in exploring proteins for which the X-ray structure is undetermined. The spectra are highly complex, but complete proton resonance assignments can be obtained by employing both COSY (Chapter 6, Section 6.3.3) and NOESY (Section 6.3.2) in a complementary fashion. The former indicates covalently bonded groups and the latter reveals spatially close proton pairs, which may actually be remote in the amino acid sequence, although of course not necessarily so.

A dramatic example of protein 2D NMR spectroscopy is shown in Fig. 7.18.[24] This is the 500 MHz proton C spectrum − shown in stacked plot form − of bovine pancreatic trypsin inhibitor, commonly abbreviated BPTI, in 90:10 $H_2O:D_2O$ solvent. It is actually a COSY spectrum and provides a "fingerprint" of the amino acid sequence. The corresponding NOESY spectrum, although providing spatial information, would be of similar appearance. The stacked plot, although time-consuming to record and less convenient for establishment of detailed correlations, provides a more striking representation of diagonal and cross-peak intensities than the contour plot.

7.3.3 Polynucleotides and Nucleic Acids

The nucleic acid chain is composed of pentose units connected by phosphate links, with purine and pyrimidine bases attached to the sugar units. The base-sugar-phosphate unit is known as a *nucleotide* unit. There are two classes of nucleic acids: those in which the sugar unit is d-ribose, called ribonucleic acids or RNA, and those in which the sugar unit is 2-deoxy-d-ribose, called deoxyribonucleic acid or DNA. A schematic segment of a DNA chain is shown below, showing the bases adenine (A), thymine (T), guanine (G), and cytosine (C) attached to the sugar units:

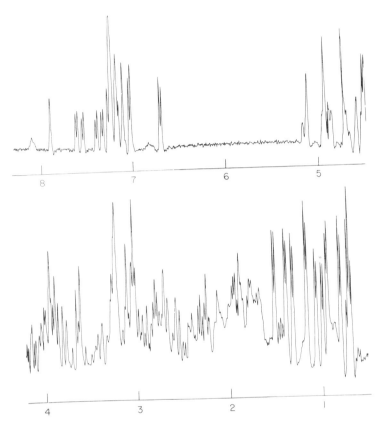

Fig. 7.17.   360 MHz proton spectrum of native carboxypeptidase
inhibitor   in   D₂O   solution.   (D. J. Patel,   private
communication.)

Segment of DNA chain

In the DNA molecule, two such chains are wound in a helical pattern about a common axis. The two chains are linked together by hydrogen bonds between the bases, the planes of which are perpendicular to the helix axis. Steric requirements are such that an adenine must be paired with a thymine at the same level in the other chain and a guanine with a cytosine. A synthetic double helical polymer may be formed in which only adenine and thymine are present, arranged in an alternating sequence and giving rise to interchain hydrogen bonds; such a polymer is designated as poly(dA—dT). In Fig. 7.19 is shown the 500 MHz proton spectrum of a helical poly(dA—dT) composed of approximately 46 base pairs (molecular weight *ca.* 32,000), observed in D_2O to avoid the very large solvent resonance seen in H_2O solutions.[25] As with proteins, the exchangeable sugar hydroxyl and NH protons have disappeared from the spectrum. The important interbase hydrogen bond protons would appear at 14 ppm in H_2O and are shown in the inset.

When the double helix is opened up to the random coil, the resonances greatly narrow and undergo shifts in position, the latter being particularly marked for the NH protons. This furnishes an important method for following and observing the details of this transition. The proton resonances are also better resolved the shorter the polynucleotide chain — as would be expected — and much information has been gained concerning the structure, stability, and binding of drugs by observation of small duplexes of known structure. Figure 7.19 represents about the limit in molecular weight at which useful NMR spectra may be obtained. At natural molecular weights — 10^8 to 10^9 — the native DNA spectra show resonances so broad as to be barely visible.

7.4 CHAIN MOTION IN MACROMOLECULES

We have seen (Chapter 5, Section 5.2.9) that as molecules become larger the principal nuclear relaxation mechanism becomes the dipole—dipole interaction. For macromolecules, both in solution and in the solid state (Chapter 8), this mechanism dominates completely for spin-1/2 nuclei and it is generally not necessary to consider any other. (For quadrupolar nuclei such as deuterium, quadrupolar relaxation dominates completely; see Chapter 8, Section 8.8.)

Synthetic polymers are useful mainly for their solid-state properties, so the solid state may be said to have greater practical importance. Nevertheless, solution measurements are often more revealing of intramolecular influences, divorced from the effects of neighboring chains. The reader is referred to reviews[26,27] for a more complete discussion.

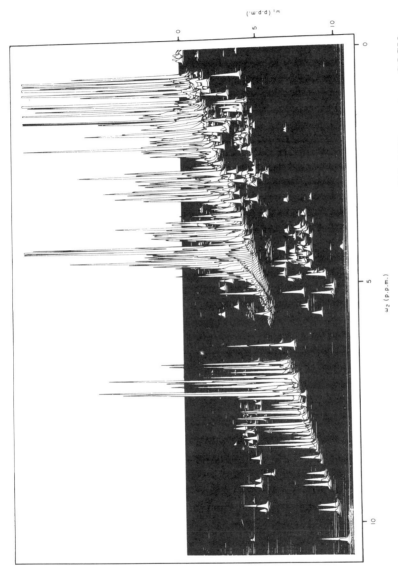

Fig. 7.18.   Two-dimensional stacked plot of absolute value 500 MHz proton COSY spectrum of a 0.02 M solution of bovine pancreatic trypsin inhibitor (BPTI) in 90:10 H₂O:D₂O solvent. [G. Wagner and K. Wüthrich, *J. Mol. Biol.* **155**, 347 (1982).]

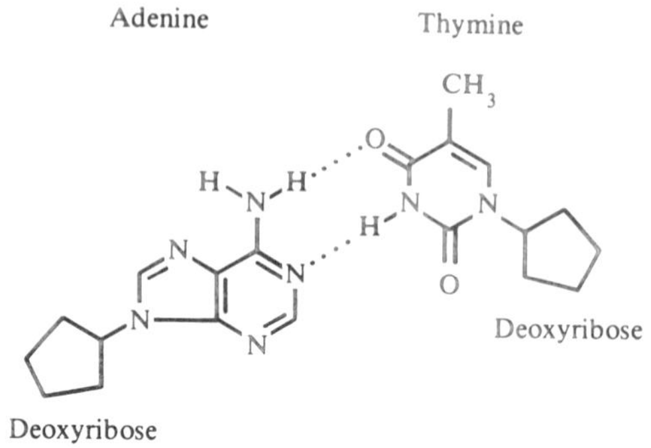
Adenine
Thymine
CH3
Deoxyribose
Deoxyribose

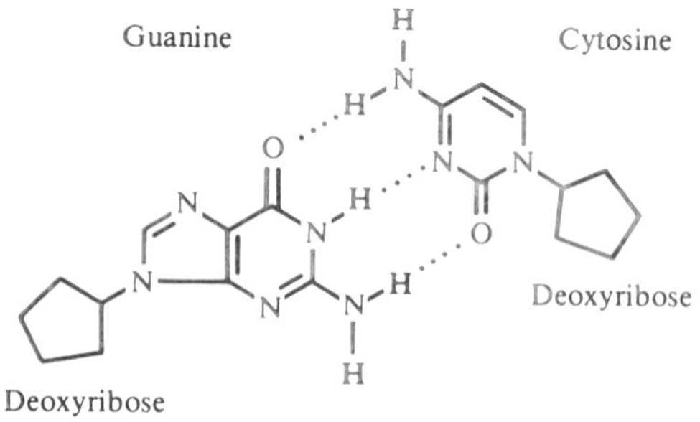
Guanine
Cytosine
Deoxyribose
Deoxyribose

We adopt here the same very simple dynamic model described in Chapter 1, Section 1.5, and Chapter 5, Section 5.2.1, i.e., a rigid sphere immersed in a viscous continuum and reoriented by small, random diffusive steps. With rare exceptions, almost all nuclear relaxation data obtained for polymers in solutions concern relaxation of carbon−13 in natural abundance by bonded protons. For the spin lattice relaxation of carbon−13 we have seen that

$$\frac{1}{T_1(\mathrm{C})} = \frac{K^2}{20}\left[J_0(\omega_{\mathrm{H}}-\omega_{\mathrm{C}}) + 3J_1(\omega_{\mathrm{C}}) + 6J_2(\omega_{\mathrm{H}}+\omega_{\mathrm{C}})\right] \quad (5.25)$$

where

$$K = \frac{\gamma_{\mathrm{C}}\gamma_{\mathrm{H}}\hbar}{r^3_{\mathrm{C-H}}} \quad (5.9)$$

and

$$J_n = \frac{2\tau_c}{1+(n\,\omega_i)^2\tau_c^2} \quad (5.10)$$

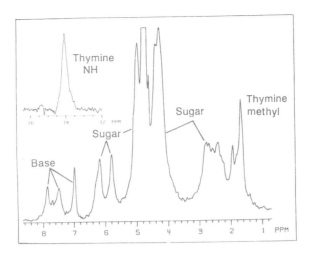

Fig. 7.19.   500 MHz proton spectrum of a helical poly $(dA-dT)$ having approximately 46 base pairs (molecular weight *ca.* 32,000) observed in D_2O.

Plots of Eq. (5.25) for two values of the magnetic field strength are shown in Fig. 5.3. For very rapid motion, when $\omega_0\tau_c \ll 1$, all J_n becomes equal to τ_c and Eq. (5.25) simplifies to

$$\frac{1}{N_H T_1} = \tau_c \frac{\gamma_H^2 \gamma_C^2 \hbar^2}{r_{C-H}^6} \tag{5.26}$$

or

$$\tau_c = \frac{4.7 \times 10^{-11}}{N_H \cdot T_1} \tag{5.28}$$

where N_H is the number of protons bonded to the observed carbon, more distant ones being neglected. We shall employ the single correlation time motional model in the subsequent discussion, although the assumption of isotropic motion clearly cannot be adequate for long chains.

We have seen that most synthetic polymers in solution give fairly well-resolved proton and carbon—13 spectra and thus must enjoy considerable segmental freedom, since rigid molecules of comparable molecular weight, such as helical polypeptides and polynucleotides (Section 7.3) give much broader resonances. It is further observed that linewidths are independent of molecular weight down to the oligomer level and very nearly independent of solution concentration and viscosity up to 25—30%. As we have seen (p.   ), solutions so viscous that they barely pour from the NMR tube give quite well-resolved spectra, another indication of the dominance of local motion. Nevertheless, polymer solutions show lines somewhat broadened compared to small-molecule counterparts. Some of this broadening, in polymers of irregular structure, represents an envelope of unresolved chemical shifts (Sec. 7.2.1). The true solution proton linewidth for many polymers is usually 3—5 Hz at moderately elevated temperatures, whereas carbon—13 resonances are somewhat narrower, typically 1—2 Hz. For even the most flexible polymer chains, it is found that $T_1 > T_2$.

The motion of flexible polymers in solution may be conveniently regarded as being composed of segmental motion and overall tumbling, a concept we have already discussed in Chapter 5, Section 5.2.9, in connection with carbon—13 relaxation in paraffinic chains. When the carbon—13 T_1 is measured as a function of the average number of monomer units per chain, it is found that it at first decreases and then levels off at a value of about 100 units. Above this relatively low molecular weight, e.g., 10,000 for polystyrene, overall tumbling becomes too slow to provide a significant contribution compared to segmental motion.

In Table 7.2 are listed a number of synthetic polymers together with the conditions of observation and the values of NT_1 for backbone carbon—13 nuclei (at 25 MHz unless otherwise indicated). In the sixth column are shown values of the correlation time τ_c calculated from Eq. (5.24b). It is found that τ_c is well on the fast-motion side of the T_1 minimum except in the case of poly(butene-1-sulfone), for which T_1 decreases with increasing temperature.

From these data we may draw these conclusions:

1. *Effect of Side Chains.* Comparison of *polyethylene* with *polypropylene*, *polybutene-1*, and *poly(vinyl chloride)* shows that the presence of side chains impedes chain mobility and that the larger the side chain the slower is the motion. A methyl group slows the chain by a factor of 2—3 and a chlorine is comparable, while ethyl and phenyl (polystyrene) groups have a further impeding effect. The main-chain conformational barrier may be regarded as composed of a symmetrical threefold torsional potential on which is superimposed nonbonded interactions between the side groups, the latter increasing with the size of the group.

It may at first seem surprising that the presence of two α substituents, as in *polyisobutylene*, *poly(vinylidene chloride)*, and *poly(methyl methacrylate)*, does not impede chain motion significantly more than a single substituent. This may be attributed to the fact that two substituents increase the crowding in *all* rotational states, including eclipsed states, without necessarily increasing the barrier between them.

2. *Effect of Heteroatoms.* The chains of *polyoxyethylene* are among the most flexible of those in Table 7.2, with a correlation time at 30°C comparable to that of polystyrene at 100°C. *Poly(propylene oxide)* shows an analogous relationship to polypropylene. *Poly(methylthiirane)* [poly(propylene sulfide)] and *poly(phenylthiirane)* show the liberating influence of a sulfur atom, while *poly(styrene peroxide)* shows the comparable effect of a peroxy link, −O−O−. These findings are consistent with the known low rotational barriers at hetero atoms, but it is not so evident why *polyoxymethylene*, with an oxygen inserted at every carbon atom, should be less flexible even than polyethylene. This may be rationalized on the basis of the energetic preference for gauche conformers in polyoxymethylene,

which tends to inhibit chain motions requiring a succession of both gauche and trans bonds intermixed.

We have seen that sulfone groups, in contrast to sulfur atoms alone, stiffen the chain very markedly, as in *poly(butene-1-sulfone)*.

3. *Effect of Double Bonds.* Although the double bond prevents rotation at the olefinic carbons, the reduced barriers at the allylic bonds more than compensate for this, and as a result, both *cis-* and *trans-*1,4-polybutadiene are substantially more flexible than polyethylene.

4. *Effect of Stereochemistry.* A dependence of chain mobility on stereochemical configuration has been observed in vinyl polymers, both for entire chains and for steric sequences in atactic chains. Syndiotactic sequences commonly appear to be more restricted than isotactic. This has been reported for *poly(methyl methacrylate)*, *polypropylene*, *polybutene-1*, and *poly(vinyl chloride)*, while for *polystyrene* the reverse is found.

TABLE 7.2

Carbon−13 T_1 Values and Approximate Correlation Times,
Calculated on the Isotropic Diffusion Model for Polymer Main-Chain Carbons in Solution*

Polymer	Solvent	Temperature (°C)	Concentration (wt.%)	NT_1, (s)	τ_c, (ns)	Reference
polyethylene	TCB[a]	110	33	2.52	0.019	m
	ODCB[b]	100	25	2.70	0.018	n
	ODCB	30[c]	25	1.24	0.040	n
polypropylene	ODCB	100	25	1.13	0.044	n
		30[d]	25	0.39	0.13	n
polyisobutylene	CDCl$_3$	30	5	0.30	0.16	o
polybutene-1	CCl$_3$CHCl$_2$	100	10	0.34[e]	0.14	p
1,4-polybutadiene						
cis	CDCl$_3$	54	20	3.00	0.016	q
trans	CDCl$_3$	54	20	2.38	0.021	q
poly(vinyl chloride)	TCB[a]	107	10	0.32	0.15	r
poly(vinylidene chloride)	HMPA-d_{18}[f]	40[g]	15	0.078[h]	0.63	s
		89	15	0.25[h]	0.20	s
poly(vinylidene fluoride)	DMF[i]	41	20	0.62	0.079	t
poly(vinyl alcohol)	Me$_2$SO	30	20	0.090	0.55	u
polyacrylonitrile	Me$_2$SO	50	20	0.15	0.33	u
polystyrene						
atactic	toluene-d_8	30	15	0.100	0.49	v
isotactic	toluene-d_8	30	15	0.090	0.55	v
poly(methyl methacrylate)						
isotactic	CDCl$_3$	38	10	0.126	0.39	w
syndiotactic	CDCl$_3$	38	10	0.080	0.62	w
atactic	DMF[i]	41	20	0.072[j]	0.81	t
poly-α-methylstyrene	CDCl$_3$	30	10	0.100	0.49	x
polyoxymethylene	HFIP[k]	30	3	0.60	0.082	y
polyoxyethylene	C$_6$D$_6$	30	5	2.80	0.018	z
poly(propylene oxide)	CDCl$_3$		5	1.00	0.049	aa
poly(methyl thiirane)	CDCl$_3$	30	10	0.86	0.057	bb
poly(phenyl thiirane)	CDCl$_3$	25	15	0.19	0.25	cc
poly(styrene peroxide)	CHCl$_3$	23	22	0.24	0.21	dd
poly(butene-1-sulfone)	CDCl$_3$	40	25	0.090	23[l]	ee

[a] 1,2,4-Trichlorobenzene.

[b] 4o-Dichlorobenzene.

[c] Extrapolated from high-temperature data with activation energy of 2.51 kcal mole^{-1}; see Heatley (Ref. 26).

[d] Extrapolated from high-temperature data with activation energy of 3.42 kcal mole^{-1}; see Heatley (Ref. 26).

* Table reprinted from F. A. Bovey and L. W. Jelinski, *J. Phys. Chem.* **89**, 571 (1985).

[e] At 22.6 MHz.

[f] Hexa(methyl-d_3) phosphoramide.

[g] Five other temperatures also reported.

[h] At 15 MHz.

[i] N,N'-Dimethylformamide.

[j] rrr sequences (racemic tetrads of CH_2 groups).

[k] Hexafluoro-2-propanol.

[l] $\overline{M}_n$ = 2020; some contribution from overall tumbling.

[m] F. C. Schilling, private communication; reported by F. A. Bovey in *Stereodynamics of Molecular Systems*, R. H. Sarma, Ed., Pergamon, Oxford, 1979.

[n] Y. Inoue, A. Nishioka, and R. Chûjô, *Makromol. Chem.* **168**, 163 (1973).

[o] F. Heatley, *Polymer* **16**, 493 (1975).

[p] F. C. Schilling, R. E. Cais, and F. A. Bovey, *Macromolecules* **11**, 325 (1978).

[q] W. Gronski and N. Murayama, *Makromol. Chem.* **177**, 3017 (1976).

[r] F. C. Schilling, *Macromolecules* **11**, 1290 (1978).

[s] K. Matsuo and W. H. Stockmayer, *Macromolecules* **14**, 544 (1981).

[t] F. A. Bovey, F. C. Schilling, T. K. Kwei, and H. L. Frisch, *Macromolecules* **10**, 559 (1977).

[u] Y. Inoue, A. Nishioka, and R. Chûjô, *J. Polym. Sci., Poly. Phys. Ed.* **11**, 2237 (1973).

[v] W. Gronski and N. Murayama, *Makromol. Chem.* **179**, 1509 (1978).

[w] J. R. Lyerla, T. T. Horikawa, and D. E. Johnson, J. Am. Chem. Soc., **99**, 2463 (1977).

[x] F. Lauprêtre, C. Noël, and L. Monnerie, *J. Polym. Sci., Polym. Phys. Ed.* **15**, 2143 (1977).

[y] G. Hermann and G. Weill, *Macromolecules* **8**, 171 (1975).

[z] F. Heatley and I. Walton, *Polymer* **17**, 1019 (1976).

[aa] F. Heatley, *Polymer* **16**, 493 (1975).

[bb] K. J. Ivin and F. Heatley, unpublished resluts reported in F. Heatley, *Prog. NMR Spectros.* **13**, 47 (1980).

[cc] R. E. Cais and F. A. Bovey, *Macromolecules* **10**, 752 (1977).

[dd] R. E. Cais and F. A. Bovey, *Macromolecules* **10**, 169 (1977).

[ee] R. E. Cais and F. A. Bovey, *Macromolecules* **10**, 757 (1977).

REFERENCES

1. F. C. Schilling, F. A. Bovey, M. D. Bruch, and S. A. Kozlowski, *Macromolecules* **18**, 1418 (1985).

2. F. A. Bovey and G. V. D. Tiers, *J. Polym. Sci.* **44**, 173 (1960); see also Ref. 1 and references therein.

3. R. C. Ferguson *Polym. Prepr.*, **8** (2), 1026 (1967).

4. R. C. Ferguson, *Trans. N.Y. Acad. Sci.* **29**, 495 (1967).

5. F. Heatley and A. Zambelli, *Macromolecules* **2**, 618 (1969).

6. F. Heatley, R. Salovey, and F. A. Bovey, *Macromolecules* **2**, 619 (1969).

7. A. Zambelli, D. E. Dorman, A. I. R. Brewster, and F. A. Bovey, *Macromolecules* **6**, 925 (1973).

8. A. E. Tonelli and F. C. Schilling, *Acc. Chem. Res.* **14**, 233 (1981).

9. F. C. Schilling, unpublished observations.

10. L. W. Jelinski, unpublished results. See also F. A. Bovey, *Chain Structure and Conformation of Macromolecules*, Academic Press, New York, 1982, pp. 105–121.

11. C. W. Wilson, III, and E. R. Santee, Jr., *J. Polym. Sci.*, Part **C81**, 97 (1965).

12. R. E. Cais and J. M. Kometani in *NMR and Macromolecules, ACS Symp. Ser.*, No. 247, J. C. Randall, ed., 1984, pp. 153–166.

13. M. D. Bruch, F. A. Bovey, and R. E. Cais, *Macromolecules* **17**, 2547 (1984).

14. C. W. Wilson, III, *J. Polym. Sci.*, Part **A1**, 1305 (1963).

15. R. Liepins, J. R. Surles, N. Morosoff, V. T. Stannett, M. L. Timmons, and J. J. Wortman, *J. Polym. Sci., Polym. Chem. Ed.* **16**, 3039 (1978).

16. R. C. Ferguson and E. G. Brame, *J. Phys. Chem.* **83**, 1379 (1979).

17. A. E. Tonelli, F. C. Schilling, and R. E. Cais, *Macromolecules* **18**, 1354 (1985).

18. R. E. Cais and J. M. Kometani, *Macromolecules* **15**, 849 (1982).

19. J. B. Kinsinger, T. Fischer, and C. W. Wilson, III, *J. Polym. Sci.*, Part **B4**, 379 (1966).

20. J. B. Kinsinger, T. Fischer, and C. W. Wilson, III, *J. Polym. Sci.*, Part **B5**, 285 (1967).

21. K. H. Hellwege, U. Johnson, and K. Kolbe, *Kolloid-Z.* **214**, 45 (1966).

22. F. A. Bovey, *High Resolution NMR of Macromolecules*, Academic Press, New York, 1972, pp. 352-256.

23. F. C. Schilling and F. A. Bovey, *Macromol. Rev.* **9** 1 (1975).

24. G. Wagner and K. Wüthrich, *J. Mol. Biol.* **155**, 347 (1982).

25. R. W. Behling and D. R. Kearns, *Biochemistry* **25**, 3335 (1986).

26. F. Heatley, *Prog. NMR Spectrosc.* **13**, 47 (1979).

27. F. A. Bovey and L. W. Jelinski, *J. Phys. Chem.* **89**, 571 (1985).

CHAPTER 8

NMR OF SOLIDS

8.1 INTRODUCTION

We have seen in earlier discussion (Chapter 1, Section 1.6; Chapter 3, Section 3.3.1; Chapter 5, Section 5.3) that the NMR spectra of solids are generally characterized by broad, featureless resonances which, while they may give significant information concerning nuclear relaxation, seem nearly devoid of structural information. It may be more correct, however, to say that the structural information — principally chemical shifts — while inherently present is masked by line broadening effects. The last 15 years have seen the development of methods by which solids can be made to yield spectra comparable in resolution to those obtained for liquid systems. The principal emphasis in this chapter will be on the description and illustration of these methods.

This will also be an appropriate place to describe *deuterium quadrupole echo* spectroscopy. This is not a high-resolution method but is uniquely powerful in providing specific geometrical and dynamic insights into the motions of molecules in the solid state.

The high-resolution NMR of solids has been the subject of many books and reviews. The books of Haeberlen,[1] Mehring,[2] and Fyfe[3] are recommended for fundamental background reading; the first two are on a somewhat more advanced mathematical level. Other reviews (of a more popular nature) are those of Griffin,[4] Harris and Packer,[5] Gray and Hill,[6] Yannoni,[7] Lyerla *et al.*,[8] Gerstein,[9,10] and Havens and Koenig.[11] An excellent discussion is also given by Harris.[12] The book edited by Komoroski[13] deals specifically with the high-resolution NMR of synthetic polymers in the solid state. An article of broader coverage on the NMR of solid polymers is that by McBrierty and Douglass,[14] while McCall[15] has reviewed the proton NMR of solid polymers.

8.2 PROTON DIPOLAR BROADENING

8.2.1 General

The local magnetic field at a carbon–13 nucleus in an organic solid is given by

$$B_{loc} = \pm \frac{h}{4\pi} \gamma_H \frac{(3 \cos^2 \theta_{C-H} - 1)}{r_{C-H}^3} \tag{8.1}$$

where γ_H is the magnetogyric ratio of the proton and r_{C-H} is the internuclear C–H distance to a nearby proton, assumed to be bonded ($r_{C-H} = 1.09$ Å). As we have seen (Chapter 1, Section 1.6), the $\pm$ sign results from the fact that the local field may add to or subtract from the laboratory field B_0 depending upon whether the neighboring proton dipole is aligned with or against the direction of B_0, an almost equal probability. The angle θ_{C-H} is defined in Fig. 8.1. If r_{C-H} and θ_{C-H} were fixed throughout the sample, as for isolated $^{13}C-{}^1H$ pairs in a single crystal, this interaction would result in a splitting of the ^{13}C resonance into equal components, the separation of which would depend on the orientation of the crystal in the magnetic field. The effect can be large. It is found from Eq. (8.1) that for $^{13}C-{}^1H$ bonded pairs oriented parallel to B_0, the splitting is 40 kHz.

Most organic solids are glassy or microcrystalline and one observes a summation over many values of θ_{C-H} and r_{C-H}, resulting in a proton

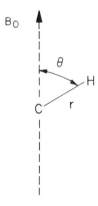

Fig. 8.1.   An internuclear $^{13}C-{}^1H$ vector of length r making an angle θ with the magnetic field B_0.

dipolar broadening of many kilohertz. We have seen in Chapter 1, Section 1.6, that rapid reorientation of the $^{13}C-{}^1H$ internuclear vectors, especially if isotropic or nearly so, will result in a narrowing of the broadened lines when the rate of reorientation exceeds the linewidth, expressed as a frequency. If the reorienting $^{13}C-{}^1H$ vectors sample all angles θ_{C-H} in a time short compared to the reciprocal of the dipolar coupling, i.e., $\ll 2 \times 10^{-5}$ s, dipolar broadening is reduced to a small value, as in liquids. In solids, such rapid isotropic tumbling is not possible, but since the term $(3 \cos^2 \theta_{C-H} - 1)$ in Eq. (8.1) is zero when θ_{C-H} equals $\cos^{-1} 3^{-1/2}$ or *ca.* 54.7°, it was long ago realized that spinning solid samples at this angle might be effective in reducing the linewidth. These early studies were devoted to proton–proton rather than proton–carbon interactions.

Lowe[16] and Andrew[17] independently and simultaneously first attempted such "magic angle" spinning (MAS) experiments but they were not really successful for this purpose because it is not mechanically feasible to design rotors that can spin faster than about 10^4 Hz, well short of the 40–50 kHz required. At higher speeds all suitable materials (metals being excluded) disintegrate under the extreme g forces generated. We shall see, however, that while MAS is not generally useful for removing proton dipolar broadening, it is very effective for collapsing the lesser broadening arising from chemical shift anisotropy.

8.2.2 High-Power Proton Decoupling

A more practical method for the removal of proton dipolar broadening is to employ a high-power proton decoupling field, i.e., *dipolar decoupling.* This is analogous to the decoupling experiment discussed in Chapter 4 (Section 4.4) for the removal of $^{13}C-{}^1H$ scalar couplings except that much greater power is required. Instead of the *ca.* 1 gauss decoupling field used for solutions, a field of *ca.* 10^{-3} tesla (*ca.* 10 gauss) must be used for the removal of direct dipole–dipole $^{13}C-{}^1H$ coupling in solids, corresponding to *ca.* 40 kHz expressed as a frequency. This approach is now standard for the observation of solid-state spectra of carbon–13 and other rare or dilute spins. We shall see several examples of its application in the subsequent discussion. It is obvious that it will not serve for the observation of high-resolution proton spectra of solids since one cannot observe the same nucleus that one is simultaneously irradiating with high power. There is a way, however, of observing such proton spectra of solids; it will be briefly discussed in Section 8.7.

8.3 CHEMICAL SHIFT ANISOTROPY

With proton dipolar decoupling alone, the resonances in a typical solid-state carbon—13 spectrum, although showing some narrowing, remain very broad, with widths of the order of 10—200 ppm (or 0.5—10 kHz at an observing frequency of 50 MHz). This broadening arises from the fact that the chemical shift of a particular carbon is *directional*, depending on the orientation of the molecule with respect to the magnetic field. We have seen (Chapter 3, Section 3.3.1) that the chemical shift is expressed as a tensor [Chapter 3, Eq. (3.3)] and that the isotropic chemical shift, observed for a rapidly tumbling molecule in solution, is one-third of the trace of this tensor:

$$\sigma_i = \frac{1}{3}(\sigma_{11} + \sigma_{22} + \sigma_{33}) \qquad (3.4)$$

Since most organic solids are amorphous or microcrystalline, their carbon chemical shifts take on a continuum of values and form the line shapes shown in Fig. 3.3. The top pattern corresponds to a chemical shift anisotropy that lacks axial symmetry, while the bottom pattern corresponds to an axially symmetric anisotropy. In the axially symmetric case, $\sigma_{22} = \sigma_{33}$ and both can be designated as $\sigma_\perp$, while $\sigma_{11} = \sigma_\parallel$. The principal values σ_{11}, σ_{22}, and σ_{33} are usually expressed in ppm, as we have seen, and can be measured directly from the singularities or discontinuities in the spectrum, as shown in Fig. 3.3. The isotropic chemical shifts are indicated by dashed lines. For axially symmetric patterns, the position of $\sigma_\parallel$ represents the observed frequency when the principal axis system is parallel with the field direction, while $\sigma_\perp$ represents the observed frequency when the orientation is perpendicular to the field.

Table 8.1 contains the principal values of the chemical shift tensors for a number of carbon types of common occurrence. They represent averaged values from a number of compounds of each kind and are taken from the extensive compilation of Duncan,[18] which also includes data and literature references for over 350 specific compounds. A smaller compilation will also be found in the book by Mehring,[2] pp. 253—256.

Spectra of a number of carbonyl compounds are shown in Fig. 8.2. These were all observed at *ca.* 77 K in the polycrystalline state by Pines *et al.*[19] The downfield pattern is that of the carbonyl group, while the narrower upfield resonances are those of methyl groups. A striking feature of these spectra is the observation that while σ_{33} and σ_{11} for the carbonyl carbons are very similar for all compounds, the isotropic chemical shifts (shown in parentheses in each spectrum and expressed with reference to

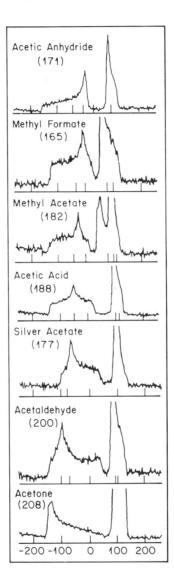

Fig. 8.2.   Carbon-13 spectra of polycrystalline compounds containing carbonyl groups. All observations at 77 K. The zero of reference is external benzene at 23°C (*ca.* 125 ppm versus TMS). [Spectra from A. Pines, M. Gibby, and J. S. Waugh, *J. Chem. Phys.* **59**, 569 (1973). See also R. G. Griffin, *Anal. Chem.* **49**, 951A (1977).]

TMS) show substantial variations, to be attributed almost entirely to variations in σ_{22} with differing substituents.

8.4 MAGIC ANGLE SPINNING

The carbon anisotropies shown in Table 8.1 and Fig. 8.2 are of much interest and repay careful experimental and theoretical study. However, in molecules of any degree of complexity there will be several such patterns that may strongly overlap, producing a broad, uninterpretable spectrum. Under these circumstances it usually becomes necessary to sacrifice the

TABLE 8.1

Principal Values of Chemical Shift Tensor
for Carbon Functionalities

Group	Formula	Chemical shift parameters				
		σ_{11}	σ_{22}	σ_{33}	σ_i	$\|\sigma_{33}-\sigma_{11}\|$
methyl	$-CH_3$	38	29	3	23	35
methylene	$-CH_2-$	54	41	16	37	38
methine	$-CH$	40	32	28	33	12
alcohol, I°	$-CH_2OH$	80	70	25	58	55
alcohol, II°	$CHOH$	90	83	42	72	48
ether	$-O-CH_3$	83	71	12	55	71
olefinic	$C=C$	223	127	34	128	189
aldehyde	$-CHO$	276	229	90	198	186
ketone	$-RCOR'$	279	245	85	203	194
carboxyl	$-COOH$	247	171	106	175	141
ester	$-COOR$	258	132	107	166	151
carboxylate	$-COO^-$	239	188	105	176	134

(*continues*)

TABLE 8.1 (Continued)

Group	Formula	\multicolumn{5}{c}{Chemical shift parameters}
		σ_{11}	σ_{22}	σ_{33}	σ_i	$\lvert\sigma_{33}-\sigma_{11}\rvert$
amide	$-CONH_2$	243	179	94	172	149
aromatic						
protonated		215	145	17	126	198
carbon-substitute		221	160	21	134	200
oxygen-substitute		227	158	71	152	156
acetylene	$-C\equiv C-$	151	151	-79	75	230
nitrile	$-C\equiv N$	227	208	-89	115	316

anisotropy information in order to retain the isotropic chemical shifts with a reasonable degree of resolution. The carbon-13 anisotropies in the last column of Table 8.1 correspond to linewidths of about 800-5000 Hz at an observing frequency of 100 MHz (proportionately higher at higher frequencies.) These frequencies are within the range attainable by mechanical spinning with air-driven rotors, such as those shown in Fig. 8.3. Under such rotation the direction cosines in Eq. (3.3) become time dependent in the rotor period. Taking the time average under rapid rotation, Eq. (3.3) becomes

$$\sigma = \frac{1}{2}\sin^2\beta(\sigma_{11}+\sigma_{22}+\sigma_{33}) + \frac{1}{2}(3\cos^2\beta-1)$$

$$\times \text{ (functions of direction cosines)} \qquad (8.2)$$

The angle β is now the angle between the rotation axis and the magnetic field direction. When β is the magic angle, $\sin^2\beta$ is $\frac{2}{3}$ and the first term becomes equal to one-third of the trace of the tensor, i.e., the isotropic chemical shift [see Eq. (3.4)]. The $3 \cos^2 \beta - 1$ term becomes zero, as we have seen. Thus, under magic angle rotation the chemical shift pattern collapses to the isotropic average. We shall discuss several examples of such "magic angle" spectra in Section 8.6, but before this we must describe one additional technique, which, while not as fundamental as dipolar decoupling and sample spinning to the recapturing of structural information from solids, is of great practical importance.

8.5 CROSS POLARIZATION

We have seen that a single 90° pulse, applied at the resonant frequency, produces a free induction decay signal. Our earlier discussion has implied that the resonant system of nuclei is a liquid, but this situation holds true for a solid as well. In solids, however, molecular motion is limited and this leads to very inefficient spin lattice relaxation. For carbon—13 nuclei one may see this by inspection of Fig. 5.3; in solids we are on the right-hand branch of the curve, since correlation times will usually be well in excess of a μs and may readily be much longer. Even in

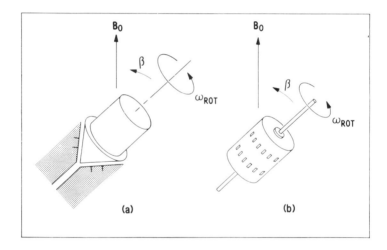

Fig. 8.3.   Two designs for magic angle sample spinners. (a) An Andrew design sample holder, rotating on air bearings within a stator (shaded), making an angle β to the magnetic field B_0. The high-pressure nitrogen jets are indicated by solid lines on the stator. (b) Lowe design rotor, supported by phosphor-bronze axles. High-pressure air or nitrogen impinges on the flutes.

solids that contain methyl groups (methyl groups generally rotate rapidly), the T_1 values for the other carbons are prohibitively long. The carbon of the methyl group cannot "communicate" its short T_1 value to other carbons by spin diffusion (mutual spin flip-flops; see Chapter 1, Section 1.6; Chapter 5, Section 5.2.1.2) as its nearest carbon−13 neighbor is on average 7Å distant. Even though the protons on the methyl carbons do communicate their short T_1 to other protons in the sample via diffusion to nearby spins, these protons cannot communicate their short T_1 back to other carbons in the sample because the carbons and protons resonate at very different frequencies.

Cross polarization takes advantage of the fact that proton spin diffusion generally causes all of the protons in a solid to have the same T_1 value, and that the proton T_1 is usually short compared to the carbon T_1 values. (For the sake of concreteness and because it is the commonest case, we shall discuss cross polarization from protons to carbons; however, these concepts apply to polarization between any abundant spin, ^{19}F, for example, and a rare spin.) Cross polarization works by effectively forcing an overlap of proton and carbon energies in the rotating frame despite the absence of such an overlap in the laboratory frame. The means of doing this was demonstrated by Hartmann and Hahn in 1962.[20] Energy transfer between nuclei with widely differing Larmor frequencies can be made to occur when

$$\gamma_C B_{1C} = \gamma_H B_{1H} \qquad (8.3)$$

This equation expresses the *Hartmann−Hahn condition*. Since γ_H is four times γ_C, the Hartmann−Hahn match occurs when the strength of the applied carbon field B_{1C} is four times the strength of the applied proton field B_{1H}. When the proton and carbon rotating frame energy levels match, polarization is transferred from the abundant protons to the rare carbon−13 nuclei. Because polarization is being transferred from the protons to the carbons, it is the shorter T_1 of the protons that dictates the repetition rate for signal averaging.

In Figs. 8.4 and 8.5 we show the pulse sequences for ^1H and ^{13}C nuclei and the behavior of the magnetization vectors in the cross-polarization experiment, the goal of which is to create a time dependence of the spins that is common to the two rotating frame systems of ^1H and ^{13}C. We first apply a 90_x° pulse to the protons [Fig. 8.4(a), Fig. 8.5, top] so that they are now along the y' axis, in the manner we have already described in previous chapters. The phase of the proton B_1 field is then shifted by 90° [Fig. 8.4(b)], i.e., the rf signal is shifted by one quarter of a wavelength.

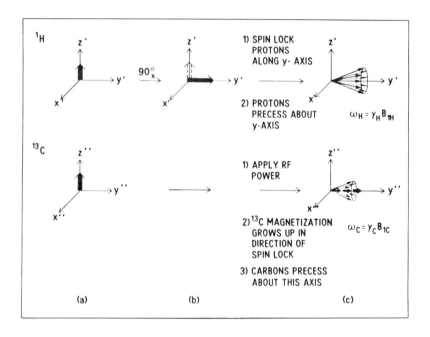

Fig. 8.4.   Vector diagram for a ^{1}H and ^{13}C double rotating frame
             cross-polarization experiment.   The carbon frame and the
             proton frame are rotating at different frequencies (see text).

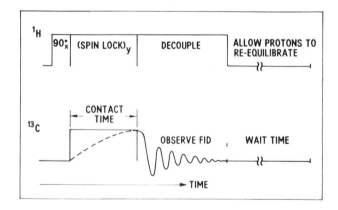

Fig. 8.5.   The ^{1}H$-^{13}$C cross-polarization pulse sequence for solid-state
             observations.

In this way, the protons are *spin-locked* along the y' axis and for the duration of the spin-locking pulse [Fig. 8.4(a); Fig. 8.5, top] they are forced to precess about the y' axis of their rotating frame with a frequency $\omega_H = \gamma_H B_{1H}$. Meanwhile, the carbons are put into contact with the protons. This is accomplished by turning the carbon field B_{1C} on during the spin-lock time (Fig. 8.5, bottom), causing the carbon magnetization to grow up in the direction of the spin-lock field [Fig. 8.4(c)]. The carbons are now precessing about their y'' axis with a frequency $W_C = \gamma_C B_{1C}$.

Figure 8.6 is intended to show in more detail the mechanism by which the polarization transfer occurs. The protons are now precessing about the B_{1H} field with frequency $\gamma_H B_{1H}$, and the carbons are simultaneously precessing with frequency $\gamma_C B_{1C}$ about the B_{1C} field. Since the Hartmann–Hahn match has been established, these frequencies are equal, and therefore the z components of both the proton and carbon magnetization must have the same time dependence, as Fig. 8.6 is meant to convey. Because of this common time dependence, mutual spin flip-flops can occur between protons and carbons. This process can be visualized as a flow of polarization or "coolth" from the abundant proton spins to the rare carbon spins. It can be shown (a good description will be found in Ref. 7) that under these circumstances there is an enhancement of the rare spin signal intensity by as much as the ratio of magnetogyric ratios of the abundant and rare spins. In the $^1H-^{13}C$ case this factor is γ_H/γ_C or 4, greater than the maximum Overhauser enhancement (Chapter 5, Section 5.2.8) of 3. This is in addition to the great time advantage of not having to deal with long carbon–13 spin lattice relaxation times in spectrum accumulation.

In order to achieve minimum linewidth in magic angle spectra, it is necessary that the spinning axis be within $0.5°$ of the magic value in order to reduce the anisotropy pattern to 1% of its static value.[7] At high magnetic field strengths, the anisotropy pattern may be as much as 10 kHz in width. Since this spinning rate is difficult to achieve, one must accept the presence of spinning sidebands located on each side of the isotropic chemical shift position by a distance equal to the spinning frequency. Examples of such sidebands will be shown and discussed in Section They are not necessarily a nuisance and may in fact be turned to good account since at relatively low spinning rates they can be used to trace out complex anisotropy patterns that are difficult to disentangle in the static spectra.[21]

8.6 MAGIC ANGLE SPECTRA

In this section we present and discuss four examples of carbon–13 magic angle spectra of solids.

8.6.1 Sucrose

In Fig. 8.7 are shown the 50.3-MHz carbon—13 spectra of crystalline *sucrose* under varied conditions.[6] Sucrose is a disaccharide with 12 nonequivalent carbon atoms:

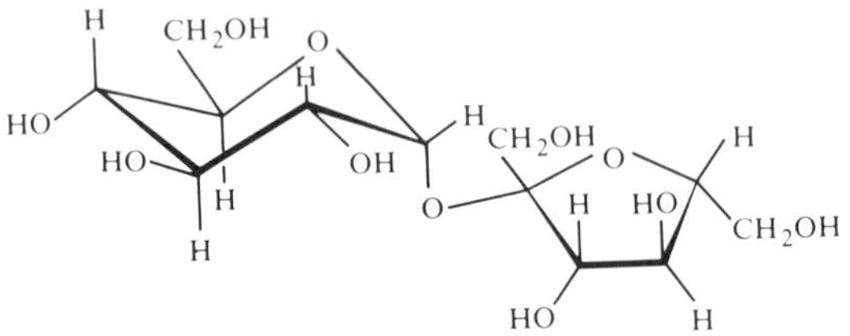

Spectrum (a) is obtained without proton decoupling, cross polarization, or spinning. It shows a half-height linewidth of *ca.* 15 kHz. Upon dipolar irradiation of the protons, the linewidth decreases to *ca.* 2 kHz [spectrum (b)]; this is mainly determined by the chemical shifts and chemical shift anisotropies of oxygen-bonded carbons (Table 8.1). Upon spinning (with cross polarization) a well-resolved spectrum is seen, as at (c), which is shown fivefold expanded at (d). Of the 12 expected resonances, 11 can be resolved. In the aqueous solution spectrum (e), all 12 are resolved since the lines are somewhat narrower. It is also observed that peak positions are not identical in the two spectra, but the differences are no greater than might be expected from a solvent effect.

8.6.2 Bisphenol A

The 50.3-MHz carbon—13 spectrum of bisphenol A (4.4'-dihydroxydiphenyl-2,2'-dimethylpropane) in the crystalline state is shown in (a) of Fig. 8.8.

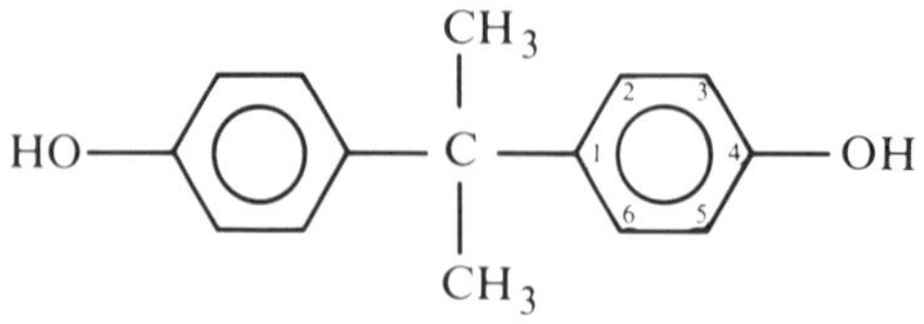

Without dipolar decoupling or spinning the spectrum is broad (*ca.* 15 kHz at the baseline), like that of sucrose [Fig. 8.7(a)], but shows some structure because of the greater difference in chemical shift between

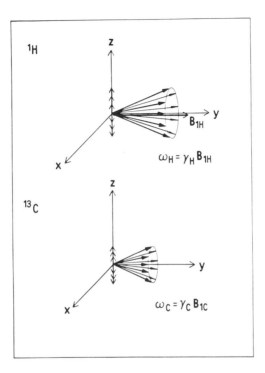

Fig. 8.6.   A more detailed representation of Fig. 8.5. The carbons are precessing with frequency $\omega_C = \gamma_C B_{1C}$ and the protons with frequency $\omega_H = \gamma_H B_{1H}$. When the Hartmann—Hahn match is achieved $(\gamma_H B_{1H} = \gamma_C B_{1C})$, the two spin systems have z components with the same frequency dependence, and mutual spin flip-flops can occur.

aromatic and aliphatic carbons. In (b) dipolar decoupling and cross polarization have been applied but without spinning; the aliphatic carbons show a discernible symmetric anisotropy pattern, but the aromatic carbons give a broader and more complex resonance. Spectrum (c) was obtained without decoupling or cross polarization but with magic angle spinning; the chemical shift anisotropy has been collapsed nearly completely but the dipolar broadening only partially, the two resonances retaining a half-height width of *ca.* 1500 Hz. In (d) all three techniques are applied. The resulting high-resolution spectrum may be compared to the solution spectrum (e). Unlike the case of sucrose, the solid and solution spectra here differ significantly: the $C_{2,6}$ and methyl resonances are split into two

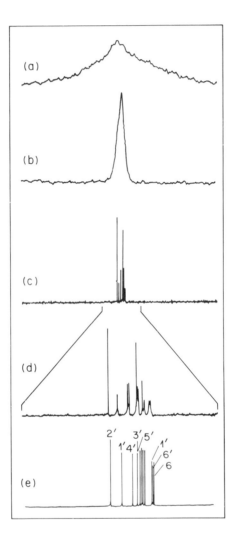

Fig. 8.7.   The 50.3 MHz carbon−13 spectra of crystalline sucrose: (a)
nonspinning and with no proton decoupling; (b) nonspinning
but with proton dipolar decoupling; (c) with magic angle
spinning, dipolar decoupling, and cross polarization; (d)
expansion of (c); (e) spectrum of aqueous solution. (From
G. A. Gray and H. D. W. Hill, *Ind. Res. and Dev.*, March
1980, p. 186.)

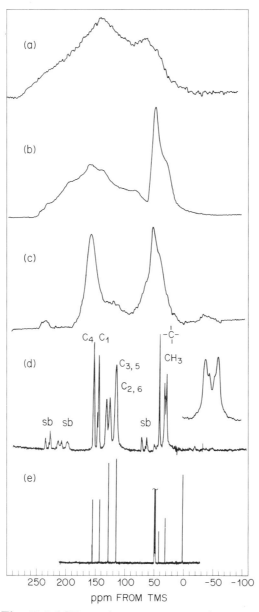

Fig. 8.8. The 50.3 MHz carbon−13 spectra of crystalline bisphenol A (4,4'-dihydroxydiphenyl-2,2'−dimethylpropane): (a) nonspinning and with no proton decoupling; (b) nonspinning but with proton dipolar decoupling; (c) with magic angle spinning and cross polarization but without

dipolar decoupling; (d) with magic angle spinning, dipolar decoupling, and cross polarization; (e) solution spectrum (10%, vol./vol., in CD_3OD, 21°C). (F. C. Schilling, unpublished observations.)

equal components, and the latter shows indications of a further splitting (see expanded spectrum in inset). The crystal structure[22] of bisphenol A reveals three molecules per unit cell in which the rings are rotated to varying degrees away from the plane bisecting the CH_3-C-CH_3 angle. There are three different sets of ϕ_1, ϕ_2 angles and $\phi_1 \neq \phi_2$ in each conformer:

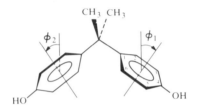

These rotations cause a nonequivalence of the methyl groups and of C_2 and C_6 within each molecule and in fact within each ring. This is probably the cause of the major splittings. The minor splittings may arise from differentiation between rings in each molecule or among the three molecules in each unit cell (or both). In any event, the MAS spectrum reveals structural information beyond that available from the solution spectrum (e).

8.6.3 Adamantane

Sucrose and bisphenol A do not rotate in the crystal but adamantane, being ball-like in form, spins readily and isotropically. The effect of this mobility on the solid-state carbon−13 spectrum is marked. In Fig. 8.9[23] (a) is the 50.3 MHz spectrum of crystalline adamantane without mechanical spinning or dipolar decoupling. The linewidth of *ca.* 1.2 kHz is reduced by over tenfold from that of sucrose. In spectrum (b) dipolar decoupling (with cross polarization) has reduced the linewidth to only *ca.* 150 Hz − or *ca.* 300 Hz at the baseline − showing that motion in the crystal also substantially collapses the chemical shift anisotropy [compare to Fig. 8.8, spectrum (b)]. In spectrum (c) we see the result of magic

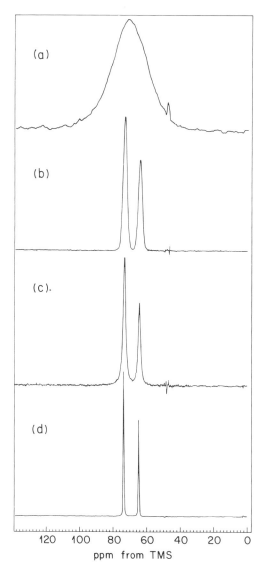

Fig. 8.9.   The 50.3 MHz carbon–13 spectra of crystalline adamantane:
(a) nonspinning and with no proton decoupling; (b)
nonspinning but with dipolar decoupling and cross
polarization; (c) with magic angle spinning but without
dipolar decoupling or cross polarization; (d) with magic angle
spinning, dipolar decoupling, and cross polarization.
(F. C. Schilling, unpublished observations.)

angle spinning alone, without dipolar decoupling (or cross polarization); the residual dipolar broadening is less than the residual chemical shift anisotropy in (b). In (d), with all three techniques applied we have reduced the linewidth to nearly the value in solution (not shown).

8.6.4 Poly(phenylene oxide)

Extensive studies of polymers in the solid state have been carried out by magic angle carbon—13 spectroscopy, and references to this work have been given in the introductory section of this chapter. An early and significant investigation is that of poly(2,6-dimethyl-1,4-phenylene oxide),

an important commercial plastic, by Schaefer and Stejskal.[24] In a way similar to what has been discussed for bisphenol-A (Section 8.6.2), the methyl groups and protons 2,6 and 3,5 on each ring are in nonequivalent environments at any instant. In solution they appear to be equivalent because of rapid bond rotation. In the solid state, however, such bond rotation may be retarded or prevented and the nonequivalence made evident. In Fig. 8.10(a) we see the spectrum with dipolar decoupling but without spinning. It exhibits the expected series of overlapping chemical shift tensor patterns. In (b) is shown the pattern upon magic angle spinning. It closely resembles the solution spectrum (c) but exhibits *six* lines rather than the expected five. The line that is doubled (also shown

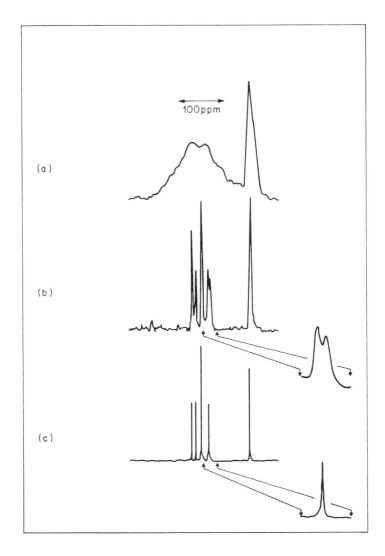

Fig. 8.10.   The 15 MHz solid-state and solution carbon—13 spectra of poly(phenylene oxide): (a) solid-state spectrum obtained with cross polarization and dipolar decoupling but without magic angle spinning; (b) same as (a) but with magic angle spinning; (c) solution carbon—13 spectrum. In (b) the splitting of the protonated ring carbon resonance is also shown in expanded form. [From J. F. Schaefer and E. A. Stejskal, *J. Am. Chem. Soc.* **98**, 1031 (1976)].

expanded) has been shown to arise from the protonated aromatic carbons. In principle, the methyl carbons and the aromatic carbons to which they are bonded should also be split into doublets, but these are not resolved.

8.6.5 Polypropylene

We have discussed the stereochemical configuration of polypropylene in Chapter 7 (Section 7.2.1.2) and have seen that the carbon chemical shifts of the solution spectra may be rationalized in terms of a γ-gauche shielding effect (see also Chapter 3, Section 3.3.3). These ideas receive striking confirmation in the solid-state magic angle carbon-13 spectra of isotactic and syndiotactic polypropylene. In the crystalline state, the isotactic polymer adopts a 3_1 helical conformation composed of alternating gauche and trans bonds: $\cdots(gt)(gt)(gt)\cdots$; this is shown at the lower left in Fig. 8.11. In this structure all CH_2, CH, and CH_3 groups occupy equivalent positions by reason of threefold symmetry and give resonances (upper left) in the same relative positions as in the solution spectrum. The syndiotactic polymer, in contrast, shows a splitting of the methylene carbon resonance (upper right), the equal components being separated by about 8 ppm. This is in conformity with the crystal structure, which has a repeating $(gg)(tt)(gg)(tt)$ helical conformation with four monomer units per turn (lower right). Here, the methylene carbons reside in two different, alternating environments, which may be termed "external" and "internal." The external CH_2 carbon experiences two trans interactions with carbons in the γ positions, whereas the internal CH_2 carbon sees two gauche carbons in the γ positions. The internal carbon should resonate upfield by approximately two gauche interaction parameters, and this is observed.

8.6.6 Polybutadiene

The observations of trans-1,4-polybutadiene in solution (Chapter 7, Section 7.2.2.1) are confirmed and extended by magic angle carbon-13 observations of the solid state. Unlike the cis polymer, which is rubbery, trans-1,4-polybutadiene is highly crystalline — about 80% — with a melting point of *ca.* 148°C. The structure of the crystals, as prepared from dilute solution, is shown in a schematic manner in Fig. 8.12. This morphology is shared by most polymer crystals. The chain axes of the macromolecules are oriented across the thickness of the lamellar platelets. The major dimensions of the crystal are manyfold greater than the thickness, as indicated. The lengths of the polymer chains are also much greater than the thickness, and they therefore must be folded many times, as shown in the magnified inset. This fold surface constitutes most of the amorphous portion of the crystalline polymer.

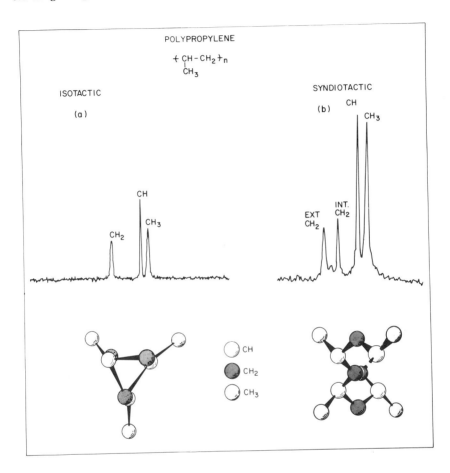

Fig. 8.11.   Solid-state carbon−13 NMR spectra of polypropylenes. (a) Solid-state spectrum of isotactic polypropylene. [From W. W. Fleming, C. A. Fyfe, R. D. Kendrick, J. R. Lyerla, Jr., H. Vanning, and C. S. Yannoni, *in* "Polymer Characterization by ESR and NMR," A. E. Woodward and F. A. Bovey, Eds., *ACS Symp. Ser.* (142), 1980, p. 212.] Below the spectrum is shown a view along the helix axis of the 3_1 helical conformation of the chains of the crystal. (b) Solid-state spectrum of syndiotactic polypropylene. [From A. Bunn, E. A. Cudby, R. K. Harris, K. J. Packer, and B. J. Say, *Chem. Comm.* **15** (1981)]. A view along the helix axis of this form is shown below the spectrum.

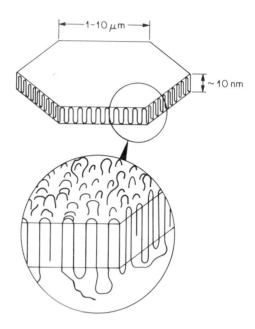

Fig. 8.12.   Schematic representation of polymer crystal lamella showing chain axes perpendicular to surface and (inset) folding of chains at surface. (Courtesy of H. D. Keith.)

In Fig. 8.13 are shown the 50.3 MHz spectra of a compacted mass of crystals prepared from heptane solution.[25] Spectrum (a) was obtained using a weak or "scalar" proton decoupling of sufficient power only to collapse the carbon–proton spin multiplets without affecting the direct dipole–dipole interactions with neighboring protons. Under these circumstances one observes only the amorphous portion of the sample, where chain motion is relatively rapid. Magic angle spinning [spectrum (b)] causes a collapse of the residual broadening, but we are still observing only the crystal surfaces. In spectrum (c) high-power dipolar decoupling with spinning allows us to observe the entire sample as narrow peaks. Both the olefinic and methylene carbon resonances are now split, the more shielded components being in the same positions as in (a) and (b). The new and larger peaks are those of the crystalline portion of the sample, where there is little or no motion. The introduction of gauche and cis conformations into the chains in the amorphous regions results in an upfield shift. The strong spinning side bands in (c) arise only from the olefinic carbons because of their much greater chemical shift anisotropy

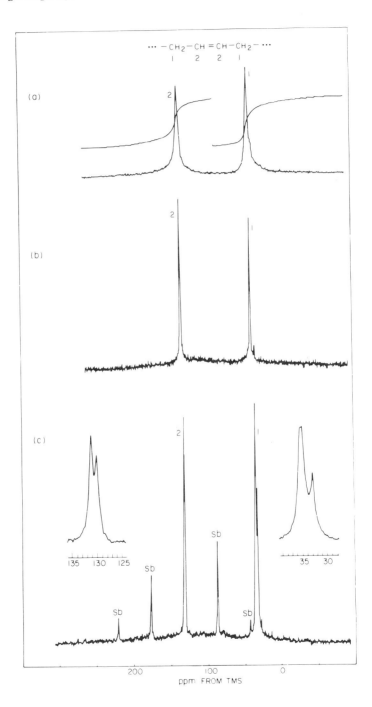

Fig. 8.13.   The 50.3 MHz carbon—13 solid-state spectra of solution-crystallized *trans*-1,4-polybutadiene: (a) with scalar proton decoupling and without spinning; (b) same as (a) but with magic angle spinning at 2.4 kHz; (c) same as (b) but with high-power proton decoupling. Insets show olefinic (left) and methylene (right) carbon resonances with expanded chemical shift scales. [From F. C. Schilling, F. A. Bovey, S. Tseng, and A. E. Woodward, *Macromolecules* **17**, 728 (1984)].

(Table 8.3), parallel behavior to that observed for the aromatic carbons of bisphenol A (Section 8.6.3). The motion of the amorphous chains is sufficient to collapse the anisotropy pattern even of the olefinic carbons to a width within the spinning frequency (*ca.* 3 kHz), as is evident in [b] (or indeed in (a)].

8.7 MULTIPLE-PULSE LINE NARROWING

We have discussed so far the attainment of high-resolution spectra of rare spins — as exemplified by carbon—13 — in the solid state. It would be desirable to be able to do the same for abundant spins such as ^{1}H and ^{19}F. It is clear that the same techniques used for carbon—13 observations are not feasible when the dipolar broadening arises from the nucleus to be observed. In addition, in the case of proton spectroscopy, the total chemical shift range is less than the narrowest linewidth one can hope to attain in rigid systems. In spectra already narrowed by molecular motion, as for example in rubbery polymers such as polybutadiene and butadiene-styrene copolymers, magic angle spinning can produce substantial line narrowing and resolution of proton chemical shifts.[26]

Multiple-pulse techniques seek to achieve line narrowing by manipulation of the nuclear magnetization rather than by mechanical motion of the sample. To understand the effect of these manipulations, one must recall that the magic angle is the angle of a cube diagonal, as shown in Fig. 8.14. The purpose of the pulse sequence, shown in Fig. 8.15, is to cause the nuclear magnetization vector to occupy in effect an average position along the cube diagonal, although it does not at any time actually occupy this position. This first and simplest of sequences for achieving this result was proposed by Waugh, Huber, and Haeberlen[27,28] and is commonly known as the WAHUHA sequence. We begin at (a) in Fig. 8.15 with the magnetization in its equilibrium state. Upon application of a 90_x° pulse (b), it is turned along the $+y'$ axis. After a time 2τ (typically of the order of 100 μs) a 90_{-x}° pulse rotates it back along the $+z'$ direction (c). Then after an interval τ a 90_y° pulse rotates it to the

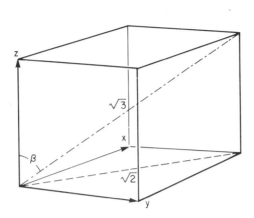

Fig. 8.14.   The magic angle as given by the diagonal of a cube: $90°$
$- (\tan^{-1} 2^{-1/2})$.

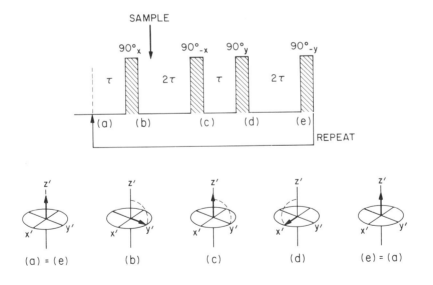

Fig. 8.15.   The WAHUHA pulse sequence (top) and the behavior of the
magnetization vector at the points indicated by the letters
(see text).

$-x'$ position as shown at (d). Then after a further interval 2τ, the magnetization is again returned to the $+z'$ alignment by a $90_{-\overset{\circ}{y}}$ pulse (e); the cycle is then repeated. The effect of this sequence is that the magnetization vector spends an equal time (2τ) along the x', y', and z' axes, which is equivalent in a time average to being along the cube diagonal or magic angle axis. At the point shown, the FID is sampled and a data point recorded. To adequately define the FID it is therefore necessary to repeat the cycle and the sampling at least a hundred times.

More complex sequences have been proposed for purposes of cancellation of errors, which arise chiefly from imperfections in pulse widths and phases.

An example of the application of this cycle is provided by the spectrum of crystalline perfluorocyclohexane

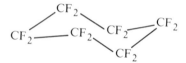

The normal ^{19}F spectrum of the polycrystalline material at $-73°C$ is shown as (a) in Fig. 8.16. At this temperature the molecule appears to rotate about the hexad axis at a rate sufficient to collapse the ^{19}F chemical shift anisotropy, but the $^{19}F-^{19}F$ dipolar interaction still gives rise to an approximately 6-kHz linewidth. When the WAHUHA cycle is applied there is a marked decrease in linewidth [spectrum (b)], which is sufficient to reveal an AB quartet arising from geminal fluorine coupling of $ca.$ 300 Hz.[29,30] Thus, molecular rotation in the crystal is not accompanied by chair-to-chair inversion, which would collapse the quartet (see Chapter 5, Section 5.4.2, for a discussion of the analogous case of cyclohexane). Ring inversion seems to proceed at about the same rate as in the liquid state or only slightly slower. The use of multipulse techniques for linewidth narrowing has been extensively reviewed.[1,2,31-34]

8.8 DEUTERIUM QUADRUPOLE ECHO NMR

8.8.1 Fundamentals

In Chapter 1, Section 1.5, and in Chapter 5, Section 5.2.4, we have seen that the interaction of molecular electric field gradients with the deuterium nucleus is a potent source of relaxation. There is also a

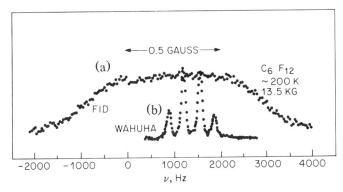

Fig. 8.16. The application of the WAHUHA cycle to polycrystalline perfluorocyclohexane at −73°C. The upper curve is the normal ^{19}F spectrum, showing dipolar $^{19}F-^{19}F$ broadening. The WAHUHA cycle narrows the linewidth sufficiently to resolve the AB quartet with geminal coupling of *ca.* 300 Hz. [From J. D. Ellett, U. Haeberlen, and J. S. Waugh, *J. Amer. Chem. Soc.* **92**, 411 (1970).]

substantial perturbation of the three Zeeman levels, which is very evident in the solid state though largely averaged out under the conditions of rapid motion characteristic of liquids. The magnitude of this quadrupolar splitting is given by

$$\Delta\nu_q = \frac{3}{4}\frac{e^2qQ}{h}(3\cos^2\theta - 1) \qquad (8.4)$$

Here (see Chapter 5, Section 5.2.4), e is the charge of the electron and e^2qQ/h is the quadrupolar coupling constant; θ is the angle that a particular carbon–deuterium bond vector makes with the direction of the external magnetic field B_0. The perturbation of the Zeeman levels is shown in Fig. 8.17(a). We again assume that the electric field gradient has cylindrical symmetry at the observed nucleus, so that η (p. 266) is zero.

We may imagine rotating a single crystal in which all the C–D bonds of a given type have the same angle θ [Fig. 8.17(b)]; in this case the line splitting $\Delta\nu q$ would change with the angle of rotation. Alternatively, if molecular motion occurred in this crystal, this motion would average the positions of these lines. In this manner the deuterium NMR line shape is averaged and can provide information about the amplitude and rate of motion occurring in a solid.

(a)

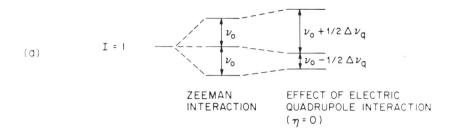

(b)

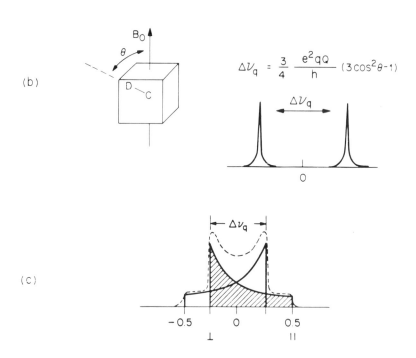

(c)

Fig. 8.17.   (a) Nuclear energy level diagram for a spin of 1. The three
quantized energy levels are degenerate in the absence of a
magnetic field. The field (i.e., the Zeeman interaction)
results in a splitting of these levels. The electric quadrupole
interaction causes a substantial perturbation of these levels.
(b) A single crystal (with only one C—D bond orientation
with respect to the field) is predicted to give two lines,
equally spaced about zero frequency. (c) The powder
average, or Pake powder pattern. [From L. W. Jelinski,
Ann. Rev. Mater. Sci. **15**, 359 (1985).]

A further complication is the fact that sizable single crystals are not common in organic chemistry and are virtually nonexistent in polymer science. Instead, one most commonly deals with polycrystalline powders or, in the case of polymers, also with glasses and rubbers. For such samples we will observe a powder average (as with chemical shift anisotropy or dipolar broadening), giving rise to the Pake doublet shown in Fig. 8.17(c). The line shape arising from one of the transitions is crosshatched. The composite line shape indicated by the dashed line is what would be observed if the C−D bond were static (with the edges softened to conform to experimental reality). Molecular motion causes averaging of this static line shape in highly specific ways. Figure 8.18 illustrates some of the line shapes that have been calculated for certain types of motion, the static pattern appearing as (a). The distance d between the singularities of this pattern is approximately 130 kHz, so that these spectra are much broader than the ordinary splittings resulting from chemical shifts and scalar couplings.

If the C−D bond forms part of a methyl group, rotation of the group causes the static line shape to collapse by a factor of $\frac{1}{2}(3\cos\theta^2-1)$, where ϕ is the angle that the C−D makes with the rotation axis, usually 109.5°. The splitting between the singularities is then $\frac{1}{3}d$ [Fig. 8.18(b)].

The spectrum shown in Fig. 8.18(c) is obtained if the C−D bond is involved in a two-site hop between two positions where the bisector of the hop angle is 54.7°, or the magic angle. The component of the molecular electric field gradient tensor perpendicular to the hop axis is unaffected by the motion and remains at $-\frac{1}{2}d$. Another component of the motionally averaged electric field gradient tensor occurs at zero frequency. The tensor must remain traceless, so the third component occurs at $\frac{1}{2}d$.

The line shape shown in Fig. 8.18(d) arises when a deuterated phenyl ring undergoes a 180° flip about the 1,4-phenylene axis. The component of the electric field gradient tensor perpendicular to the flip axis is unaffected by the motion and remains at $-\frac{1}{2}d$. Another component of the tensor is averaged by 120°, and occurs at $-\frac{1}{8}d$. The third component must therefore be at $\frac{5}{8}d$ so that the tensor remains traceless. It should also be noted that a deuteron on the flip axis, i.e., in the para position, is unaffected by the 180° flip.

Another possible mode of motion for an aromatic ring is shown in Fig. 8.18(e). In this case the phenyl ring undergoes rotational diffusion by random steps about the 1,4-phenylene axis. It is found that the entire

(a) STATIC C-D

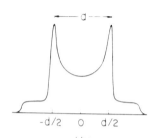

(b) ROTATING METHYL

(c) TWO-SITE HOP

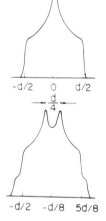

(d) 180° PHENYL RING FLIP

180°

(e) FREE DIFFUSION OF PHENYL RING

Fig. 8.18.   Calculated solid-state deuterium NMR line shapes for (a) a static C—D bond; (b) a rotating methyl group; (c) a two-site hop whose bisector forms the magic angle with the C—D bond; (d) a 180° phenyl ring flip; and (e) free diffusion of a phenyl ring about the 1,4 axis.

pattern is averaged by a factor of $\frac{1}{8}$, as the C—D bond makes an angle of 60° with the rotation axis.

These examples illustrate the effect of motions that are fast on the deuterium NMR time scale, i.e., with correlation times less than *ca.* 10^{-7}s. The effect of motions that are intermediate on this scale — with correlation times less than 10^{-3}s but greater than 10^{-7} — is of particular interest. In this regime the frequency of a single NMR transition will fluctuate between two values in a manner analogous to chemical exchange in high-resolution NMR spectra. Line shapes can be calculated for such intermediate motions,[2,35-37] and these yield rate constants which, when plotted against inverse temperature, afford Arrhenius activation energies.

One of the seeming disadvantages of deuterium is its very low natural abundance of only 0.0156% (Chapter 1, Table 1.1). This is actually an advantage, however, because it permits one to specifically label a particular portion of the molecule being studied without having to be concerned about the natural background from other parts of the molecule.

8.8.2 Experimental Observation of Solid-State Deuterium Spectra

Because of their great breadth — up to *ca.* 250 kHz — solid-state deuterium spectra present special problems with respect to both hardware and software and cannot be acquired with the equipment and methods described in Chapter 2 for spin-$\frac{1}{2}$ nuclei. It is necessary for the receiver to have sufficient bandwidth and for the transmitter to be powerful enough to generate an accurate 90° pulse over the entire spectral region. The probe must be capable of withstanding this high rf power. An additional consideration is the rate of digitization of the signal. Parts of the deuterium FID are very broad and thus decay very rapidly. It should be recalled that a 10 Hz signal in the time domain would decay to $1/e$ of its original intensity in 30 ms. A pattern of 120 kHz breadth would decay in less than a microsecond. Very rapid digitization, of the order of 10^9 s^{-1}, is necessary to adequately capture such a signal. Finally, one must consider that such an FID will decay in less than the time required by the receiver to recover from the high-power pulse. This problem is solved by software, using a quadrupolar echo pulse sequence (not shown), which refocuses the magnetization while the receiver recovers from the pulse, in effect "buying time" while the system settles down. It preserves the inhomogeneously broadened deuterium powder pattern. The first 90° pulse is followed, after a delay of 25–30 μs, by a second 90° pulse which has been phase shifted by 90°, and a second time delay. The length of the second time delay is adjusted so digitization of the FID begins at exactly the top of the quadrupolar echo maximum.

8.8.3 Phenyl Ring Rotation in Crystalline Enkephalin

[Leu^5] enkephalin is a natural pentapeptide with this structure:

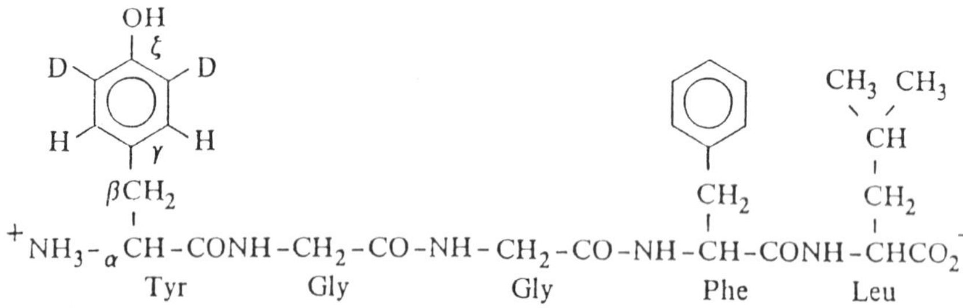

The tyrosyl ring was deuterated at the positions shown in order to study its motions in the crystal,[38] of which the most important is motion about the $C_\beta - C_\gamma$ bond, resulting in rotation about the $\gamma-\zeta$ axis. If closely packed about with sterically demanding neighboring groups, the ring motion might be limited to small librations. If freer, the ring might be able to rotate in a diffusional manner or to execute 180° jumps, as described in Section 8.8.1. In Fig. 8.19 the experimental observations are shown (a) at −20°, 25°, 68°, and 101°C; (b) and (c) are simulated spectra calculated on two bases. The theoretical spectra in (c) are calculated as if the spectra were normal Bloch free induction decays, acquired in the usual way immediately following the pulse. In fact, the echo technique results in some distortions of the FID at intermediate exchange rates, because at some crystal orientations in the microcrystalline sample the effective spin–spin relaxation time T_2 is longer than at others (see Refs. 36 and 37 for more detailed explanation). The series of theoretical spectra at (b) have been corrected for this distortion and give a good match to the observed spectra. Both sets of spectra were calculated on the assumption of 180° jumps of the ring, since it is evident that the experimental spectra do not correspond to the free diffusional case [Fig. 8.18(e)]. At −20°C the flip rate is less than 10^3 s^{-1}, while at 101°C the rate is approximately 10^6 s^{-1}.

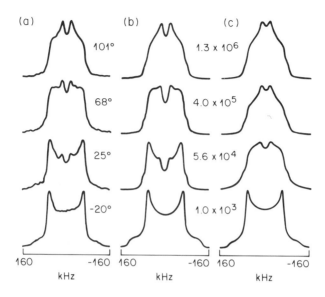

Fig. 8.19. (a) Observed 45.1-MHz deuterium quadrupole echo spectra of polycrystalline [Leu[5]] enkephalin substituted with deuterium in 3 and 5 positions of the tyrosyl ring (see text, p. 438) as a function of temperature, obtained with 30,000 to 50,000 scans. (b) Simulations of the spectra using a 180° jump model including corrections for intensity distortions produced by the quadrupole echo. The jump rate at each temperature is indicated, expressed as s^{-1}. (c) Simulations at the same jump rate but without the echo distortion corrections, i.e., as if the signal were a normal Bloch decay. The differences between (b) and (c) are particularly marked at 25° and 68°C — intermediate jump rates. [From D. M. Rice *et al.*, *J. Am. Chem. Soc.* **103**, 7707 (1981).]

8.8.4 Poly(butylene terephthalate)

Deuterium spectroscopy of selectively labeled poly(butylene terephthalate), a partially crystalline polyester resin has been used[39,40] to establish the nature of the motion about the three central bonds in the aliphatic chain. Earlier carbon–13 relaxation and chemical shift anisotropy considerations indicated that the terephthalate carbonyl groups can be considered to be static compared to motions occurring in the aliphatic region. In addition, the latter is the shortest sequence that can undergo motion about three bonds, with the terephthalate groups acting as molecular anchors to restrict longer range motional modes.

In Fig. 8.20 are shown the observed and calculated solid-state deuterium spectra of this labeled polyester as a function of temperature, together with the rate constants deduced from the line shape. In going

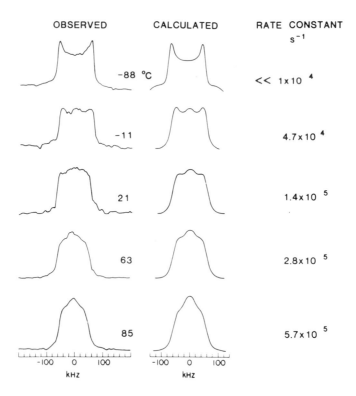

Fig. 8.20.   Deuterium spectra of the central CD_2 groups of the aliphatic segment of poly(butylene terephthalate). The observed spectra are on the left at the temperatures indicated. The calculated spectra on the right are for the hopping rate shown. [From L. W. Jelinski, J. J. Dumais, and A. K. Engel, *Macromolecules* **16**, 492 (1983).]

from $-88\,°C$, to $+85\,°C$, the pattern progresses from that of a static system to that of a powder pattern representing fairly fast motion. The calculated spectra were based on a two-site gauche—trans hop model (Fig. 8.21) in which the C—D bond vector moves between sites separated by a dihedral angle of $103\,°$. This is substantially less than the $120\,°$ value corresponding to the projection of tetrahedral bond angles, but is substantiated by X-ray results. The observed motion is believed to represent cooperative gauche—trans conformational transitions, such as shown in Fig. 8.22, first proposed by Helfand.[41-43] It is significant that the activation energy derived from the temperature dependence of the rate constants in Fig. 8.20 is 5.8 kcal mole^{-1}, which is greater than that for the gauche—trans transition of an isolated bond — approximately 4.0 kcal mole^{-1} — but less than that for two such transitions occurring exactly simultaneously, *ca.* 8.0 kcal-mole^{-1}. It is clearly impossible for such a two-site hop to occur in isolation, as this would necessitate sweeping a large portion of the chain through a highly viscous medium. The Helfand proposals (Fig. 8.22) picture a pairwise quasi-simultaneous occurrence of gauche-trans transitions that do not require any major translation of the chain.

By labeling the phenylene rings with deuterium, it has been found[44] that those in the amorphous phase of this semicrystalline polymer undergo $180\,°$ ring flips at ambient temperatures, whereas those in the crystalline phase do not.

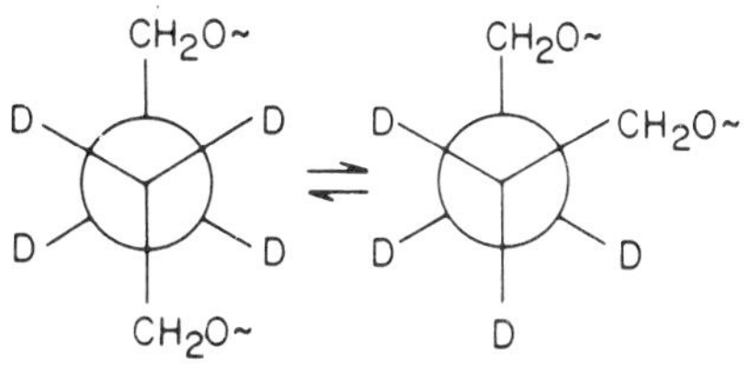

Fig. 8.21.   Newman projections of the gauche-trans bond rotation and C—D bond hopping process in the central CD_2 groups of the aliphatic portion of poly(butylene terephthalate) in the solid state. [From L. W. Jelinski, J. J. Dumais, and A. K. Engel, *Macromolecules* **16**, 492 (1983).]

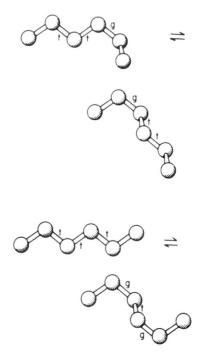

Fig. 8.22.   Helfand proposal for counter-rotation about second-neighbor parallel bonds. [From E. Helfand, Refs. 41–43.)

REFERENCES

1. U. Haeberlen, "High Resolution NMR in Solids: Selective Averaging," *Adv. Magn. Reson.*, Suppl. 1 (1976).

2. M. Mehring, *High Resolution NMR in Solids*, 2nd ed., Springer-Verlag, Berlin and New York, 1983.

3. C. A. Fyfe, *Solid State NMR for Chemists*, C. F. C. Press, Guelph, Ontario, Canada, 1983.

4. R. G. Griffin, *Anal. Chem.* **49**, 951A (1977).

5. R. K. Harris and K. J. Packer, *Eur. Spectrosc. News* **21**, 37 (1978).

6. G. A. Gray and H. D. W. Hill, *Industrial Res. and Dev.*, March 1980, p. 136.

7. C. S. Yannoni, *Acc. Chem. Res.* **15**, 201 (1982).

8. J. R. Lyerla, C. S. Yannoni, and C. A. Fyfe, *Acc. Chem. Res.* **15**, 208 (1982).

9. B. C. Gerstein, *Anal. Chem.* **55**, 781A (1983).

10. B. C. Gerstein, *Anal. Chem.* **55**, 899A (1983).

11. J. R. Havens and J. L. Koenig, *Appl. Spectrosc.*, **37**, 226 (1983).

12. R. K. Harris, *Nuclear Magnetic Resonance Spectroscopy*, Pitman, London, 1983; Chapter 6.

13. R. A. Komoroski, Ed. *High Resolution NMR Spectroscopy of Synthetic Polymers in Bulk*, VCH Publishers, Dearfield Beach, Florida, 1986.

14. V. J. McBrierty and D. C. Douglass, *Phys. Rep.* **63**, 61 (1980).

15. D. W. McCall, *Acc. Chem. Res.* **4**, 223 (1971).

16. I. J. Lowe, *Phys. Rev. Lett.* **2**, 285 (1959).

17. E. R. Andrew, A. Bradbury, and R. G. Eades, *Nature* **183**, 1802 (1959).

18. T. M. Duncan, *J. Phys. and Chem. Ref. Data* **00**, 000 (1987).

19. A. Pines, M. Gibby, and J. S. Waugh, *J. Chem. Phys.* **59**, 569 (1973).

20. S. R. Hartmann and E. L. Hahn, *Phys. Rev.* **128**, 2042 (1962).

21. L. W. Jelinski *Macromolecules* **14**, 1341 (1981) and references therein.

22. V. K. Bel'ski, N. Yu. Chernikova, V. K. Rotaru, and M. M. Kurchinin, *Sov. Phys. Crystallogr.* **28** (4), 405 (1983).

23. F. C. Schilling, unpublished observations. See also K. W. Zilm, D. W. Alderman, and D. M. Grant, *J. Mag. Res.* **30**, 563 (1978).

24. J. F. Schaefer and E. A. Stejskal, *J. Am. Chem. Soc.* **98**, 1031 (1976).

25. F. C. Schilling, F. A. Bovey, S. Tseng, and A. E. Woodward, *Macromolecules*, **17**, 728 (1984).

26. H. Benoit and M. Rabii, *Chem. Phys. Lett.* **21**, 466 (1973).

27. J. S. Waugh, L. M. Huber, and V. Haeberlen, *Phys. Rev. Lett.* **20**, 180 (1968).

28. J. D. Ellett, V. Haeberlen, and J. S. Waugh, *J. Amer. Chem. Soc.* **92**, 411 (1970).

29. G. V. D. Tiers, *Proc. Chem. Soc.* **1960**, 389.

30. H. S. Gutowsky and F. Chen, *J. Phys. Chem.* **69**, 3216 (1965).

31. P. Mansfield, *Phil. Trans. Roy. Soc. London* **A299**, 479 (1981).

32. U. Haeberlen, *Phil. Trans. Roy. Soc. London* **A299**, 497 (1981).

33. B. C. Gerstein, *Phil. Trans. Roy. Soc. London* **A299**, 505 (1981).

34. T. M. Duncan and C. Dybowski, *Surf. Sci. Rep.* **1**, 157 (1981).

35. U. Pschorn and H. W. Spiess, *J. Magn. Reson.* **39**, 217 (1980).

36. H. W. Spiess and H. Sillescu, *J. Magn. Reson.* **42**, 381 (1981).

37. L. W. Jelinski, *Ann. Rev. Mater. Si* **15**, 359 (1985).

38. D. M. Rice, R. J. Wittebort, R. G. Griffin, E. Meirovitch, E. R. Stimson, Y. C. Meinwald, J. H. Freed, and H. A. Scheraga, *J. Am. Chem. Soc.* **103**, 7707 (1981).

39. L. W. Jelinski, J. J. Dumais, and A. K. Engel, *Macromolecules* **16**, 492 (1983).

40. L. W. Jelinski, J. J. Dumais, P. L. Watnick, A. K. Engel, and M. D. Sefcik, *Macromolecules* **16**, 409 (1983).

41. E. Helfand, *J. Chem. Phys.* **54**, 4651 (1971).

42. E. Helfand, *J. Polym. Sci., Polym. Symp.* **73**, 39 (1985).

43. E. Helfand, *Science* **226**, 647 (1984).

44. A. L. Cholli, J. J. Dumais, A. K. Engel, and L. W. Jelinski, *Macromolecules* **17**, 2399 (1984).

CHAPTER 9

SPECIAL TOPICS

9.1 LESS COMMON NUCLEI (F. A. BOVEY)

Of the many nuclei (see Appendix A) that are capable in principle of giving a high resolution NMR spectrum, only those shown in Table 1.1 have received major attention. Of these, we shall discuss here ^{19}F, ^{14}N, ^{15}N, and ^{31}P since these nuclei correspond to large classes of organic compounds.

9.1.1 Fluorine

9.1.1.1 ^{19}F Chemical Shifts

The magnetic moment of ^{19}F is only slightly smaller than that of ^1H, so its relative sensitivity is high. (Relative sensitivities of nuclei are approximately proportional to the cubes of the ratios of magnetic moments at constant field strength.) It has a spin of 1/2 and, therefore, experiences no complications from quadrupole relaxation. Because of the greater polarizability of its electron cloud, ^{19}F, in common with most other nuclei, exhibits a much greater range of chemical shifts than does ^1H. This greater range often makes possible the discrimination of subtler environmental differences.

The understanding of fluorine shieldings is in a very incomplete state. It is probable that the origin of variations in fluorine shielding is mainly in the paramagnetic term of Ramsey's nuclear shielding equation (Chapter 3, Eq. 3.3). Shielding anisotropies of neighboring groups would be expected to have at most only relatively small effects and to be of the same magnitude as for ^1H, but it appears that this may not necessarily be the case. Dispersion forces from neighboring substituents and from solvent molecules may be expected to play an important part and even to reverse trends anticipated from substituent electronegativity alone.

The least shielded ^{19}F nuclei are those attached directly to N and S (entries 1–5, Table 9.1; in this table ^{19}F chemical shifts are expressed on the Φ scale of Filipovich and Tiers.[1]) Accumulation of halogen atoms on the same carbon also results in marked deshielding (entries 6–9). Most of the generalizations that describe relative shielding in hydrocarbon groups

TABLE 9.1

Selected Values of Fluorine Chemical Shifts[a]

Entry number	Compound	Φ^* (ppm)	Concentration (%)
1	NF_3	147	*ca.* 3
2	SOF_2	75.68	16
3	$C_6H_5SO_2F$	65.464(Φ)	0
		65.514	20
4	SF_6	57.617(Φ)	0
		57.42	*ca.* 10
5	SO_2F_2	33.17	*ca.* 10
6	CBr_3F	7.388(Φ)	0
		7.043	80
7	CCl_3F	0.000	neat
8	CCl_2F_2	−6.848	2
9	$CClF_3$	−28.1	*ca.* 3
10	$(CF_3)_3N$	−55.969(Φ)	0
		−55.978	30
11	$CF_3CF_2CF_{2\,(1)}I$	$F_{(1)}$:−60.470	10
12	CF_4	−62.3	*ca.* 3
13	$C_6H_5CF_3$	−63.732(Φ)	0
		−63.370	40
14	PF_5	−67.5	*ca.* 5
15	$CCl_2F \cdot CCl_2F$	−67.775(Φ)	0
		−67.834	20
16	$(CF_{3\,(2)})_3CF_{(1)}$	$F_{(2)}$:−74.625	2
17	CF_3CO_2H	−76.530(Φ)	0
		−76.542	20
18	$CF_3CF_2CF_2CF_2CF_2CF_2CF_{3\,(1)}$	−81.60	20
19	$CF_3CF_2CF_2CF_{3\,(1)}$	−81.85	*ca.* 4
20	$[CF_3CF_2CF_{2\,(1)}]_3N$	−85.19	5
21	POF_3	−90.7	*ca.* 5
22		$F_{(1)}$:−93.30	*ca.* 40
23		$F_{(1)}$:−101.89	25
24	$CF_3CF_2CF_{2\,(2)}CF_{2\,(1)}CN$	$F_{(1)}$:−105.764	20
25		$F_{(2)}$:−107.10	*ca.* 40

TABLE 9.1 (continued)   439

26	$C_6H_5 \cdot CF_{2\,(1)}CF_2CF_3$	$F_{(1)}$:-112.545	8
27	C_6H_5F	-113.12	0
		-113.23	40
28	(cyclobutane: H_2, F_2)	-114.49	15
29	Cl, $F_{(2)}$, $F_{(3)}\!>\!C=C\!<\!F_{(1)}$	$F_{(2)}$:-119.33	25
30	$CF_3CF_{2\,(2')}CF_{2\,(3')}CF_{2\,(4)}CF_{2\,(3)}CF_{2\,(2)}CF_3$	$F_{(4)}$:-122.2	20
31		$F_{(3)}$:-123.07	20
32		$F_{(2)}$:-126.66	20
33	$CF_3CF_2CF_{2\,(2)}CF_3$	$F_{(2)}$:-127.66	ca. 4
34	$CF_2 = CF_2$	$-132.837(\Phi)$	0
35	$Cyclo$-C_6F_{12}	$-133.25(\Phi)$	0
		-133.23	20
36		-134.916	0
		-135.15	ca. 70
37	$n-C_6F_{13}CF_{2\,(1)}H$	-137.62	20
38		$F_{(3)}$:-143.90	25
39	SiF_4	-164.8	ca. 5
40	(cyclohexane, $F_{(eq)}$ / $F_{(ax)}$)	$F_{(eq)}$:-165.46^b	25
		$F_{(ax)}$:-185.96^b	
		$F_{(avg)}$:-175.0^c	
41	$[CF_{3\,(2)}]_3CF_{(1)}$	$F_{(1)}$:-190.2	2
42		$F_{(3)}$:-192.78	ca. 40
43	$CH_3(CH_2)_4CH_2F$	-219.02	0
		-219.00	40

[a] Expressed on Φ scale; units are Φ^* unless Φ is indicated (see text).
[b] At $-90°C$.
[c] At $25°C$.

have no applicability to fluorocarbon chains. Thus, in a saturated fluorocarbon, the CF_3 end groups are substantially deshielded as compared to the CF_2 groups (entries 18 and 30; 19 and 32); fluorines attached to unsaturated carbons vary over a wide shielding range (entries 22, 23, 25, 29, 34, 38, 42); and those attached to a carbon bearing a substituent other than fluorine may be markedly shielded as compared to fluorines on saturated chains (entries 37, 40). Most shielded are lone fluorine atoms at the ends of saturated hydrocarbon chains (entry 43); even a single proton on a fluorinated chain produces marked shielding of the fluorine nuclei on the same carbon (compare 37 and 18), while, in contrast, protons on adjacent carbons cause deshielding (compare entries 28 and 36). So far, very little progress has been made toward even a qualitative rationalization of these observations.

A significant but only very roughly linear, correlation has been observed between the shielding of ^{19}F nuclei at meta and at para positions on benzene rings and the Hammett σ constants of the group $R^{2,3}$:

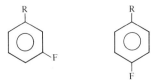

The shielding decreases as the σ_m and σ_p of R increase, the range of chemical shifts produced by para substituents being greater than that produced by meta substituents. Taft[3] has shown that the σ parameter may be separated into an inductive component σ_I and a resonance contribution σ_R; σ_I shows a much higher correlation with ^{19}F shielding at meta and para positions than does σ.

Fluorine chemical shifts have been extensively reviewed. In the series *Annual Reports on NMR Spectroscopy*, reports have appeared in Volumes 1,[4] 3,[5] 5A,[6] 6B,[7] 10B,[8] and 14.[9] A massive review of the subject is that of Emsley and Phillips.[10] A 113-page discussion of the theory (such as it is) and empirical correlations of ^{19}F chemical shifts is followed by a 410-page appendix in which chemical shifts are listed, divided into 43 structural types.

The most widely (but not universally) accepted scale for ^{19}F chemical shifts is that of Filipovich and Tiers.[1] They proposed using CCl_3F, a volatile liquid, as both reference and solvent in recognition of the fact that fluorine chemical shifts may be strongly dependent on solvent. They further proposed that chemical shifts on this scale be denoted Φ values if extrapolated to infinite dilution and Φ^* values if not, as is most commonly the case. In Table 9.1 ^{19}F chemical shifts are shown expressed on this scale. In some cases, both Φ and Φ^* values are indicated. Dilution effects are seldom very large. Chemical shifts corresponding to shielding less than that of CCl_3F are given positive signs and those more shielded are given negative signs.

9.1.1.2 ^{19}F-^{19}F Coupling

^{19}F-^{19}F couplings show unexpected behavior, particularly in highly fluorinated groups. Vicinal couplings in perfluoroalkyl chains are often nearly zero (entries 3 and 5, Table 9.2), while those through four and five bonds may be of order of 5—15 Hz (entries 1—3). In many instances, these stronger, long-range couplings make the vicinal couplings indeterminate through "virtual coupling" effects (Chapter 4, Section 4.2.3), the result being that only the sum of the vicinal coupling and the long-range coupling can be directly observed (entries 1, 2, 4).

Since ^{19}F-^{19}F (and ^{19}F-^{1}H) couplings are transmitted to an observable degree through a larger number of intervening bonds than ^{1}H-^{1}H couplings, ^{19}F multiplets are frequently quite complex and may appear poorly resolved and broad because of many closely spaced transitions. This effect is illustrated by the spectrum of perfluoroheptane (Fig. 9.1). The

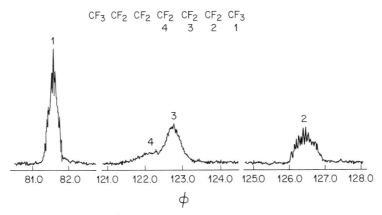

Fig. 9.1   56.4 MHz ^{19}F spectrum of perfluoroheptane, observed as 50% (vol./vol.) solution CCl_3F; see text. (F. P. Hood, private communication.)

TABLE 9.2

Some Representative ^{19}F-^{19}F Couplings

Compound number	Compound	J (Hz)		Reference
1	$CF_{3(2)} — CF_{2(1)}$ $\searrow$ $N — CF_{3(3)}$ $CF_{3(2')} — CF_{2(1')}$ $\nearrow$	$J_{12} + J_{12'}$:	10.2	a
		J_{13}:	15.8	
		J_{23}:	6.8	
		$J_{11'}$:	*ca.* 9	
2	$[CF_{3(2)}CF_{2(1)}]_2NCF_{2(1')}CF_{2(2')}$	$J_{12} + J_{12'}$:	13.6	a
3	$CF_{3(2)}CF_{2(1)}N[CF_{3(3)}]_2$	J_{12}:	$\leqslant 1$	b
		J_{13}:	16	
		J_{23}:	6	
4	$CF_{3(2)}CF_{2(1)}OCF_{2(1')}CF_{3(2')}$	$J_{12} + J_{12'}$:	3.4	a
5	$[CF_{3(2)}]_3CF_{(1)}$	J_{12}:	1.4	c
6	$CF_{3(2)}CF_{2(1)}H$	J_{12}:	2.8	d
7	$CF_{3(2)}CF_{(1)}H_2$	J_{12}:	15.5	d
8	$CF_{3(3)}CF_{2(2)}CHF_{2(1)}$	J_{12}:	4.5	e
		J_{13}:	7.3	
9	$CF_{3(3)}CF_{2(2)}CH_2F_{(1)}$	J_{12}:	15.2	e
		J_{13}:	7.9	
10	$CF_{2(3)}Cl \cdot CF_{2(2)} \cdot CH_2F_{(1)}$	J_{12}:	15.1	e
		J_{13}:	7.7	
		J_{23}:	3.9	
11	$CF_{2(3)}Br \cdot CF_{2(2)}CH_2F_{(1)}$	J_{12}:	15.5	e
		J_{13}:	7.7	
		J_{23}:	3.9	
12	$CF_{2(2a2b)}Br \cdot CHF_{(1)}Br$	$J_{1,2a}, J_{1,2b}$:	21, 24	f
		$J_{2a, 2b}$:	174	
13	$CF_{2(2a2b)}Br \cdot CHF_{(1)}Cl$	$J_{1,2a}, J_{1,2b}$:	18, 18	g, h
		$J_{2a, 2b}$:	177	
14	$CF_{2(2a2b)}Br \cdot CF_{(1)}ClBr$	$J_{1,2a}, J_{1,2b}$:	13, 14	i
		J_{2ab}:	159	
15	$CF_{2(2a2b)}(SiCl_3)CF_{(1)}ClH$	$J_{1,2a}, J_{1,2b}$:	16.8, 16.8	j
		$J_{2a, 2b}$:	343	
16	$CF_3CF_{2(2a2b)}CF_{(1)}ICl$	$J_{2a, 2b}$:	270.4	k
17	$F_{(2)}$ $>$ $C = C <$ Cl $F_{(1)}$ $F_{(3)}$	J_{12}:	76	l
		J_{13}:	56	
		J_{23}:	116	
18	F $>$ $C = C <$ F Cl Cl	J_{cis}:	37.5	m

TABLE 9.2 (continued) 443

19		J_{trans}:	129.57	m

| 20 | | J_{12}:
J_{13}:
J_{23}: | 87
33
119 | n |

| 21 | | J_{12}:
J_{13}:
J_{23}:
J_{14}:
J_{24}:
J_{34}: | 57
39
116
8
22
13 | n |

| 22 | | J_{gem}: | 157 | o |

| 23 | | J_{gem}: | 187 | p |

| 24 | | J_{gem}: | 249 | p |

| 25 | | J_{gem}: | 192 | p |

| 26 | | J_{2-gem}:
J_{3-gem}: | 230
224 | o |

| 27 | | J_{gem}: | 284 | q |

28

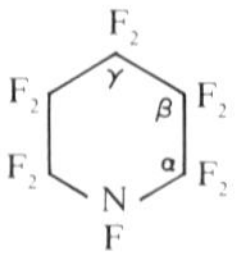

F$_2$
F$_2$ γ F$_2$
β
F$_2$ α F$_2$
N F$_2$
F

$J_{\alpha-gem}$: 185   r
$J_{\beta-gem}$: 278
$J_{\gamma-gem}$: 284

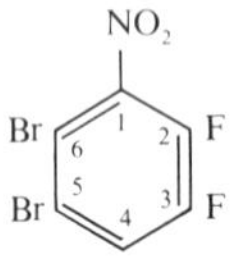

29

NO$_2$
Br — 1, 2 — F
6
5, 3
Br — 4 — F

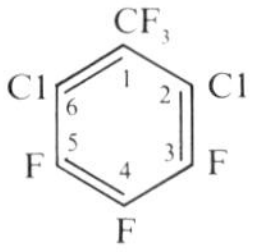

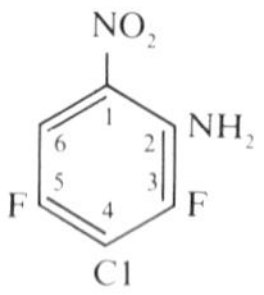

J_{23}: 20.2   s

30

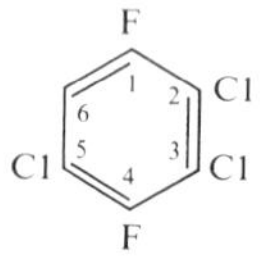

CF$_3$
Cl — 1, 2 — Cl
6
5, 3
F — 4 — F
F

J_{35}: 1.9   s

31:

NO$_2$
1, 2 — NH$_2$
6
5, 3
F — 4 — F
Cl

J_{35}: 3.1   s

32

F
1, 2 — Cl
6
Cl — 5, 3 — Cl
4
F

J_{14}: 12.0   s

33

NO$_2$
F — 1, 2 — Cl
6
5, 3
4 — F
Cl

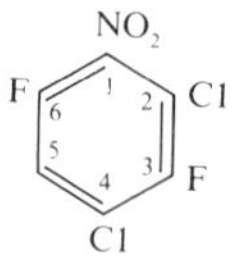

J_{36}: 14.4   s

Key to references:

a. L. Petrakis and C. H. Sederholm, *J. Chem. Phys.* **35**, 1243 (1961); for compounds 1, 2, and 4, the spectral interpretation is made ambiguous by "virtual coupling"; see J. I. Musher, *J. Chem. Phys.* **36**, 1086 (1962); and Chapter IV, Sec. VII, B.

b. A. Saika and H. S. Gutowsky, *J. Am. Chem. Soc.* **78**, 4818 (1956).

c. C. A. Reilly, *J. Chem. Phys.* **25**, 604 (1956).

d. D. D. Elleman, L. C. Brown, and D. Williams, *J. Mol. Spectrosc.* **7**, 307 (1961).

e. D. D. Elleman, L. C. Brown, and D. Williams, *J. Mol. Spectrosc.* **7**, 322 (1961).

f. J. Lee and L. H. Sutcliffe, *Trans. Faraday Soc.* **55**, 880 (1959).

g. J. Lee and L. H. Sutcliffe, *Trans. Faraday Soc.* **54**, 308 (1958).

h. J. N. Shoolery and B. Crawford, *J. Mol. Spectrosc.* **1**, 270 (1957).

i. J. J. Drysdale and W. D. Phillips, *J. Am. Chem. Soc.* **79**, 319 (1957).

j. R. K. Harris, private communication, quoted in J. W. Emsley, J. Feeney, and L. H. Sutcliffe, *High Resolution Nuclear Magnetic Resonance Spectroscopy,* Pergamon, Oxford, 1966, p. 886.

k. L. M. Crapo and C. H. Sederholm, *J. Chem. Phys.* **33**, 1583 (1960).

l. Author's observations; see Fig. 4.9. See also H. M. McConnell, C. A. Reilly, and A. D. McLean, *J. Chem. Phys.* **24**, 479 (1956).

m. G. V. D. Tiers and P. C. Lauterbur, *J. Chem. Phys.* **36**, 1110 (1962).

n. H. M. McConnell, C. A. Reilly, and A. D. McLean, *J. Chem. Phys.* **24**, 479 (1956).

o. W. D. Phillips, *J. Chem. Phys.* **25**, 949 (1956).

p. J. B. Lambert and J. D. Roberts, *J. Am. Chem. Soc.* **87**, 3884 (1965).

q. G. V. D. Tiers, *Proc. Chem. Soc.* **1960**, 389.

r. L. W. Reeves and E. J. Wells, *Discussions Faraday Soc.* **34**, 177 (1962).

s. H. S. Gutowsky, C. H. Holm, A. Saika, and G. A. Williams, *J. Am. Chem. Soc.* **79**, 4596 (1957).

CF_3 group resonance appears as a triplet of triplets at 81.60 Φ^*, the central $CF_2[F_{(4)}]$ as a broad, unresolved multiplet at 122.2 Φ^*, the $F_{(3)}$, $F_{(5)}$ resonance as a somewhat narrower but poorly resolved peak at 122.7 Φ^*, and the $F_{(2)}$ $F_{(6)}$ resonance as a complex multiplet at 126.7 Φ^*.

The substitution of other atoms for ^{19}F markedly increases vicinal ^{19}F-^{19}F couplings. This is not a simple function of electronegativity or of bulkiness for protons (entries 7—11, Table 9.2) are nearly as effective as halogens (12—15).

The geometrical dependence of ^{19}F-^{19}F couplings is not so well worked out as for proton—proton couplings, but it is clear that very strong effects may occur. Harris and Sheppard[11] deduced from the ^{13}C sideband spectra of $CBrF_2 \cdot {}^{13}CBrF_2$ that the gauche coupling is about 12 Hz and the trans coupling about 1.5 Hz. More direct indications are obtained from the study of substituted ethanes in which the C-C rotational barrier is high enough to permit the observation of separate spectra of the rotamers at low temperatures (to *ca.* $-100°C$). Thompson *et al.*[12] found that for the meso isomer of $CFClBr \cdot CFClBr$, the coupling in the gauche rotamers (which are mirror images) is 21.5 Hz; the fluorine nuclei are geometrically equivalent in the trans rotamer and in all rotamers of the *d* isomer:

Newmark and Sederholm[13] have extended these studies to other ethanes and have found for $CF_2Br \cdot CFBr_2$:

$$J_g = -16.1 \text{ Hz} \qquad\qquad J_g = -21.5 \text{ Hz}$$
$$J_t = -18.4 \text{ Hz}$$
$$J_{gem} = 168.7 \text{ Hz}$$

Comparison of this result with that of Harris and Sheppard indicates that replacement of ^{19}F by other halogens increases J_g somewhat and J_t very markedly. For $CF_2Br \cdot CCl_2Br$, J_{gem} is 155.8 Hz for the mirror image conformers:

It is indeterminate for the third conformer as the fluorine nuclei are equivalent.

Vicinal couplings across double bonds are very much larger than in saturated chains and are reminiscent of proton—proton couplings in that J_{trans} is larger than J_{cis} (entries 18—21, Table 9.2). J_{gem}, however, is much larger than for protons and exceeds J_{cis} in magnitude. In perfluoropropene (entry 21) the cis-like coupling of CF_3 to $^{19}F_{(2)}$ is larger than the trans-like coupling to $^{19}F_{(1)}$, again reminiscent of the corresponding proton—proton couplings. The three-bond coupling J_{34} is intermediate in magnitude between the four-bond couplings.

Very marked geometrical dependence of four-bond ^{19}F-^{19}F couplings has been observed in perfluoroacrylyl fluoride,[14] which exhibits two distinguishable conformers at --105°C:

$$\begin{array}{cc}
\underset{F_{(4)}}{\overset{F_{(3)}}{>}}C=C\underset{\underset{F_{(1)}}{\diagdown}C=O}{\overset{F_{(2)}}{\diagup}} & \underset{F_{(4)}}{\overset{F_{(3)}}{>}}C=C\underset{\underset{O\diagup}{\diagdown}C-F_{(1)}}{\overset{F_{(2)}}{\diagup}}
\end{array}$$

	I	II
J_{12}:	34.4 Hz	31.1 Hz
J_{13}:	<2 Hz	41.6 Hz
J_{14}:	84.5 Hz	<2 Hz
J_{23}:	33.7 Hz	36.2 Hz
J_{24}:	111.0 Hz	117.0 Hz
J_{34}:	4.0 Hz	<2 Hz

The vicinal coupling J_{12} is here only rather weakly dependent on conformation, but J_{13} and J_{14} are strongly dependent. It is also found that J_{23} and J_{34} are larger at room temperature — 36.8 Hz and 6.0 Hz, respectively — than for either of the conformers at low temperature. It is thus evident that the couplings in each conformer are temperature dependent; this behavior of ^{19}F couplings has been observed in several other molecules.

Geminal ^{19}F-^{19}F couplings in rings of varying size are shown in entries 22—28, Table 9.2; the largest values are shown by fluorine nuclei on tetrahedral carbon atoms in six-membered rings (27,28).

Ortho and *meta* ^{19}F-^{19}F couplings in aromatic rings (entries 29–33, Table 9.2) are much smaller than the couplings across the same number of bonds in olefins. *Para* couplings (five bonds) are of intermediate magnitude between *ortho* and *meta*.

The signs of ^{19}F-^{19}F couplings, most of which have been determined by double resonance (see Chapter 4, Section 4.4), are summarized in the following structural formulas. These signs are based on the assumption that $J_{13_{CH}}$ is positive and the observation (see Section 9.1.1.4) that $J_{13_{CH}}$ and $J_{13_{C-F19}}$ have opposite signs:

$$
\begin{array}{ccccc}
\underset{F}{\overset{F}{\diagdown}}C\diagup & \underset{F}{\overset{F}{\diagdown}}C= & \overset{F}{\diagdown}C-C\underset{\diagdown}{\diagup}^{F} & F-\overset{C}{\diagdown}C=C\overset{}{\diagup} \\
+ & + & - & - \\
& & \text{\textit{(gauche and trans)}} &
\end{array}
$$

$$
\begin{array}{ccc}
\overset{F}{\diagdown}C=C\overset{\diagup F}{\diagdown} & \overset{F}{\diagdown}C=C\overset{\diagdown}{\diagup_F} & \overset{F}{\diagdown}C-C-C\overset{\diagup F}{\diagdown} \\
+ & + & + \\
F-\overset{}{\diagdown}C=C\overset{\diagup F}{\diagdown} & F-\overset{}{\diagdown}C=C\overset{\diagdown}{\diagup_F} & \overset{F}{\diagdown}C=C\overset{}{\diagdown}C-C-F \\
+ & + & +
\end{array}
$$

(ortho-difluorobenzene structure) −
(meta-difluorobenzene structure) −
(para-difluorobenzene structure) +

9.1.1.3 ^{19}F-^{1}H Coupling

^{19}F-^{1}H couplings are more "normal" than ^{19}F-^{19}F couplings in that they tend to fall off in magnitude monotonically with increasing numbers of intervening bonds. They also exhibit very strong geòmetrical dependencies, which can upset this correlation. Vicinal and geminal ^{19}F-^{1}H couplings for a representative group of substituted alkanes are given in Table 9.3 (entries 1–20). Geminal couplings are all in the range of *ca.* 45–55 Hz, but have been observed to increase to as much as 75.6 Hz in CHF_2X with an increase in electronegativity of X.[13] Vicinal couplings usually have less than half the magnitude of geminal couplings. Compounds 1–3 (Table 9.3) show a marked decrease in the vicinal coupling as the number of fluorine atoms on the group CH_nF_{3-n} is increased. This trend is even more marked in compounds 5 and 6. The series of compounds 2, 11, 10, 3 indicate a decrease in vicinal coupling in CH_3CF_2X as the electronegativity of X is increased. In the ethanes 4 and 12–16, the observed couplings are, of course, dependent upon conformational preferences and cannot be directly compared with the others. For $CF_2Br \cdot CHBrCl$, the following values have been found[13] by analysis of the low-temperature rotamer spectra:

I	II	III

$J_{H-F \text{ gauche}}$: <2 Hz $J_{H-F \text{ gauche}}$: <2 Hz $J_{H-F \text{ gauche}}$: 1.8, 3.4 Hz

$J_{H-F \text{ trans}}$: 18.9 Hz $J_{H-F \text{ trans}}$: 18.5 Hz (not assigned)

$(J_{F-F \text{ gem}}$: 160.8 Hz) $(J_{F-F \text{ gem}}$: 159.0 Hz)

The substituted propanes (entries 17–20) seem, in general, to follow the same trends as the ethanes; in compounds 19 and 20 a measurable ^{19}F-^1H coupling is transmitted through four bonds.

The form of the dependence of vicinal ^{19}F-^1H couplings on the dihedral angle is still not entirely clear. There is evidence that it is similar to that of ^1H-^1H couplings, i.e., that there are maxima at 0° and 180° and a minimum near 90°, in agreement with theoretical expectation.[15,16] Williamson et al.[17] measured the vicinal couplings in a number of rigid cyclic compounds, among them entry 32 in Table 9.3. They found that when the dihedral angle was supposedly near 0° (as for $J_{F-H(a)}$, entry 32), J_{F-H} varied from 10.55 to 30.8 Hz. When the angle was 60°, values of 3.8 and 6.0 Hz were observed. This is consistent with the low values of J_{gauche} indicated above for the rotamers of $CF_2Br \cdot CHBrCl$ and with the small values of $J_{H_{(eq)}F_{(ax)}}$, $J_{H_{(ax)}F_{(eq)}}$, and $J_{H_{(eq)}F_{(eq)}}$ derived from the low-temperature spectra of the conformers of cyclohexyl fluoride (entry 36). When the angle was increased to 120°, a coupling of 12.2 Hz was observed; at 132° ($J_{F-H_{(b)}}$, entry 32) it decreased to 6.3 Hz. Williamson et al. did not report a coupling corresponding to a dihedral angle of 180°; the results for $CF_2Br \cdot CHBrCl$ indicate a marked increase to ca. 18 Hz, but $J_{H_{(ax)}F_{(ax)}}$ in the equatorial cyclohexyl fluoride conformer has the much larger value of 43.5 Hz.[18] It remains to be determined whether this inconsistency arises chiefly because of substituent effects (which from entries 1–11 can be clearly seen to play a major role), or because the dihedral angles deviate from what they are presumed to be.

TABLE 9.3

Some Representative ^{19}F-^1H Couplings

Compound number	Compound	J (Hz)		Ref.
1	$CH_{3(2)}CH_{2(1)}F$	$J_{H_{(1)}-F}$:	47.5	a, b
		$J_{H_{(2)}-F}$:	25.7	
2	$CH_{3(2)}CH_{(1)}F_2$	$J_{H_{(1)}-F}$:	57.2	a
		$J_{H_{(2)}-F}$:	20.8	
3	CH_3CF_3	J_{H-F}:	12.8	a, c
4	$CF_{2(2)}H_{(2)}CF_{2(1)}H_{(1)}$	$J_{H_{(1)}-F_{(1)}}$:	52.1	a
		$J_{H_{(1)}-F_{(2)}}$:	4.8	
5	$CF_{3(2)}CF_{(1)}H_{2(1)}$	$J_{H_{(1)}-F_{(1)}}$:	45.5	a
		$J_{H_{(1)}-F_{(2)}}$:	8.0	
6	$CF_{3(2)}CF_{2(1)}H_{(1)}$	$J_{H_{(1)}-F_{(1)}}$:	52.6	a
		$J_{H_{(1)}-F_{(2)}}$:	2.6	
7	CF_3CH_2Cl	J_{H-F}:	8.5	c
8	CF_3CH_2Br	J_{H-F}:	9.0	c
9	$CF_3CH_{2(1)}OH$	$J_{H_{(1)}-F}$:	8.9	c
10	CH_3CF_2Cl	J_{H-F}:	15.0	a
11	CH_3CF_2Br	J_{H-F}:	15.9	a
12	$CH_2Cl \cdot CF_2Cl$	J_{H-F}:	6, 16 (AA'XX')	d
13	$CH_2Cl \cdot CF_2Br$	J_{H-F}:	3, 21 (AA'XX')	e
14	$CH_2Br \cdot CF_2Br$	J_{H-F}:	4, 22 (AA'XX')	d
15	$CF_{2(2a,2b)}Br \cdot CH_{(1)}F_{(1)}Br$	$J_{H_{(1)}-F_{(1)}}$:	48	e
		$J_{H_{(1)}-F_{(2a)}}$,		
		$J_{H_{(1)}-F_{(2b)}}$:	3.2, 9.1	
16	$CF_{2(2a,2b)}Br \cdot CH_{(1)}F_{(1)}Cl$	$J_{H_{(1)}-F_{(1)}}$:	48	d, f
		$J_{H_{(1)}-F_{(2a)}}$,		
		$J_{H_{(1)}-F_{(2b)}}$:	3.5, 6.3	
17	$CF_{3(3)}CF_{2(2)}CH_{2(1)}F_{(1)}$	$J_{H_{(1)}-F_{(1)}}$:	46.0	a
		$J_{H_{(1)}-F_{(2)}}$:	11.7	
18	$CF_{3(3)}CF_{2(2)}CH_{(1)}F_{2(1)}$	$J_{H_{(1)}-F_{(1)}}$:	52.1	a
		$J_{H_{(1)}-F_{(2)}}$:	4.5	
19	$CF_{2(3)}Cl \cdot CF_{2(2)}CH_{2(1)}F_{(1)}$	$J_{H_{(1)}-F_{(1)}}$:	45.9	a
		$J_{H_{(1)}-F_{(2)}}$:	11.8	
		$J_{H_{(1)}-F_{(3)}}$:	1.0	
20	$CF_{2(3)}Br \cdot CF_{2(2)}CH_{2(1)}F_{(1)}$	$J_{H_{(1)}-F_{(1)}}$:	46.0	a
		$J_{H_{(1)}-F_{(2)}}$:	11.7	
		$J_{H_{(1)}-F_{(3)}}$:	0.9	

TABLE 9.3 (continued) 451

21	$H_2C=CCl_2$ (H, F on left carbon; Cl, Cl on right)	J_{H-F}: 81	g
22	(H, H on left; F, F on right)	$J_{H-F(cis)}$: ca.1 $J_{H-F(trans)}$: 34	g
23	(H, F on left; H, F on right)	$J_{H-F(trans)}$: 20.4 $J_{H-F(gem)}$: 72.7	h
24	(H, F on left; F, H on right)	$J_{H-F(cis)}$: 4.4 $J_{H-F(gem)}$: 74.3	h
25	(F, F on left; H, F on right)	$J_{H-F(cis)}$: <3 $J_{H-F(trans)}$: 12 $J_{H-F(gem)}$: 72	g
26	(H, H on left; F, H on right)	$J_{H-F(cis)}$: 20.1 $J_{H-F(trans)}$: 52.4 $J_{H-F(gem)}$: 84.7	i
27	(H, H on left; F, Cl on right)	$J_{H-F(cis)}$: 8 $J_{H-F(trans)}$: 40	j
28	(F, F on left; H, Cl on right)	$J_{H-F(cis)}$: <3 $J_{H-F(trans)}$: 13	g
29	($CH_{3(3)}$, $H_{(2)}$ on left; F, $H_{(1)}$ on right)	$J_{F-H_{(1)}}$: 89.9 $J_{F-H_{(2)}}$: 41.8 $J_{F-H_{(3)}}$: 2.6	j
30	($CH_{3(3)}$, $H_{(2)}$ on left; $H_{(1)}$, F on right)	$J_{F-H_{(1)}}$: 84.8 $J_{F-H_{(2)}}$: 19.9 $J_{F-H_{(3)}}$: 3.3	j
31	$F-C\equiv C-H$	J_{F-H}: 21	k

TABLE 9.3 (continued)

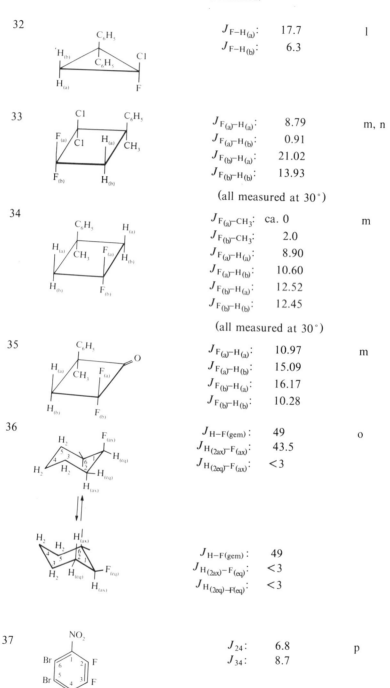

32

$J_{F-H_{(a)}}$:   17.7   l

$J_{F-H_{(b)}}$:   6.3

33

$J_{F_{(a)}-H_{(a)}}$:   8.79   m, n

$J_{F_{(a)}-H_{(b)}}$:   0.91

$J_{F_{(b)}-H_{(a)}}$:   21.02

$J_{F_{(b)}-H_{(b)}}$:   13.93

(all measured at 30°)

34

$J_{F_{(a)}-CH_3}$:   ca. 0   m

$J_{F_{(b)}-CH_3}$:   2.0

$J_{F_{(a)}-H_{(a)}}$:   8.90

$J_{F_{(a)}-H_{(b)}}$:   10.60

$J_{F_{(b)}-H_{(a)}}$:   12.52

$J_{F_{(b)}-H_{(b)}}$:   12.45

(all measured at 30°)

35

$J_{F_{(a)}-H_{(a)}}$:   10.97   m

$J_{F_{(a)}-H_{(b)}}$:   15.09

$J_{F_{(b)}-H_{(a)}}$:   16.17

$J_{F_{(b)}-H_{(b)}}$:   10.28

36

$J_{H-F(gem)}$:   49   o

$J_{H_{(2ax)}-F_{(ax)}}$:   43.5

$J_{H_{(2eq)}-F_{(ax)}}$:   <3

$J_{H-F(gem)}$:   49

$J_{H_{(2ax)}-F_{(eq)}}$:   <3

$J_{H_{(2eq)}-F(eq)}$:   <3

37

J_{24}:   6.8   p

J_{34}:   8.7

452

TABLE 9.3 (continued)

38		J_{36}:	-2.2	p
		J_{56}:	9.4	
39		J_{16}:	8.4	p
		J_{46}:	6.3	
40		J_{35}:	6.3	p
		J_{56}:	8.2	
41		$J_{24}(J_{46})$:	7.0	q
		$J_{34}(J_{45})$:	8.6	
42		$J_{F-CH_{3(a)}}$:	1.2	r
		$J_{F-CH_{3(b)}}$:	ca. 0	

a. D. D. Elleman, L. C. Brown, and D. Williams, *J. Mol. Spectrosc.* **7**, 307 (1961).

b. S. L. Stafford and J. D. Baldeschwieler, *J. Am. Chem. Soc.* **83**, 4473 (1961).

c. N. Muller and D. T. Carr, *J. Phys. Chem.* **67**, 112 (1963).

d. J. N. Shoolery and B. L. Crawford, *J. Mol. Spectrosc.* **1**, 270 (1957).

e. J. Lee and L. H. Sutcliffe, *Trans. Faraday Soc.* **55**, 880 (1959).

f. J. Lee and L. H. Sutcliffe, *Trans. Faraday Soc.* **54**, 308 (1958).

g. H. M. McConnell, C. A. Reilly, and A. D. McLean, *J. Chem. Phys.* **24**, 479 (1956).

h. G. W. Flynn, M. Matsushima, N. C. Craig, and J. D. Baldeschwieler, *J. Chem. Phys.* **38**, 2295 (1963).

i. C. N. Banwell and N. Sheppard, *Proc. Roy. Soc.* **A263**, 136 (1961).

j. R. A. Beaudet and J. D. Baldeschwieler, *J. Mol. Spectrosc.* **9**, 30 (1962).

k. W. J. Middleton and W. H. Sharkey, *J. Am. Chem. Soc.* **81**, 803 (1959).

l. K. L. Williamson, Y-F. Li, F. H. Hall, and S. Swager, *J. Am. Chem. Soc.* **88**, 5678 (1966).

m. J. B. Lambert and J. D. Roberts, *J. Am. Chem. Soc.* **87**, 3891 (1965).

n. M. Takahashi, D. R. Davies, and J. D. Roberts, *J. Am. Chem. Soc.* **84**, 2935 (1962).

o. F. A. Bovey, E. W. Anderson, F. P. Hood, and R. L. Kornegay, *J. Chem. Phys.* **40**, 3099 (1964).

p. H. S. Gutowsky, C. H. Holm, A. Saika, and G. A. Williams, *J. Am. Chem. Soc.* **79**, 4596 (1957).

q. T. Schaefer, *Can. J. Chem.* **37**, 882 (1959).

r. A. H. Lewin, *J. Am. Chem. Soc.* **86**, 2303 (1964).

In alkenes (entries 21–30), geminal ^{19}F-^{1}H couplings do not vary widely, but the vicinal couplings are sensitive to the presence of substituents. Trans couplings are always much larger than cis, just as for vicinal ^{1}H-^{1}H and ^{19}F-^{19}F couplings. Comparison of entries 22–26 shows that the accumulation of fluorine substituents depresses both $J_{H-F(cis)}$, and $J_{H-F(trans)}$. In the 1-fluoropropenes (entries 29 and 30), the four-bond ^{19}F-^{1}H couplings are considerably larger than in the saturated chains (19 and 20). The coupling in fluoroacetylene (entry 31) is comparable to the trans ^{19}F-^{1}H coupling in *cis*-1,2-difluoroethylene (entry 23).

^{19}F-^{1}H couplings in *gem*-difluorocyclobutanes have been measured by Takahashi *et al.*[19] and Lambert and Davis.[20] A selection of the data are included in Table 9.3 as entries 33–35. The four couplings shown were measured at 30°C; they also measured these couplings at −50° and 100°C. It was found that the couplings of the cyclobutanone, entry (35), were insensitive to temperature, which is consistent with a single, probably nearly planar conformation. Substantial changes (*ca.* 1–2 Hz) were observed over this temperature range in the couplings of entry 34 and smaller but measurable changes in those of entry 33. It is also noteworthy that the trans couplings differ markedly in these latter two compounds. This behavior was interpreted in terms of changing populations of conformers. For entry 34, an equilibrium between two nonplanar forms, differing only moderately in energy, was proposed; for entry 33, in which the trans couplings differ greatly, one nonplanar form was believed to predominate at low temperature, with increasing participation of a planar form at higher temperature. These conclusions were based on the assumption of Karplus-like vicinal ^{19}F-^{1}H coupling dependence.

A few examples of ^{19}F-^{1}H couplings in benzene rings are provided by entries 37—42. *Ortho* and *meta* couplings are similar in magnitude, *ortho* couplings being generally somewhat larger. *Para* couplings are usually about 2.2 ± 0.2 Hz.

Appreciable ^{19}F-^{1}H couplings can be transmitted over five and even six intermediate bonds if certain rather special geometrical requirements are met. The couplings of ring fluorine nuclei to angular methyl groups have been studied by Cross and Landis.[21,22] It is found, for example, that the 19-H resonance in 6β-fluoro steroids [(a) below] appears as a doublet, whereas in the 6α-fluoro isomer (b) no observable coupling occurs:

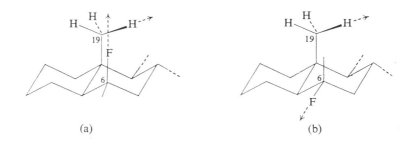

(a)   (b)

Again, in the $5\beta,6\beta$-difluorocyclopropyl steroid (c), the methyl group resonance is a doublet, indicating that only one of the fluorine nuclei is appreciably coupled:

(c)

These findings were summarized in an empirical rule which states that ^{19}F-^{1}H coupling will be observed only when a vector directed along the C-F bond and originating at the carbon atom can converge upon and intersect a similar vector directed along a C-H bond of the methyl group. This condition is satisfied for structure (a) but not for structure (b) above, and for $F_{(a)}$ but not $F_{(b)}$ in structure (c). In compound 33 (Table 9.3) it is found that only one of the two fluorine nuclei is observably coupled to the methyl group, and this is believed to be $F_{(b)}$. This is in accord with the rule provided the conformation

is fairly strongly preferred, and this, in turn, is consistent with the other observed couplings, as we have seen.

An observable ^{19}F-^{1}H coupling is transmitted through six bonds in compound 42, Table 9.3. Here, the partial double-bond character of the C-N and aromatic C-C bonds probably enhances the effect. Only one of the methyl groups is split by ^{19}F couplings, and it is probably the one cis to the phenyl group. This again is in accord with the vector rule.

Information concerning signs of ^{19}F-^{1}H couplings may be summarized as follows, adopting the convention discussed on p. 448:

9.1.1.4 ^{19}F-^{13}C Coupling

A selected group of carbon-fluorine couplings are given in Table 9.4. Direct, one-bond couplings are found to be negative in those molecules for which they have been determined. Accumulation of fluorines tends to decrease the magnitude of the coupling (entries 1−6), but accumulation of protons has even more effect (7−10). Couplings to trigonal carbons (entries 11 and 12) are comparable to tetrahedral couplings, while those of aromatic carbons (entries 17−24) are somewhat reduced and show at most a weak dependence on the electronegativity of *para* substituents.

Two-bond couplings (entries 11, 14, 15−17, 19) are one-fifth to one-tenth of one-bond couplings in magnitude and may be of opposite sign even in closely related molecules (entries 15, 16).

The theoretical interpretation of ^{19}F couplings has not been notably successful. It has been reviewed by Emsley, Phillips, and Wray,[23] who also present an extensive table of couplings reported up to 1972.

9.1.2 Nitrogen

9.1.2.1 ^{14}N and ^{15}N

Next to hydrogen and carbon, nitrogen is the most important of the elements composing organic and bioorganic molecules, occurring both in skeletons and functional groups. Unfortunately, the abundant species, ^{14}N (99.63%), is inconvenient to observe because of its nuclear electric quandrupole. We have seen several instances of ^{14}N quadrupolar relaxation in previous discussion (Chapter 4, Section 4.4.3.1; Chapter 5, Sections 5.24, 5.29). Sometimes this is beneficial; for example, it makes possible the observation of protons directly bonded to nitrogen with resolution sufficient to resolve vicinal ^{1}H-^{1}H couplings in amides and peptides.[24] For the observation of ^{14}N itself, however, it is not helpful since it causes severe line broadenings, sometimes by as much as 10^3 Hz, which of course prevents the resolution of structural detail.

With the introduction of FT NMR spectroscopy, attention turned to ^{15}N, which has a spin of 1/2 and therefore presents no problem from quadrupolar line broadenings. Although its observing sensitivity in natural abundance is only 0.357% of that of ^{14}N (and *ca.* 4 ppm compared to the proton), this drawback is in part overcome by the large nuclear Overhauser effect ($\eta = -4.93$) under proton decoupling. This NOE, added to the normal signal intensity of +1.0, results in a maximum inverted signal of approximately −4.0 intensity. For large molecules, however, this enhancement will diminish and the signal intensity may pass through zero to positive values. This may also happen if the ^{15}N is relaxed by mechanisms other than dipole−dipole interaction with the bonded proton.

TABLE 9.4

Some Representative ^{19}F-^{13}C Couplings

Compound number	Compound	^1J (Hz)	^2J (Hz)	^nJ (Hz)	Ref.
1	CFBr$_3$	372			a
2	CF$_2$Br$_2$	358			b
3	CF$_3$Br	324			b
4	CFCl$_3$	337			b
5	CF$_2$Cl$_2$	325			b
6	CF$_3$Cl	299			b
7	CF$_4$	257			b
8	CF$_3$H	272			b
9	CF$_2$H$_2$	232			b
10	CH$_3$F	−157.5			c
11		−291.0	+54.5		d
12	CF$_2$ = CD$_2$	287			b
13	CF$_3$COOH	283			e
14	CF$_3$CF$_3$	281.3	46.0		f
15		298	−25.8		b
16		−299.67	+26.94		g
17		252	17.5		h
18		260			i

TABLE 9.4 (continued)

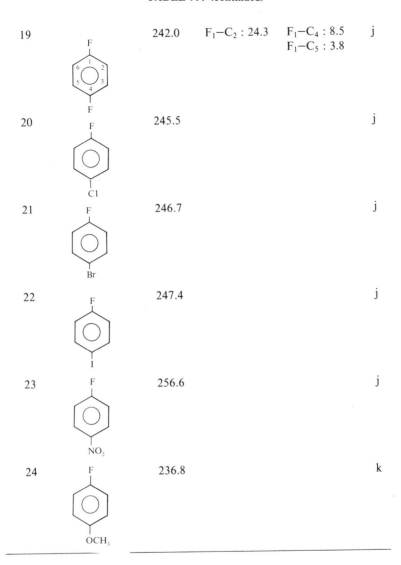

19		242.0	F_1-C_2 : 24.3	F_1-C_4 : 8.5	j
				F_1-C_5 : 3.8	
20		245.5			j
21		246.7			j
22		247.4			j
23		256.6			j
24		236.8			k

Key to references:

a. W. D. Phillips, *Determination of Organic Structures by Physical Methods*, F. C. Nachod and W. D. Phillips, Eds., Vol. 2, Academic Press, New York, 1962.

b. N. Muller and D. T. Carr, *J. Phys. Chem.* **67**, 112 (1963).

c. S. G. Frankiss, *J. Phys. Chem.* **67**, **752** (1963).

d. R. Kaiser and A. Saika, *Mol. Phys.* **15**, 221 (1968).

e. G. V. D. Tiers, *J. Phys. Soc. Jpn.* **15**, 354 (1960).

f. R. E. Graves and R. A. Newmark, *J. Chem. Phys.* **47**, 3681 (1967).

g. R. K. Harris, and V. J. Robinson, *J. Mag. Res.* **1**, 362 (1969).

h. T. F. Page, *Mol. Phys.* **13**, 523 (1967).

i J. P. Imbaud, Compt. rend. **261**, 5442 (1965).

j. F. J. Weigert and J. D. Roberts, *J. Amer. Chem. Soc.* **93**, 2361 (1971).

k. S. Mohanty and P. Venkateswarlu, *Mol. Phys.* **12**, 277 (1967).

The NMR of nitrogen-15 has been reviewed by Martin *et al.*,[25] Witanowski *et al.*,[26a], Stefaniak *et al.*[26b], Levy and Lichter,[27] and von Philipsborn and Müller.[28] The 1981 monograph of Witanowski *et al.*[26a] contains 156 tables of ^{15}N chemical shifts and couplings, while that of Stefaniak *et al.*[26b] contains 163 tables. The review of von Philipsborn and Müller[28] is recommended as a general survey. It contains a discussion of the 2D NMR of ^{15}N and the use of special pulse sequences for sensitivity enhancement.

9.1.2.2 ^{15}N Chemical Shifts

^{15}N chemical shifts extend over a range of nearly 1000 ppm, from alkylamines (most shielded) to nitroso compounds (least shielded). This is about five times the range usually observed for carbon—13 shieldings (Chapter 3, Section 3.3.2), with correspondingly great discrimination of structural features. The reference compound now most commonly employed is *nitromethane*; on this scale, NH_3 (neat), also proposed and used in early work, appears at -381.9 ppm. Following the practice already described (Chapter 2, Section 2.6.3; Chapter 3, Section 3.3.2), values more shielded than the reference are given negative signs and those less shielded are given positive signs. The result is that the majority of ^{15}N chemical shifts for organic compounds are negative. Some workers in this field have considered this undesirable, and in fact the opposite sign convention (but still based on nitromethane) has been adopted in the extensive tables of Ref. 26. However, it seems to be generally acceptable.[25] The use of this scale is illustrated in Fig. 9.2, taken from Ref. 28.

Nitrogen—15 and carbon—13 are similar in electronic structure and bonding and their chemical shifts generally show parallel dependence on chemical structure. Table 9.5 shows the ^{15}N chemical shifts (referred to

TABLE 9.5

Selected Values of ^{15}N Chemical Shifts

Entry number	Compound	ppm vs. $CH_3^{15}NO_2$	Solvent; conditions
Miscellaneous			
1	CH_3NO_2	0.000	neat
2	NH_3	−381.9	neat
3	$NH_4^+Cl^-$	−352.9	sat. aq. sol.
n-Alkylamines:			
4	CH_3NH_2	−378.7	neat
		−377.3	2M in CH_3OH
5	$(CH_3)_2NH$	−369.5	2M in CH_3OH
6	$(CH_3)_3N$	−363.1	2M in CH_3OH
7	$CH_3NH_3^+$	−361.8	2M in CH_3OH
8	$(CH_3)_2NH_2^+$	−356.6	1M in CH_3OH
9	$(CH_3)_3NH^+$	−349.4	1M in CH_3OH
10	$(CH_3)_4N^+$	−336.7	sat. aq. sol.
11	$CH_3CH_2NH_2$	−355.4	2M in CH_3OH
12	$CH_3(CH_2)_2NH_2$	−359.6	2M in CH_3OH
13	$CH_3(CH_2)_3NH_2$	−359.4	2M in CH_3OH
14	$(CH_3)_2CH\ NH_2$	−338.1	2M in CH_3OH
15	$(CH_3)_2CH\ CH_2NH_2$	−362.7	2M in CH_3OH
16	$(CH_3CH_2)_2NH$	−333.7	2M in CH_3OH
17	$(CH_3CH_2)_3N$	−332.0	2M in CH_3OH
18	$c-C_6H_{11}-NH_2$	−340.4	2M in CH_3OH
19	$CH_3(CH_2)_2NH_3^+$	−349.0	1M in CH_3OH
20	$CH_3(CH_2)_3NH_3^+$	−348.8	1M in CH_3OH

TABLE 9.5 (continued)

21	$(CH_3)_2CH\ NH_3^+$	−334.0	1M in CH_3OH
Cyclic amines:			
22	NH (aziridine)	−388.7	4−5 M in $CDCl_3$
23	N CH_3	−379.5	4−5 M in $CDCl_3$
24	NCH_2CH_3	−363.8	4−5 M in $CDCl_3$
25	HN (azetidine)	−354.9	4−5 M in $CDCl_3$
26	NCH_3	−348.7	4−5 M in CDI_3
27	NCH_2CH_3	−335.2	4−5 M in $CDCl_3$
28	NH (pyrrolidine)	−343.5	H_2O, 0.5 M NH_4NO_3
29	N–CH_3	−339.5	H_2O, 0.5M NH_4NO_3
30	NH (piperidine)	−343.2	H_2O, 0.5M NH_4NO_3
31	N–CH_3	−343.5	H_2O, 0.5M NH_4NO_3
32	NH_2^+	−337.5	82:18 CH_3Cl_3:CH_3OH

TABLE 9.5 (continued)

Anilines:

33	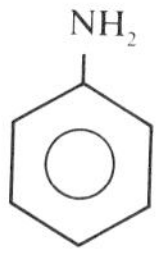 (structure: aniline, NH$_2$ on benzene)	-322.3	acetone
34	4-F	-325.7	acetone
35	4-Cl	-322.8	acetone
36	4-Br	-321.5	acetone
37	4-I	-321.6	DMSO
38	4-NO$_2$	-307.4	acetone
39	4-OCH$_3$	-328.2	DMSO
40	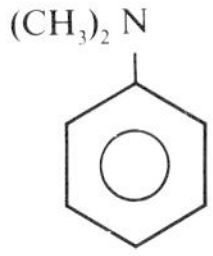 (structure: (CH$_3$)$_2$N on benzene)	-339.8	neat

Hydrazines: (chemical shifts in left to right order)

41	H$_2$N$\cdot$NH$_2$	-334.8	neat
42	CH$_3$HN$\cdot$NH$_2$	$-328.0, -305.5$	neat
43	(CH$_3$)$_2$N$\cdot$NH$_2$	$-322.7, -281.4$	neat
44	CH$_3$HN$\cdot$NHCH$_3$	-306.6	neat
45	(CH$_3$)$_2$N$\cdot$N(CH$_3$)$_2$	-303.6	neat

Ureas and amides

46	(H$_2$N)$_2$CO	-302.6	DMSO
47	CH$_3$NH$-$C $=$ O$\cdot$NH$_2$	NH:-310.2 NH$_2$:-304.7	DMSO
48	CH$_3$NH$-$C $=$ O$\cdot$NH CH$_3$	-316.7	DMSO
49	(CH$_3$)$_2$N$-$C $=$ O$\cdot$OCH$_3$	-315.7	neat
50	H$-$C $=$ O$\cdot$NH$_2$	-267.8	neat and in H$_2$O
51	H$-$C $=$ O$\cdot$NHCH$_3$	cis -273.8 trans -271.8	neat

TABLE 9.5 (continued)

52	$H-C = O \cdot N(CH_3)_2$	-277.01	neat
53	$CH_3-C = O \cdot NH_2$	-276.8	CDCl$_3$
54	$CH_3-C = O \cdot NHCH_3$	-272.8 (trans?)	neat
55	$CH_3-C = O \cdot N(CH_3)_2$	-281.6	neat
56		-260.4	H_2O
57		-260.7	H_2O
58		-257.8	H_2O
59		-267.4	CF$_3$COOH
60		LL -254.5 DL -254.9	CF$_3$COOH

Linear peptides (chemical shifts in left to right order):

61	H−Gly−Gly−OH	$-354.2, -272.4$	HCOOH
62	(Gly)$_n$	-272.2	HCOOH
63	tert.−BuOCO−Gly−Gly−OH	$-302.3\ -274.0$	DMSO
64	H−Ala−Ala−OH	$-340.6\ -257.8$	DMSO
65	tert.−BuOCO−Ala−Ala−OH	$-287.9, -262.2$	DMSO

TABLE 9.5 (continued)

Cyano, azide, cyanate, isocyanate

66	HCN	-129	neat
67	CH_3CN	-135.8_3	neat
68	CH_3NC	-219.6	neat
69	$CH_3-N=C=0$	-365.4_2	neat
70	$C_6H_5 \cdot N=C=0$	$-388._3$	neat
71	$H-N=N \equiv N$	$-324.5, -134.1, -178.6$	Et_2O
72	$CH_3-N=N \equiv N$	$-321.7, -130.2, -171.5$	CD_2Cl_2

Heterocyclics:

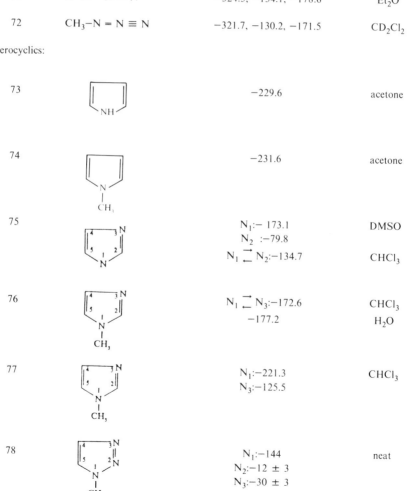

73		-229.6	acetone
74		-231.6	acetone
75		$N_1 :- 173.1$ $N_2 :-79.8$ $N_1 \rightleftarrows N_2 :-134.7$	DMSO CHCl$_3$
76		$N_1 \rightleftarrows N_3 :-172.6$ -177.2	CHCl$_3$ H$_2$O
77		$N_1 :-221.3$ $N_3 :-125.5$	CHCl$_3$
78		$N_1 :-144$ $N_2 :-12 \pm 3$ $N_3 :-30 \pm 3$	neat

TABLE 9.5 (continued)

79		N_1:-154.2 N_2:-11.7 N_3:$+10.6$ N_4:-52.1	CDCl$_3$
80		N_1:-75.0 N_2:-104.9 N_3:-2.4 N_4:-50.3	CDCl$_3$
81		N_1:-98.8 N_2:$+4.4$ N_3:-42.9 N_5:-68.7	CDCl$_3$
82		-124 ± 1	CCl$_4$
83		-62.0 -69.2	neat CHCl$_3$
84		-167.8	CHCl$_3$
85		-85.4	neat
86		-46.3	DMSO
87		$+20.3$	DMSO

TABLE 9.5 (continued)

88	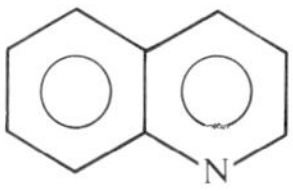 (quinoline structure)	−66.9	acetone

Azo and nitroso

89	$C_6H_5 \cdot \underset{\underset{O}{\downarrow}}{N} = N \cdot C_6H_5$	cis NO:−36.0	CDCl₃
		N:−19.8	
90	$C_6H_5 \cdot \overset{+}{N} \equiv N$	$\overset{+}{-N} \equiv \;:-156.4$	in H₂O with
		$\equiv$ N:−63.4	BF_4^- counterion
91	tert.−Bu−N = O	+578 ± 3	neat
92	$C_6H_5 \cdot N = O$	+529.4	acetone
93	NO_2^-	+227.6	H₂O

Oximes

94	$H_2C = N \diagup^{OH}$	−2.2	20 mole %, H₂O
95	(CH₃)(H)C=N−OH structures	cis −30.3	CHCl₃
		trans −34.6	
96	(CH₃CH₂)(H)C=N−OH structures	cis −32.8	CHCl₃
		trans −34.4	

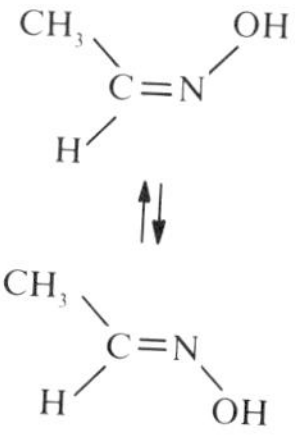

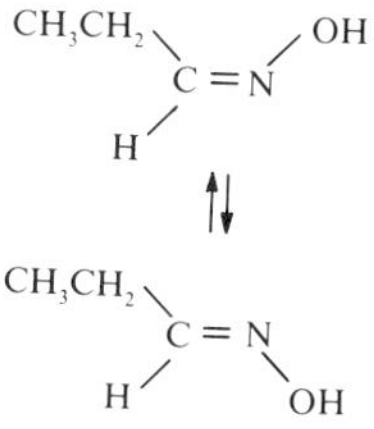

TABLE 9.5 (continued)

97	CH₃\C=N\OH (CH₃)	−45.9	CHCl₃
98	=NOH (cyclobutanone oxime)	−52.4	CHCl₃
99	=NOH (cyclopentanone oxime)	−53.0	CHCl₃
100	=NOH (cyclohexanone oxime)	−52.6	CHCl₃
101	=NOH (cycloheptanone oxime)	−46.8	CHCl₃

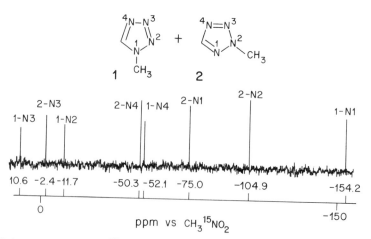

Fig. 9.2   40.6 MHz ^{15}N spectrum of 1-methyltetrazole (1) and 2-methyltetrazole (2); 200 mg in 2.5 mL of CDCl₃; 75 mg Cr(acac). [Adapted from W. von Philipsborn and R. Müller, *Angew. Chem.* **25**, 383 (1986).]

CH_3NO_2) of representative organic nitrogen compounds arranged by classes rather than in a rigorous sequence of chemical shifts. It should be pointed out that a similar listing of ^{14}N chemical shifts (which would be less complete) would show the same values within a few tenths of a ppm since the primary isotope effect is very small. In relation to the nature of the N bonding orbitals it is found that, in general, shielding decreases in the order $sp^3 > sp > sp^2$. Electronegative substituents cause deshielding, as expected.

Protonation of alkylamines causes strong deshielding of NH_3 (by 29 ppm; items 2 and 3), less so in alkylamines (by *ca.* 13 ± 3; items 7, 8, 9, 19, 20, 21). In contrast, protonation of pyridines (items 83, 84) causes shielding by *ca.* 100 ppm.

There is a relatively small but fairly regular effect of carbon substitution and chain length in alkyl amines. Thus, one methyl group causes an approximately 5 ppm increase of shielding of ammonia, while two and three methyls cause a slightly larger but comparable increase of *ca.* 6 ppm per methyl (items 1, 4, 5, 6). The addition of carbons β to the nitrogen has a marked effect (item 11 versus item 4), whereas a third carbon shows a shielding effect (item 12 versus item 11) reminiscent of the carbon γ effect.

As in cyclopropanes and oxiranes, aziridines show marked ^{15}N shielding compared to the corresponding open-chain compounds. Thus, in aziridine itself (item 22) the nitrogen is 19.2 ppm more shielded than in dimethylamine (item 5). The four-membered (item 25) and five-membered (item 29) rings are deshielded in comparison to dimethylamine, but this trend does not continue in larger rings.

In *para*-substituted anilines, ^{15}N shielding shows little effect from the presence of halogens (items 34–37), but NO_2 and OCH_3 show a marked effect in the expected direction (items 38, 39). In heterocyclic rings, ^{15}N shielding is strongly dependent on ring position. This is seen in items 75–81 and is strikingly illustrated by the N-methyl tetrazole spectra in Fig. 9.2 (items 79 and 80).

In peptides (items 61–65) ^{15}N chemical shifts are similar to those of amides (items 50–58), as expected, but are much more shielded at the N–terminus.

^{14}N and ^{15}N chemical shifts may be observed in the solid state in the same manner as for carbon–13 (Chapter 8, Section 8.6). Schaefer *et al.*[29] have used magic angle spinning and cross polarization to detect amide and amino groups in ^{15}N-labeled soybean plants.

9.1.2.3 ^{14}N and ^{15}N Couplings

We have seen (Chapter 4, Section 4.4.3.1, Chapter 5, Section 5.2.4) that direct (one-bond) ^{14}N—proton J couplings, which are approximately 50—60 Hz, are usually not observed because the ^{14}N quadrupole causes decoupling. (The ^{1}H and ^{14}N multiplicity is preserved in symmetrical environments such as ammonium ions.) The corresponding ^{15}N couplings are readily observable as proton or nitrogen multiplicity, the magnitude of which is calculable from the relationship (Chapter 4, Section 4.1):

$$J(^{15}\text{N} - \text{X}) = J(^{14}\text{N} - \text{X})\, \frac{\gamma_{15_\text{N}}}{\gamma_{14_\text{N}}} \qquad (9.1\text{a})$$

$$= -1.40 \cdot J(^{14}\text{N} - \text{X}) \qquad (9.1\text{b})$$

Selected values of ^{15}N couplings are given in Table 9.6. One-bond ^{15}N-^{1}H couplings are expected from Eq. 9.1 to be negative and to be *ca.* 70—90 Hz. They are the largest of nitrogen couplings and, like ^{13}C couplings, are markedly dependent on the *s* character of the bonding orbitals. Thus, ammonia and alkylamines (sp^3) are −60 to −75 Hz whereas arylamines (sp^2) are −80 to −90 Hz and pyrroles and pyridinium ion are *ca.* −95 Hz. Amides (sp^2) are *ca.* −90 Hz. Nitrilium ions, $\text{R} - \text{C} \equiv \overset{+}{\text{N}} - \text{H}$, which may be assumed to be *sp*, exhibit $^{1}J_{\text{N–H}}$ values of *ca.* −135 Hz.

Two-bond ^{15}N-^{1}H couplings are weaker than one-bond couplings, usually by an order of magnitude, but are more sensitive to structure and may be either positive or negative. Thus, in methylamine (item 3) the nitrogen-methyl proton coupling is −1.0 Hz, whereas the insertion of a carbonyl group, as in N-methylformamide (item 10), increases the coupling to −14.1 Hz; in the latter molecule, the coupling to the methyl protons is +1.4 Hz. Positive charge decreases two-bond couplings: in pyridine, the coupling to H_2 and H_6 is reduced from −10.93 to −3.01 Hz when the nitrogen is protonated (items 18 and 19), and a similar effect is observed in quinoline (items 23 and 24).

Three-bond ^{15}N-^{1}H couplings appear to offer the possibility of measurement of torsional angles in the same manner as ^{1}H-^{1}H couplings, but as they exhibit a narrow range of absolute values — *ca.* 1-5 Hz — and may be of either sign their application has been rather limited. There is a marked effect of positive charge on nitrogen and it is in the opposite direction to what we have seen for two-bond couplings. These trends may be observed in pyridine (items 18 and 19).

TABLE 9.6

Selected Values of ^{15}N Couplings

Entry number	Compound	J (Hz)	Solvent; conditions
1	NH_3	$^{15}N-^1H: -61.2$	neat
2	NH_4^+	$^{15}N-^1H : -73.2$	H_2O
3	CH_3NH_3	$^1J : {}^{15}N-^1H : -75.6$	H_2O
		$^2J : {}^{15}N-^1H(CH_3) : -1.0$ $^1J : {}^{15}N-^{13}C : -4.5$	
4	$C_6H_5 \cdot NH_2$	$^{15}N-^1H : -78.0$ $^1J : {}^{15}N-^{13}C : (-)10.9$	CCl_4 $CDCl_3$
5	$C_6H_5 \cdot \overset{+}{N}H_3$	$^{15}N-^1H : -73.5$	
6	$C_6H_5 \cdot NHCH_3$	$^{15}N-^1H : -78.0$	acetone-d_6
7	$C_6H_5 \cdot \overset{+}{N}H_2CH_3$	$^{15}N-^1H : -76.0$	
8	[structure: H₁–¹³C(=O)–N(H₂)(H₃)]	$^{15}N-H_1 : (-) 16.4$ $^{15}N-H_2 : (-) 89.7$ $^{15}N-H_3 : (-)86.4$	acetone

H_2O |
9	[structure: CH₃–¹³C(=O)–N(H₁)(H₂)]	$^{15}N-H_1 : (-) 89$ $^{15}N-H_2 :$ $^{15}N-^{13}C : (-) 14.8$	H_2O
10	[structure: H₁–¹³C(=O)–N(H₂)(CH₃)]	$^{15}N-H_1 : (-) 14.1$ $^{15}N-H_2 : (-)93.6$ $^{15}N-H(CH_3) : (+)1.4$	H_2O
11	[structure: H₁–¹³C(=O)–N(CH₃₍₂₎)(CH₃₍₃₎)]	$^{15}N-H_1 : \pm15.6$ $^{15}N-H(CH_{3(2)}) : \mp1.1$ $^{15}N-H(CH_{3(3)}) : \mp1.2$	neat
12	$(H_2H)_2C = O$	$^{15}N-H_1 : -90.3$ $^{15}N-^{13}C : -20.2$	D_2O

TABLE 9.6 (continued)

13

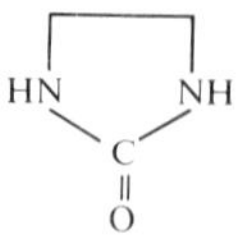

$^{15}N-H_1$: -92.2

DMSO

14

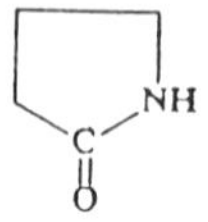

$^{15}N-H_1$:-91

H_2O

15

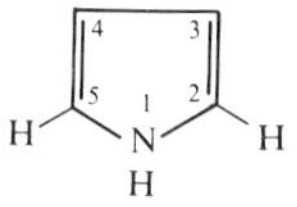

$^{15}N-H_1$:-96.4
$^{15}N-H_{2,5}$: -5.36
$^{15}N-H_{3,4}$: -4.55
^1J : $^{15}N-^{13}C$: -10.3

benzene

16

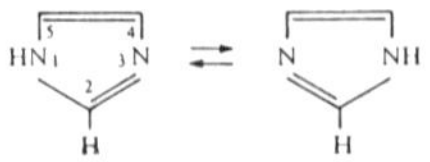

^2J : $^{15}N-H_2$: -9.6
$^{15}N-H_{4,5}$: -7.2
^3J : $^{15}N-H_{5,4}$: -2.5
^1J : $^{15}N-^{13}C_2$: -6.9
: $^{15}N-^{13}C_{4,5}$: -5.9

H_2O

17

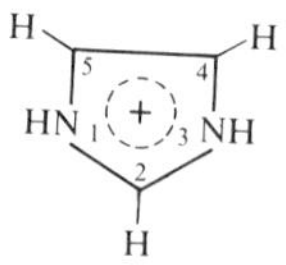

^2J : $^{15}N-H_2$: -5.5
$^{15}N-H_{4,5}$: -4.0 to -4.6
^1J : $^{15}N-C_2$: -16.2
$^{15}N-C_{4,5}$: -10.6

H_2O

18

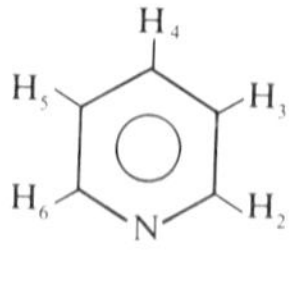

^2J : $^{15}N-H_{2,6}$: -10.93
^3J : $^{15}N-H_{3,5}$: -1.48
^4J : $^{15}N-H_4$: $+0.27$
^1J : $^{15}N-C_{2,6}$: $+0.62$
^2J : $^{15}N-C_{3,5}$: $+2.53$
^3J : $^{15}N-C_4$: -3.85

acetone-d_6

19

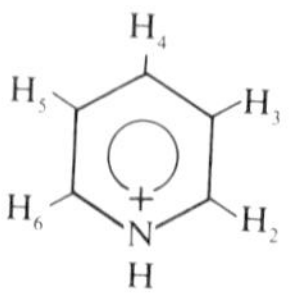

^1J : $^{15}N-H$: -96.3
^2J : $^{15}N-H_{2,6}$: -3.01
^3J : $^{15}N-H_{3,5}$: -3.98
^4J : $^{15}N-H_4$: $+0.69$
^1J : $^{15}N-C_{2,6}$: -11.85
^2J : $^{15}N-C_{3,5}$: $+2.01$
^3J : $^{15}N-C_4$: -5.30

CF$_3$COOH

TABLE 9.6 (continued)

20		$^2J : {}^{15}N-H_{2,6} : +0.47$ $^3J : {}^{15}N-H_{3,5} : -5.32$ $^4J : {}^{15}N-H_4 : +1.11$ $^1J : {}^{15}N-C_{2,6} : -15.2$ $^2J : {}^{15}N-C_{3,5} : +1.43$ $^3J : {}^{15}N-C_4 : -5.17$	CDCl$_3$
21		$^2J : {}^{15}N-H : -14.7$	nematic liq. crystal
22	$NH_2CH_2COO^-$ $\overset{+}{N}H_3CH_2COO^-$ $\overset{+}{N}H_3CH_2COOH$	$^1J : {}^{15}N-{}^{13}C : 4.6$ $^1J : {}^{15}N-{}^{13}C : 6.4$ $^1J : {}^{15}N-{}^{13}C : 7.3$	H$_2$O
23		$^2J : {}^{15}N-H_2 : -11.0$	H$_2$O
24		$^1J : {}^{15}N-H_1 : -96.0$ $^2J : {}^{15}N-H_2 : -2.0$	H$_2$O
25	$H-C \equiv \overset{+}{N}H$	$^1J : {}^{15}N-H : -134$	
26	$CH_3-C \equiv \overset{+}{N}H$	$^1J : {}^{15}N-H : -136$	
27	$C_6H_5 \cdot C \equiv \overset{+}{N}H$	$^1J : {}^{15}N-H : -136$	
28	CH_3NO_2	$^1J : {}^{15}N-{}^{13}C : -10.5$	
29	$CH_3C \equiv N$	$^1J : {}^{15}N-{}^{13}C : -17.5$ $^2J : {}^{15}N-{}^{13}C(CH_3) : +3.0$	
30		$^1J : {}^{15}N-{}^{13}C : -77.5$	

The observation of ^{15}N-^{13}C couplings generally requires enrichment in at least one of these rare nuclei. One-bond ^{15}N-^{13}C couplings are usually negative and 2–20 Hz in magnitude (items 3, 4, 12, 15, 16, 17, 18, 19, 20, 22, 27, 28, 29), although in exceptional cases (item 30) they may be as large as −77.5 Hz. The presence of unsaturation on the carbon (items 12, 28) or of a charge on the nitrogen (items 18, 19) tends to increase the coupling. Two-bond ^{15}N-^{13}C couplings tend to be weaker than one-bond couplings and may have either sign. Three-bond couplings are usually negative and weak, seldom exceeding 5 Hz; they have not proved broadly useful for establishing torsional angles. In pyridine, pyridinium ion, and pyridine oxide (items 18, 19, 20), the three-bond ^{15}N-^{13}C couplings are larger than the two-bond couplings.

9.1.3 Phosphorus

^{31}P is the only naturally occurring isotope of phosphorus. It has a spin of 1/2, a magnetic moment of 1.1305, a magnetogyric ratio of 1.0829×10^8, and a relative sensitivity of 0.0664 compared to the proton. It is thus relatively easy to observe with modern instrumentation. In an 11.75 T magnetic field (500 MHz for protons) it resonates at 202 MHz. (See Chapter 1, Table 1.1; Appendix A.) Relaxation times of ^{31}P nuclei, while longer than for protons, are not so long as to impose serious limitations on pulse repetition rates and ^{31}P is therefore relatively well suited for study. Its occurrence in nucleotides makes it relevant to biophysical studies both *in vitro* and even *in vivo*. For these reasons, the NMR literature is large.[30–42]

The chemical shift range of ^{31}P is over 600 ppm. The standard of reference is external 85% aqueous phosphoric acid as zero; as we have seen for other nuclei, ^{31}P nuclei less shielded than this are given positive values and those more shielded are given negative values; in earlier literature, this convention may be reversed in sign. A brief selection from the very large store of available data is shown in Table 9.7, together with some ^{31}P-^{1}H, ^{31}P-^{13}C, and a few other phosphorus couplings.

Trivalent phosphorus compounds cover a range of 700 ppm, from +210 to −490 ppm; the range of quadruply connected phosphorus is considerably smaller, from +130 to −115 ppm; pentavalent phosphorus is in the range of about +30 to −100 ppm; while the chemical shifts of the few known structures based on sixfold phosphorus are high, being −135 to −300 ppm. In Fig. 9.3 is shown a schematic summary of these chemical shift ranges, showing the effects of bonding hybridization and of neighboring bonded atoms of varying electronegativity. The most shielded of all phosphorus-containing molecules is phosphorus itself, P_4, a cage structure in which the ^{31}P must be thought of as trivalent (item 1); it is more shielded than

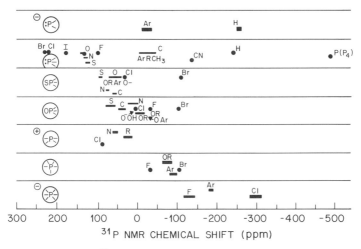

Fig. 9.3   Range of ^{31}P chemical shifts for substituted phosphorus compounds. Each group shown is equivalently substituted about phosphorus. Filled circles designate specific compounds. [Adapted from D. G. Gorenstein, *Progr. Nucl. Mag. Res. Spectros.*, J. W. Emsley, J. Feeney, and L. H. Sutcliffe, Eds., **16**, 1 (1983).]

^{31}P in PH_3 (item 6) and diphosphine (item 13), which are of lower symmetry. It will be noted that the substitution of any group other than H causes very strong deshielding of trivalent phosphorus (line 2 of the diagram in Fig. 9.3); halogens cause the greatest effect, but not in order of their electronegativity. Theoretical interpretation of these chemical shift effects has been reviewed in Refs. 31 and 42.

Specific phosphorus couplings are shown in Table 9.7. These are presented by classes in a more systematic way in Tables 9.8, 9.9, and 9.10, taken from Gorenstein.[41,42] Some couplings for certain specific compounds are also shown in these tables. One-bond $^{31}P-^{1}H$ couplings are positive and show some indication of varying with the degree of *s* hybridization of

TABLE 9.7

Selected Values of ^{31}P Chemical Shifts and Coupling Constants

Entry number	Compound	Chemical shift vs. 85% aq. H_3PO_4 as zero	J (Hz)
1	P_4	−488	
2	PF_3	+97.0	P−H : 1400
3	PCl_3	+215 to +220	
4	PBr_3	+222−227	
5	PI_3	+178	
6	PH_3	−238 to −241	P−H : 180−182
7	$P(CH_3)Cl_2$	+190 to 193	P−H : 16−16.9
8	$P(CH_3)_2Cl$	+92 to 96	
9	$P(CH_3)_3$	−61 to 62	P−H : 2.66
10	$PH_2(CH_3)$	−163.5	P−H : 210
11	$PH(CH_3)_2$	−98.5 to 99.5	P−H : 188 to 210 P−C−H : ±3.6
12	$PH_2(C_6H_5)$	−122.3	P−H : 199.5 P−H (ortho) : 6.72 P−H (para) : 1.56
13	$H_2P−PH_2$	−204.0 (−80°C)	P−H : 186.5 P−P−H : 11.9 P−P : 108.2
14	$P(CH_2CH_3)_3$	−19 to 20.4	P−C−H : ±13.7 P−C−C−H : ±0.5
15	$P(n−C_3H_7)_3$	−33	
16	$P(n−C_4H_9)_3$	−32.3 to 33.4	
17	$P(C_6H_5)_3$	−5.6 to 8	P−C (ipso) + 12.51 P−C (ortho) −19.65 P−C (meta) −6.80 P−C (para) ±0.33

TABLE 9.7 (continued)

Selected Values of ^{31}P Chemical Shifts and Coupling Constants

Entry number	Compound	Chemical shift vs. 85% aq. H_3PO_4 as zero	J (Hz)
18	$P(OCH_3)_3$	+140 to 141	
19	$P(OCH_3)_2OC_6H_5$	+135.2	P−O−C−H : 9.7
20	$P(OCH_2CH_3)_3$	+137 to 140	
21	$P(O-n-C_3H_7)_3$	+137.9	
22	$P(OCH_2CH = CH_2)_3$	+138.5	
23	$P(O-t-Bu)_3$	+138.3	
24	$P(OC_6H_5)_3$	+125 to 129	
25	$O = PF_3$	−35.5	P−F : 1080, − 1055
26	$O = PCl_3$	+1.9 to +5.4	
27	$O = PBr_3$	−102 to −104.3	
28	$O = P(CH_3)F_2$	+26.8 to 27.4	P−F : 1270, 1090
29	$O = P(CH_3)Cl_2$	+43.5 to 44.5	
30	$O = P(CH_3)Br_2$	+8.5	P−C−H : 14.7
31	$O = P(CH_3)_2F$	−66.3	P−F : 990
32	$O = P(CH_3)_2Cl$	−62.8 to 68.3	P−C−H : 19
33	$O = P(CH_3)_2Br$	−50.7	
34		+65.8	
35			P−C$_1$: 66.8 P−C$_2$: 7.9

TABLE 9.7 (continued)

Selected Values of ^{31}P Chemical Shifts and Coupling Constants

Entry number	Compound	Chemical shift vs. 85% aq. H_3PO_4 as zero	J (Hz)
36			P–C$_{2,6}$: 63 P–C$_4$: 35 P–C$_7$: 4 P–C–C$_4$–H : 28
37	O = P(OCH$_3$)Cl$_2$	+5.6	
38	O = P(OCH$_2$CH$_3$)F$_2$	+20.9 to 21.2	P–F : 1010, 1015
39	O = P(OC$_2$H$_5$)Cl$_2$	+3.4 to 6.4	
40	O = PH$_2$(OCH$_3$)	+19.0	P–H : 575 P–O–C–H : 13
41	O = PH$_2$(OC$_2$H$_5$)	+15.0	P–H : 567
42	O = PH(OCH$_3$)$_2$		P–^{17}O : 220 P–O–C–H : 88
43	O = P(CH$_3$)(OCH$_3$)$_2$	+32.3 to 32.6	
44			P–C$_1$: 88.6 P–C$_2$: 10.7 P–C$_3$: 12.8 P–C$_4$: 3.0 P–C(CH$_3$) : 56.8
45		−88.0 to 89.0	−90° : P–C$_{(eq)}$: 113.0 +80° : P–C$_{(eq/ax)}$: 88.0

TABLE 9.7 (continued)

Selected Values of ^{31}P Chemical Shifts and Coupling Constants

Entry number	Compound	Chemical shift vs. 85% aq. H_3PO_4 as zero	J (Hz)
46	adenosine-3'-triphosphate, (ATP) pH 7.40	P_α : -11.45	$P_\alpha-P_\beta$: 19.75
		P_β : -22.66	$P_\beta-P_\gamma$: 19.75
		P_γ : -7.33	

| 48 | adenosine-3'-diphosphate (ADP) | P_α : -11.13 | $P_\alpha-P_\beta$: 23.1 |
| | | P_β : -6.96 | |

the bond, but this relationship is by no means as exact as is found for ^{13}C-^1H couplings (Chapter 4, Section 4.3.6.1). ^{31}P-^{13}C couplings are weaker while ^{31}P-^{19}F and ^{31}P-^{31}P couplings are an order of magnitude stronger than ^{31}P-^1H couplings. Two-bond couplings, ^{31}P-C-^1H, may be of either

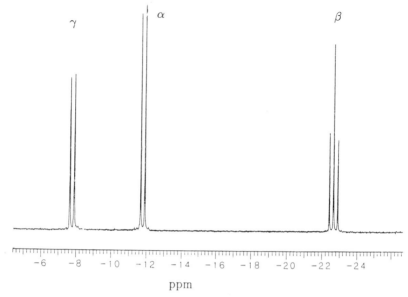

Fig. 9.4   73.8 MHz ^{31}P spectrum of adenosine triphosphate (ATP) in
            aqueous solution at pH 7.0 (F. C. Schilling, private
            communication.)

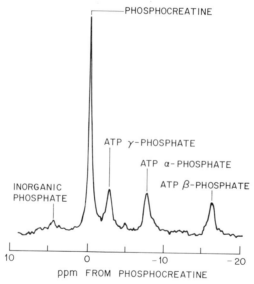

Fig. 9.5   73.8 MHz ^{31}P spectrum obtained on a rat using a surface coil
            placed on the live animal. (J. A. Ackerman, private
            communication.)

TABLE 9.8

One-Bond Phosphorus Spin–Spin Coupling Constants $^1J_{PX}$ [a]

Structural class (or structure)	^1J (Hz)
P(II)	
PH_2^-	139
P(III)	
$>$PH	180–225
PH_3	+182
$P(CH_3)_2H$	+191.6
$>$P–C	0–45
$P(CH_3)_3$	−13.5 to −14.0
$P(CH_3)Cl_2$	45
P(IV)	
$>$$\overset{+}{P}$H	490–600
$\overset{+}{P}H_4$	548
$\overset{+}{P}H(CH_3)_3$	+505.5
$>$P(O)H	460–1030
$>$P(M)F	1000–1400

TABLE 9.8 (continued)

One-Bond Phosphorus Spin–Spin Coupling Constants $^1J_{PX}$ [a]

Structural class (or structure)	^1J (Hz)
(M = O,S)	
P(O)F$_3$	1055–1080
P(S)F$_3$	1170–1184
F$_2$P(O)OP(O)F$_2$	−1063
P(V)	
>PH (with vertical bonds)	700–1000
PF$_4$H	1075
P(CH$_3$)F$_3$H	850
P(VI)	
PF$_6$$^-$	706
>PF	820–1450
PF$_3$	1400–1450
P$_2$F$_4$	−1194
>P–P<	100–400
H$_2$PPH$_2$	−108
F$_2$PPF$_2$	−227.4
>PC	50–305

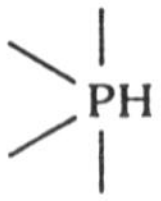

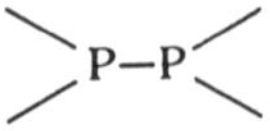

TABLE 9.8 (continued)

One-Bond Phosphorus Spin–Spin Coupling Constants $^1J_{PX}$ [a]

Structural class (or structure)	^1J (Hz)
$\overset{+}{P}(CH_3)_4$	+56
$P(S)(CH_3)_3$	+56.1
$P(O)(OC_2H_5)_2C \equiv CCH_3$	304
$\diagdown\!\!\!\!/ P(S)H$	490–650
$\diagdown\!\!\!\!/ \overset{\mid}{\underset{\mid}{P}}\!\!-\!F$	530–1100
PF_5	938
$P(CH_3)_2F_2$	541

[a] From Gorenstein, Refs. 41 and 42.

TABLE 9.9

Two-Bond Phosphorus Spin–Spin Coupling Constants $^2J_{PX}$ [a]

Structural class (or structure)	^2J (Hz)
P(III)	
$\diagdown\!\!\!\!/ PCH$	0–18
$P(CH_3)_3$	+2.7
$P(C_2H_5)_3$	−0.5
$P(Cl)_2CH_3$	17.5, 17.7

TABLE 9.9 (continued)

Two-Bond Phosphorus Spin–Spin Coupling Constants $^2J_{PX}$ [a]

Structural class (or structure)	^2J (Hz)
$\diagup\hspace{-0.4em}\diagdown$PCF	40–149
$\diagup\hspace{-0.4em}\diagdown$P(CF$_3$)$_3$	85.5
$\diagup\hspace{-0.4em}\diagdown$PNC	
P[N(CH$_3$)$_2$]$_3$	17
$\diagup\hspace{-0.4em}\diagdown$PNH	13–28
P(IV)	
P(O)CH	7–30
P(O)(CH$_3$)$_3$	−12.8, −13.4
P(O)(CH$_3$)F$_2$	19, 19.9
$\diagup\hspace{-0.4em}\diagdown\overset{+}{P}$CH	12–18
$\overset{+}{P}$(CH$_3$)$_4$	−14.6
PH$_3$CH$_3$	−17.6
$\diagup\hspace{-0.4em}\diagdown$PCC	0–40
$\overset{+}{P}$(C$_2$H$_5$)$_4$	−4.3

TABLE 9.9 (continued)

Two-Bond Phosphorus Spin–Spin Coupling Constants $^2J_{PX}$ [a]

Structural class (or structure)	^2J (Hz)
$\diagup\!\!\!\!\diagdown$ POP	0–23
$F_2(O)POP(O)F_2$	0–4
P(V)	
$\diagdown\!\!\!\!\diagup$ PCH	10–18
$P(CH_3)_3F_2$	17.2
$P(CH_3)F_3H$	17.6–18.0
P(VI)	
$\diagup\!\!\!\!\diagdown$ PCF	130–160
$P(CF_3)F_5^-$	145
$\diagdown$ PCC	12–20
$P(C_2H_5)_3$	+14.1
$P(C_6H_5)_3$	19.6
$\diagdown$ POC	10–12
$P(OCH_3)_3$	+10.0
$P(OC_2H_5)_3$	+11.3

TABLE 9.9 (continued)

Two-Bond Phosphorus Spin–Spin Coupling Constants $^2J_{PX}$ [a]

Structural class (or structure)	^2J (Hz)
>PXP (X = S, C)	70–90
>P(S)CH	11–15
P(S)(CH$_3$)$_3$	−13.0
P(S)(CH$_3$)$_2$H	14.4
>P(O)CF	100–130
P(O)(CF$_3$)$_3$	113.4
>POC	~6
P(O)(OCH$_3$)$_3$	−5.8
P(S)(OCH$_3$)$_3$	−5.6
>PSC	
P(S)(SCH$_3$)$_3$	<10
>PCF	124–193
P(CF$_3$)Cl$_4$	154

[a] From Gorenstein, Refs. 41 and 42.

TABLE 9.10

Three-Bond Phosphorus Spin–Spin Coupling Constants $^3J_{PX}$ [a]

Structural class (or structure)	^3J (Hz)
P(III)	
>POCH	0–15
$P(OCH_3)_3$	10.8–11.8
$P(OC_2H_5)_3$	7.3–8.0
>PCCH	10–16
$P(C_2H_5)_3$	+13.7
$P(C_2H_5)(C_6H_5)_2$	16.5
	12.5
>PCCC	
$P(C_4H_9)_3$	12.5
P(IV)	
>POCH$^+$	7–11
$P(OCH_3)_4^+$	11.2
>PCCH$^+$	15–22
$P(C_2H_5)_4^+$	+18.1

TABLE 9.10 (continued)

Three-Bond Phosphorus Spin–Spin Coupling Constants $^3J_{PX}$ [a]

Structural class (or structure)	^3J (Hz)
$\overset{+}{P}(C_2H_5)_3H$	+20.0
P(O)OCH	0–13
$P(O)(OCH_3)_3$	10.2–11.4
$\diagdown\diagup P(M)CCH$	14–25
(M = O, S)	
$P(O)(C_2H_5)_3$	+18.0
$P(S)(t-C_4H_9)_3$	+14.0
P(V)	
$\diagdown\diagup\overset{\mid}{P}CCH_{\mid}$	20–27
$P(C_2H_5)F_3H$	26.4
$\diagdown\diagup\overset{\mid}{P}SCH_{\mid}$	13–25
$P(SCH_3)F_4$	25.2
$\diagdown\diagup\overset{\mid}{P}NCH_{\mid}$	3–14
$P[N(CH_3)_2]_3$	8.8–9.0
$P(Cl)_2N(CH_3)_2$	12.9–13.0
PNCH	2–15

[a] From Gorenstein, Refs. 41 and 42.

sign. Three-bond couplings, $^1P-O-C-^1H$, show the expected Karplus-like angular dependence, the coupling being of the order of 20 Hz when ϕ is *ca.* 180°. This relationship in $^{31}P-O-C-^1H$ and, $^{31}P-O-C-^{13}C$, fragments has been used to determine the conformation about the ribose-phosphate backbone of nucleotides and nucleic acids in solution.

^{31}P NMR spectroscopy has become of great importance not only in the study of polynucleotide structure but also in the observation of metabolic processes *in vitro* and *in vivo*. In Fig. 9.4 is shown the 73.8 MHz ^{31}P spectrum of adenosine triphosphate (ATP) in aqueous solution at pH 7.0[43]:

The splitting of the α, β, and γ phosphorus resonances arises from the ^{31}P-^{31}P couplings (tabulated in Table 9.4). It is possible to observe this and other phosphorus-containing metabolites *in vivo* as well. In Fig. 9.5 the 73.8 MHz phosphorus spectrum is obtained on a rat[44] using a surface coil placed on the live animal. The ATP resonances are broadened but readily detectable. One may observe in addition to inorganic phosphate the ^{31}P resonance of another important metabolite, phosphocreatine:

9.2 NMR IMAGING (L. W. JELINSKI)

9.2.1 Introduction

NMR imaging, also called MRI or *magnetic resonance imaging*, has become a valuable medical diagnostic tool. Medical applications of imaging evolved rapidly from Lauterbur's demonstration in 1973[45] of a two-dimensional NMR image from two water-containing 1-mm tubes whose centers were separated by 3.2 mm. Lauterbur proposed the term *zeugmatography* to describe this technique. From the Greek word *zeugma* meaning *that which is used for joining*, the term refers to bringing together the radio frequency field necessary to produce the NMR signal and the magnetic field gradients that produce the spatial encoding of this NMR signal.

NMR imaging is a noninvasive technique that generates images of "slices" of the human body, such as the one in Fig. 9.6. Since the magnet must be large enough to accommodate even obese humans, NMR imaging spectrometers for medical applications are massive. Figure 9.7 is a photograph of one such system. This figure shows a patient positioned on a tray that can be slid into the bore of the horizontal magnet. NMR images get their contrast from the differences in water concentration in various parts of the sample and because different tissues and tumors have different relaxation times (T_1 and T_2). For example, the proton content increases in going from bone to blood to muscle to fat to cerebrospinal fluid.[46] Furthermore, the relaxation times in these materials are very different. In general, the amount of water and the relaxation times of tumors are different from the surrounding tissue, also affording contrast.[47]

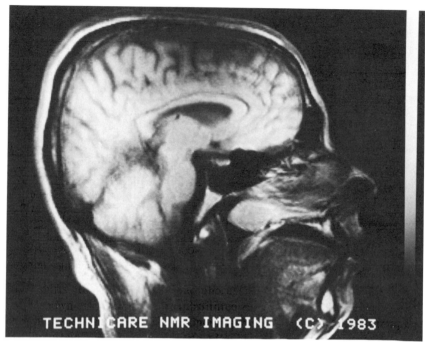

Fig. 9.6   NMR image of a human head. This is a *sagittal* image, or one
that runs along a plane parallel to the plane through the nose
and back of the head. *Coronal* images are parallel to the plane
running from ear to ear. (Courtesy of Technicare.)

NMR imaging uses many of the concepts that we have already
encountered, with one major difference. So far we have concentrated on
homogeneity — homogeneous magnetic fields and homogeneous B_1 fields.
These methods give average information about the composition of
ostensibly homogeneous materials. For example, two capillaries of water
placed side by side in a homogeneous magnetic field would produce a
composite signal, a single line whose frequency, ω, is given by $\omega = \gamma B_o$ [see
Fig. 9.8(a)].[46] (Controlled test samples such as these capillaries are often
called *phantoms*.) However, if instead of a homogeneous magnetic field,
the two capillaries were placed in a *magnetic field gradient* (produced by
the coils shown in [Fig. 9.8(b)], the water in capillary A would resonate at
a frequency different from that of capillary B. The frequency of peak A
would be given by $\omega_A = \gamma B_A$, where B_A is the magnetic field at the
position of capillary A. (A similar condition holds for capillary B.)
Magnetic field gradients produce the spatial discrimination necessary for
magnetic resonance imaging. The distance between two objects can be

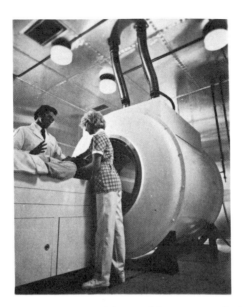

Fig. 9.7   NMR imaging machine.   (Courtesy of General Electric Company.)

determined if the field gradient is linear and if the rate at which it changes is known.

We can see that this simple analysis readily falls apart if we were to position the capillaries differently in the magnetic field [see for example, Fig. 9.8(c)]. In practice, one wants the x, y, and z coordinates of a particular volume element, and in order to avoid the confusion illustrated in Fig. 9.8c, various gradients along x, y, and z are used to produce images.

9.2.2 Imaging Methods

There are many techniques for obtaining an NMR image, and the interested reader is referred to Ref. 48 for a comprehensive review. Here we will present the rudiments of some of the more basic methods. These methods can be divided into four general classes according to the volume element that they sample.[49] This classification scheme is illustrated in Fig. 9.9. The methods range from building an image out of a number of single point measurements [Fig. 9.9(a)] to sequential measurement of a number of lines [Fig. 9.9(b)] to sequential build-up of a series of planar images [Fig. 9.9(c)] to obtaining the entire image at once [Fig. 9.9(d)] and then reconstructing the desired slice for visualization purposes.

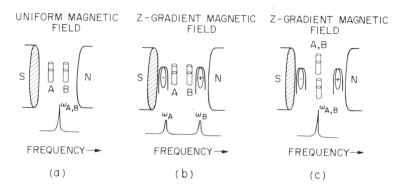

Fig. 9.8   (a) Two capillaries of water side-by-side in a homogeneous magnetic field produce a single line. In the presence of a field gradient these two capillaries produce separate lines (b) unless the capillaries experience the same gradient (c). [Adapted from J. D. Roberts, *Eng. & Sci.*, 10 (January 1986).]

9.2.2.1 Sequential Point Measurements

We have already seen that surface coils [Chapter 2, Fig. 2.13(c)] can provide information about the area immediately beneath them. The fall-off of the B_1 homogeneity of these coils determines the volume that they sample. Outside the region of B_1 homogeneity, the sample receives inadequate radio frequency excitation, and the signals from parts of the sample lying outside of this region are markedly diminished. In general the B_1 field inhomogeneity of these coils is such that they are selective for hemispheric regions of the sample that are within approximately the

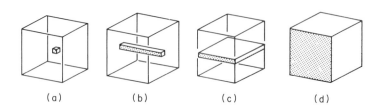

Fig. 9.9   Classification of NMR imaging methods according to the volume they sample: (a) sequential point measurement; (b) line measurement; (c) plane measurement; (d) volume measurement, or 3-D zeugmatography. [Adapted from *NMR: A Perspective on Imaging*, General Electric Corporation, Medical Systems Operations, Milwaukee, Wisconsin (1982).]

circumference of the coil and extending to a depth corresponding to one radius of the coil. Although various pulse sequences have been proposed to make these volumes more selective,[50] surface coils are generally best for observing chemical shift spectra (see Chapter 3) of localized regions of tissue. A two dimensional image plane could be built up by physically moving the surface coil over the surface of the sample.

A somewhat different technique involves creating high order magnetic field gradients so that the magnetic field is homogeneous over only a very small volume. In the rest of the region , the magnetic field homogeneity falls off very steeply, and thus signals from objects outside the sensitive region are extremely broad. It then becomes necessary to move the sample about in the magnetic field in order to build up the image scan.

Moving the coil around on the surface of the sample or moving the sample around within the magnetic field are tedious and inefficient ways to obtain an image. The *sensitive point* method[51] overcomes these drawbacks. This method relies on time-dependent magnetic field gradients along the x, y, and z directions. They produce a sensitive point that can be scanned along the sample by changing the magnetic field gradients. The way this works can be understood by looking first at the effect of one time-dependent magnetic field gradient, given by

$$(x - x_0)G_x \cos(\Omega_x t) \qquad (9.2)$$

where x_0 describes the position, G_x the x-gradient amplitude, and Ω_x the frequency. This produces time dependence on the NMR signals from all regions of the sample except where $x = x_0$, the region where the time-dependent gradient vanishes. A special pulse sequence, to provide a steady state free precession signal, then excites the signal that arises from the sensitive plane, described by $x = x_0$. When gradients in the y and z directions are also added to this scheme, the time independent signal arises from the sensitive point given by (x_0, y_0, z_0). This sensitive point can then be moved about in the sample by changing the strengths of the magnetic field gradients.

Although not computationally demanding, the sequential point methods suffer from inefficiency and have been largely supplanted by planar and volume methods. The foregoing description is useful, however, for pedagogical purposes.

9.2.2.2 Line Methods

We have seen that one may build an NMR scan from an assembly of sensitive points. There are also a number of ways to obtain a scan by the sequential scanning of lines. We shall describe only one of them here, as these, too, have been largely superseded by the more efficient planar and volume methods.

One of the sequential line methods involves an extension of the sensitive point technique.[52,53] The sensitive line is generated by creating two (instead of three, as in the sensitive point method) time dependent magnetic field gradients. Let us assume that they are along the x and y directions. As before, the NMR signal will be time-dependent from all regions except where $x = x_0$ and $y = y_0$. This provides a line that runs along z. In conjunction with the steady-state free precession pulse sequence, this produces a spectrum along the line defined by x_0, y_0. Discrimination of the signal along this line is accomplished by imposing a *static* magnetic field gradient along the z direction, thereby producing spatial encoding along this line. The scan is then built up by moving the position of this line across the sample.

9.2.2.3 Planar Methods

Lauterbur used the *projection reconstruction* method in his original report on capillaries of water.[45] This principle has been used extensively in other imaging methods, most notably in X-ray computed tomography (CT scans). A series of one-dimensional projections are obtained by incrementally rotating the magnetic field gradient through 180°.[54] The two-dimensional image is then reconstructed from these one-dimensional images using standard projection reconstruction algorithms borrowed from CT methodology. Figure 9.10 illustrates this process. In Fig. 9.10(a), the magnetic field B_0 is parallel to the magnetic field gradient G, which produces the one dimensional image shown at the bottom of the figure. The gradient G is then rotated to other positions and other one-dimensional images such as the ones in (b) and (c) are obtained. The spatial arrangement of the objects is then back-calculated to produce the image shown on the right-hand side of the figure.

Up until now we have ignored the problem of *slice selection*. Lauterbur's projection reconstruction image relied on the receiver coil geometry for crude control of the slab or slice thickness. Slice selection is now generally controlled by a tailored, frequency-selective radio frequency pulse that is applied in the presence of a field gradient perpendicular to the plane of the desired slice [see Fig. 9.11(a)]. Both the strength of the magnetic field gradient and the frequency bandwidth (see Chapter 1, Section 1.4, and Chapter 2, Section 2.4.3) of the selective 90° radio frequency pulse control the slice thickness. Figure 9.12 illustrates these effects.[55] In (a) the frequency bandwidth, $\Delta\omega_0$, is kept constant and the gradient strength is varied from a steep gradient (illustrated by the line

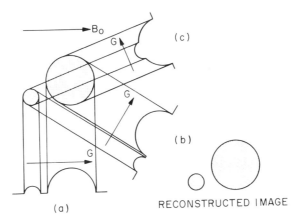

Fig. 9.10   Method of obtaining one dimensional images for the
             projection-reconstruction technique. The magnetic field
             gradient, G, is rotated about the sample. When G is parallel to
             the magnetic field, B_0, the image in (a) is produced. Other one
             dimensional images are shown in (b) and (c). The
             reconstructed image is shown.

A), which gives a slab of thickness Δz_A, to a lower gradient (the line B),
which produces a thicker slice. If the gradient is kept constant
[Fig. 9.12(b)], a more selective radio frequency pulse $(\Delta\omega_A)$ produces a
thinner slice (Δz_A).

Having described methods for slice selection, we address another class
of planar methods of imaging. These include two dimensional NMR
Fourier zeugmatography,[56,57] rotating frame zeugmatography,[58] the echo
planar method,[59] and spin warp imaging.[60] All of these methods bear a
formal similarity to the two dimensional NMR methods that we have
already seen (Chapter 6) in that they involve preparation, evolution, and
acquisition time periods. It is beyond the scope of this book to describe all
of these methods, so we shall concentrate on one of them, the spin warp
technique. A description of this method illustrates the principles of phase
encoding, which are common to many two- and three-dimensional
techniques.

A simplified version of the spin warp technique is shown in Fig. 9.11.
The preparation period consists of selective excitation of the desired plane
using a tailored 90° radio frequency pulse [Fig. 9.11(a)]. Phase encoding
is produced by applying a gradient in the y direction. This gradient causes

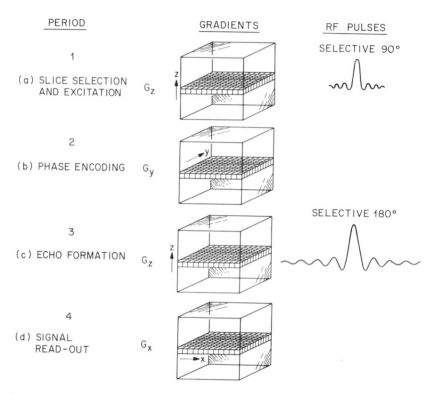

Fig. 9.11   Basic time periods in measuring an NMR image. (Adapted
from F. W. Wehrli, J. R. MacFall, and T. H. Newton in
Advanced Imaging Techniques, T. H. Newton and D. G. Potts,
Eds., Clavadel Press, San Anselmo, California, 1983,
pp. 81—117.)

the nuclei across the sample to precess at frequencies that are
characteristic of their location with respect to the G_y gradient. (For now
we shall assume that there is only one chemical species, i.e., water,
contributing to the image.) At the end of this period, each of the volume
elements has accumulated a phase angle that is directly related to its
position in the gradient. Figure 9.13 illustrates the principle of phase
encoding for three volume elements A, B, and C in a G_y gradient. At time
$t = 0$, the nuclei in these volume elements have been polarized by the
selective excitation pulse and lie in the $x\,y$ plane. Since the nuclei in these
volumes experience different magnetic fields due to the G_y field gradient,
they precess at different frequencies. At the end of the G_y gradient at
$t = \Delta t_y$, the magnetic moments from each of the volume elements have
precessed through an angle related to their position in the G_y gradient.

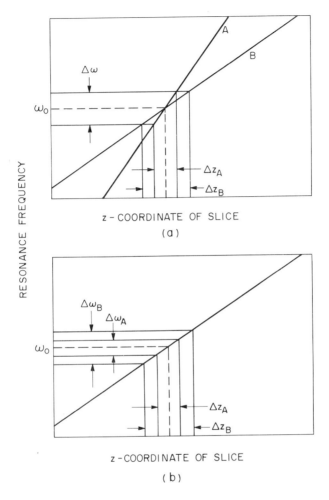

z – COORDINATE OF SLICE

(a)

z – COORDINATE OF SLICE

(b)

Fig. 9.12   Illustration of the effect on the slice thickness of variations in the strength of the magnetic field gradient (a) and the bandwidth of the selective radio frequency pulse (b).[55] In (a), the steeper field gradient A produces the narrower slab, Δz_A. In (b), the narrower bandwidth pulse $\Delta\omega_A$ produces the thinner slice.

The third period shown in Fig. 9.11(c) is composed of a selective $180°$ pulse (again, with the G_z gradient on), which accomplishes the second part of the T_2 $(90° - \tau - 180°)$ spin echo pulse sequence (see Section 5.3). Because of this $180°$ pulse, an image obtained with this pulse sequence would be called a T_2 *image*. A T_1 *image* could be obtained instead by

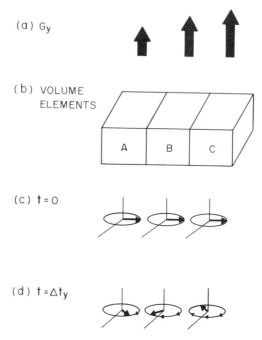

(a) G_y

(b) VOLUME
ELEMENTS

A B C

(c) $t = 0$

(d) $t = \Delta t_y$

Fig. 9.13   In the presence of field gradients, G_y, the volume elements A,
B, and C will initially have identical phases (c), but at the end
of the gradient period, the spins in these elements will have
accumulated different phases (d). (Adapted from F. W.
Wehrli, J. R. MacFall, and T. H. Newton in *Advanced
Imaging Techniques*, T. H. Newton and D. G. Potts, Eds.,
Clavadel Press, San Anselmo, California, 1983, pp. 81−117.)

performing a nonselective 180° pulse prior to the slice selection part of this
sequence. The last part of the image acquisition process [Fig. 9.11(d)]
involves signal readout, this time in the presence of the third gradient.

In order to obtain an image of, for example, 128 × 128 picture
elements, or *pixels*, it is necessary to obtain 128 different free induction
decay signals, each one differing from the others by an incremental value
of the phase encoding gradient, G_y. These FIDs are then transformed with
respect to the read-out coordinate, x, then transposed and Fourier
transformed with respect to the phase encoding gradient in a fashion
analogous to the two-dimensional Fourier transform techniques described
in Chapter 6 (see Fig. 6.1).

The description of the spin warp technique given in Fig. 9.11 shows only the main features of the select — encode — read sequence. A more complete description is shown in Fig. 9.14. The tailored 90° excitation

The description of the spin warp technique given in Fig. 9.11 shows only the main features of the select — encode — read sequence. A more complete description is shown in Fig. 9.14. The tailored 90° excitation pulse for slice selection can be of a $(\sin x)/x$ form convoluted with a gaussian. The G_z gradient is on during this time. Immediately following the slice selection, the G_z gradient is inverted, producing time reversal, which rephases the spins. An inverted x gradient is also applied and acts to advance the position of the echo along the time axis and away from the time of the pulses. Simultaneously, the G_y phase encoding gradient is applied. (The amplitude of this gradient is incremented for each successive scan.) A 180° selective radio frequency pulse is then applied to the G_z-selected slice. The purpose of this pulse is to provide the Hahn echo (see Chapter 5, Section 5.3) that will provide the T_2 contrast necessary for the image. The G_x gradient is then turned on during the time that the spin echo is being detected.

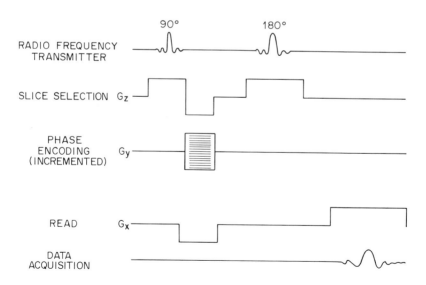

Fig. 9.14   Pulse sequence for a T_2 image acquired with the spin warp technique.

The relatively long T_1 of water in biological tissues (hundreds of milliseconds) usually governs the repetition rate of the data acquisition, and therefore the rate at which the G_y gradient is incremented. *Multislice imaging* is a way around this. This method involves cycling through scans of different slices at the same value of G_y while waiting for the protons in the first slice to relax. This method affords an enormous time savings and makes the difference between clinically reasonable and clinically infeasible NMR images.

9.2.2.4 Three-Dimensional Techniques

Data can be collected simultaneously from the entire imaging volume using three-dimensional techniques.[57] The methods for performing this are based on an extension of two dimensional imaging. Instead of selecting a slice with a slice-selective pulse, the entire volume is excited. A variable phase encoding gradient, G_z, is applied to separate the protons along the z axis, and a G_y gradient is simultaneously applied to differentiate the protons along the y direction. At the end of this phase-encoding period, the signal from each *voxel* (the three-dimensional equivalent of a pixel) element has a phase that is related to its position with respect to the y and z axes. The signal is then read out with a G_x static gradient. Fourier transformation is then performed with respect to G_x and then with respect to the two phase encoding gradients, G_y and G_z. Although computationally intensive, this method produces data that can be reconfigured to trace out the shape and size of tumors or other objects in three dimensions.

9.2.3 Relaxation and Contrast Enhancement in Imaging

9.2.3.1 Water Content and Relaxation

The protons in biological tissue exist in widely diverse environments. For example, there are protons in water, bone, fibrous proteins that make up the skin, lipids, blood, and DNA. Of these components, the protons in water and lipids are the primary contributors to the NMR image. The protons in bone and large proteins, because they have long correlation times, generally produce signals that are so broad that they are not observed in the NMR image. Although there are differences in the proton content in various parts of the body, the density of protons is not the major factor controlling the relative contrast observed in an image. Instead, most of the contrast in an NMR image depends on differences in the T_1 and T_2 of various tissues.

We have already examined the fundamental processes leading to nuclear relaxation (see Chapter 1, Section 1.7, and Chapter 5, Section 5.3). These fundamental principles apply to relaxation in NMR imaging.

For example, of the two major contributors to the NMR image signal, we would expect that water would have a longer T_1 than the lipids, since the water is a smaller molecule and will have a shorter correlation time. (Recall that relaxation is most efficient when $\omega\tau_c = 1$; see Fig. 1.8.) However, the water molecules in biological tissue have relaxation times that are far shorter (i.e., relaxation is more efficient) than is observed for bulk or free water. This shortening of the relaxation time arises because part of the water exists in a "bound" state, where it is associated with protein or DNA macromolecules. This temporarily immobilized water exchanges with the bulk water, in effect producing an averaged relaxation time for the water in a particular tissue or tumor. Therefore, the T_1 of water in biological tissues is of the order of hundreds of milliseconds, rather than the order of seconds, as it is in bulk or free water. It is thought that the exchange between free and bound water is affected in certain pathological conditions, perhaps because proteins are present in differing amounts, or because different proteins are produced.[61] Differences in water exchange lead to differences in T_1, which lead to differences in the contrast of an image.

Recall also that T_2 processes involve the loss of transverse magnetization (M_x and M_y), and that this decay in magnetization happens because the nuclear magnetic moments get out of phase with each other. We have already seen that many processes contribute to this dephasing, most notably magnetic field inhomogeneities. These magnetic field inhomogeneities can arise from the static magnetic field or from field inhomogeneities generated by the tissues themselves. We have already seen (Chapter 5, Section 5.3) that T_2 relaxation is most efficient for large molecules with long correlation times. As is the case with T_1 relaxation, the T_2 for bound water is much shorter than for free water, thereby affording image contrast.

The standard pulse sequences we have already seen [T_1 by inversion-recovery (Chapter 5, Fig. 5.5), partial saturation, and T_2 by spin echo (Chapter 5, Section 5.4.7.)] can be used with slice-selection pulses and gradients to produce what are known as "T_1" and "T_2" images. The pulse sequence diagram in Fig. 9.14 illustrates how a T_2 image would be obtained.

Much research has been performed to determine the conditions for obtaining optimum contrast in medical images. The interested reader is referred to Ref. 55 for an excellent overview of the subject.

9.2.3.2 Contrast Enhancement Agents

We have already seen that relaxation reagents, usually paramagnetic materials, are added to samples for high resolution NMR to enhance T_2

relaxation (see Chapter 2, Section 2.7.1). Relaxation reagents, generally known in medical imaging as *contrast agents*, can be employed to enhance the appearance of images and to facilitate diagnoses.[62] Most of these reagents are based on paramagnetic ions such as manganese or gadolinium, or on stable nitroxide free radicals. The ideal reagent is tissue specific, nontoxic, and rapidly excreted from the body without undergoing metabolism to other materials. Such reagents have been used, for example, to determine whether or not the blood—brain barrier is intact. Nitroxides have been shown to accumulate where there is damage to the blood—brain barrier and they produce enhanced relaxation in these regions.[62]

Monoclonal antibodies with covalent or chelated paramagnetic ions have been employed as tissue-seeking contrast enhancement reagents.[62]

9.2.4 Imaging of Flow

It has long been recognized that NMR spectroscopy can be used to measure flow, and these principles have now been extended to NMR imaging. The basic idea is that the slice-selection pulse excites only those spins in the slice, as before. However, some fraction of these saturated spins flow out of the excitation slice by the time the signal is detected. In their place in the slice are "fresh" spins — spins that have not been saturated by the initial radio frequency pulse (see Fig. 9.15). The contrast

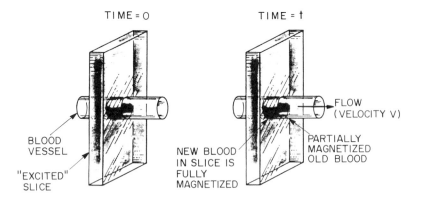

(a)   (b)

Fig. 9.15   (a) At time = 0, the slice selection pulse "tags" the nuclei in the slice. (b) When the signal is detected at a later time t both the fresh and "tagged" blood in the imaging slice will be detected. (Adapted from F. W. Wehrli, J. R. MacFall, and T. H. Newton in *Advanced Imaging Techniques*, T. H. Newton and D. G. Potts, Eds., Clavadel Press, San Anselmo, California, 1983, pp. 81—117.)

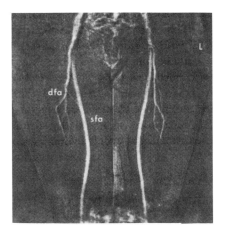

Fig. 9.16   Flow image of the thighs of a normal subject: sfa, superficial
femoral artery; dfa, deep femoral artery. [Adapted from V. J.
Wedeen, R. A. Meuli, R. R. Edelman, S. C. Geller, L. R.
Frank, T. J. Brady, and B. R. Rosen, *Science* **230**, 946 (1985).]

depends on the blood flow rate and upon the time between the slice
selection pulse and the "read" pulse. If new blood is flowing into the slice,
it will have a greater signal strength than the blood that was originally
there, as the new blood is fully magnetized, whereas the old blood,
depending on the time between pulses, is still partially saturated.

The foregoing is a very simplified description of the use of NMR to
measure flow. Actual experiments are generally more complicated and use
electrocardiographic gating of signal acquisition and signal subtraction
techniques together with velocity dependent phase shifts.[63] These velocity
dependent phase shifts are caused by the x gradient pulses, which
simultaneously perform the spatial encoding in x. Using these methods,
vessels as small as $1 - 2$ mm in diameter can be detected in a 50-cm field
of view in less than 15 min. Figure 9.16 shows a flow image of the thighs
of a human, where the superficial femoral artery (sfa) and the deep
femoral artery (dfa) are clearly seen. These NMR measurements of flow
are completely noninvasive and do not require arterial catheterization as
does Roengten angiography. Furthermore, they can be used in cranial
regions, where other techniques, such as ultrasonography, provide
insufficient resolution.[63]

9.2.5 Chemical Shift Imaging

So far in our discussion of NMR imaging we have assumed that we are observing one chemical shift, i.e., that of water. In reality, the major contributors to the NMR image, water and lipids, have chemical shifts that differ by several parts per million. These differences in chemical shifts will manifest themselves as differences in spatial position in an NMR image. At low magnetic field strengths (less than about 1 tesla) these so-called *chemical shift artifacts* are not a problem, since the lipid and water components produce a composite signal. However, at higher magnetic field strengths, the signals from water and from lipid will produce "ghosts." A number of techniques have been proposed to avoid the problems of chemical shift artifacts.

It is also possible to exploit these chemical shift differences in a method known as *chemical shift imaging*. This is, in essence, a form of four-dimensional imaging, with three of the dimensions being spatial and the fourth being the chemical shift.

There are several techniques that will produce a chemical shift image. For example, the projection reconstruction method can be used to obtain a chemical shift spectrum for each point in the projection.[48] Each point is obtained with a selective excitation pulse, applied in the presence of an appropriate field gradient. The signal is then detected in the absence of a field gradient. In another more complicated method, a slice is selected with an appropriately shaped radio frequency pulse, a nonselective 90° pulse is applied, and the G_x, G_y, and G_z gradients are used to spatially encode the signal. The free induction decay signal is then recorded in the absence of a field gradient. The free induction decay is then transformed with respect to the gradients to give the three-dimensional spatial information and with respect to the acquisition time to give the fourth dimension of frequency (i.e., chemical shift).

A more detailed description of chemical shift imaging is beyond the scope of this section, and the interested reader is directed to Ref. 64 for more details about this and other imaging methods. However, it is important to note that techniques such as chemical shift imaging may become potentially useful techniques in the diagnosis of certain disease states.[65]

9.2.6 Imaging of Solids and Materials

Up until now our description of NMR imaging has centered about imaging of protons, and in particular, the protons from water and from lipids. Protons from these substances generally have narrow linewidths, owing to their nearly isotropic motion. The lines are narrow because this rapid motion averages out the local dipolar fields. As we have already seen (Chapter 8, Section 8.2), these fields are very large in solids, but can be removed by applying a field that is large compared to the interaction strength (i.e., dipolar decoupling). Because these fields can be as large as 5 G, a gradient greater than 50 G/cm would be needed to achieve a resolution of 1 mm.[66] One way to circumvent this problem in NMR imaging is to use multiple-pulse line narrowing (Chapter 8, Section 8.7), or to use it in combination with multiple quantum NMR.[66] An alternate method involves eliminating the line broadening effects by observing the free induction decay signal at a fixed time while the experiment is repeated, each time with incremented field gradient steps.[67] This latter method has been applied to solid adamantane.

There are certain problems in materials science that can be solved without going to these line-narrowing techniques. For example, NMR imaging can be used to determine the uptake patterns of water in epoxy resins, follow the time dependence of polymer curing (going from a liquid or mobile material to a solid), or monitor the distribution of oil in a solid sample.

9.3 SOLVENT SUPPRESSION (P. A. MIRAU AND F. A. BOVEY)

9.3.1 Introduction

It may happen in solution NMR that the resonance of the solvent itself, which is usually present in great molar excess over the solute, is inconveniently large, and may mask solute resonances of interest or cause dynamic range problems (see Chapter 2, Section 2.5.2). In carbon−13 spectroscopy this is seldom a difficulty because of the large range of chemical shifts; if a particular solvent has interfering resonances one chooses another. The most serious problems arise in the study of biomolecules, where H_2O and D_2O are often necessarily the solvent of choice. Since H_2O is *ca.* 55 molar, its proton resonance, appearing near 5 ppm, overwhelms all others. Even when employing D_2O, the residual HDO peak is likely to be inconveniently large and in addition one loses the rapidly exchangeable solute proton resonances. It is therefore often desirable to either (a) saturate the solvent resonance or (b) selectively excite the solute resonances. There have been many proposals for accomplishing solvent resonance suppression by one or the other of these approaches, and we shall discuss some of the more effective procedures.

9.3.2 Solvent Saturation

The simplest scheme is to apply a proton field using the proton decoupler for a sufficient time and at an appropriate power level [see

Chapter 1, Eq. (1.50)] to saturate the solvent resonance. The decoupler is then shut off, a 90° pulse applied, and the solute spectrum acquired. This will not work if solute resonances of interest are under or very close to the solvent peak, or if such resonances represent protons that are in exchange with the solvent. In the latter case, the solvent saturation will be transferred to the solute in the manner we have already described (Chapter 5, Section 5.4.7.1.) It is in general best to use minimum field strength for the shortest time that gives adequate suppression.

9.3.3 Solvent Nulling and Selective Excitation

9.3.3.1 WEFT

The acronym stands for *W*ater *E*liminated *F*ourier *T*ransform NMR spectroscopy.[68] This method is based on the inversion-recovery technique for the measurement of the spin lattice relaxation time, T_1. We have seen (Chapter 5, Section 5.2.7.2) that this involves the inversion of the magnetization — in this case both the solvent and solute resonances — by a 180° pulse followed after an interval τ by a 90° sampling pulse to measure the regrowth of the magnetization toward equilibrium. We have further seen (Eq. 5.36) that when $\tau = \ln 2 T_1$ (solvent), or 0.693 T_1 (solvent) the recovering solvent magnetization passes through zero. If the solute T_1 is substantially shorter than that of the solvent, a 90° pulse at this time, followed by acquisition of the signal, will provide a spectrum of the solute without that of the solvent. After an interval of *ca.* $5T_1$ (solvent), the pulse sequence may be repeated.

Actually, it has been found that for HDO, it is not necessary in practice that T_1 (solvent) be greatly in excess of T_1 (solute); the condition T_1 (solvent) $\geqslant 1.4 \ T_1$ (solute) has been found sufficient.[69]

In Fig. 9.17 is shown the 100 MHz proton spectra of threonine, $CH_3CHOHCH(NH_3^+)CO_2^-$, in D_2O[68] (details in the figure caption). Spectrum (a) is the normal spectrum with the strongly dominant HDO resonance at *ca.* 5.2 ppm; in (b) is the WEFT spectrum, exhibiting an at least 1000-fold suppression of the solvent resonance.

9.3.3.2 The Redfield "2-1-4" Pulse

A frequently employed method for the selective excitation of solute resonances without exciting the solvent is the Redfield "2-1-4" pulse.[70,71] This is an example of a *composite* pulse, i.e., one in which the phase of B_1 is shifted during the pulse without the insertion of intervals. In order to understand the operation of this pulse one must realize that even in the absence of a resonating nucleus the rf pulse itself has a Fourier transform. For a square-wave pulse this is of the form sinc x, i.e., $(\sin x)/x$, a function having a central maximum with an infinite train of diminishing wiggles on each side:

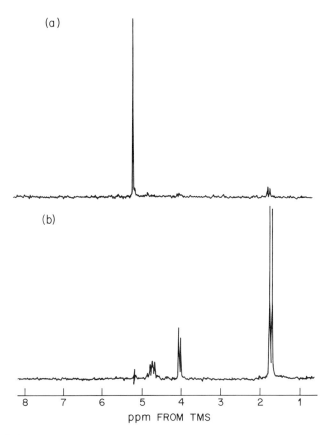

Fig. 9.17   Solvent resonance suppression by the WEFT sequence. The
solute is threonine, $CH_3CHOH-CH(NH_3^+)CO_2^-$ dissolved in
D_2O. (a) Normal proton spectrum (100 MHz) with HDO
resonance at *ca.* 5.2 ppm; T_1 (solvent) = 15 s; spectral width
1000 Hz; (b) WEFT spectrum, acquisition time t_{ac} is 4 s and
pulse interval τ is 6.3 s. (From *Practical NMR Spectroscopy*
by M. L. Martin, J.-J. Delpuech, and G. J. Martin, Heyden,
London, 1980, p. 278.)

In Fig. 9.18(b) the curve marked "long pulse" represents the central
portion of such a function. The interval from the center to the first null,
corresponding to the difference between the transmitter frequency ν_0 and
(if properly adjusted) the solvent frequency ν_s, is given by $1/\tau$, the
reciprocal of the pulse width τ. If, for example, τ is 500 μs (a rather long
or "soft" pulse), this frequency is 2000 Hz. A nucleus resonating at

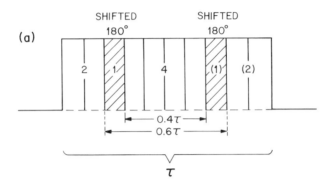

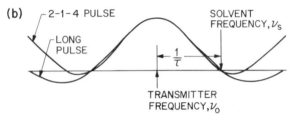

Fig. 9.18   (a) Composition of Redfield "2—1—4" pulse; τ is its total
duration, here represented as divided into 10 equal segments.
(b) Comparison of the 2—1—4 pulse and long pulse, viewed in
terms of the waveform of the Fourier transform. [From A. G.
Redfield, S. D. Kunz, and E. K. Ralph, *J. Magn. Reson.* **19**,
114 (1975).]

2000 Hz from ν_0 will not experience any net flipping by the rf field and so
will not appear in the spectrum. The disadvantage of a simple "soft" pulse
is that the sinc x function crosses the null rather abruptly; this means that
careful adjustment of ν_0 and τ is required in order to achieve effective
suppression of a very strong solvent resonance.

The Redfield "2-1-4" pulse is shown schematically in Fig. 9.18(a).
Following the first 0.2 segment of τ, the rf phase is shifted by 180°.
The phase is then shifted back for another 0.2 τ interval, bringing us to the
center of the pulse. This series of events is then repeated. The Fourier
transform of this composite pulse is shown by the function marked "2-1-
4 pulse." It is similar to that of the simple pulse but with the important
difference that the null is broader, making optimum adjustments much
easier.

Another approach is to replace the "soft" pulse by a sequence of hard pulses having the same sinc x excitation pattern. The simplest of these is the Jump-and-Return pulse sequence, consisting of $45°-\tau-45°$-acquire. This has an effect comparable to that of the Redfield pulse, but is easier to implement because the B_1 power and pulse widths are fixed and only the delay needs to be tuned for maximum suppression.

It should be realized that with these techniques solute resonances at or near the solvent resonance, e.g., H_2O or HDO, will also inevitably be suppressed. Nevertheless, the solvent suppression of at least 1000-fold that may be achieved means that one may be able to observe small solute resonances within 50–250 Hz of the solvent resonance. These techniques do not depend on T_1 differences between solute and solvent.

9.3.3.3 WATR and SWATTR

The methods designated by these acronyms make it possible to eliminate the water proton resonance almost completely and so permit observation of solute peaks not only near it but under it. A common feature of both methods is the addition of a salt, usually ammonium chloride, to produce a shortening of the water T_2. At a pH of 6.5 the T_2 of 1% HDO in D_2O, normally about 5 s, is reduced 160-fold by the presence of 0.5 M NH_4Cl[72] because of the moderately rapid exchange of water protons with the less shielded protons of the ammonium ion [see Chapter (5, Eq. 5.73)]. The proton spectrum is collected using the Carr−Purcell−Meiboom−Gill pulse sequence (Chapter 5, Section 5.3): $90_x°-(\tau-180_y°-\tau_n)$-acquire. (The delays τ are of the order 0.1–0.4 s.) By making the time between the $90_x°$ pulse and the acquisition of the FID long compared to T_2, the HDO resonance may be completely eliminated. This method is designated by the acronym WATR, standing for *W*ater *A*ttenuation by T_2 *R*elaxation[72], and is effective for solutes of small or moderate size.

For large molecules with short T_2's (0.005-0.020 s), this method is not feasible because the solute resonances tend to be eliminated along with that of the solvent. In such cases, Mirau[73] has proposed an alternative scheme (SWATTR), which also uses chemical exchange with added salt to shorten the solvent proton T_2, but employs *selective* pulses at the water frequency to null the H_2O signal. The resonances of the solute are then observed with a high-powered, nonselective $90_x°$ pulse.

The operation of this pulse sequence is shown in Fig. 9.19. In panel (a) are represented the nonselective $90_x°$ solute observe pulse (top) and the selective $90°-180°$ water nulling sequence (below), which is the same as in a T_2 experiment and in WATR. [If the period between the $180_y°$ pulses is short compared to $1/J$ of the observed solute protons, the phases of

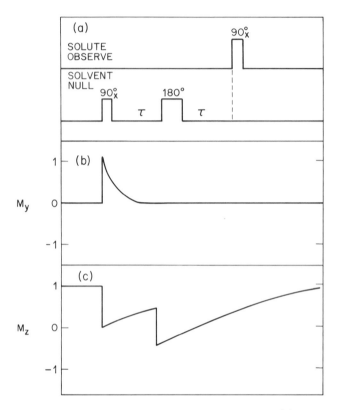

Fig. 9.19   SWATTR procedure for solvent suppression. (a) Proton pulse
sequence for solute observation and solvent nulling; (b)
behavior of solvent y and z magnetization components as a
function solvent nulling pulses. [From P. A. Mirau, *J. Magn.
Reson.* **73**, 123 (1987).]

peaks are not modulated by the J couplings as in the 2D J-resolved
experiment (Chapter 6, Section 6.2).] In (b) and (c) are shown the effects
of the water nulling sequence on the y and z components of the water
proton magnetization. Following the 90_x° selective pulse [(a), top], the
water magnetization component M_y decays to zero in a time of the order
of T_2 (b) — *ca.* 0.02 s — provided the interval τ is longer than this. In
practice, it is sufficiently long in comparison to the water T_1 that the water
M_z will regrow as shown (c) and must be inverted by the 180° pulse. The
90_x° observe pulse is applied just as the solvent M_z crosses the null [curve
(c)], and so the solvent is eliminated from the solute spectrum.

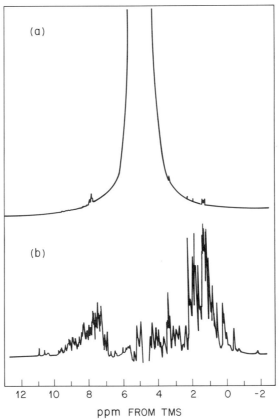

Fig. 9.20   500 MHz proton spectrum of HEW lysozyme; 0.005 M
            lysozyme, 0.35 M NH₄Cl at pH 6.6 with 5% D₂O. (a)
            Without solvent suppression; (b) with SWATTR solvent
            suppression applied. [From P. A. Mirau, *J. Magn. Reson.* **73**,
            123 (1987).]

   In Fig. 9.20 an example of the application of this technique is shown;
(a) is the normal 500-MHz proton spectrum of hen egg-white (HEW)
lysozyme, a globular protein of 14,600 molecular weight, 0.005 M in 95:5
$H_2O:D_2O$ at 20°C and pH 6, with 0.35 M ammonium chloride. The
solvent proton spectrum dominates so strongly that the protein spectrum
can barely be seen. In (b) the SWATTR sequence has been applied using
16 transients.

   Both WATR and SWATTR are also suitable for 2D spectroscopy.[72,73]

9.4 ZERO FIELD NMR (F. A. BOVEY)

9.4.1 Introduction

We have described at many points in previous chapters the benefits of performing NMR experiments in magnetic fields of the greatest strength possible. Observing sensitivity is proportional to the resonance frequency ν_0, which in turn is proportional to B_0. Sensitivity is also proportional to the separation of Zeeman levels, likewise proportional to B_0. Sensitivity thus increases as B_0^2. Another major reason for employing large B_0 is the fact that the separations of chemical shifts are also proportional to B_0, in both the liquid and solid states; this of course is of the greatest importance for the determination of chemical structure, particularly of large and complex molecules. What advantage then can there be in operating without a magnetic field?

In one way the existence of a magnetic field is a detriment, since it establishes a direction within the sample. We have seen (Chapter 1, Section 1.6; Chapter 8, Section 8.2) that in glassy or polycrystalline solids the random orientation of the molecules with respect to the field direction gives rise to powder patterns. These in combination with the mathematically similar chemical shift anisotropy patterns broaden the resonances so greatly as to mask all isotropic chemical shifts and scalar couplings. We have also described the experimental means for collapsing these patterns in order to recover the isotropic chemical shifts. If there were no magnetic field and the direction within the sample that it establishes, these patterns could not exist, as all molecular orientations would be equivalent. Experimental methods have accordingly been developed that allow one to capture the best of both worlds: the full sensitivity of high fields combined with the resolved fine structure, which becomes observable when molecular orientation is without effect.

9.4.2 The Basic Experiment

In Fig. 9.21 is shown the idealized behavior of the magnetic field B_0 and the response of the nuclear magnetization M_z.[74,75] The decrease of B_0 to zero takes place in a time short compared to T_2 so that the spins retain their original high-field alignment and extent of polarization but now precess only in their own local fields — of the order of a few gauss. Their magnetization undergoes coherent oscillations in these local fields; these are allowed to proceed for a period t_1. They are then sampled by reapplication of B_0 and detected at high field. (The zero field oscillations are themselves of too low frequency to be detected directly.) By incrementing t_1 a time domain spectrum is generated that is then Fourier transformed to the frequency domain in the usual manner.

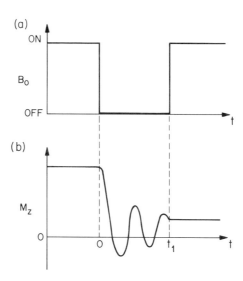

Fig. 9.21   Idealized scheme for field cycling to achieve zero field. (a)
Laboratory field B_0 as a function of time. (b) The
corresponding profile of the nuclear magnetization, M_z. At the
time marked 0, B_0 is instantly shut off. The magnetization
begins slow oscillations in the local magnetic fields of the nuclei
and is allowed to evolve for a time t_1. To sample M_z, B_0 is
then instantly reapplied and the evolution terminated. The
cycling is repeated for a number of equally spaced increments
of t_1 and the resulting series of values of M_z constitute the zero
field FID. This is then transformed to the frequency domain.
[From D. B. Zax, A. Bielecki, K. W. Zilm, A. Pines, and D. P.
Weitekamp, *J. Chem. Phys.* **83**, 4877 (1985).]

In practice, it is not possible to switch off a field of several teslas in
times of the order of a microsecond, as required. An acceptable substitute
procedure is to transfer the sample mechanically into an intermediate field
B_1 of *ca.* 0.01 T, followed by switching to zero field by means of a
cancelling field B_2 which may be very rapidly applied. The apparatus and
magnetic field (as a function of time) are diagrammed in somewhat
idealized form in Fig. 9.22. The reapplication of the field takes a course
that is just the reverse of its decrease. The incremented time t_1 is the time
spent at zero field, as shown.

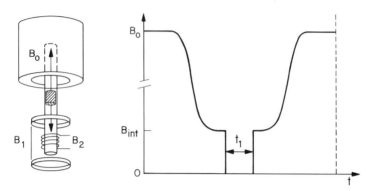

Fig. 9.22   Schematic diagram of the field cycling experiment as actually
carried out. At left is the main magnet coil generating field B_0.
The sample (cross-hatched) is moved along the glass tube by
air pressure. The larger of the two coils generates the
intermediate field B_1, while the smaller coil generates the
bucking field B_2, which cancels B_1 to generate the zero field.
[From D. B. Zax, A. Bielecki, K. W. Zilm, A. Pines, and D. P.
Weitekamp, *J. Chem. Phys.*, **83** 4877 (1985).]

9.4.3 Form of the Zero Field Spectrum

The evolution of the zero field spectrum under the influence of local
fields is not simple to describe. The reader is referred to Ref. 74 for a full
treatment. One result of such a treatment is that the number of lines in
the frequency-domain spectrum is found to be given by

$$N = 2^n(2^n - 1), \quad n = 2, 4, 6, \ldots \tag{9.3a}$$

and

$$N = 2^{n-1}(2^{n-1} - 1) + 1, \quad n = 3, 5, 7, \ldots \tag{9.3b}$$

where n is the number of interacting spin-1/2 nuclei. Thus, only two
nuclei can in principle give rise to as many as 12 lines and four nuclei can
give 240 lines. In practice, molecular symmetry (i.e., equality of bond
length) and limited resolution greatly reduce this number. In spin-1/2
spectra, the basic parameters that determine the line pattern are
internuclear distances, together with the averaging effects of molecular
motion, if any. Chemical shifts lose their meaning and scalar couplings are
generally unresolved as the spectra are commonly of the order of
50–100 kHz in width.

A representative spin-1/2 spectrum is that of 1,2,3,4-tetrachloronaphthalenebis(hexachlorocyclopentadiene) adduct:

The structural question at issue here is the orientation of the protons on the central ring, involving the further questions of whether this ring is chair, boat, or planar and whether the protons are cis or trans (or equatorial versus axial). The observed spectrum is shown in Fig. 9.23(b). In Fig. 9.24 are shown six possible structures (these clearly do not exhaust the possibilities), together with the predicted zero field spectra. The differences between these arise from differing interproton distances. The best match is given by structure (f), the simulated spectrum that resembles Fig. 9.23(b) very closely. A C_2 axis of symmetry, which interconverts the two innermost (1 and 1') and the two outermost (2 and 2') protons has been assumed in this simulation and therefore only 4 distances are needed: $r_{11'} = 2.38$ Å, $r_{12} = 2.22$ Å, $r_{12'} = 4.34$ Å and $r_{22'} = 5.01$ Å. (Distances to neighboring molecules are not included as these are too great to affect the simulation.)

Many other intriguing zero field experiments have been described (Refs. 74 and 75), and it promises to be a method of considerable interest for the solid state. Heteronuclear spin systems, for example, exhibit the interesting property that differences in the magnetogyric ratio, however large, vanish, yet the different spin populations may be separately manipulated by irradiation at their resonant frequencies while still in B_0.

REFERENCES

1. G. Filipovich and G. V. D. Tiers, *J. Phys. Chem.* **63**, 761 (1959).

2. H. S. Gutowsky, D. W. McCall, B. R. McGarvey, and L. H. Meyer, *J. Am. Chem. Soc.* **74**, 4809 (1952).

(a)

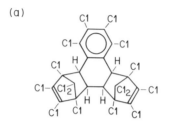

(b)

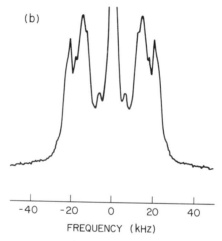

-40   -20   0   20   40

FREQUENCY (kHz)

Fig. 9.23   (a)   1,2,3,4-tetrachloronaphthalene-bis(hexachlorocyclopentadiene) adduct.   (b) The experimental zero field proton
spectrum of microcrystalline (a).   [From D. B. Zax,
A. Bielecki, K. W. Zilm, A. Pines, and D. P. Weitekamp, *J.
Chem. Phys.* **83**, 4877 (1985).]

3.   R. W. Taft, *J. Am. Chem. Soc.* **79**, 1045 (1957).

4.   E. F. Mooney and P. H. Winson, *Annual Reports on NMR
Spectroscopy*, E. F. Mooney, Ed., Vol. 1, Academic Press, New York
1968, pp. 244–312.

5.   K. Jones and E. F. Mooney, *Annual Reports on NMR Spectroscopy*,
E. F. Mooney, Ed., Vol. 3, Academic Press, New York, 1970,
pp. 261-422.

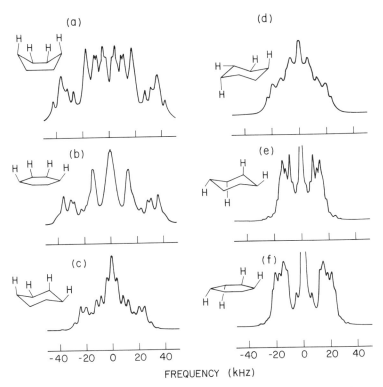

Fig. 9.24   Simulated zero field spectra for six possible configurations of the central ring in 1,2,3,4-tetrachloronaphthalene-bis(hexachlorocyclopentadiene) adduct. The spectrum of structure (f) most closely matches the observed spectrum of Fig. 9.23(b). [From D. B. Zax, A. Bielecki, K. W. Zilm, A. Pines, and D. P. Weitekamp, *J. Chem. Phys.* **83**, 4877 (1985).]

6.   R. Fields, *Annual Reports on NMR Spectroscopy*, E. F. Mooney, Ed., Vol. 5A, Academic Press, New York, 1972, pp. 99–304.

7.   L. Cavalli, *Annual Reports on NMR Spectroscopy*, E. F. Moooney, Ed., Vol. 6B, Academic Press, New York, 1976, pp. 43–228.

8.   V. Wray, *Annual Reports on NMR Spectroscopy*, E. F. Mooney, Ed., Vol. 10B, Academic Press, New York, 1980, pp. 3–507.

9.   V. Wray, *Annual Reports on NMR Spectroscopy*, G. A. Webb, Ed., Vol. 13, Academic Press, New York, 1983, pp. 3–406.

10. J. W. Emsley and L. Phillips, *Prog. in Nucl. Magn. Reson. Spectrosc.* **7** (1971).

11. R. K. Harris and N. Sheppard, *Trans. Faraday Soc.* **59**, 606 (1963).

12. D. S. Thompson, R. A. Newmark, and C. H. Sederholm, *J. Chem. Phys.* **37**, 411 (1962).

13. R. A. Newmark and C. H. Sederholm, *J. Chem. Phys.* **43**, 602 (1965).

14. W. S. Brey and K. C. Ramsey, *J. Chem. Phys.* **39**, 844 (1963).

15. M. S. Gopinathan and P. T. Narasimhan, *Mol. Phys.* **21**, 1141 (1971).

16. G. Govil, *Mol. Phys.* **21**, 953 (1971).

17. K. L. Williamson, Y. F. Li, F. H. Hall, and S. Swager, *J. Am. Chem. Soc.* **88**, 5678 (1966).

18. F. A. Bovey, E. W. Anderson, F. P. Hood, and R. L. Kornegay, *J. Chem. Phys.* **40**, 3099 (1964).

19. A. M. Takahashi, D. R. Davis, and J. D. Roberts, *J. Am. Chem. Soc.* **84**, 2935 (1962).

20. J. B. Lambert and J. D. Roberts, *J. Am. Chem. Soc.* **87**, 3891 (1965).

21. A. D. Cross and P. W. Landis, *J. Am. Chem. Soc.* **84**, 1736, 3784 (1962).

22. A. D. Cross and P. W. Landis, *J. Am. Chem. Soc.* **86**, 4005 (1964); A. D. Cross, *J. Am. Chem. Soc.* **86**, 4011.

23. J. W. Emsley, L. Phillips, and V. Wray, *Prog. in Nucl. Magn. Res. Spectrosc.* **10**, 85 (1977). This review has also been issued as a separate volume, Pergamon Press, 1977.

24. G. V. D. Tiers and F. A. Bovey, *J. Phys. Chem.* **63**, 302 (1959).

25. G. J. Martin, M. L. Martin, and J.-P. Gouesnard, "^{15}N-NMR Spectroscopy" in *NMR Basic Principles and Progress*, Vol. 18, P. Diehl, E. Fluck, and R. Kosfeld, Eds., 1981.

26a. M. Witanowski, L. Stefaniak, and G. A. Webb, *Annual Reports on NMR Spectroscopy* **11B** (1981).

26b. L. Stefaniak, G. A. Webb, and M. Witanowski, *Annual Reports on NMR Spectroscopy* **18** (1986).

27. G. C. Levy and R. L. Lichter, *Nitrogen-15 Nuclear Magnetic Resonance Spectroscopy*, Wiley, New York, 1979.

28. W. von Philipsborn and R. Müller, *Angew. Chem.* **25**, 383 (1983).

29. J. Schaefer, R. A. McKay, and E. O. Stejskal, *J. Mag. Reson.* **38**, 361 (1980).

30. G. Mavel, *Prog. in Nucl. Magn. Res. Spectros.* **1**, 251 (1966), J. W. Emsley, J. Feeney, and L. H. Sutcliffe, Eds., **1**, 251 (1966).

31. M. M. Crutchfield, C. H. Dungan, L. H. Letcher, V. Mark, and J. R. Van Wazer, *Topics in Phosphorus Chemistry*, M. Grayson and E. F. Griffin, Eds., **5**, 1967.

32. G. Mavel, *Annual Reports on NMR Spectroscopy*, E. F. Mooney, Ed., **5b**, 1 (1973).

33. C. T. Burt, T. Glonek, and M. Brany, *Science*, **195**, 145 (1977).

34. D. G. Gadian, G. K. Radda, R. E. Richards, P. J. Seeley in *Biological Application of Magnetic Resonance*, R. G. Shulman, Ed., Academic Press, New York, 1979, p. 463.

35. D. G. Gadian, *Nuclear Magnetic Resonance and Its Applications to Living Systems*, Oxford Univ. Press (Clarendon), London and New York, 1982..

36. D. P. Hollis, *Bull. Magn. Reson.* **1**, 27 (1979).

37. K. Ugurbil, R. G. Shulman, and T. R. Brown, *Biological Applications of Magnetic Resonance*, R. G. Shulman, Ed. Academic Press, New York, 1979, p. 537.

38. M. Cohn and B. D. N. Rao, *Bull. Magn. Reson.* **1**, 38 (1979).

39. I. K. O'Neill and C. P. Richards, *Annual Reports on NMR Spectroscopy* **10a**, 1 (1980).

40. O. Jardetzky and G. C. K. Roberts, *NMR in Molecular Biology*, Academic Press, New York, 1981, p. 681.

41. D. G. Gorenstein, *P-31 NMR: Principles and Applications*, Academic Press, New York, 1981.

42. D. G. Gorenstein in *Prog. in NMR Spectrosc.*, J. W. Emsley, J. Feeney, and L. H. Sutcliffe, Eds., **16**, 1 (1983).

43. F. C. Schilling, private communication.

44. J. A. Ackerman, private communication.

45. P. C. Lauterbur, *Nature (London)* **242**, 190 (1973).

46. J. D. Roberts, *Eng. & Sci.*, 10 (January 1986).

47. R. Damadian, *Science* **171**, 1151 (1971).

48. P. A. Bottomly, *Rev. Sci. Instrum.* **53**, 1319 (1982).

49. *NMR: A Perspective on Imaging*, General Electric Corporation, Medical Systems Operations, Milwaukee, Wisconsin, 1982.

50. M. R. Bendall and R. E. Gordon, *J. Magn. Reson.* **53**, 365 (1983).

51. W. S. Hinshaw, *J. Appl. Phys.* **47**, 3709 (1976).

52. P. A. Bottomley, *Cancer Res.* **39**, 468 (1979).

53. W. S. Hinshaw, P. A. Bottomley, and G. N. Holland, *Nature (London)* **270**, 722 (1977).

54. I. L. Pykett, *Sci. Am.* **246**, 78 (1982).

55. F. W. Wehrli, J. R. MacFall, and T. H. Newton in *Advanced Imaging Techniques*, T. H. Newton and D. G. Potts, Eds., Clavadel Press, San Anselmo, California, 1983, pp. 81–117.

56. A. Kumar, I. Welti, and R. R. Ernst, *Naturwissenschaften* **62**, 34 (1975).

57. A. Kumar, I. Welti, and R. R. Ernst, *J. Magn. Reson.* **18**, 69 (1975).

58. D. I. Hoult, *J. Magn. Reson.* **33**, 183 (1979).

59. P. Mansfield and I. L. Pykett, *J. Magn. Reson.* **29**, 355 (1978).

60. W. A. Edelstein, J. M. S. Hutchison, G. Johnson, and T. Redpath, *Phys. Med. Biol.* **25**, 751 (1980).

61. E. T. Fossel, J. M. Carr, and J. McDonagh, *New England J. Med.* **315**, 1369 (1986).

62. T. F. Budinger and P. C. Lauterbur, *Science* **226**, 288 (1984).

63. V. J. Wedeen, R. A. Meuli, R. R. Edelman, S. C. Geller, L. R. Frank, T. J. Brady, and B. R. Rosen, *Science* **230**, 946 (1985).

64. R. E. Steiner and G. K. Radda, *Brit. Med. Bull.* **40**, 113 (1984).

65. G. K. Radda, *Science* **233**, 640 (1986).

66. A. N. Garroway, J. Baum, M. G. Munowitz, and A. Pines, *J. Magn. Reson.* **60**, 337 (1984).

67. S. Emid and J. H. N. Creyghton, *Physica* **128B**, 81 (1985).

68. S. L. Patt and B. D. Sykes, *J. Chem. Phys.* **56**, 3182 (1972).

69. F. W. Benz, J. Feeney, and G. C. K. Roberts, *J. Magn. Reson.* **8**, 114 (1972).

70. M. L. Martin, J.-J. Delpuech, and G. J. Martin, *Practical NMR Spectroscopy*, Heyden, London, 1980, p. 278.

71. A. G. Redfield, S. D. Kunz, and E. K. Ralph, *J. Magn. Reson.* **19**, 114 (1975).

72. D. L. Rabenstein, S. Fan, and T. T. Nakashima, *J. Magn. Reson.* **64**, 541 (1985).

73. P. A. Mirau, *J. Magn. Reson.* **73**, 123 (1987).

74. D. B. Zax, A. Bielecki, K. W. Zilm, A. Pines, and D. P. Weitekamp, *J. Chem. Phys.* **83**, 4877 (1985).

75. A. M. Thayer and A. Pines, *Acc. Chem. Res.* **20**, 47 (1987).

Appendix A

TABLE OF NUCLEAR PROPERTIES

NUCLEAR PROPERTIES[a]

Isotope (* indicates radioactive)	NMR frequency (Mc/sec for a 10 kgauss field)	Natural abundance (%)	Relative sensitivity for equal number of nuclei		Magnetic moment μ (multiples of the nuclear magneton, $eh/4\pi\,M_p c$)	Spin I (multiples of $h/2\pi$)	Electric quadrupole moment Q (multiples of $e \times 10^{-24}$ cm²)
			At constant field	At constant frequency			
^{1}n*	29.165		0.322	0.685	-1.9130	1/2	
^{1}H	42.577	99.9844	1.000	1.000	2.79270	1/2	
^{2}H	6.536	1.56×10^{-2}	9.64×10^{-3}	0.409	0.85738	1	2.77×10^{-3}
^{3}H*	45.414		1.21	1.07	2.9788	1/2	
^{3}He	32.434	10^{-5} to 10^{-7}	0.443	0.762	-2.1274	1/2	
^{6}Li	6.265	7.43	8.51×10^{-3}	0.392	0.82191	1	4.6×10^{-4}
^{7}Li	16.547	92.57	0.294	1.94	-3.2560	3/2	-4.2×10^{-2}
^{9}Be	5.983	100	1.39×10^{-2}	0.703	-1.1774	3/2	2×10^{-2}
^{10}B	4.575	18.83	1.99×10^{-2}	1.72	1.8006	3	0.111
^{11}B	13.660	81.17	0.165	1.60	2.6880	3/2	3.55×10^{-2}
^{13}C	10.705	1.108	1.59×10^{-2}	0.251	0.70216	1/2	
^{14}N	3.076	99.635	1.01×10^{-3}	0.193	0.40357	1	2×10^{-2}
^{15}N	4.315	0.365	1.04×10^{-3}	0.101	-0.28304	1/2	
^{17}O	5.772	3.7×10^{-2}	2.91×10^{-2}	1.58	-1.8930	5/2	-4×10^{-3}
^{19}F	40.055	100	0.834	0.941	2.6273	1/2	
^{21}Ne		0.257				$\geqq 3/2$	
^{22}Na*	4.434		1.81×10^{-2}	1.67	1.745	3	
^{23}Na	11.262	100	9.27×10^{-2}	1.32	2.2161	3/2	0.1
^{25}Mg	2.606	10.05	2.68×10^{-2}	0.714	-0.85471	5/2	
^{27}Al	11.094	100	0.207	3.04	3.6385	5/2	0.149
^{29}Si	8.460	4.70	7.85×10^{-2}	0.199	-0.55477	1/2	
^{31}P	17.235	100	6.64×10^{-2}	0.405	1.1305	1/2	-6.4×10^{-2}
^{33}S	3.266	0.74	2.26×10^{-2}	0.384	0.64274	3/2	
^{35}S*	5.08		8.50×10^{-3}	0.599	1.00	3/2	4.5×10^{-2}
^{35}Cl	4.172	75.4	4.71×10^{-3}	0.490	0.82089	3/2	-7.97×10^{-2}
^{36}Cl*	4.893		1.21×10^{-2}	0.919	1.2838	2	-1.68×10^{-2}
^{37}Cl	3.472	24.6	2.72×10^{-3}	0.408	0.68329	3/2	-6.21×10^{-2}

Isotope							
^{39}K	1.987	93.08	5.08×10^{-4}	0.233	0.39094	3/2	
^{40}K*	2.470	1.19×10^{-2}	5.21×10^{-3}	1.55	−1.296	4	
^{41}K	1.092	6.91	8.39×10^{-5}	0.128	0.21453	3/2	
^{43}Ca	2.865	0.13	6.39×10^{-2}	1.41	−1.3153	7/2	
^{45}Sc	10.343	100	0.301	5.10	4.7491	7/2	
^{47}Ti	2.400	7.75	2.10×10^{-3}	0.659	−0.78712	5/2	
^{49}Ti	2.401	5.51	3.76×10^{-3}	1.19	−1.1023	7/2	
^{50}V	4.245	0.24	5.53×10^{-2}	5.58	3.3413	6	
^{51}V	11.193	~100	0.383	5.53	5.1392	7/2	
^{53}Cr	2.406	9.54	1.0×10^{-4}	0.29	−0.4735	3/2	0.3
^{55}Mn	10.553	100	0.178	2.89	3.4610	5/2	0.5
^{57}Fe		2.245			≦0.05		
^{57}Co*	10.0		0.274	4.95	4.6	7/2	
^{58}Co*	13.3		0.25	2.5	3.5	2	
^{59}Co	10.103	100	0.281	4.83	4.6388	7/2	0.5
^{60}Co*	4.6		5×10^{-2}	4.3	3.0	5?	
^{61}Ni		1.25			<0.25		
^{63}Cu	11.285	69.09	9.38×10^{-2}	1.33	2.2206	3/2	−0.15
^{65}Cu	12.090	30.91	0.116	1.42	2.3790	3/2	−0.14
^{67}Zn	2.635	4.12	2.86×10^{-3}	0.730	0.8735	5/2	
^{69}Ga	10.218	60.2	6.93×10^{-2}	1.201	2.0108	3/2	0.2318
^{71}Ga	12.984	39.8	0.142	1.525	2.5549	3/2	0.1461
^{73}Ge	1.485	7.61	1.40×10^{-3}	1.15	−0.8768	9/2	−0.2
^{75}As	7.292	100	2.51×10^{-2}	0.856	1.4349	3/2	0.3
^{77}Se	8.131	7.50	6.97×10^{-3}	0.191	0.5333	1/2	
^{79}Se*	2.210		2.94×10^{-3}	1.12	−1.015	7/2	
^{79}Br	10.667	50.57	7.86×10^{-2}	1.26	2.0990	3/2	0.9
^{81}Br	11.498	49.43	9.84×10^{-2}	1.35	2.2626	3/2	0.33
^{83}Kr	1.64	11.55	1.89×10^{-3}	1.27	−0.968	9/2	0.28
^{85}Rb	4.111	72.8	1.05×10^{-2}	1.13	1.3483	5/2	0.15
^{87}Rb	13.932	27.2	0.177	1.64	2.7415	3/2	0.31
^{87}Sr	1.845	7.02	2.69×10^{-3}	1.43	−1.0893	9/2	0.15
^{89}Y	2.086	100	1.17×10^{-4}	4.90×10^{-2}	−0.1368	1/2	
^{91}Zr	4.0	11.23	9.4×10^{-3}	1.04	−1.3	5/2	

NUCLEAR PROPERTIES[a] (*Continued*)

Isotope (* indicates radioactive)	NMR frequency (Mc/sec for a 10 kgauss field)	Natural abundance (%)	Relative sensitivity for equal number of nuclei		Magnetic moment μ (multiples of the nuclear magneton, $eh/4\pi M_p c$)	Spin I (multiples of $h/2\pi$)	Electric quadrupole moment Q (multiples of $e \times 10^{-24}$ cm^2)
			At constant field	At constant frequency			
^{93}Nb	10.407	100	0.482	8.06	6.1435	9/2	-0.4 ± 0.3
^{95}Mo	2.774	15.78	3.22×10^{-3}	0.761	-0.9099	5/2	
^{97}Mo	2.833	9.60	3.42×10^{-3}	0.776	-0.9290	5/2	
^{99}Tc*	9.583		0.376	7.43	5.6572	9/2	0.3
^{99}Ru		12.81				3	
^{101}Ru		16.98				5/2	
^{103}Rh	1.340	100	3.12×10^{-5}	3.15×10^{-2}	-0.0879	1/2	
^{105}Pd	1.74	22.23	7.79×10^{-4}	0.47	-0.57	5/2	
^{107}Ag	1.722	51.35	6.69×10^{-5}	4.03×10^{-2}	-0.1130	1/2	
^{109}Ag	1.981	48.65	1.01×10^{-4}	4.66×10^{-2}	-0.1299	1/2	
^{111}Cd	9.028	12.86	9.54×10^{-3}	0.212	-0.5922	1/2	
^{113}Cd	9.444	12.34	1.09×10^{-2}	0.222	-0.6195	1/2	
^{113}In	9.310	4.16	0.345	7.22	5.4960	9/2	1.144
^{115}In*	9.329	95.84	0.348	7.23	5.5072	9/2	1.161
^{115}Sn	13.22	0.35	3.50×10^{-2}	0.327	-0.9132	1/2	
^{117}Sn	15.77	7.67	4.53×10^{-2}	0.356	-0.9949	1/2	
^{119}Sn	15.87	8.68	5.18×10^{-2}	0.373	-1.0409	1/2	
^{121}Sb	10.19	57.25	0.160	2.79	3.3417	5/2	-0.8
^{123}Sb	15.518	42.75	4.57×10^{-2}	2.72	2.5334	7/2	-1.0
^{123}Te	11.59	0.89	1.80×10^{-2}	0.262	-0.7319	1/2	
^{125}Te	13.45	7.03	3.16×10^{-2}	0.316	-0.8824	1/2	
^{127}I	8.519	100	9.35×10^{-2}	2.33	2.7939	5/2	-0.75
^{129}I*	5.669		4.96×10^{-2}	2.80	2.6030	7/2	-0.43
^{129}Xe	11.78	26.24	2.12×10^{-2}	0.277	-0.7726	1/2	
^{131}Xe	3.490	21.24	2.77×10^{-3}	0.410	0.6868	3/2	-0.12
^{133}Cs	5.585	100	4.74×10^{-2}	2.75	2.5642	7/2	≤ 0.3

Isotope						Spin	
^{134}Cs*	5.64		6.21×10^{-2}	3.53	2.96	4	
^{135}Cs*	5.94		5.70×10^{-2}	2.94	2.727	7/2	
^{137}Cs*	6.19		6.44×10^{-2}	3.05	2.84	7/2	
^{135}Ba	4.25	6.59	4.99×10^{-3}	0.499	0.837	3/2	
^{137}Ba	4.76	11.32	6.97×10^{-3}	0.559	0.936	3/2	
^{138}La*	5.617	0.089	9.18×10^{-2}	2.64	3.6844	5	2.7
^{139}La	6.014	99.911	5.92×10^{-2}	2.97	2.7615	7/2	0.9
^{141}Ce*	0.35		1.1×10^{-5}	0.17	0.16	7/2	
^{141}Pr	11.3	100	0.234	3.18	3.8	5/2	-5.4×10^{-2}
^{143}Nd	2.2	12.20	2.81×10^{-3}	1.07	-1.1	7/2	$\leqq 1.2$
^{145}Nd	1.4	8.30	6.70×10^{-4}	0.666	-0.69	7/2	$\leqq 1.2$
^{147}Sm	1.47	15.07	8.8×10^{-4}	0.725	-0.68	7/2	0.72
^{149}Sm	1.19	13.84	4.7×10^{-4}	0.591	-0.55	7/2	0.72
^{151}Eu	10	47.77	0.168	2.84	3.4	5/2	~1.2
^{153}Eu	4.6	52.23	1.45×10^{-2}	1.25	1.5	5/2	~2.5
^{155}Gd		14.68			-0.19	(7/2)	
^{157}Gd		15.64			-0.33	(7/2)	
^{159}Tb		100				3/2	
^{161}Dy		18.73				7/2	
^{163}Dy		24.97				7/2	
^{165}Ho		100				7/2	
^{167}Er		22.82				7/2	
^{169}Tm		100				1/2	~10
^{171}Yb	6.9	14.27	4.19×10^{-3}	0.161	0.45	1/2	
^{173}Yb	1.98	16.08	1.18×10^{-3}	0.543	-0.65	5/2	3.9
^{175}Lu	5.7	97.40	4.94×10^{-2}	2.79	2.6	7/2	5.9
^{176}Lu*		2.60			4.2	$\geqq 7$	6–8
^{177}Hf		18.39				1/2 or 3/2	
^{179}Hf		13.78				1/2 or 3/2	
^{181}Ta	4.6	100	2.60×10^{-2}	2.26	2.1	7/2	6.5
^{183}W	1.75	14.28	6.98×10^{-5}	4.12	0.115	1/2	
^{185}Re	9.586	37.07	0.133	2.63	3.1437	5/2	2.8
^{187}Re	9.684	62.93	0.137	2.65	3.1760	5/2	2.6
^{189}Os	3.307	16.1	2.24×10^{-3}	0.385	0.6507	3/2	2.0

NUCLEAR PROPERTIES[a] (Continued)

Isotope (* indicates radioactive)	NMR frequency (Mc/sec for a 10 kgauss field)	Natural abundance (%)	Relative sensitivity for equal number of nuclei		Magnetic moment μ (multiples of the nuclear magneton, $eh/4\pi M_p c$)	Spin I (multiples of $h/2\pi$)	Electric quadrupole moment Q (multiples of $e \times 10^{-24}$ cm^2)
			At constant field	At constant frequency			
^{191}Ir	0.81	38.5	3.5×10^{-5}	9.5×10^{-2}	0.16	3/2	~1.2
^{193}Ir	0.86	61.5	4.2×10^{-5}	0.104	0.17	3/2	~1.0
^{195}Pt	9.153	33.7	9.94×10^{-3}	0.215	0.6004	1/2 $\pm$	
^{197}Au	0.691	100	2.14×10^{-5}	8.1×10^{-2}	0.136	3/2	0.56
^{199}Hg	7.612	16.86	5.72×10^{-3}	0.179	0.4993	1/2	
^{201}Hg	3.08	13.24	1.90×10^{-3}	0.362	-0.607	3/2	0.5
^{203}Tl	24.33	29.52	0.187	0.571	1.5960	1/2	
^{205}Tl	24.57	70.48	0.192	0.577	1.6114	1/2	
^{207}Pb	8.899	21.11	9.13×10^{-3}	0.209	0.5837	1/2	
^{209}Bi	6.842	100	0.137	5.30	4.0389	9/2	-0.4
^{235}U*		0.71				5/2	
^{237}Np*	~20		1.0	5.0	6 ± 2.5	5/2	
^{239}Pu*	6.1		2.9×10^{-3}	0.14	0.4	1/2	
^{241}Pu*	4.3		1.2×10^{-2}	1.2	1.4	5/2	
Free electron	27,994		2.85×10^8	658	-1836	1/2	

[a] Reproduced by permission of Varian, Palo Alto, California.

I. AB₂

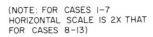

(NOTE: FOR CASES 1-7
HORIZONTAL SCALE IS 2X THAT
FOR CASES 8-13)

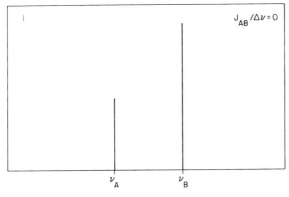

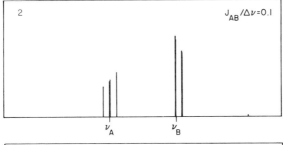

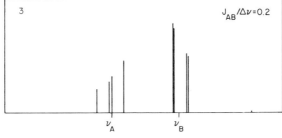

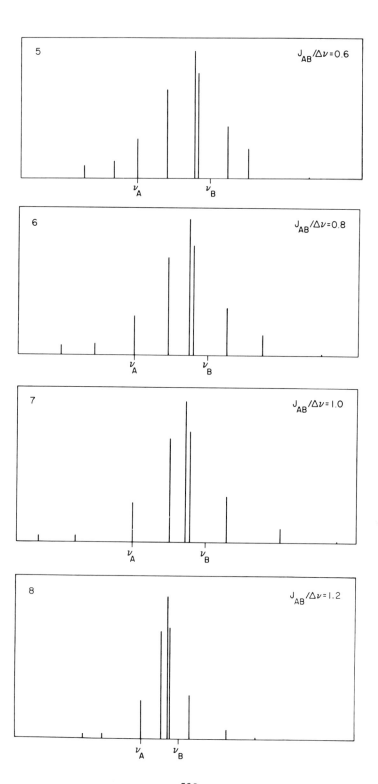

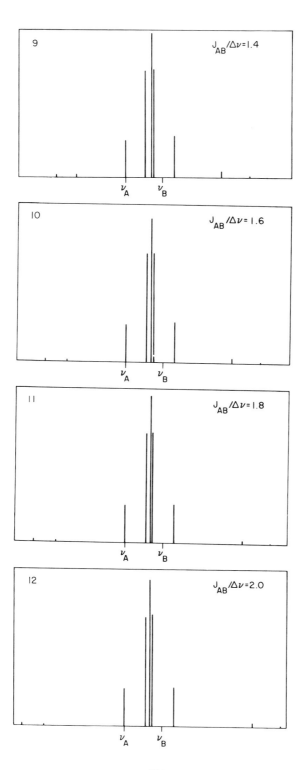

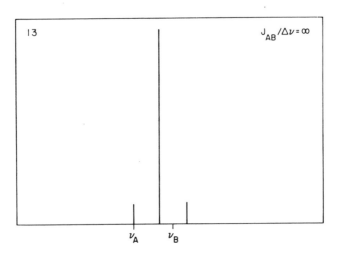

II. AB₃

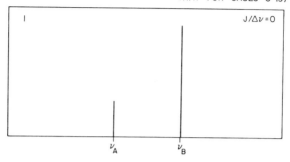

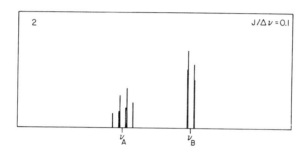

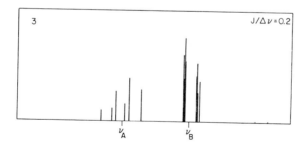

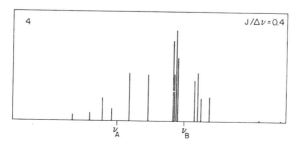

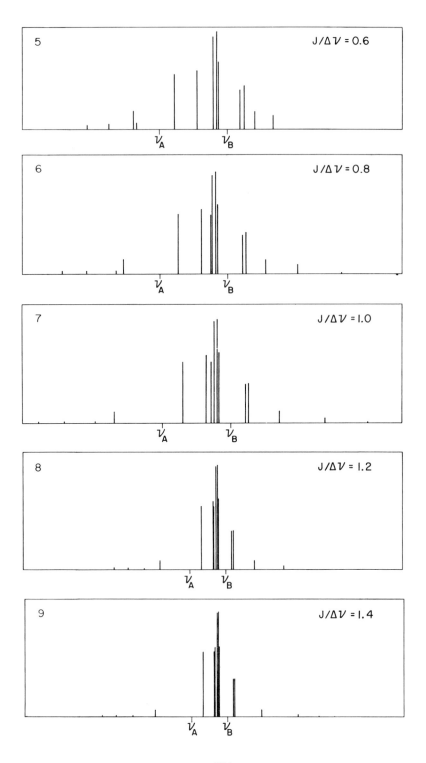

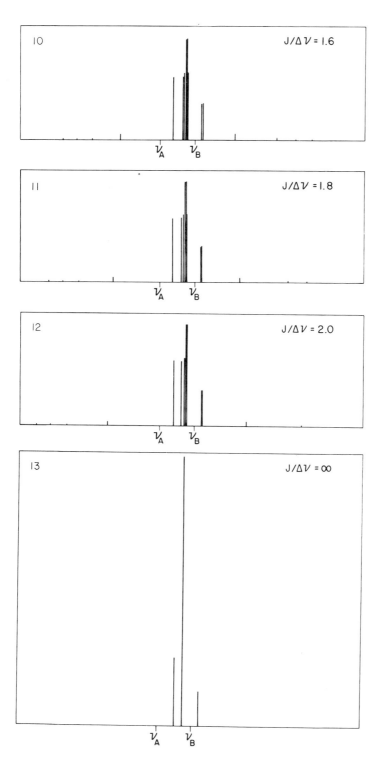

III. ABX

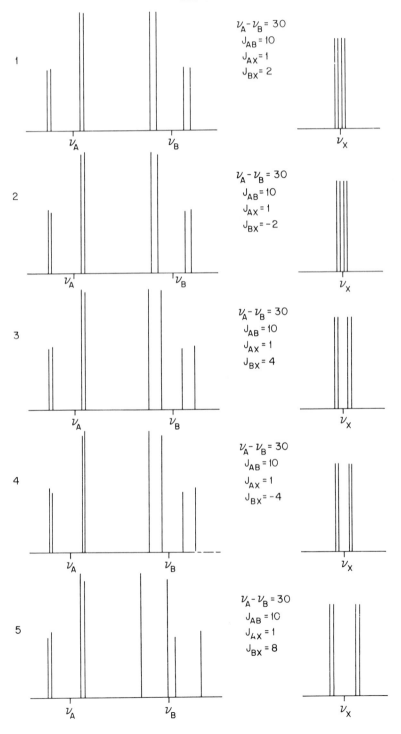

$\nu_A - \nu_B = 30$
$J_{AB} = 10$
$J_{AX} = 1$
$J_{BX} = 2$

1

$\nu_A - \nu_B = 30$
$J_{AB} = 10$
$J_{AX} = 1$
$J_{BX} = -2$

2

$\nu_A - \nu_B = 30$
$J_{AB} = 10$
$J_{AX} = 1$
$J_{BX} = 4$

3

$\nu_A - \nu_B = 30$
$J_{AB} = 10$
$J_{AX} = 1$
$J_{BX} = -4$

4

$\nu_A - \nu_B = 30$
$J_{AB} = 10$
$J_{AX} = 1$
$J_{BX} = 8$

5

536

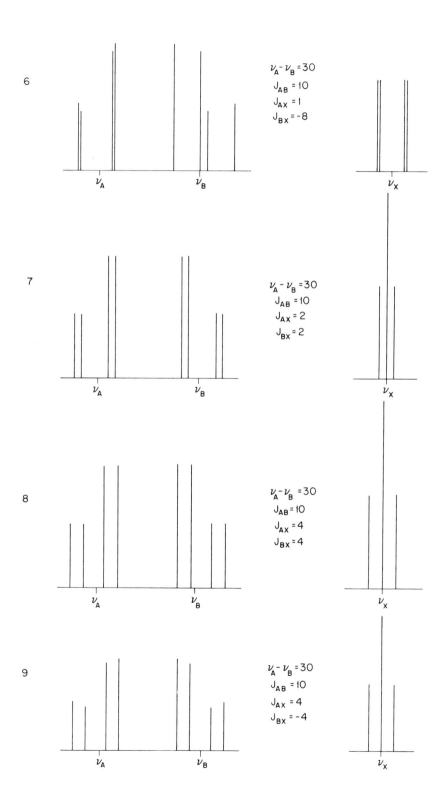

6

$\nu_A - \nu_B = 30$
$J_{AB} = 10$
$J_{AX} = 1$
$J_{BX} = -8$

ν_A ν_B

ν_X

7

$\nu_A - \nu_B = 30$
$J_{AB} = 10$
$J_{AX} = 2$
$J_{BX} = 2$

ν_A ν_B

ν_X

8

$\nu_A - \nu_B = 30$
$J_{AB} = 10$
$J_{AX} = 4$
$J_{BX} = 4$

ν_A ν_B

ν_X

9

$\nu_A - \nu_B = 30$
$J_{AB} = 10$
$J_{AX} = 4$
$J_{BX} = -4$

ν_A ν_B

ν_X

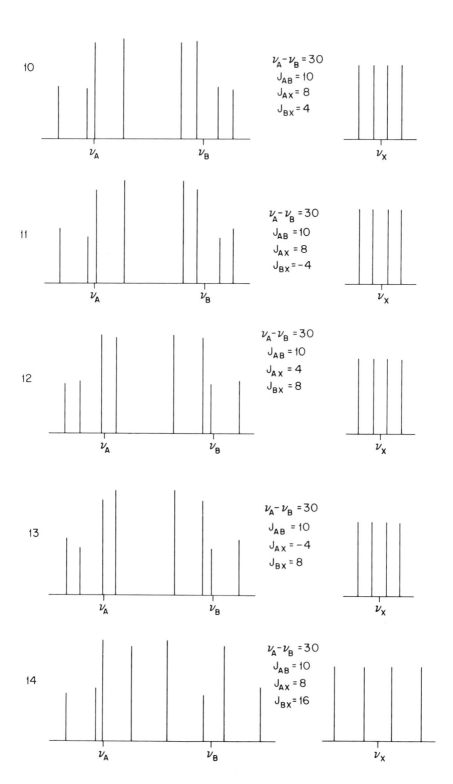

10

$\nu_A - \nu_B = 30$
$J_{AB} = 10$
$J_{AX} = 8$
$J_{BX} = 4$

ν_A ν_B ν_X

11

$\nu_A - \nu_B = 30$
$J_{AB} = 10$
$J_{AX} = 8$
$J_{BX} = -4$

ν_A ν_B ν_X

12

$\nu_A - \nu_B = 30$
$J_{AB} = 10$
$J_{AX} = 4$
$J_{BX} = 8$

ν_A ν_B ν_X

13

$\nu_A - \nu_B = 30$
$J_{AB} = 10$
$J_{AX} = -4$
$J_{BX} = 8$

ν_A ν_B ν_X

14

$\nu_A - \nu_B = 30$
$J_{AB} = 10$
$J_{AX} = 8$
$J_{BX} = 16$

ν_A ν_B ν_X

538

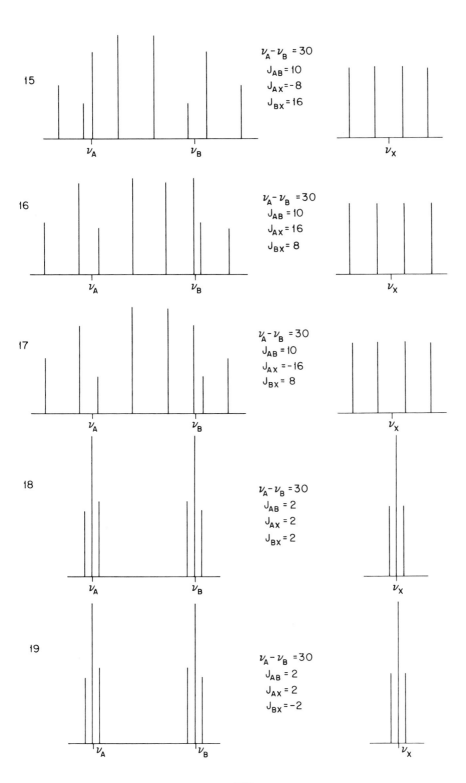

15    $\nu_A - \nu_B = 30$
      $J_{AB} = 10$
      $J_{AX} = -8$
      $J_{BX} = 16$

      ν_A ν_B ν_X

16    $\nu_A - \nu_B = 30$
      $J_{AB} = 10$
      $J_{AX} = 16$
      $J_{BX} = 8$

      ν_A ν_B ν_X

17    $\nu_A - \nu_B = 30$
      $J_{AB} = 10$
      $J_{AX} = -16$
      $J_{BX} = 8$

      ν_A ν_B ν_X

18    $\nu_A - \nu_B = 30$
      $J_{AB} = 2$
      $J_{AX} = 2$
      $J_{BX} = 2$

      ν_A ν_B ν_X

19    $\nu_A - \nu_B = 30$
      $J_{AB} = 2$
      $J_{AX} = 2$
      $J_{BX} = -2$

      ν_A ν_B ν_X

APPENDIX B

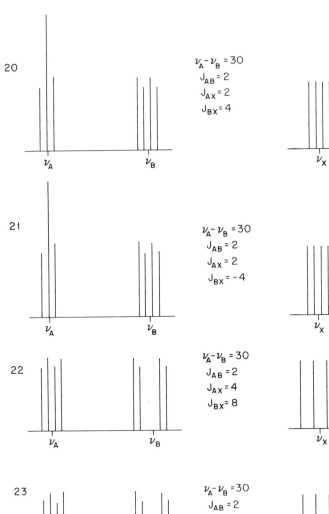

20

$\nu_A - \nu_B = 30$
$J_{AB} = 2$
$J_{AX} = 2$
$J_{BX} = 4$

ν_A ν_B ν_X

21

$\nu_A - \nu_B = 30$
$J_{AB} = 2$
$J_{AX} = 2$
$J_{BX} = -4$

ν_A ν_B ν_X

22

$\nu_A - \nu_B = 30$
$J_{AB} = 2$
$J_{AX} = 4$
$J_{BX} = 8$

ν_A ν_B ν_X

23

$\nu_A - \nu_B = 30$
$J_{AB} = 2$
$J_{AX} = 4$
$J_{BX} = -8$

ν_A ν_B ν_X

24

$\nu_A - \nu_B = 30$
$J_{AB} = 2$
$J_{AX} = 8$
$J_{BX} = 4$

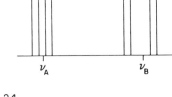
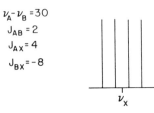
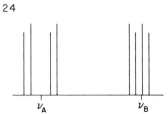
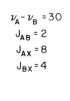
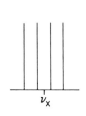

ν_A ν_B ν_X

25
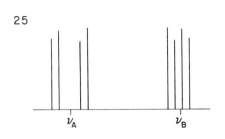

$\nu_A - \nu_B = 30$
$J_{AB} = 2$
$J_{AX} = 8$
$J_{BX} = -4$
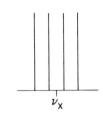

26
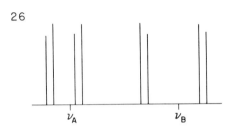

$\nu_A - \nu_B = 30$
$J_{AB} = 2$
$J_{AX} = 8$
$J_{BX} = 16$
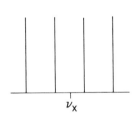

27
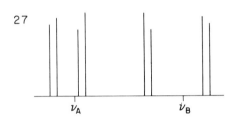

$\nu_A - \nu_B = 30$
$J_{AB} = 2$
$J_{AX} = -8$
$J_{BX} = 16$
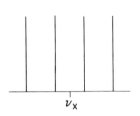

28
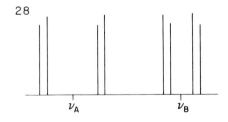

$\nu_A - \nu_B = 30$
$J_{AB} = 2$
$J_{AX} = 16$
$J_{BX} = 8$
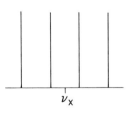

29
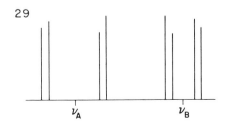

$\nu_A - \nu_B = 30$
$J_{AB} = 2$
$J_{AX} = 16$
$J_{BX} = -8$
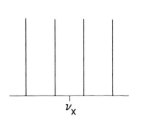

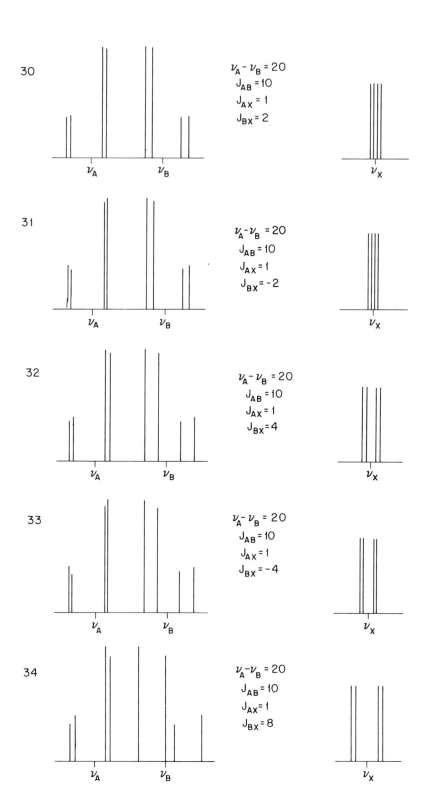

30

$\nu_A - \nu_B = 20$
$J_{AB} = 10$
$J_{AX} = 1$
$J_{BX} = 2$

ν_A ν_B ν_X

31

$\nu_A - \nu_B = 20$
$J_{AB} = 10$
$J_{AX} = 1$
$J_{BX} = -2$

ν_A ν_B ν_X

32

$\nu_A - \nu_B = 20$
$J_{AB} = 10$
$J_{AX} = 1$
$J_{BX} = 4$

ν_A ν_B ν_X

33

$\nu_A - \nu_B = 20$
$J_{AB} = 10$
$J_{AX} = 1$
$J_{BX} = -4$

ν_A ν_B ν_X

34

$\nu_A - \nu_B = 20$
$J_{AB} = 10$
$J_{AX} = 1$
$J_{BX} = 8$

ν_A ν_B ν_X

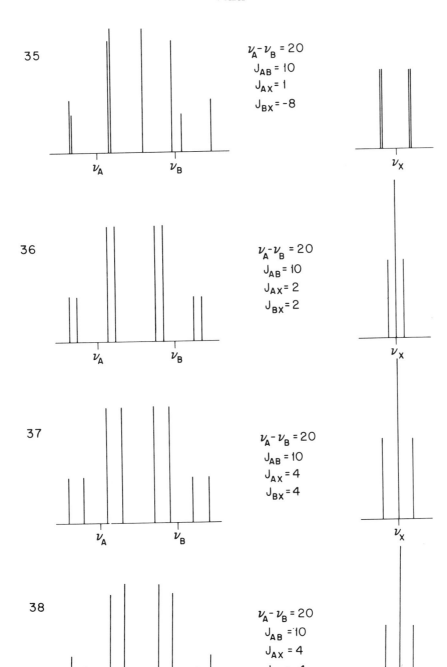

35   $\nu_A - \nu_B = 20$
     $J_{AB} = 10$
     $J_{AX} = 1$
     $J_{BX} = -8$

36   $\nu_A - \nu_B = 20$
     $J_{AB} = 10$
     $J_{AX} = 2$
     $J_{BX} = 2$

37   $\nu_A - \nu_B = 20$
     $J_{AB} = 10$
     $J_{AX} = 4$
     $J_{BX} = 4$

38   $\nu_A - \nu_B = 20$
     $J_{AB} = 10$
     $J_{AX} = 4$
     $J_{BX} = -4$

39

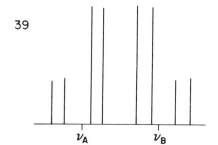

$$\nu_A - \nu_B = 20$$
$$J_{AB} = 10$$
$$J_{AX} = 3$$
$$J_{BX} = 4$$

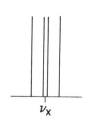

40

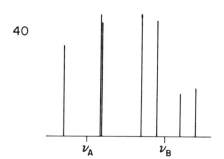

$$\nu_A - \nu_B = 20$$
$$J_{AB} = 10$$
$$J_{AX} = 0$$
$$J_{BX} = 4$$

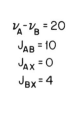

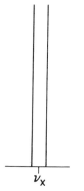

41

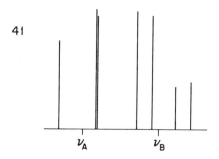

$$\nu_A - \nu_B = 20$$
$$J_{AB} = 10$$
$$J_{AX} = 0$$
$$J_{BX} = -4$$

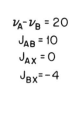

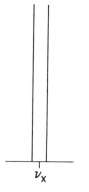

42

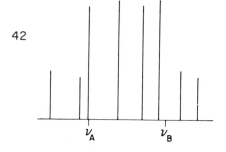

$$\nu_A - \nu_B = 20$$
$$J_{AB} = 10$$
$$J_{AX} = 8$$
$$J_{BX} = 4$$

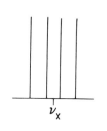

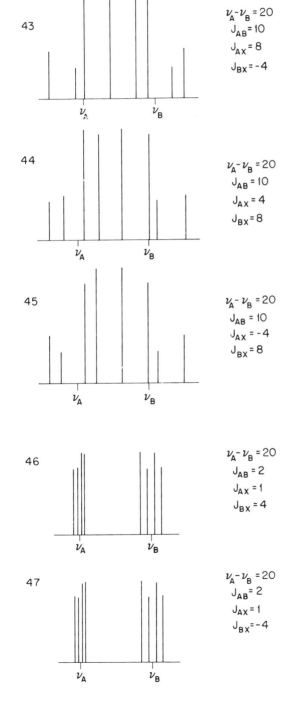

43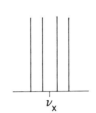

$\nu_A - \nu_B = 20$
$J_{AB} = 10$
$J_{AX} = 8$
$J_{BX} = -4$

44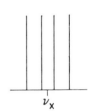

$\nu_A - \nu_B = 20$
$J_{AB} = 10$
$J_{AX} = 4$
$J_{BX} = 8$

45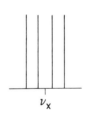

$\nu_A - \nu_B = 20$
$J_{AB} = 10$
$J_{AX} = -4$
$J_{BX} = 8$

46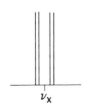

$\nu_A - \nu_B = 20$
$J_{AB} = 2$
$J_{AX} = 1$
$J_{BX} = 4$

47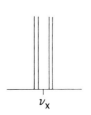

$\nu_A - \nu_B = 20$
$J_{AB} = 2$
$J_{AX} = 1$
$J_{BX} = -4$

48

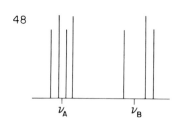

$$\nu_A - \nu_B = 20$$
$$J_{AB} = 2$$
$$J_{AX} = 4$$
$$J_{BX} = 8$$

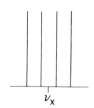

49

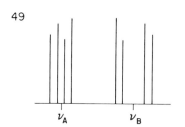

$$\nu_A - \nu_B = 20$$
$$J_{AB} = 2$$
$$J_{AX} = 4$$
$$J_{BX} = -8$$

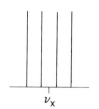

50

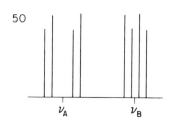

$$\nu_A - \nu_B = 20$$
$$J_{AB} = 2$$
$$J_{AX} = 8$$
$$J_{BX} = 4$$

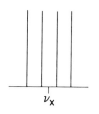

51

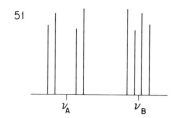

$$\nu_A - \nu_B = 20$$
$$J_{AB} = 2$$
$$J_{AX} = 8$$
$$J_{BX} = -4$$

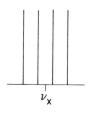

52

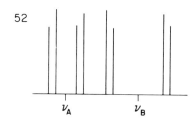

$$\nu_A - \nu_B = 20$$
$$J_{AB} = 2$$
$$J_{AX} = 8$$
$$J_{BX} = 16$$

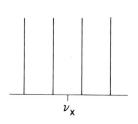

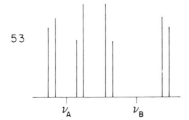

53

$\nu_A - \nu_B = 20$
$J_{AB} = 2$
$J_{AX} = -8$
$J_{BX} = 16$

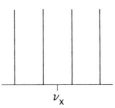

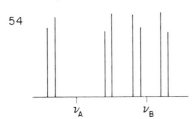

54

$\nu_A - \nu_B = 20$
$J_{AB} = 2$
$J_{AX} = 16$
$J_{BX} = 8$

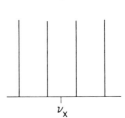

55

$\nu_A - \nu_B = 20$
$J_{AB} = 2$
$J_{AX} = 16$
$J_{BX} = -8$

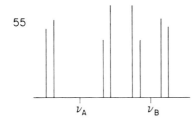

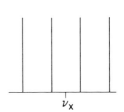

56

$\nu_A - \nu_B = 10$
$J_{AB} = 10$
$J_{AX} = 1$
$J_{BX} = 2$

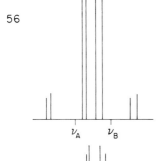

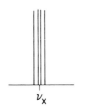

57

$\nu_A - \nu_B = 10$
$J_{AB} = 10$
$J_{AX} = 1$
$J_{BX} = -2$

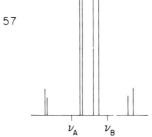

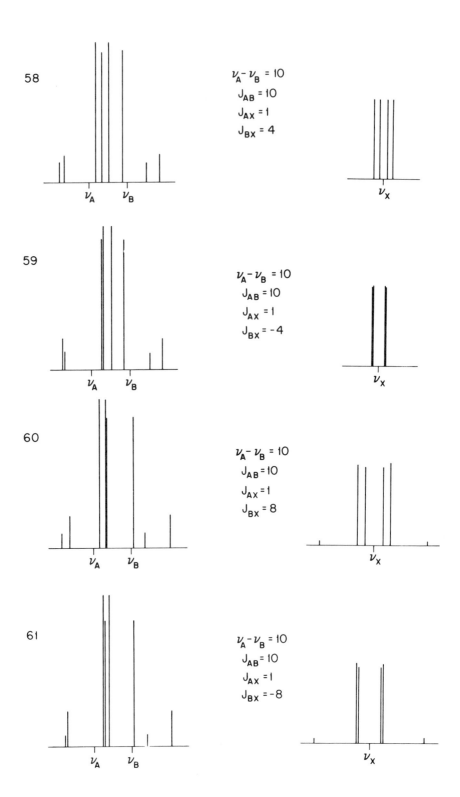

58

$\nu_A - \nu_B = 10$
$J_{AB} = 10$
$J_{AX} = 1$
$J_{BX} = 4$

ν_A ν_B

ν_X

59

$\nu_A - \nu_B = 10$
$J_{AB} = 10$
$J_{AX} = 1$
$J_{BX} = -4$

ν_A ν_B

ν_X

60

$\nu_A - \nu_B = 10$
$J_{AB} = 10$
$J_{AX} = 1$
$J_{BX} = 8$

ν_A ν_B

ν_X

61

$\nu_A - \nu_B = 10$
$J_{AB} = 10$
$J_{AX} = 1$
$J_{BX} = -8$

ν_A ν_B

ν_X

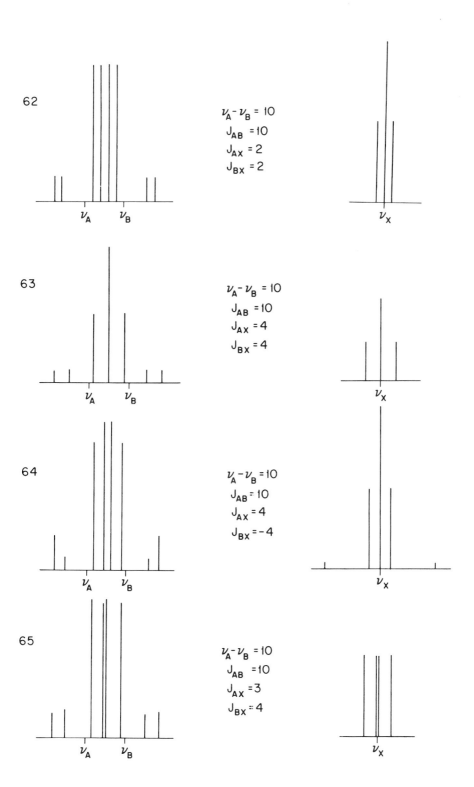

62

$\nu_A - \nu_B = 10$
$J_{AB} = 10$
$J_{AX} = 2$
$J_{BX} = 2$

ν_A ν_B

ν_X

63

$\nu_A - \nu_B = 10$
$J_{AB} = 10$
$J_{AX} = 4$
$J_{BX} = 4$

ν_A ν_B

ν_X

64

$\nu_A - \nu_B = 10$
$J_{AB} = 10$
$J_{AX} = 4$
$J_{BX} = -4$

ν_A ν_B

ν_X

65

$\nu_A - \nu_B = 10$
$J_{AB} = 10$
$J_{AX} = 3$
$J_{BX} = 4$

ν_A ν_B

ν_X

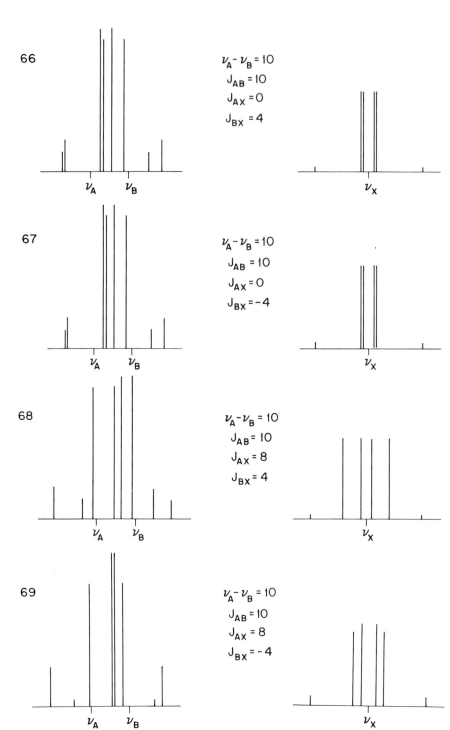

66

$\nu_A - \nu_B = 10$
$J_{AB} = 10$
$J_{AX} = 0$
$J_{BX} = 4$

ν_A ν_B ν_X

67

$\nu_A - \nu_B = 10$
$J_{AB} = 10$
$J_{AX} = 0$
$J_{BX} = -4$

ν_A ν_B ν_X

68

$\nu_A - \nu_B = 10$
$J_{AB} = 10$
$J_{AX} = 8$
$J_{BX} = 4$

ν_A ν_B ν_X

69

$\nu_A - \nu_B = 10$
$J_{AB} = 10$
$J_{AX} = 8$
$J_{BX} = -4$

ν_A ν_B ν_X

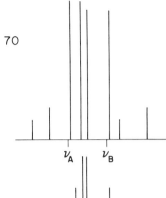

70

$\nu_A - \nu_B = 10$
$J_{AB} = 10$
$J_{AX} = 4$
$J_{BX} = 8$

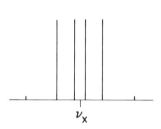

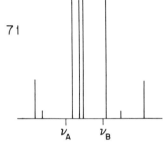

71

$\nu_A - \nu_B = 10$
$J_{AB} = 10$
$J_{AX} = -4$
$J_{BX} = 8$

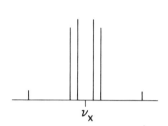

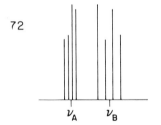

72

$\nu_A - \nu_B = 10$
$J_{AB} = 2$
$J_{AX} = 1$
$J_{BX} = 4$

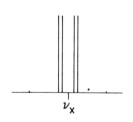

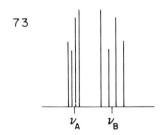

73

$\nu_A - \nu_B = 10$
$J_{AB} = 2$
$J_{AX} = 1$
$J_{BX} = -4$

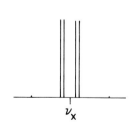

74

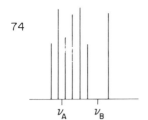

$$\nu_A - \nu_B = 10$$
$$J_{AB} = 2$$
$$J_{AX} = 4$$
$$J_{BX} = 8$$

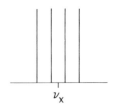

75

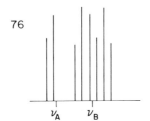

$$\nu_A - \nu_B = 10$$
$$J_{AB} = 2$$
$$J_{AX} = 4$$
$$J_{BX} = -8$$

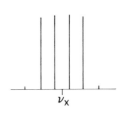

76

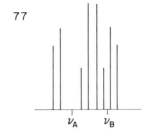

$$\nu_A - \nu_B = 10$$
$$J_{AB} = 2$$
$$J_{AX} = 8$$
$$J_{BX} = 4$$

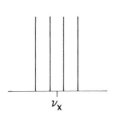

77

$$\nu_A - \nu_B = 10$$
$$J_{AB} = 2$$
$$J_{AX} = 8$$
$$J_{BX} = -4$$

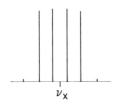

78

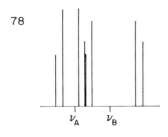

$$\nu_A - \nu_B = 10$$
$$J_{AB} = 2$$
$$J_{AX} = 8$$
$$J_{BX} = 16$$

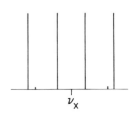

79

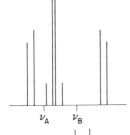

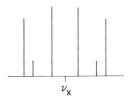

$\nu_A - \nu_B = 10$
$J_{AB} = 2$
$J_{AX} = -8$
$J_{BX} = 16$

80

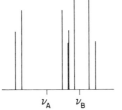

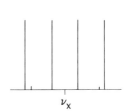

$\nu_A - \nu_B = 10$
$J_{AB} = 2$
$J_{AX} = 16$
$J_{BX} = 8$

81

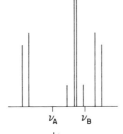

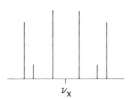

$\nu_A - \nu_B = 10$
$J_{AB} = 2$
$J_{AX} = 16$
$J_{BX} = -8$

82

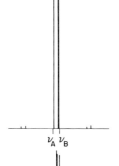

$\nu_A - \nu_B = 2$
$J_{AB} = 10$
$J_{AX} = 1$
$J_{BX} = 2$

83

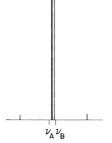

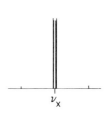

$\nu_A - \nu_B = 2$
$J_{AB} = 10$
$J_{AX} = 1$
$J_{BX} = -2$

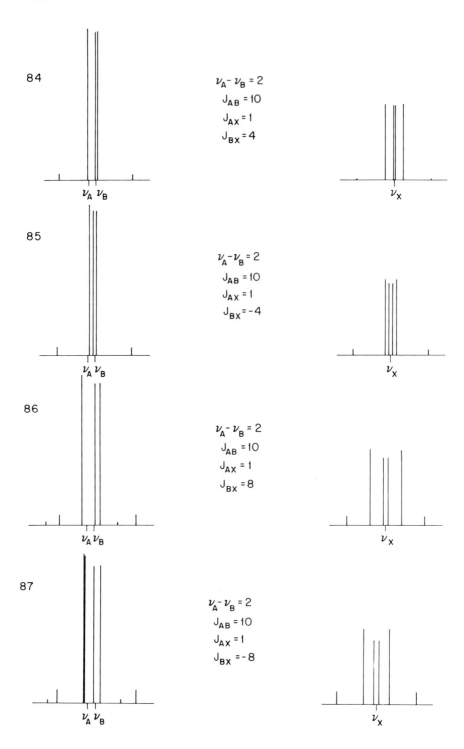

84

$\nu_A - \nu_B = 2$
$J_{AB} = 10$
$J_{AX} = 1$
$J_{BX} = 4$

$\nu_A \ \nu_B$ ν_X

85

$\nu_A - \nu_B = 2$
$J_{AB} = 10$
$J_{AX} = 1$
$J_{BX} = -4$

$\nu_A \ \nu_B$ ν_X

86

$\nu_A - \nu_B = 2$
$J_{AB} = 10$
$J_{AX} = 1$
$J_{BX} = 8$

$\nu_A \ \nu_B$ ν_X

87

$\nu_A - \nu_B = 2$
$J_{AB} = 10$
$J_{AX} = 1$
$J_{BX} = -8$

$\nu_A \ \nu_B$ ν_X

88

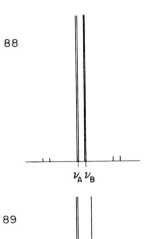

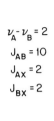

$\nu_A - \nu_B = 2$
$J_{AB} = 10$
$J_{AX} = 2$
$J_{BX} = 2$

89

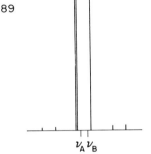

$\nu_A - \nu_B = 2$
$J_{AB} = 10$
$J_{AX} = 4$
$J_{BX} = 4$

90

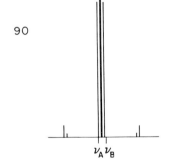

$\nu_A - \nu_B = 2$
$J_{AB} = 10$
$J_{AX} = 4$
$J_{BX} = -4$

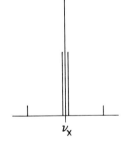

91

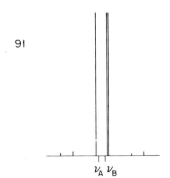

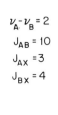

$\nu_A - \nu_B = 2$
$J_{AB} = 10$
$J_{AX} = 3$
$J_{BX} = 4$

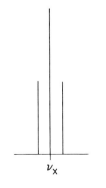

APPENDIX B

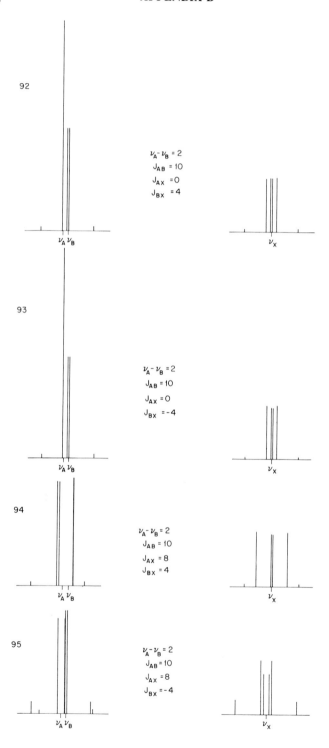

92

$\nu_A - \nu_B = 2$
$J_{AB} = 10$
$J_{AX} = 0$
$J_{BX} = 4$

ν_A ν_B

ν_X

93

$\nu_A - \nu_B = 2$
$J_{AB} = 10$
$J_{AX} = 0$
$J_{BX} = -4$

ν_A ν_B

ν_X

94

$\nu_A - \nu_B = 2$
$J_{AB} = 10$
$J_{AX} = 8$
$J_{BX} = 4$

ν_A ν_B

ν_X

95

$\nu_A - \nu_B = 2$
$J_{AB} = 10$
$J_{AX} = 8$
$J_{BX} = -4$

ν_A ν_B

ν_X

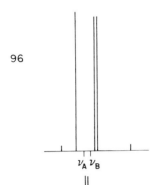

96

$\nu_A - \nu_B = 2$
$J_{AB} = 10$
$J_{AX} = 4$
$J_{BX} = 8$

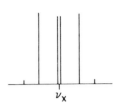

ν_A ν_B ν_X

97

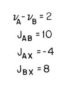

$\nu_A - \nu_B = 2$
$J_{AB} = 10$
$J_{AX} = -4$
$J_{BX} = 8$

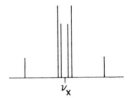

ν_A ν_B ν_X

98

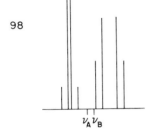

$\nu_A - \nu_B = 2$
$J_{AB} = 2$
$J_{AX} = 8$
$J_{BX} = 16$

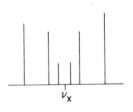

ν_A ν_B ν_X

99

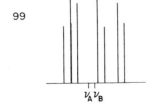

$\nu_A - \nu_B = 2$
$J_{AB} = 2$
$J_{AX} = -8$
$J_{BX} = 16$

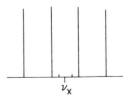

ν_A ν_B ν_X

100

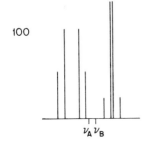

$\nu_A - \nu_B = 2$
$J_{AB} = 2$
$J_{AX} = 16$
$J_{BX} = 8$

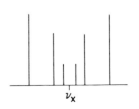

ν_A ν_B ν_X

101

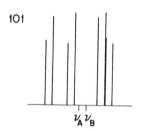

$$\nu_A - \nu_B = 2$$
$$J_{AB} = 2$$
$$J_{AX} = 16$$
$$J_{BX} = -8$$

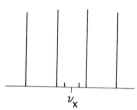

102

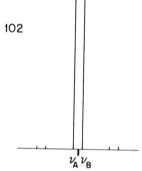

$$\nu_A - \nu_B = 0$$
$$J_{AB} = 10$$
$$J_{AX} = 1$$
$$J_{BX} = 4$$

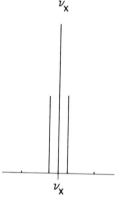

103

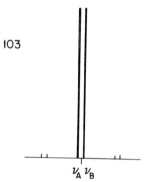

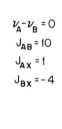

$$\nu_A - \nu_B = 0$$
$$J_{AB} = 10$$
$$J_{AX} = 1$$
$$J_{BX} = -4$$

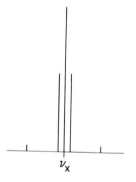

104

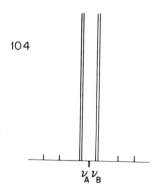

$$\nu_A - \nu_B = 0$$
$$J_{AB} = 10$$
$$J_{AX} = 1$$
$$J_{BX} = 8$$

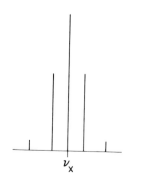

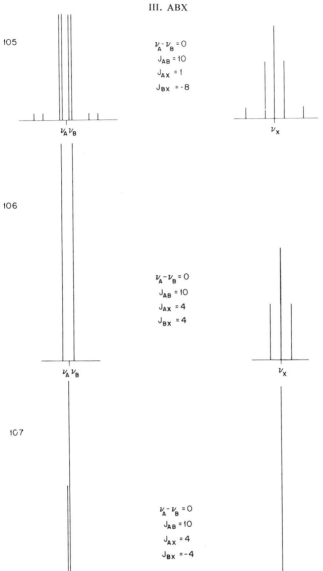

105     $\nu_A - \nu_B = 0$
$J_{AB} = 10$
$J_{AX} = 1$
$J_{BX} = -8$

$\nu_A \nu_B$ ν_X

106     $\nu_A - \nu_B = 0$
$J_{AB} = 10$
$J_{AX} = 4$
$J_{BX} = 4$

$\nu_A \nu_B$ ν_X

107     $\nu_A - \nu_B = 0$
$J_{AB} = 10$
$J_{AX} = 4$
$J_{BX} = -4$

$\nu_A \nu_B$ ν_X

108     $\nu_A - \nu_B = 0$
$J_{AB} = 10$
$J_{AX} = 0$
$J_{BX} = 4$

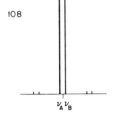
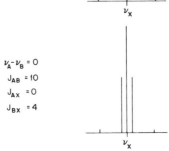

$\nu_A \nu_B$ ν_X

109

$\nu_A - \nu_B = 0$

$J_{AB} = 2$

$J_{AX} = 1$

$J_{BX} = 4$

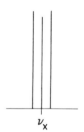

$\nu_A \nu_B$ ν_X

110

$\nu_A - \nu_B = 0$

$J_{AB} = 2$

$J_{AX} = 1$

$J_{BX} = 8$

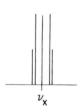

$\nu_A \nu_B$ ν_X

111

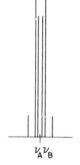

$\nu_A - \nu_B = 0$

$J_{AB} = 2$

$J_{AX} = 0$

$J_{BX} = 4$

$\nu_A \nu_B$ ν_X

112

$\nu_A - \nu_B = 0$

$J_{AB} = 0$

$J_{AX} = 1$

$J_{BX} = 4$

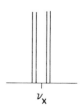

$\nu_A \nu_B$ ν_X

IV. ABC

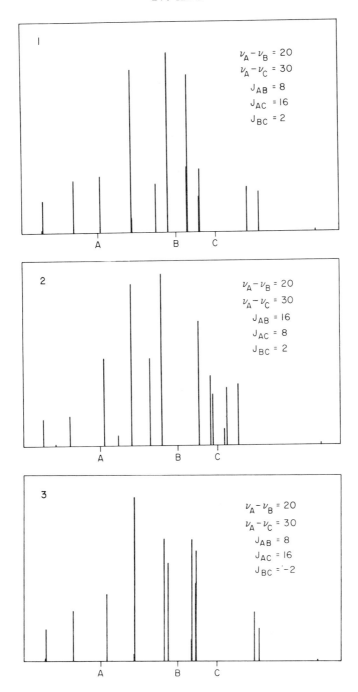

1

$\nu_A - \nu_B = 20$
$\nu_A - \nu_C = 30$
$J_{AB} = 8$
$J_{AC} = 16$
$J_{BC} = 2$

A   B   C

2

$\nu_A - \nu_B = 20$
$\nu_A - \nu_C = 30$
$J_{AB} = 16$
$J_{AC} = 8$
$J_{BC} = 2$

A   B   C

3

$\nu_A - \nu_B = 20$
$\nu_A - \nu_C = 30$
$J_{AB} = 8$
$J_{AC} = 16$
$J_{BC} = -2$

A   B   C

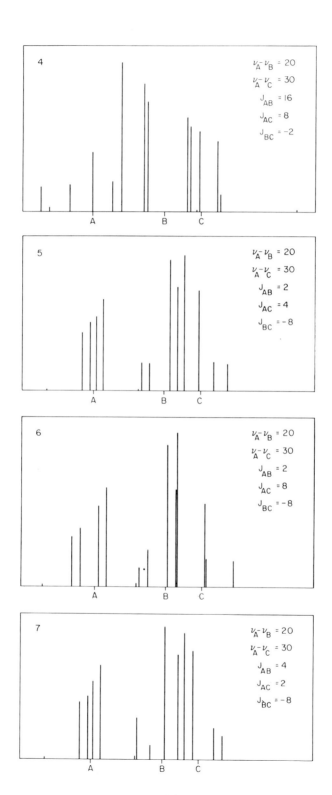

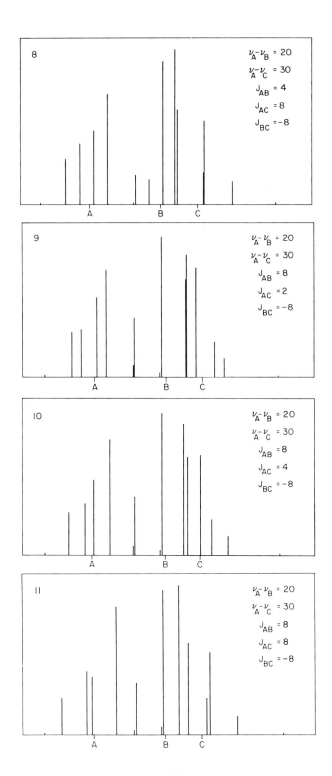

563

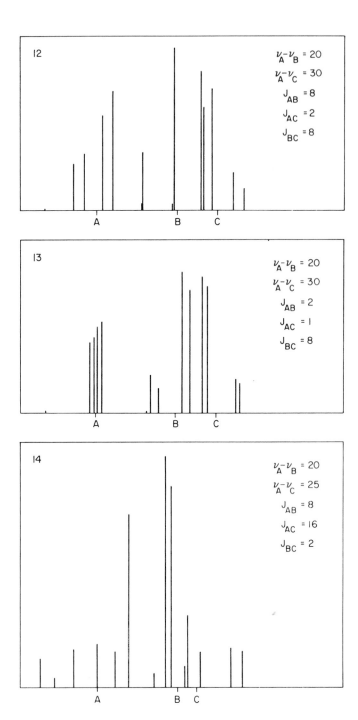

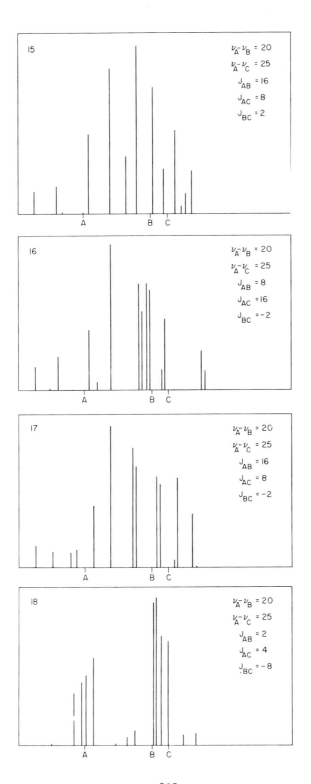

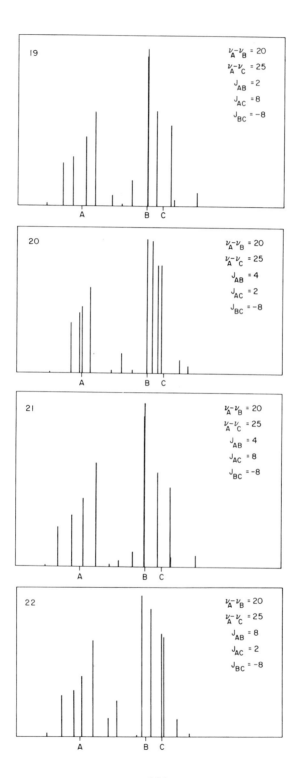

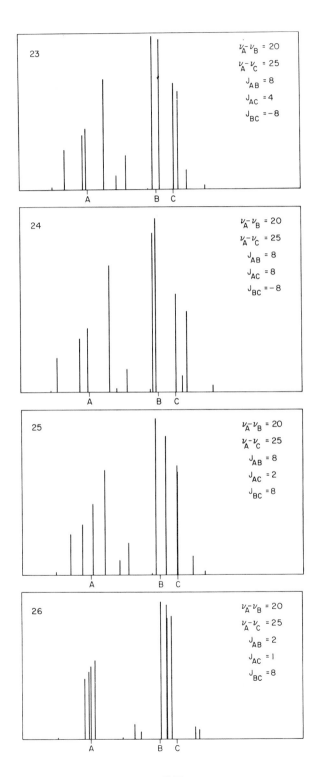

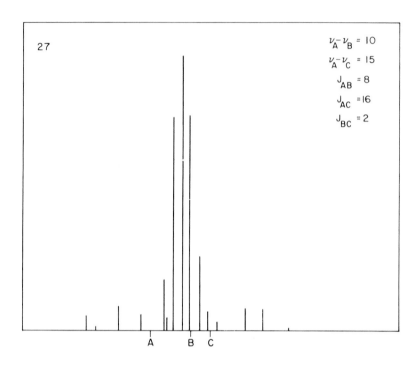

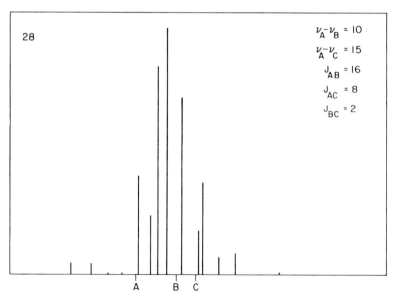

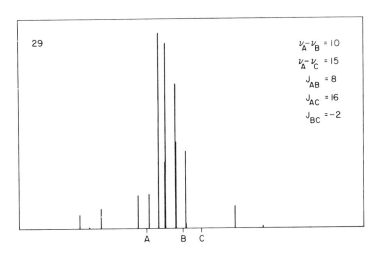

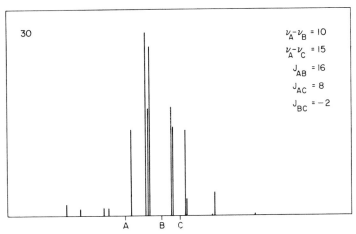

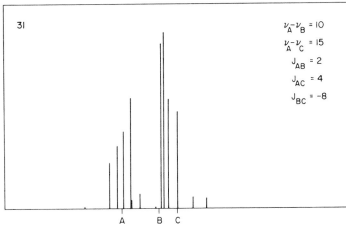

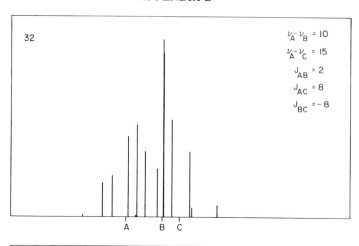

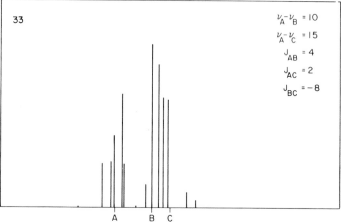

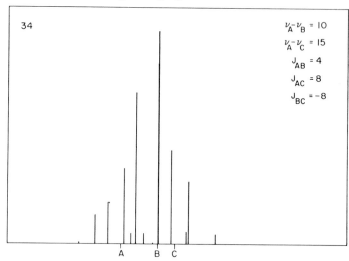

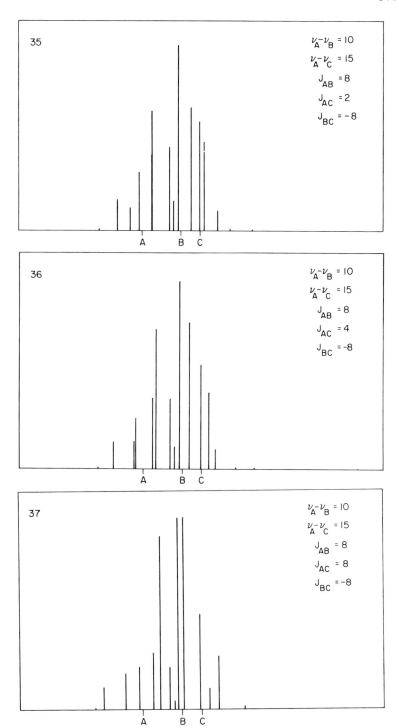

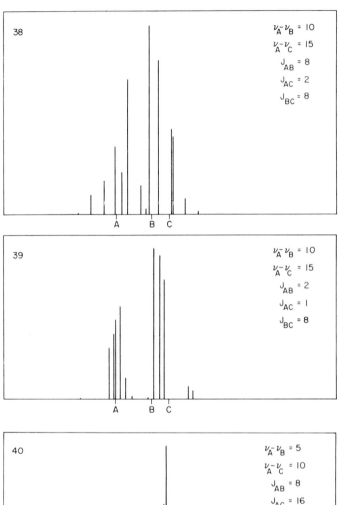

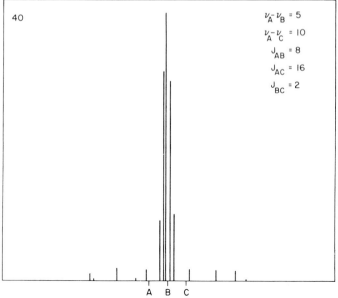

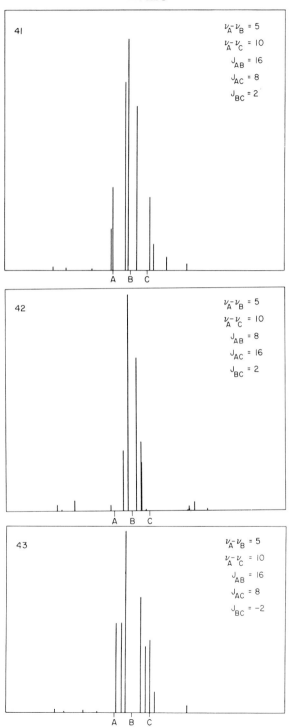

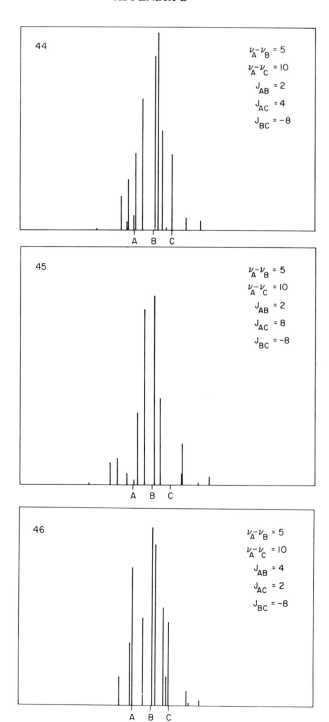

44

$\nu_A - \nu_B = 5$

$\nu_A - \nu_C = 10$

$J_{AB} = 2$

$J_{AC} = 4$

$J_{BC} = -8$

A   B   C

45

$\nu_A - \nu_B = 5$

$\nu_A - \nu_C = 10$

$J_{AB} = 2$

$J_{AC} = 8$

$J_{BC} = -8$

A   B   C

46

$\nu_A - \nu_B = 5$

$\nu_A - \nu_C = 10$

$J_{AB} = 4$

$J_{AC} = 2$

$J_{BC} = -8$

A   B   C

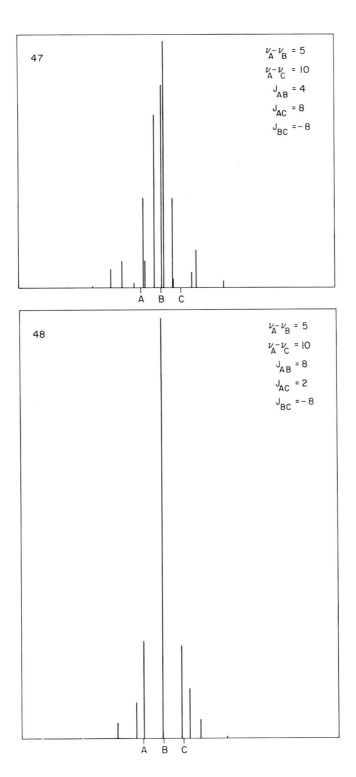

47

$$\nu_A - \nu_B = 5$$
$$\nu_A - \nu_C = 10$$
$$J_{AB} = 4$$
$$J_{AC} = 8$$
$$J_{BC} = -8$$

A B C

48

$$\nu_A - \nu_B = 5$$
$$\nu_A - \nu_C = 10$$
$$J_{AB} = 8$$
$$J_{AC} = 2$$
$$J_{BC} = -8$$

A B C

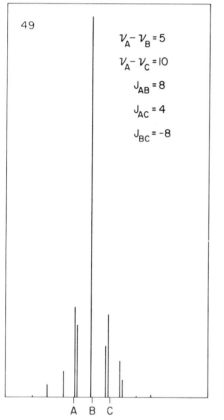

49

$$\nu_A - \nu_B = 5$$
$$\nu_A - \nu_C = 10$$
$$J_{AB} = 8$$
$$J_{AC} = 4$$
$$J_{BC} = -8$$

A B C

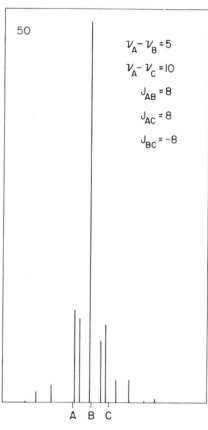

50

$$\nu_A - \nu_B = 5$$
$$\nu_A - \nu_C = 10$$
$$J_{AB} = 8$$
$$J_{AC} = 8$$
$$J_{BC} = -8$$

A B C

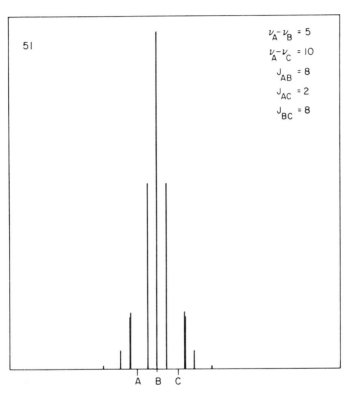

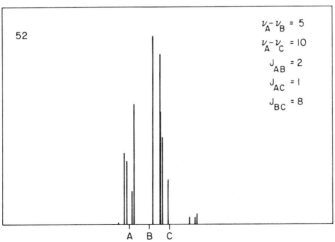

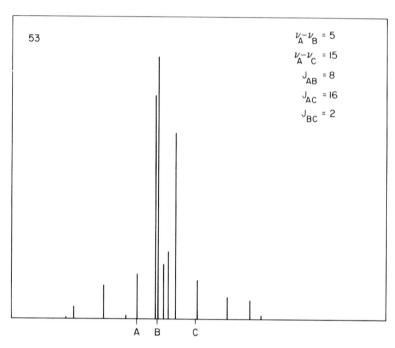

53

$\nu_A - \nu_B = 5$

$\nu_A - \nu_C = 15$

$J_{AB} = 8$

$J_{AC} = 16$

$J_{BC} = 2$

A B   C

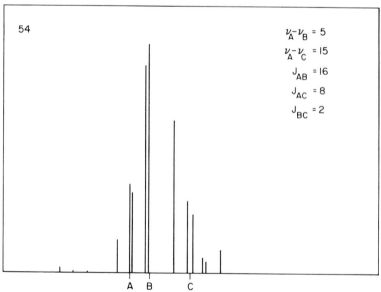

54

$\nu_A - \nu_B = 5$

$\nu_A - \nu_C = 15$

$J_{AB} = 16$

$J_{AC} = 8$

$J_{BC} = 2$

A B   C

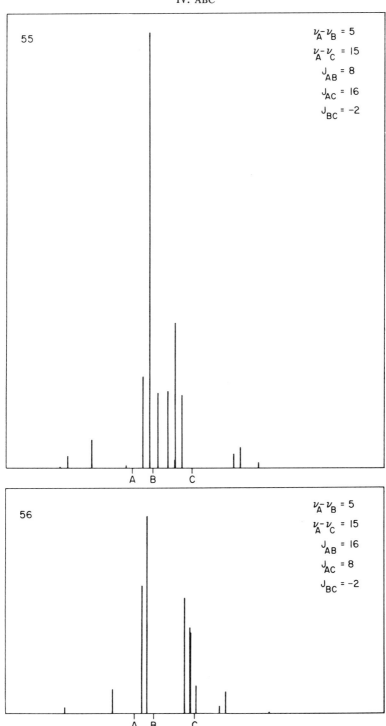

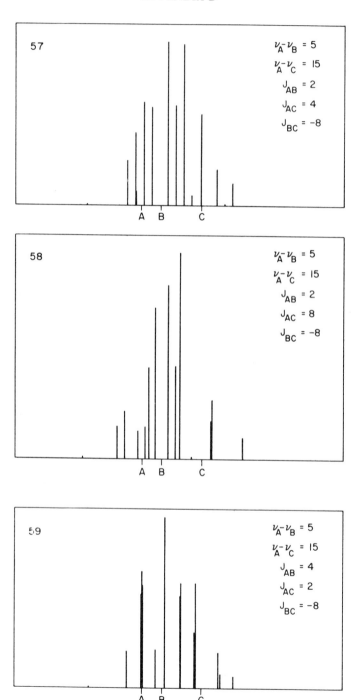

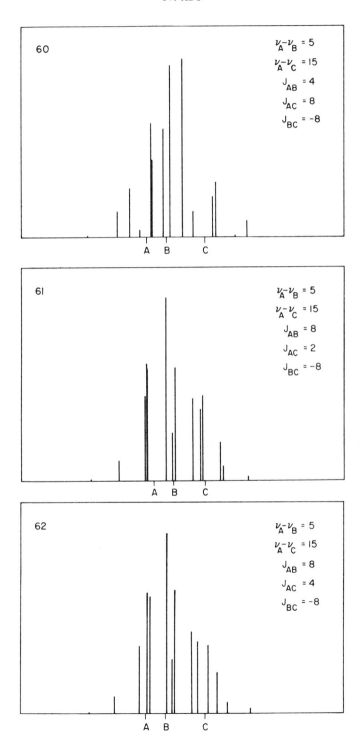

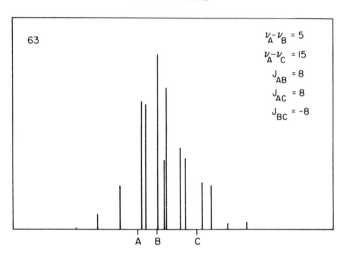

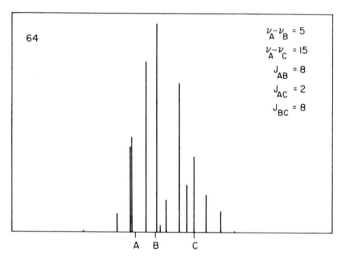

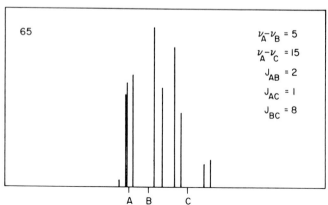

IV. ABC

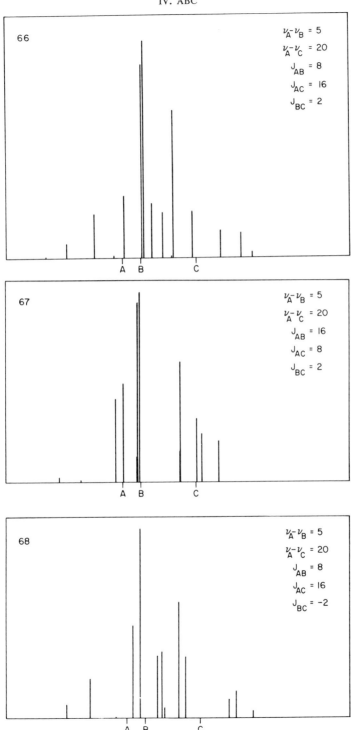

66

$\nu_A - \nu_B = 5$

$\nu_A - \nu_C = 20$

$J_{AB} = 8$

$J_{AC} = 16$

$J_{BC} = 2$

67

$\nu_A - \nu_B = 5$

$\nu_A - \nu_C = 20$

$J_{AB} = 16$

$J_{AC} = 8$

$J_{BC} = 2$

68

$\nu_A - \nu_B = 5$

$\nu_A - \nu_C = 20$

$J_{AB} = 8$

$J_{AC} = 16$

$J_{BC} = -2$

583

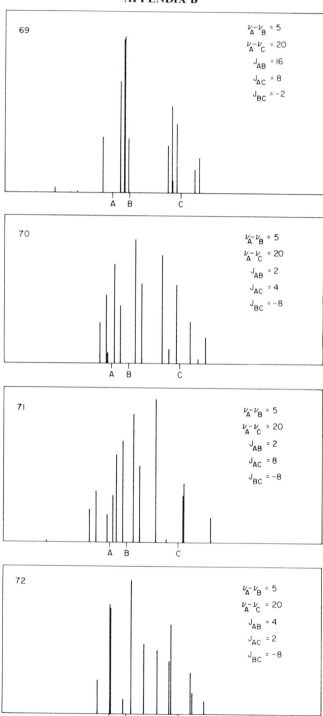

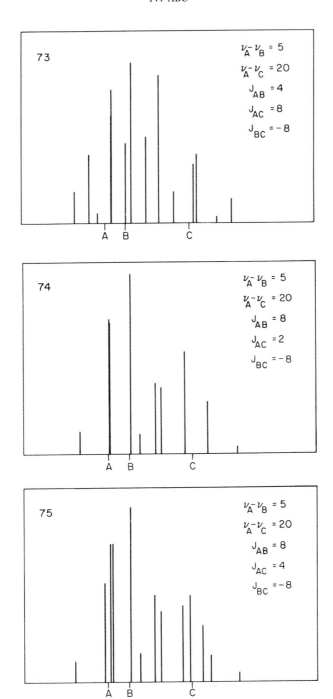

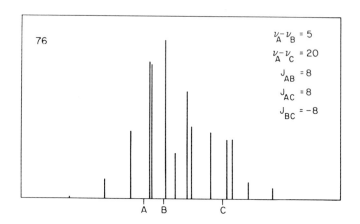

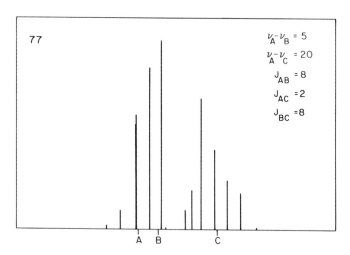

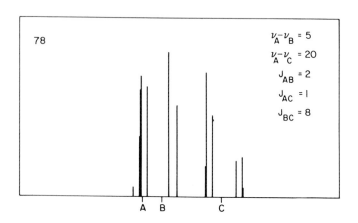

V. AA′BB′

GROUP I. $J_{AA} = 0$ $K = 0$

$J_{BB} = 0$ $M = 0$

$J_{AB} = 10$ $N = 20$

$J_{AB'} = 10$ $L = 0$

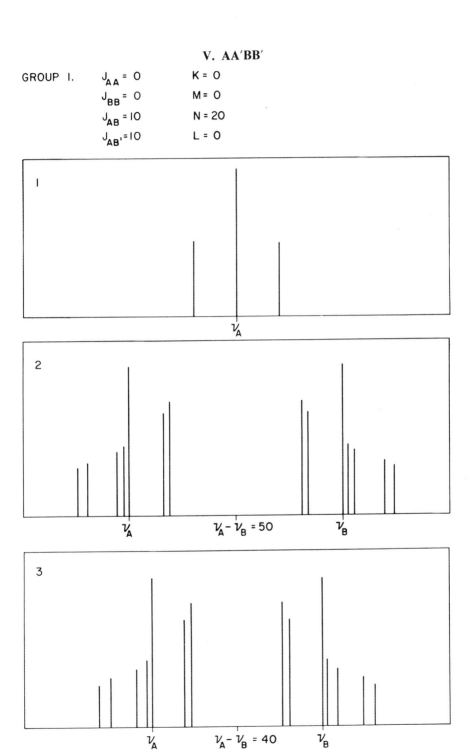

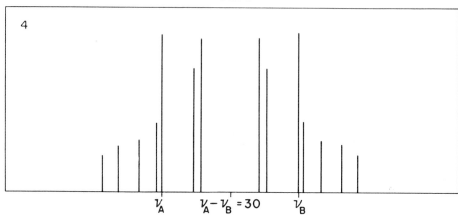

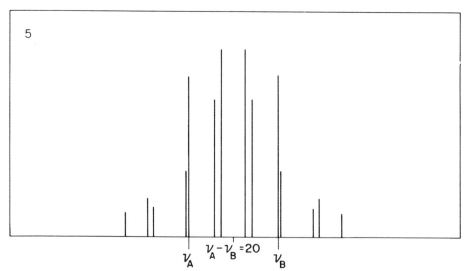

GROUP 2. J_{AA} = 0.5   K = 8
             J_{BB} = 7.5   M = –7
             J_{AB} = 8.5   N = 11
             $J_{AB'}$ = 2.5   L = 6

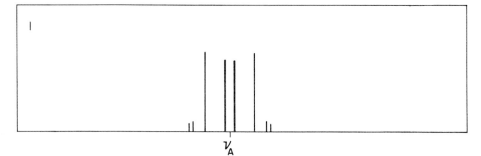

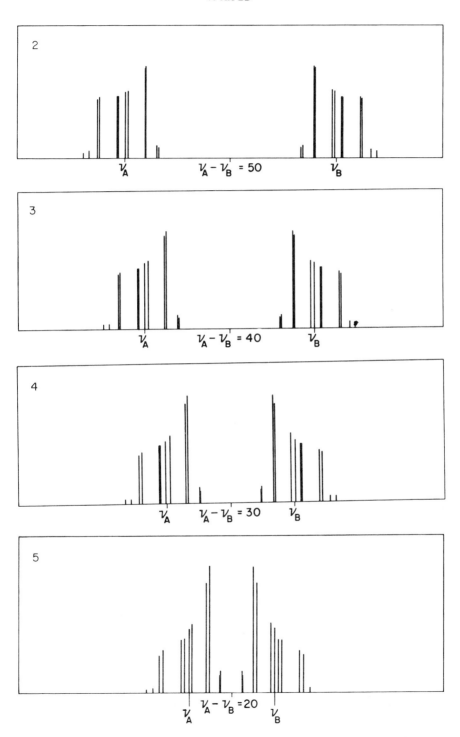

APPENDIX B

GROUP 3a. J_{AA} = 2.5   K = 5
          J_{BB} = 2.5   M = 0
          J_{AB} = 8.5   N = 9
          $J_{AB'}$ = 0.5   L = 8

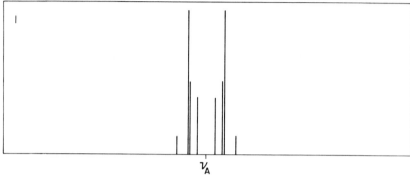

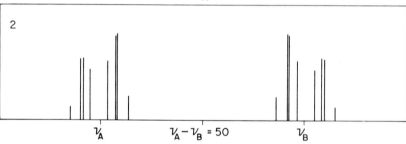

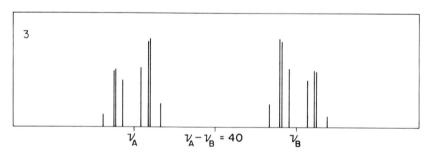

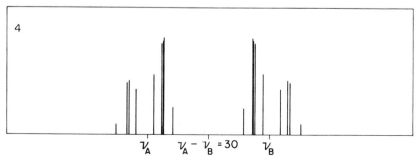

V. AA′BB′

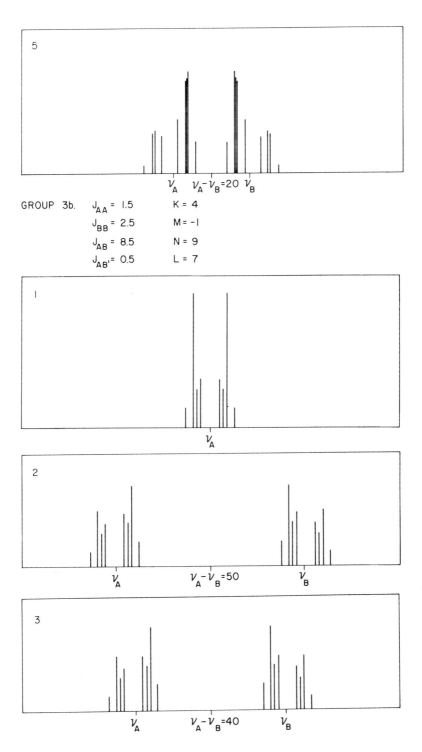

GROUP 3b.　J_{AA} = 1.5　　K = 4
　　　　　J_{BB} = 2.5　　M = -1
　　　　　J_{AB} = 8.5　　N = 9
　　　　　$J_{AB'}$= 0.5　　L = 7

591

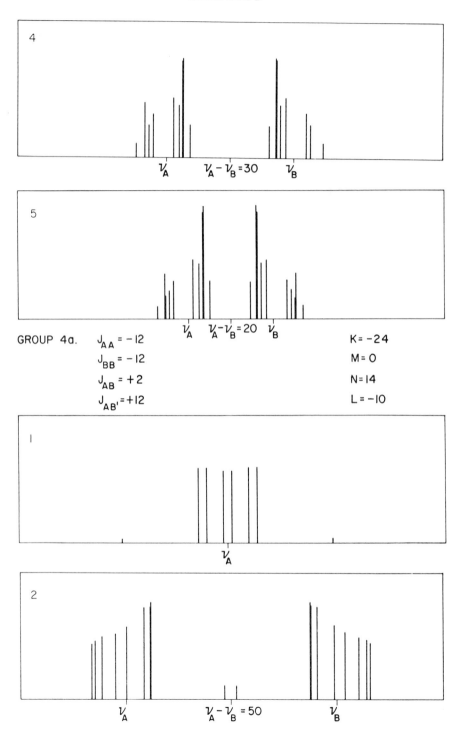

GROUP 4a.

$J_{AA} = -12$

$J_{BB} = -12$

$J_{AB} = +2$

$J_{AB'} = +12$

$K = -24$

$M = 0$

$N = 14$

$L = -10$

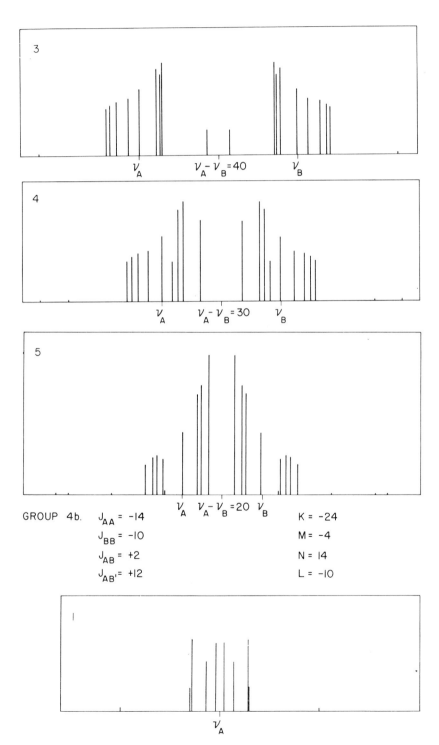

GROUP 4b.

$J_{AA} = -14$

$J_{BB} = -10$

$J_{AB} = +2$

$J_{AB'} = +12$

$K = -24$

$M = -4$

$N = 14$

$L = -10$

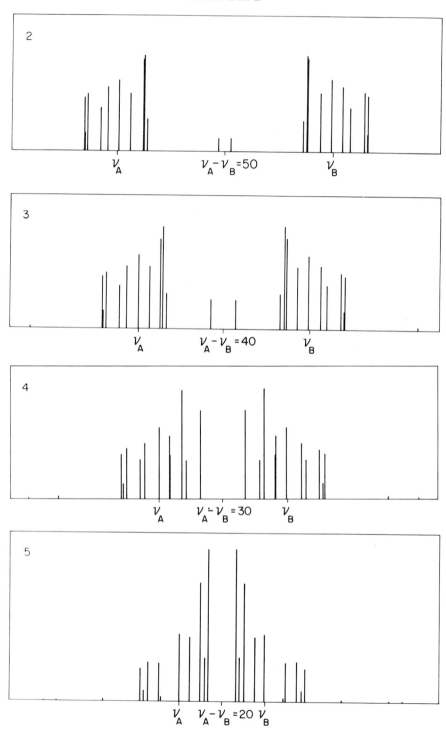

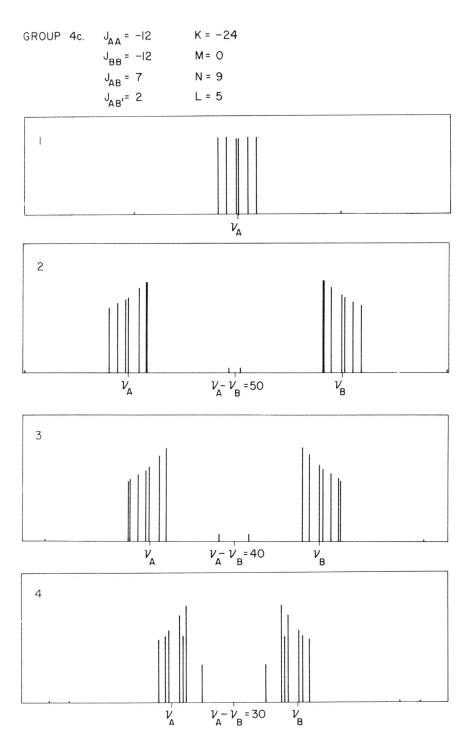

GROUP 4c. $J_{AA} = -12$ $K = -24$
            $J_{BB} = -12$ $M = 0$
            $J_{AB} = 7$ $N = 9$
            $J_{AB'} = 2$ $L = 5$

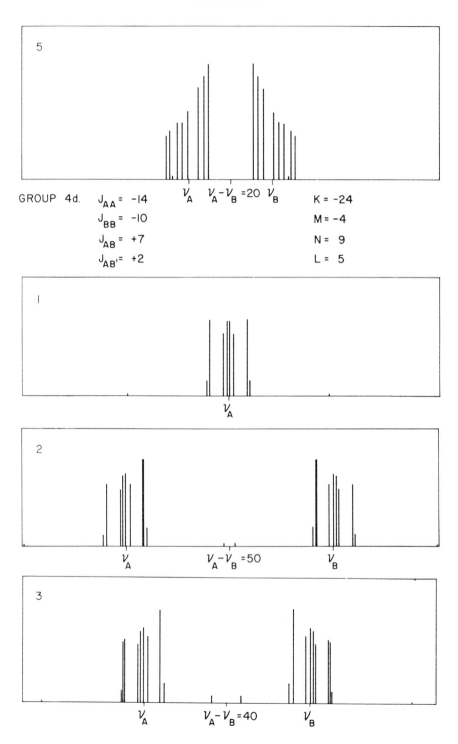

GROUP 4d. $J_{AA} = -14$

$J_{BB} = -10$

$J_{AB} = +7$

$J_{AB'} = +2$

$K = -24$

$M = -4$

$N = 9$

$L = 5$

V. AA′BB′

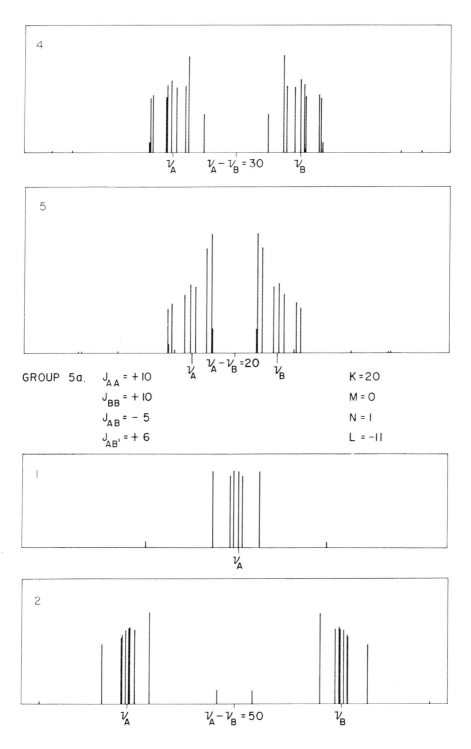

GROUP 5a. $J_{AA} = +10$ $K = 20$
            $J_{BB} = +10$ $M = 0$
            $J_{AB} = -5$ $N = 1$
            $J_{AB'} = +6$ $L = -11$

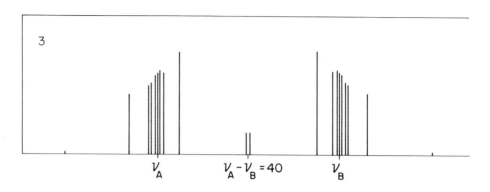

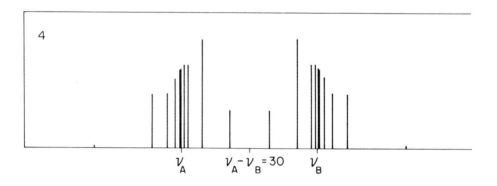

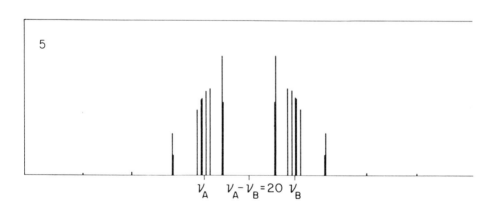

V. AA'BB'

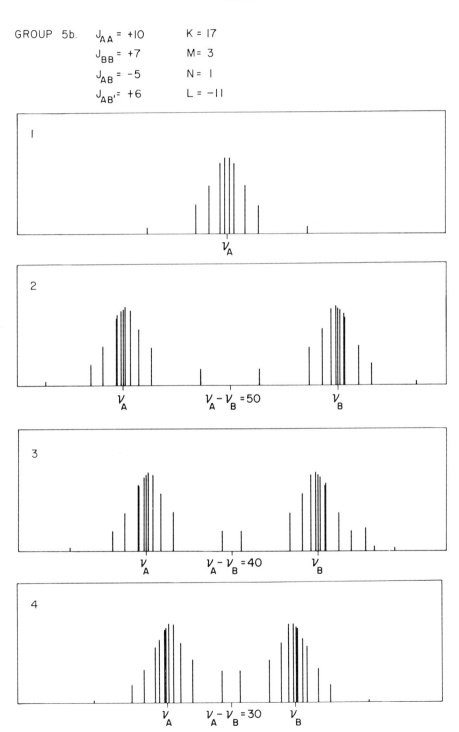

GROUP 5b. $J_{AA} = +10$ K = 17

                 $J_{BB} = +7$ M = 3

                 $J_{AB} = -5$ N = 1

                 $J_{AB'} = +6$ L = -11

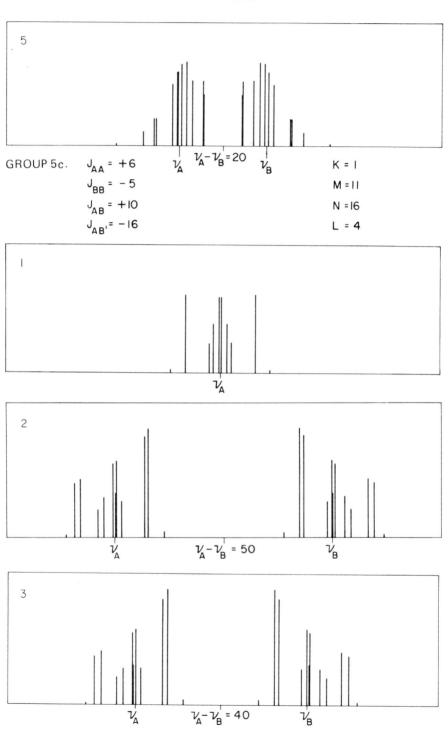

GROUP 5c. $J_{AA} = +6$ $K = 1$
$J_{BB} = -5$ $M = 11$
$J_{AB} = +10$ $N = 16$
$J_{AB'} = -16$ $L = 4$

$\nu_A - \nu_B = 20$

$\nu_A - \nu_B = 50$

$\nu_A - \nu_B = 40$

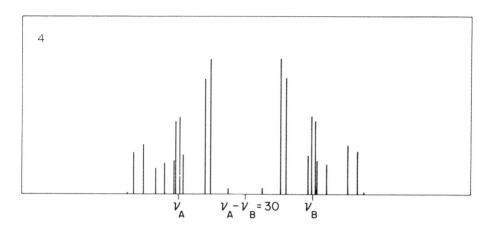

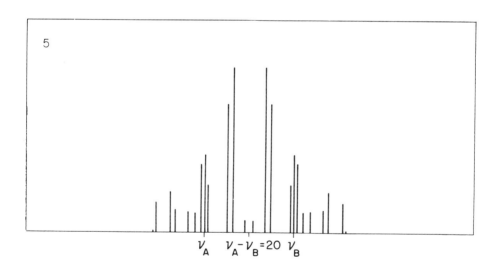

VI. A₂B₃

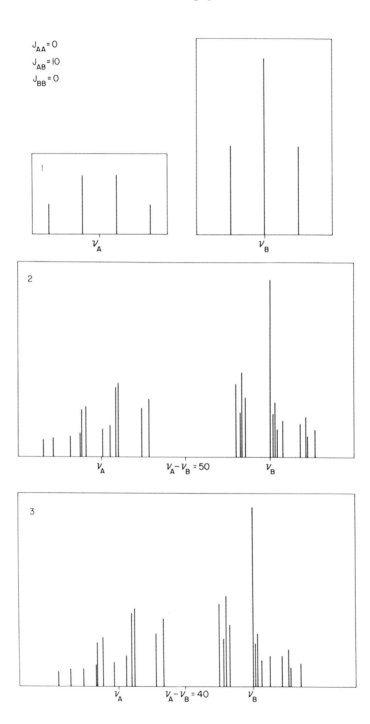

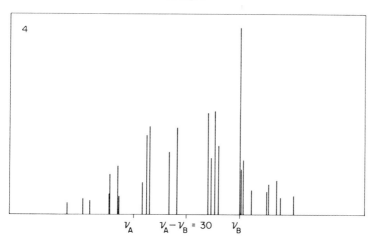

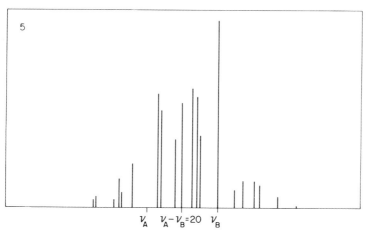

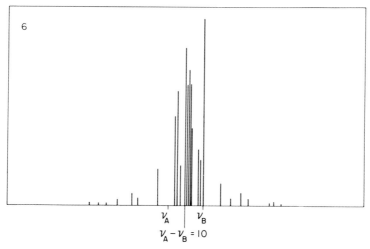

Appendix C

PROTON SPIN-SPIN COUPLING CONSTANTS

In this tabulated material, proton-proton coupling constants are given for specific compounds and for typical structures. For specific compounds, a literature reference is given in the fourth column; this refers to the bibliography at the end of the table. For typical structures, the range of measured values is indicated. For both, the protons involved are specifically indicated in the third column, where necessary. The sign of the coupling is given as + or as − for specific compounds when this is known. If not actually measured but strongly indicated by analogy, the sign is given as (+) or (−). If not determined and not strongly indicated by analogy, the sign is indicated as ±.

For olefinic compounds (entries 102–136), geminal, vicinal, and allylic couplings are given together for convenience.

A. Geminal Couplings (Geminal Couplings for Ethylenic Compounds are Given in Section B)

No.	Compound or structural type	J_{gem} (cps)	Ref.[a]
1	CH_4	$(-)12.4$	1–3
2	CH_3CCl_3	$(-)13.0$	1–3
3	CH_3I	$(-)\ 9.2$	4
4	CH_3Br	$(-)10.2$	5
5	CH_3Cl	$(-)10.8$	4
6	CH_3F	$(-)\ 9.6$	4
7	CH_2Cl_2	$(-)\ 7.5$	4
8	CH_3CN	$(-)16.9$	1–3
9	$NC—CH_2—CN$	$(-)20.4$	1–3
10	CH_3CO_2H	$(-)14.6$	1–3
11	CH_3NO_2	$(-)13.2$	1
12	CH_3COCH_3	$(-)14.9$	1–3
13	CH_3OH	$(-)10.8$	4, 6
14	$C_6H_5 \cdot CH_3$	$(-)14.4$	1
15	![pyridine with CH₃] N⟩—CH_3	$(-)14.5$	7
16	$(CH_3{-}\overset{H_{(a)}}{\underset{H_{(b)}}{C}}{-}O)_2SO$	$H_{(a)}—H_{(b)}: -10.45$	8
17	$(CH_3{-}\overset{H_{(a)}}{\underset{H_{(b)}}{C}}{-}O)_2CHCH_3$	$H_{(a)}—H_{(b)}: -9.3$	8
18	$\overset{H}{\underset{H}{>}}C{=}N{\underset{OH}{\diagdown}}$	$(-)7.63{-}9.95$ (solvent dependent)	9, 10
19	$\overset{H}{\underset{H}{>}}C{=}N{\underset{OCH_3}{\diagdown}}$	$(-)6.96{-}9.22$ (solvent dependent)	9, 10
20	$\overset{H}{\underset{H}{>}}C{=}O$	$+40.22{-}42.42$ (solvent dependent)	11
21	![cyclopropane with 2 H]	$-3.9{-}8.8$	—
22	![cyclopropane with 2 CH₃ and 3 H]	-4.5	12

[a] Key to references can be found on page 626.

No.	Compound or structural type	J_{gem} (cps)	Ref.
23	Cl, Cl	-6.0	12
24	O	$+5.4–6.3$	—
25	O	$+5.5$	13
26	CH_3, O	$+5.4$	14, 16
27	CN, O	$+5.53$	15
28	CH_3CO, O	$+5.78$	15
29	HO_2C, O	$+6.28$	15
30	C_6H_5, O	$+5.65$	15
31	NH	$+1.5$	13
32	CH_3, NH	$+2.0, +3.4$	16
33	CH_3, S	$|J| < 0.4$	17

A. GEMINAL COUPLINGS—(*Continued*)

No.	Compound or structural type	J_{gem} (cps)	Ref.
34		$H_{(a)}$—$H_{(a')}$: -10.92 $H_{(b)}$—$H_{(b')}$: -15.31	18
35		$H_{(a)}$—$H_{(b)}$: $(-)13.7$	19
36		$H_{(2)}$: -17.3 $H_{(3)}$ -15.5	20
37		-21.5	1–3
38		± 18.2 to ± 19.8	—
39		± 1.5	—
40		$H_{(a)}$—$H_{(b)}$: -8.3	21
41		$H_{(3)}$ *cis*: $(-)14.23$ $H_{(3)}$ *trans*: $(-)14.06$ $H_{(5)}$ *cis*: -12.50 $H_{(5)}$ *trans*: -12.69	22

607

No.	Compound or structural type	J_{gem} (cps)	Ref.
42		-10 $(-)12.6$	23 24
43		$H_{(ax)}$—$H_{(eb)}$: -12.2	25
44		$H_{(a)}$—$H_{(b)}$: $(-)5.4$ to $(-)8.4$	26
45		$H_{(a)}$—$H_{(b)}$: $(-)5.4$	26
46		$H_{(a)}$—$H_{(b)}$: $(-)6.6$	26
47		$H_{(a)}$—$H_{(b)}$: $(-)8.4$	26

No.	Compound or structural type	J_{gem} (cps)	Ref.
48		$H_{(6)}$: -12.5	27
49		$H_{(1)}$: -13.5	27
50		$H_{(2)}$: -14.0	27
51		ca. $+10$	28
52		$H_{(a)}$—$H_{(b)}$: $(-)20$	23
53		CH_2: $(-)14.9$	29
54	 (fixed conformation)	$H_{(a)}$—$H_{(b)}$: -12.0	30

B. VICINAL COUPLINGS

No.	Compound or structural type	J_{vic}	Ref.
55	CH_3CH_2-	$(+)4.7$ to ca. $(+)9.0$	—
56	CH_3CH_3	$+8.0$	31, 32
57	$CH_3CH_2O^+(Et)_2$	$(+)4.7$	33
58	$CH_3CH_2B(Et)_3$	$(+)6.8$	34
59	$CH_3CH_2OCH_2CH_3$	$(+)6.97$	35
60	CH_3CH_2OAc	$(+)7.12$	35
61	$(CH_3CH_2)_3N$	$(+)7.13$	35
62	$CH_3CH_2 \cdot C\equiv CH$	$(+)7.2$	36
63	CH_3CH_2Cl	$(+)7.23$	35
64	$CH_3CH_2CH_3$	$(+)7.26$	37
65	$CH_3CH_2COCH=CH_2$	$(+)7.32$	35
66	CH_3CH_2Br	$(+)7.33$	35
67	$CH_3CH_2N(Et)_2$	$(+)7.4$	34
68	$CH_3CH_2N^+(Et)_3$	$(+)7.4$	34
69	CH_3CH_2I	$(+)7.45$	35
70	$(CH_3CH_2)_4C$	$(+)7.53$	35
71	CH_3CH_2CN	$(+)7.60$	35
72	$CH_3CH_2 \cdot C_6H_5$	$(+)7.62$	35
73	$(CH_3CH_2)_4Si$	$(+)8.0$	36
74	CH_3CH_2Li	$(+)8.90$	38
75	CH_3CH_2MgBr	$(+)8.96$	38
76	$(CH_3)_2CH-$	$(+)6.1$ to $(+)7.0$	—
77	$(CH_3)_2CHOH$	$(+)6.2$	39, 40
78	$(CH_3)_2CHCl$	$(+)6.4$	39, 40
79	$(CH_3)_2CHBr$	$(+)6.5$	39, 40
80	$(CH_3)_2CHI$	$(+)6.6$	39, 40
81	$(CH_3)_2CH \cdot C_6H_5$	$(+)6.9$	39, 40
82	$(CH_3)_2CH \cdot CHO$	$(+)7.0$	39, 40
83	$(CH_3)_3CH$	$(+)6.8$	41
84	CH_3CHCl_2	$(+)6.10$	28, 42
85	CH_3CHBr_2	$(+)6.35$	28, 42
86	CH_3CHF_2	$(+)4.5$	45
87		$J_t: (+)8\text{–}12$ $J_g: (+)2\text{–}4$	—
88	$CHOHCO_2H$ $\mid$ $CHDCO_2H$	$erythro: (+)7.1$ $threo: (+)4.3$	44
89	$Cl \cdot CH_2CH_2 \cdot Cl$	J_{avg} (neat): $(+)6.8 \pm 0.1$ $J_t: (+)18.0 \pm 4.0$ $J_g: (+) 9.2 \pm 1.5$ $[6.8]$	28
90	$Br \cdot CH_2CH_2 \cdot Br$	J_{avg} (neat): $(+)8.30 \pm 0.15$ $J_t: (+)15.3 \pm 1.0$ $J_g: (+) 3.4 \pm 1.0$ $[7.4]$	28

B.   VICINAL COUPLINGS—(*Continued*)

No.	Compound or structural type	J_{vic}	Ref.
91	Cl_2CHCH_2Cl	J_{avg} (neat): $(+)5.9 \pm 0.1$ J_t: $(+)10.5 \pm 3.0$ J_g: $(+) 3.0 \pm 1.5$ [5.5]	28
92	Br_2CHCH_2Br	J_{avg} (neat): $(+)6.65 \pm 0.13$	28
93	$Cl_2CHCHCl_2$	J_{avg} (neat): $(+)3.06 \pm 0.12$ J_t: $(+)16.35 \pm 0.80$ J_g: $(+) 2.01 \pm 0.08$ [8.1]	28
94	$Br_2CHCHBr_2$	J_{avg} (neat): $(+)2.92 \pm 0.12$	14, 28
95	$CHCl_2CHF_2$	J_{avg} (neat): $(+)3.09 \pm 0.03$ J_t: $(+)10.25 \pm 0.40$ J_g: $(+) 2.01 \pm 0.09$ [4.8]	45

meso: structure with Cl, Cl on top; CH3 and CH3 ends; H, H on bottom

No.		J_{vic}	Ref.
96	*meso*	H—H: $+7.8$–8.8 (solvent dependent)	46
	dl	H—H: $(+)3.0$–3.1 (solvent dependent)	46
97	*meso*	H—H: $(+)6.26$–7.39 (solvent dependent)	46, 47
	dl	H—H: $(+)3.28$–3.45 (solvent dependent)	46, 47
98		$+10.7$	48
99		$+4.6$, $+10.9$	49

No.	Compound or structural type	J_{vic}	Ref.

100 *meso:*

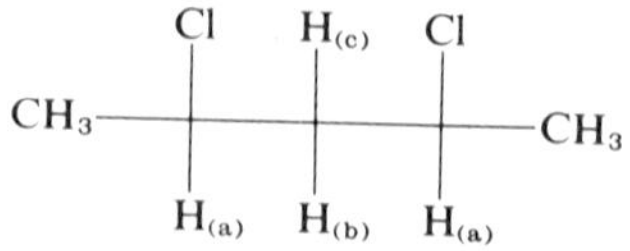

$H_{(a)}$—$H_{(b)}$: ca. (+)7.0
$H_{(a)}$—$H_{(c)}$: ca. (+)7.0

50–52

dl:

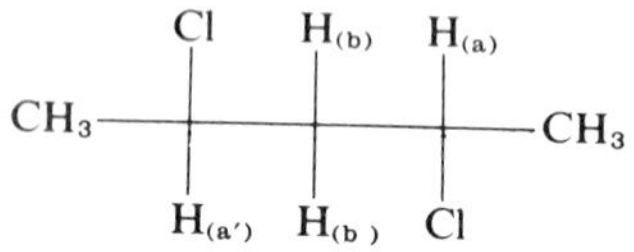

$H_{(a)}$—$H_{(b)}$: (+)2.0–2.9
$H_{(a')}$—$H_{(b')}$: (+)10.6–11.4

50–52

101 *meso:*

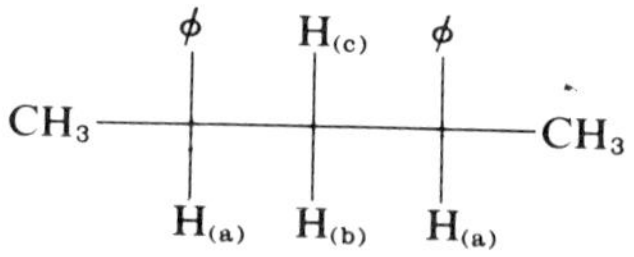

$H_{(a)}$—$H_{(b)} \simeq H_{(a)}$—$H_{(c)}$:
(+)7.45
$H_{(a)}$—CH_3: (+)6.8

53

dl:

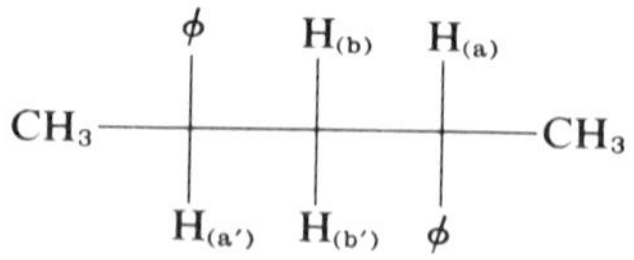

(at 35°; highly temperature
dependent)

$H_{(a)}$—$H_{(b)}$: (+)9.75, (+)5.05
J_t: (+)12.5 ± 0.1
J_g: (+)1.5 ± 0.1

53

C. OLEFINIC COUPLINGS

No.	Compound or structural type	J_{olef}	Ref.

102

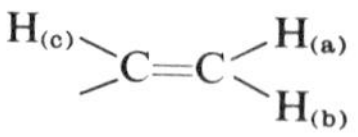

$H_{(a)}$—$H_{(b)}$: − 3.2–+7.4
$H_{(a)}$—$H_{(c)}$: + 4.65–19.3
$H_{(b)}$—$H_{(c)}$: +12.75–23.9

—

103

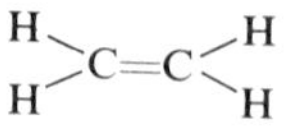

cis: +11.5 ± 0.1*
trans: +19.0 ± 0.1*
gem: + 2.3 ± 0.2*
(asterisk = avg of 3 reports)

31, 32, 54

104

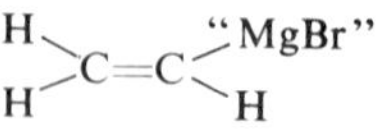

cis: +17.2
trans: +22.1
gem: + 7.4

38

105

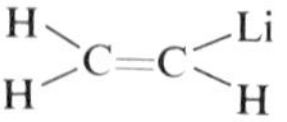

cis: +19.3
trans: +23.9
gem: + 7.1

55

106

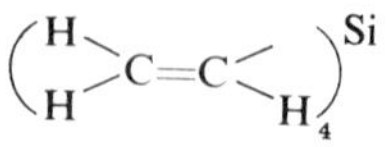

cis: +14.17; +15.33
trans: +19.95; +20.47
gem: + 4.38; +2.70

56, 57

107

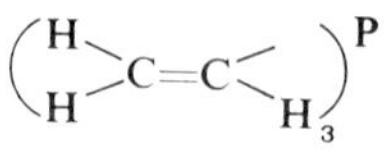

cis: +11.76
trans: +18.37
gem: + 2.02

58

C. OLEFINIC COUPLINGS—(*Continued*)

No.	Compound or structural type	J_{olef}	Ref.
108	$H_2C=CH-SCH_3$	*cis*: +10.3 *trans*: +16.4 *gem*: − 0.3	59
109	$H_2C=CH-CO_2H$	*cis*: +10.2 *trans*: +17.2 *gem*: + 1.7	60
110	$H_2C=CH-CO_2C_2H_5$	*cis*: +11.7 *trans*: +19.0 *gem*: +0.24	60
111	$H_2C=CH-COCH_3$	*cis*: +10.77 *trans*: +17.80 *gem*: + 1.10	56
112	$H_2C=CH-CN$	*cis*: +11.75 *trans*: +17.92 *gem*: + 0.91	57
113	$H_2C=CH-C_6H_5$	*cis*: +10.7 *trans*: +17.5 *gem*: + 1.3	14
114	$H_2C=CH-CH_3$	*cis*: +10.02 *trans*: +16.81 *gem*: + 2.08	61
115	$H_2C=CH-CH_2F$	*cis*: +10.66 *trans*: +17.21 *gem*: + 1.54	62
116	$H_2C=CH-CHF_2$	*cis*: +10.85 *trans*: +17.49 *gem*: + 0.67	62, 63
117	$H_2C=CH-CF_3$	*cis*: +11.13 *trans*: +17.49 *gem*: − 0.15	62, 63
118	$H_2C=CH-CH_2Cl$	*cis*: +10.11 *trans*: +16.92 *gem*: + 1.27	64
119	$H_2C=CH-CHCl_2$	*cis*: +10.14 *trans*: +16.80 *gem*: + 0.22	63
120	$H_2C=CH-CCl_3$	*cis*: +10.07 *trans*: +16.05 *gem*: − 0.42	65
121	$H_2C=CH-CH_2Br$	*cis*: + 9.95 *trans*: +16.78 *gem*: + 1.22	64
122	$H_2C=CH-CH_2I$	*cis*: + 9.69 *trans*: +16.49 *gem*: + 1.25	64

No.	Compound or structural type	J_{olef}	Ref.
123	H—C=C(OCH$_3$)—H (H,H / OCH$_3$,H)	*cis*: + 7.0 *trans*: +14.1 *gem*: − 2.0	59
124	H$_2$C=CHF	*cis*: + 4.65 *trans*: +12.75 *gem*: − 3.2	66
125	H$_2$C=CHCl	*cis*: + 7.4 *trans*: +14.8 *gem*: − 1.3	14, 66
126	H$_2$C=CHBr	*cis*: + 7.1 *trans*: +15.2 *gem*: − 1.8	66
127	H,H / EtO$_2$C,CO$_2$Et (C=C)	*cis*: (+)11.9	67
128	EtO$_2$C,H / H,CO$_2$Et (C=C)	*trans*: (+)15.5	67
129	Cl,Cl / H,H (C=C)	*cis*: (+)5.2	67
130	Cl,H / H,Cl (C=C)	*trans*: (+)12.2	67
131	CH$_{3(3)}$,CH$_3$O$_2$C / H$_{(2)}$,H$_{(1)}$ (C=C)	H$_{(1)}$—H$_{(2)}$: (+)2.0 H$_{(1)}$—H$_{(3)}$: (−)1.0 H$_{(2)}$—H$_{(3)}$: (−)2.0	24
132	H$_{(a)}$,Cl / H$_{(b)}$ C=C; C=C Cl,H$_{(b')}$ / H$_{(a)}$	H$_{(a)}$—H$_{(b)}$: + 7.28 [H$_{(b)}$—H$_{(b')}$: +10.36]	68
133	Cl,H$_{(a')}$ / H$_{(b)}$ C=C; C=C H$_{(b')}$,H$_{(a)}$ / Cl	H$_{(a)}$—H$_{(b)}$: +13.12 [H$_{(b)}$—H$_{(b')}$: +11.24]	68
134	H$_{(3)}$,H$_{(2)}$ C=C H$_{(4)}$—C(CH$_3$)$_2$ / H$_{(1)}$	H$_{(1)}$—H$_{(2)}$: +10.37 H$_{(1)}$—H$_{(3)}$: +17.22 H$_{(2)}$—H$_{(3)}$: + 1.74 H$_{(1)}$—H$_{(4)}$: + 6.41 H$_{(2)}$—H$_{(4)}$: − 1.17 H$_{(3)}$—H$_{(4)}$: − 1.43	64
135	H$_{(3)}$,H$_{(2)}$ C=C H$_{(4)}$—C(*t*-Bu)$_2$ / H$_{(1)}$	H$_{(1)}$—H$_{(2)}$: + 9.97 H$_{(1)}$—H$_{(3)}$: +17.01 H$_{(2)}$—H$_{(3)}$: + 2.63 H$_{(1)}$—H$_{(4)}$: +10.65 H$_{(2)}$—H$_{(4)}$: − 0.10 H$_{(3)}$—H$_{(4)}$: − 0.63	64
136	H,H C=C H$_{(1')}$; C=C H$_{(1)}$,H$_{(2)}$ / H$_{(3)}$	H$_{(1)}$—H$_{(2)}$: +10.17 H$_{(1)}$—H$_{(3)}$: +17.05 H$_{(2)}$—H$_{(3)}$: + 1.74 H$_{(1)}$—H$_{(1')}$: +10.41	69

D. ADDITIONAL VICINAL COUPLINGS

No.	Compound or structural type	J_{vic}	Ref.
137	HC≡CH	± 9.8 ± 9.53	31 32
138	(structure: three-membered ring with H substituents)	cis: $+7.4$–11.2 $trans$: $+3.6$–8.6	—
139	(structure: three-membered ring with H substituents)	cis: $+9.5$ $trans$: $+5.5$	70
140	CO_2H (structure with $H_{(c)}$, $H_{(a)}$, $H_{(c')}$, $H_{(b)}$, $H_{(b')}$)	$H_{(a)}$—$H_{(b)}$: $+8.0$ $H_{(a)}$—$H_{(c)}$: $+4.6$ $H_{(b)}$—$H_{(b')}$: $+10.5$ $H_{(c)}$—$H_{(c')}$: $+11.0$ $H_{(b)}$—$H_{(c)}$: $+7.5$	71
141	NH_2 (structure with $H_{(c)}$, $H_{(a)}$, $H_{(c')}$, $H_{(b)}$, $H_{(b')}$)	$H_{(a)}$—$H_{(b)}$: $(+)6.6$ $H_{(a)}$—$H_{(c)}$: $(+)3.6$ $H_{(b)}$—$H_{(b')}$: $(+)12.5$ $H_{(c)}$—$H_{(c')}$: $(+)12.5$ $H_{(b)}$—$H_{(c)}$: $(+)7.5$	72
142	Br (structure with $H_{(c)}$, $H_{(a)}$, $H_{(c')}$, $H_{(b)}$, $H_{(b')}$)	$H_{(a)}$—$H_{(b)}$: $+7.30$ $H_{(a)}$—$H_{(c)}$: $+3.92$ $H_{(b)}$—$H_{(b')}$: $+10.5$ $H_{(c)}$—$H_{(c')}$: $+10.3$ $H_{(b)}$—$H_{(c)}$: $+6.6$	71
143	CH_3 (structure with CH_3 and H substituents)	H cis: $+9.2$ H $trans$: $+5.4$	12
144	Cl (structure with Cl and H substituents)	H cis: $+11.2$ H $trans$: $+8.0$	12
145	O (oxirane-type structure with H substituents)	cis: $+4.0$ to $+5.0$ $trans$: $+1.9$ to $+2.5$	—
146	O (oxirane-type structure with H substituents)	cis: $(+)4.45$ $trans$: $(+)3.10$	13

No.	Compound or structural type	J_{vic}	Ref.
147		*cis*: (+)5.4 *trans*: (+)2.6	14, 16
148		*cis*: +4.23 *trans*: +2.51	15
149		*cis*: +5.21 *trans*: +1.95	15
150		*cis*: +7.15 *trans*: (+)5.65	13
151		*cis*: +6.3 *trans*: +5.4	17
152		*cis*: +6.0 *trans*: (+)3.1	13
153		*vicinal*: *cis*: +10.4 *trans*: +4.9 crossing: ca. 0.0	70
154		*cis*: ±6.3, ±10.6 *trans*: ±5.9	73
155		*cis*: ±9.4 *trans*: +4.5, +9.3	74

D. ADDITIONAL VICINAL COUPLINGS—(*Continued*)

No.	Compound or structural type	J_{vic}	Ref.
156		*cis*: or *trans*: $+7.67$, $+11.16$ (unassigned)	18
157		$H_{(1)}$—$H_{(2)}$: $+3.9$ $H_{(2)}$—$H_{(3)}$: $+6.3$ $H_{(3)}$—$H_{(4)}$: $+1.6$	75
158		$H_{(2)}$—$H_{(3a)}$: $+10.48$ $H_{(2)}$—$H_{(3b)}$: $+3.84$ $H_{(3a)}$—$H_{(4)}$: $+4.71$ $H_{(3b)}$—$H_{(4)}$: $+2.09$ $H_{(4)}$—$H_{(5a)}$: $+4.57$ $H_{(4)}$—$H_{(5b)}$: $+0.94$	22
159		$H_{(2)}$—$H_{(3a)}$: $+7.66$ $H_{(2)}$—$H_{(3b)}$: $+10.44$ $H_{(3a)}$—$H_{(4)}$: $+1.41$ $H_{(3b)}$—$H_{(4)}$: $+4.31$ $H_{(4)}$—$H_{(5a)}$: 1.22 $H_{(4)}$—$H_{(5b)}$: $+4.09$	22
160		$H_{(ax)}$—$H_{(ax)}$: $+10$–12.5 $H_{(ax)}$—$H_{(eq)}$: $+2.3$–5.5 $H_{(eq)}$—$H_{(eq)}$: ca. $+2.7$	—
161		$H_{(ax)}$—$H_{(ax)}$: $(+)11.8$ $H_{(ax)}$—$H_{(eq)}$: ca. $(+)3.9$ $H_{(eq)}$—$H_{(eq)}$: ca. $(+)3.9$	24
162		$H_{(ax)}$—$H_{(eq)}$: $+2.5$–3.2 [$H_{(eq)}$—$H_{(eq)}$: ca. $(+)2.5$]	—

No.	Compound or structural type	J_{vic}	Ref.

163

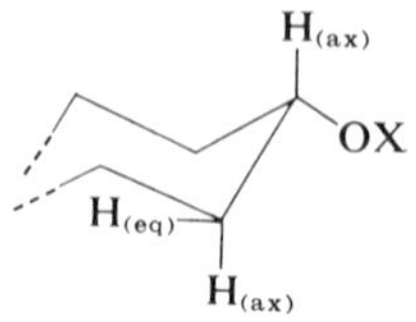

$H_{(ax)}$—$H_{(eq)}$: $(+)5.5 \pm 1.0$ —
$[H_{(ax)}$—$H_{(ax)}$: ca. $(+)10]$

164

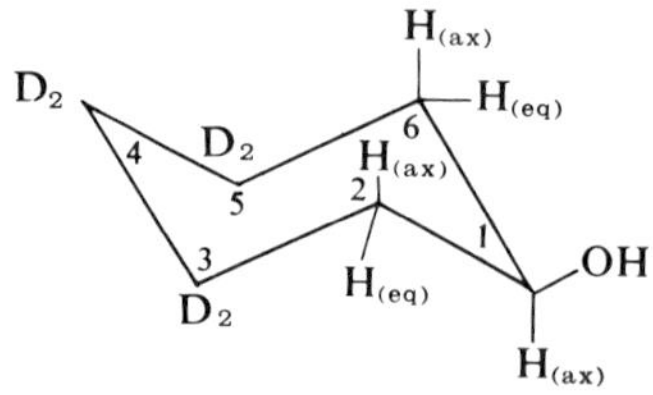

(preferred conformation)

$H_{(1)}$—$H_{(2eq)}$: $+4.08$ 25
$H_{(1)}$—$H_{(2ax)}$: $+10.17$

165

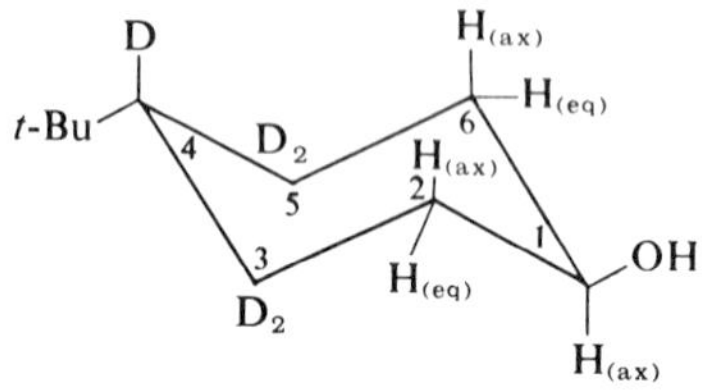

$H_{(1)}$—$H_{(2eq)}$: $+4.31$ 25
$H_{(1)}$—$H_{(2ax)}$: $+11.07$

166

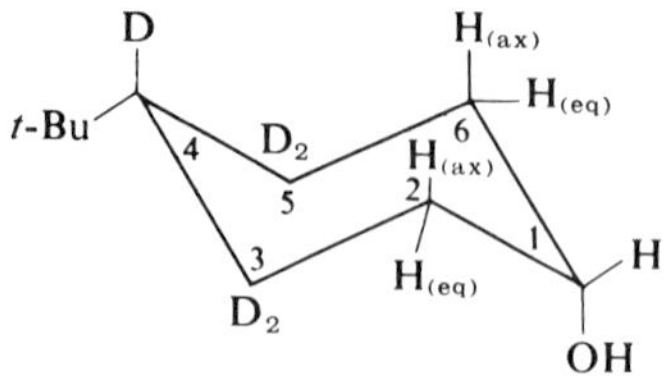

$H_{(1)}$—$H_{(2eq)}$: $+2.72$ 25
$H_{(1)}$—$H_{(2ax)}$: $+3.00$

167

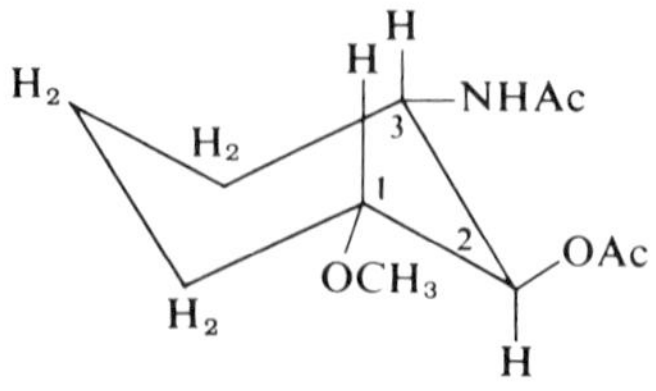

$H_{(1)}$—$H_{(2)}$: $(+)9.1$ 76
$H_{(2)}$—$H_{(3)}$: $(+)9.1$

168

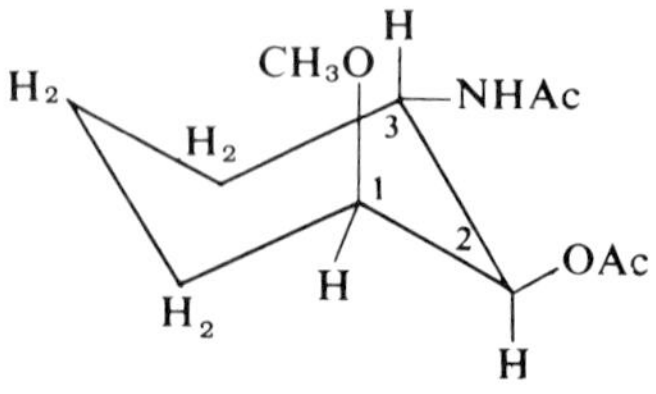

$H_{(1)}$—$H_{(2)}$: $(+)2.3$ 76
$H_{(2)}$—$H_{(3)}$: $(+)10.4$

D. ADDITIONAL VICINAL COUPLINGS—(*Continued*)

No.	Compound or structural type	J_{vic}	Ref.
169	(equilibrates with equienergy mirror-image conformer when $D_{(1)} = H$)	$H_{(2)}$—$H_{(3)}$ (avg'd value): (+)3.8, (+)7.1 (in CHCl$_3$)	77
170	(locked conformation)	$H_{(2)}$—$H_{(3eq)}$: (+)4.5 $H_{(2)}$—$H_{(3ax)}$: (+)11.0 (in CHCl$_3$)	77
171		$H_{(2ax)}$—$H_{(3ax)}$ = (+)10.0 [$H_{(4ax)}$—$H_{(3ax)}$ = (+)10.0] $H_{(2eq)}$—$H_{(3ax)}$ = (+)5.0 [$H_{(4eq)}$—$H_{(3ax)}$ = (+)5.0]	78
172		$H_{(11ax)}$—$H_{(12eq)}$: +2.5 $H_{(11eq)}$—$H_{(12eq)}$: +2.5	78

(171 and 172 are portions of
5-β-pregnan-3α,12α-diol-20-one diacetate)

173		J_{56} (*cis*) X = F: (+)1.2 X = Cl: (+)2.4 X = Br: (+)2.8 X = I: (+)4.3	79

No.	Compound or structural type	J_{vic}	Ref.
174		J_{56} (*trans*) X = F: (+)10.4 X = Cl: (+)11.4 X = Br: (+)11.8 X = I: (+)11.9	79
175		$H_{(4)}$—$H_{(5)}$: (+)8.1	80
176		$H_{(4)}$—$H_{(5)}$: (+)4.2	80
177		$H_{(ax)}$—$H_{(ax)}$: (+)9.42 $H_{(ax)}$—$H_{(eq)}$: (+)2.72 [$H_{(eq)}$—$H_{(eq)}$: ca. (+)2.7]	28
178		$H_{(4eq)}$—$H_{(5ax)}$: (+)7.02 $H_{(4eq)}$—$H_{(5eq)}$: (+)0.58 $H_{(5eq)}$—$H_{(6ax)}$: (+)3.06 $H_{(5ax)}$—$H_{(6ax)}$: (+)11.3	81
179		$H_{(5ax)}$—$H_{(6ax)}$: (+)10.7 $H_{(5eq)}$—$H_{(6ax)}$: (+)2.6	81

D. ADDITIONAL VICINAL COUPLINGS —(*Continued*)

No.	Compound or structural type	J_{vic}		Ref.
180		*cis*: *trans*:	$+3.2$ $+13.5$	82
181		$H_{(5)}$—$H_{(6)}$: $H_{(5)}$—$H_{(7)}$: $H_{(6)}$—$H_{(8)}$: $H_{(7)}$—$H_{(8)}$:	$+2.4$ $+1.8$ $+9.8$ $+3.6$	29
182		$H_{(a)}$—$H_{(a')}$: $H_{(b)}$—$H_{(b')}$: $H_{(a)}$—$H_{(b')}$:	$+12.3$ $+3.2$ $+4.0$	30
183		$H_{(2)}$—$H_{(3)}$: $H_{(3)}$—$H_{(4)}$:	$(+)7.7$ ca. 0	83
184		$H_{(2)}$—$H_{(3)}$: $H_{(3)}$—$H_{(4)}$:	$(+)2.3$ $(+)4.0$	83
185		$H_{(2)}$—$H_{(3)}$: $H_{(3)}$—$H_{(4)}$:	$(+)2.2$ ca. 0	83

No.	Compound or structural type	J_{vic}	Ref.
186		$H_{(2)}$—$H_{(3)}$: $(+)8.9$ $H_{(3)}$—$H_{(4)}$: $(+)4.4$	83
187		<2	84
188		$(+)2.5$–4.2	—
189		$H_{(1)}$—$H_{(2)}$: $(+)5.3$ $H_{(2)}$—$H_{(3)}$: $(+)2.1$	85
190		$H_{(1)}$—$H_{(2)}$: $(+)8.8$ $H_{(2)}$—$H_{(3)}$: $(+)3.1$	85
191		$H_{(1)}$—$H_{(2)}$: $(+)10.8$ $H_{(2)}$—$H_{(3)}$: $(+)5.7$	85
192		$H_{(1)}$—$H_{(2)}$: $(+)10.3$ $H_{(2)}$—$H_{(3)}$: $(+)7.8$	85

No.	Compound or structural type	J_{arom}	Ref.
193		*ortho*: +6.5–9.4 *meta*: +0.8–3.0 *para*: +0.4–1.0	—
194		*ortho*: 7.5 *meta*: 1.4 *ortho*: 7.68 *meta*: 1.42 *para*: 0.60	86 87
195		$H_{(A)}$—$H_{(B)}$: +7.1–8.1 (For R = NO_2: 8.4) $H_{(B)}$—$H_{(B')}$: +1.1–1.7 (For R = NO_2: 8.0) $H_{(A)}$—$H_{(B')}$: +1.1–1.7 $H_{(A)}$—$H_{(A')}$: +0.3–0.6 (For R = NO_2: 0.1)	—
196		$H_{(A)}$—$H_{(B)}$: +1.8–1.9 (for R = NO_2: +2.2) $H_{(A)}$—$H_{(C)}$: +0.3–0.6 $H_{(B)}$—$H_{(C)}$: +7.8–8.1 (For R = NO_2: +8.3)	—
197		$H_{(A)}$—$H_{(B)}$: +8.5–8.7 (For R = CH_3, R′ = I: +7.9) (For R or R′ = NO_2: +8.9–9.0) $H_{(A)}$—$H_{(A')}$, $H_{(B)}$—$H_{(B')}$ mostly: +2.3–2.7 (extreme range: +2.1–3.0) $H_{(A')}$—$H_{(B)}$: +0.3–0.5	88 88
198		(+)1.7–(+)2.2	—
199		(+)2.3	89
200		(+)2.5	89

No.	Compound or structural type	J_{arom}	Ref.
201		$(+)2.8$	89
202		$H_{(A)}$—$H_{(B)}$: $(+)8.1$ $H_{(B)}$—$H_{(B')}$: $(+)6.4$ $H_{(A)}$—$H_{(B')}$: $(+)1.1$ $H_{(A)}$—$H_{(A')}$: —	90
203		$H_{(A)}$—$H_{(B)}$: $(+)8.3$ $H_{(B)}$—$H_{(B')}$: $(+)6.5$ $H_{(A)}$—$H_{(B')}$: $(+)1.2$ $H_{(A)}$—$H_{(A')}$: —	90
204		$H_{(A)}$—$H_{(B)}$: $(+)7.6$ $H_{(B)}$—$H_{(B')}$: —	90

F. Heterocyclic Ring Couplings

No.	Compound or structural type	J_{hetero}	Ref.
205		$H_{(2)}$—$H_{(3)}$: $(+)4.5$–5.5 $H_{(3)}$—$H_{(4)}$: $(+)7.5$–8.2 $H_{(2)}$—$H_{(5)}$: $(+)0.7$–1.0 $H_{(3)}$—$H_{(5)}$: $(+)0.3$–1.6 $H_{(2)}$—$H_{(6)}$: $(+)0.0$–0.4	—
206		$H_{(2)}$—$H_{(3)}$: $(+)5.5$ $H_{(3)}$—$H_{(4)}$: $(+)7.5$ $H_{(2)}$—$H_{(5)}$: $(+)0.9$ $H_{(3)}$—$H_{(5)}$: $(+)1.6$ $H_{(2)}$—$H_{(6)}$: $(+)0.4$	91
207		$H_{(3)}$—$H_{(4)}$: $(+)7.5$–7.9 $H_{(4)}$—$H_{(5)}$: $(+)7.0$–8.0 $H_{(5)}$—$H_{(6)}$: $(+)4.5$–4.8 $H_{(3)}$—$H_{(5)}$: $(+)0.3$–1.2 $H_{(4)}$—$H_{(6)}$: $(+)1.7$–2.0 $H_{(3)}$—$H_{(6)}$: $(+)0.75$–0.9	—
208		$H_{(2)}$—$H_{(4)}$: ca. 0.0 $H_{(2)}$—$H_{(5)}$: $(+)0.7$–0.9 $H_{(2)}$—$H_{(6)}$: $(+)0.0$–0.3 $H_{(4)}$—$H_{(5)}$: $(+)7.8$–8.2 $H_{(4)}$—$H_{(6)}$: $(+)1.5$–1.8 $H_{(5)}$—$H_{(6)}$: $(+)4.7$–4.9	—

No.	Compound or structural type	J_{hetero}	Ref.
209		$H_{(2)}$—$H_{(3)}$: $(+)1.4$–2.0 $H_{(3)}$—$H_{(4)}$: $(+)3.1$–3.8 $H_{(2)}$—$H_{(4)}$: $(+)0.7$–1.0	—
210		$H_{(2)}$—$H_{(3)}$: $(+)1.4$ $H_{(2)}$—$H_{(4)}$: $(+)1.2$ $H_{(3)}$—$H_{(4)}$: —	92
211		$H_{(3)}$—$H_{(4)}$: $(+)3.2$–3.8 $H_{(4)}$—$H_{(5)}$: $(+)1.7$–1.9 $H_{(3)}$—$H_{(5)}$: $(+)0.4$–0.9	—
212		$H_{(3)}$—$H_{(4)}$: $+3.5$ $H_{(4)}$—$H_{(5)}$: $+1.8$ $H_{(3)}$—$H_{(5)}$: $+0.8$	93
213		$H_{(2)}$—$H_{(3)}$: $(+)4.7$–5.5 $H_{(3)}$—$H_{(4)}$: $(+)3.3$–4.0 $H_{(2)}$—$H_{(4)}$: $(+)1.3$–1.5 $H_{(2)}$—$H_{(5)}$: $(+)2.7$–3.2	—
214		$H_{(2)}$—$H_{(3)}$: $+4.7$ $H_{(3)}$—$H_{(4)}$: $+3.35$ $H_{(2)}$—$H_{(4)}$: $+1.0$ $H_{(2)}$—$H_{(5)}$: $+2.85$	94
215		$H_{(3)}$—$H_{(4)}$: $+3.4$–4.0 $H_{(4)}$—$H_{(5)}$: $+4.5$–5.2 $H_{(3)}$—$H_{(5)}$: $+1.1$–1.7	—
216		$H_{(2)}$—$H_{(4)}$: $+1.2$–1.5 $H_{(2)}$—$H_{(5)}$: $+2.7$–3.2 $H_{(4)}$—$H_{(5)}$: 4.9–5.4	—
217		$H_{(1)}$—$H_{(2)}$: ca. $(+)3$ $H_{(1)}$—$H_{(3)}$: ca. $(+)2$ $H_{(2)}$—$H_{(3)}$: $(+)2.4$–3.1 $H_{(3)}$—$H_{(4)}$: $(+)3.4$–3.8 $H_{(2)}$—$H_{(4)}$: $(+)1.3$–1.5 $H_{(2)}$—$H_{(5)}$: ca. $(+)2$	—
218		$H_{(1)}$—$H_{(2)}$: $(+)2.43$ $H_{(1)}$—$H_{(3)}$: $(+)2.43$ $H_{(2)}$—$H_{(3)}$: $(+)2.63$ $H_{(3)}$—$H_{(4)}$: $(+)3.42$ $H_{(2)}$—$H_{(4)}$: $(+)1.44$ $H_{(2)}$—$H_{(5)}$: ca. $(+)2$	95

^a Key to references:

1. M. Barfield and D. M. Grant, *J. Am. Chem. Soc.* **83**, 4726 (1961).
2. M. Barfield and D. M. Grant, *J. Chem. Phys.* **36**, 2054 (1962).
3. M. Barfield and D. M. Grant, *J. Am. Chem. Soc.* **85**, 1901 (1963).
4. H. J. Bernstein and N. Sheppard, *J. Chem. Phys.* **37**, 3012 (1962).
5. P. J. Black and M. L. Heffernan, *Australian J. Chem.* **15**, 862 (1962).
6. J. F. Bagli, P. E. Marand, and R. Gaudry, *J. Org. Chem.* **28**, 1207 (1963).
7. D. P. Biddiscombe, E. F. G. Herington, J. J. Lawrenson, and J. F. Martin, *J. Chem. Soc.* **1963**, 444.
8. F. Kaplan and J. D. Roberts, *J. Am. Chem. Soc.* **84**, 1053 (1962).
9. B. L. Shapiro, S. J. Ebersole, and R. M. Kopchik, *J. Mol. Spectry.* **11**, 201 (1963).
10. B. L. Shapiro, S. J. Ebersole, G. J. Karabatsos, F. M. Vane, and S. L. Manatt, *J. Am. Chem. Soc.* **85**, 4041 (1963).
11. B. L. Shapiro, R. M. Kopchik, and S. J. Ebersole, *J. Chem. Phys.* **39**, 3154 (1963).
12. D. J. Patel, M. E. H. Howden, and J. D. Roberts, *J. Am. Chem. Soc.* **85**, 3218 (1963).
13. F. S. Mortimer, *J. Mol. Spectry.* **5**, 199 (1960).
14. F. A. Bovey, *Chem. Eng. News.* **43**, 98 (1965).
15. C. A. Reilly and J. Swalen, *J. Chem. Phys.* **32**, 1378 (1960).
16. H. S. Gutowsky, M. Karplus, and D. M. Grant, *J. Chem. Phys.* **31**, 1278 (1959).
17. J. I. Musher and R. G. Gordon, *J. Chem. Phys.* **36**, 3097 (1962).
18. K. L. Servis and J. D. Roberts, *J. Phys. Chem.* **67**, 2885 (1963).
19. K. Takahashi, T. Stone, Y. Matsuki, and G. Hazato, *Bull. Chem. Soc. Japan* **36**, 108 (1963).
20. E. Lustig, private communication quoted by A. A. Bothner-By. (see footnote 4, Chapter V).
21. R. R. Fraser, R. U. Lemieux, and J. D. Stevens, *J. Am. Chem. Soc.* **83**, 3901 (1961).
22. R. J. Abraham and K. A. McLauchlan, *Mol. Phys.* **5**, 195 (1962).
23. F. A. Bovey and F. P. Hood, unpublished observations.
24. N. Muller and P. J. Schultz, *J. Phys. Chem.* **68**, 2026 (1964).
25. F. A. L. Anet, *J. Am. Chem. Soc.* **84**, 1053 (1962).
26. K. B. Wiberg, B. R. Lowry, and B. J. Nist, *J. Am. Chem. Soc.* **84**, 1594 (1962).
27. D. H. Williams and N. S. Bhacca, *Chem. Ind. (London)* **1965**, 506 (1965).
28. N. Sheppard and J. J. Turner, *Proc. Roy. Soc.* **A252**, 506 (1959).
29. J. S. Webb, R. W. Broschard, D. B. Cosulick, J. H. Mowat, and J. E. Lancaster, *J. Am. Chem. Soc.* **84**, 3183 (1962).
30. H. S. Gutowsky and C. Juan, *J. Chem. Phys.* **37**, 120 (1962); *Discussions Faraday Soc.* **34**, 52 (1962).
31. D. M. Graham and C. E. Holloway, *Can. J. Chem.* **41**, 2114 (1963).
32. R. M. Lynden-Bell and N. Sheppard, *Proc. Roy. Soc.* **A269**, 385 (1962).
33. S. Brownstein, B. C. Smith, G. Ehrlich, and A. W. Laubengayer, *J. Am. Chem. Soc.* **81**, 3826 (1959).
34. A. G. Massey, E. W. Randall, and D. Shaw, *Spectrochim. Acta* **20**, 379 (1964).
35. S. E. Ebersole, S. Castellano, and A. A. Bothner-By, *J. Phys. Chem.* **68**, 3420 (1964).
36. P. T. Narasimhan and M. T. Rogers, *J. Am. Chem. Soc.* **82**, 5983 (1960).
37. D. R. Whitman, L. Onsager, M. Saunders, and H. E. Dubb, *J. Chem. Phys.* **32**, 67, (1960).
38. G. Fraenkel, D. G. Adams, and J. Williams, *Tetrahedron Letters* **1963**, 767.
39. A. A. Bothner-By and R. E. Glick, *J. Chem. Phys.* **25**, 362 (1956).
40. J. Ranft, *Ann. Physik* [7] **10**, 1 (1962).
41. J. S. Waugh and F. W. Dobbs, *J. Chem. Phys.* **31**, 1235 (1959).

42. R. J. Abraham and K. G. R. Pachler, *Mol. Phys.* **7**, 165 (1964).
43. G. W. Flynn and J. D. Baldeschwieler, *J. Chem. Phys.* **37**, 2907 (1962).
44. F. A. L. Anet, *J. Am. Chem. Soc.* **82**, 994 (1960).
45. H. S. Gutowsky, G. G. Belford, and P. E. McMahon, *J. Chem. Phys.* **36**, 3353 (1962).
46. A. A. Bothner-By and C. Naar-Colin, *J. Am. Chem. Soc.* **84**, 743 (1962).
47. F. A. L. Anet, *J. Am. Chem. Soc.* **84**, 747 (1962).
48. R. Freeman and K. G. R. Pachler, *Mol. Phys.* **5**, 85 (1962).
49. R. Freeman, K. A. McLauchlan, J. I. Musher, and K. G. R. Pachler, *Mol. Phys.* **5**, 321 (1962).
50. D. Doskocilova, *J. Polymer Sci.* **B2**, 421 (1964).
51. D. Doskocilova and B. Schneider, *Collection Czech. Chem. Commun.* **29**, 2290 (1964).
52. T. Shimanouchi, M. Tasumi, and Y. Abe, *Makromol. Chem.* **86**, 34 (1965).
53. F. A. Bovey, F. P. Hood, E. W. Anderson, and L. C. Snyder, *J. Chem. Phys.* **42**, 3900 (1965).
54. G. S. Reddy and J. H. Goldstein, *J. Mol. Spectry.* **8**, 475 (1962).
55. C. S. Johnson, M. A. Weiner, J. S. Waugh, and D. Seyferth, *J. Am. Chem. Soc.* **83**, 1306 (1961).
56. S. Castellano and J. S. Waugh, *J. Chem. Phys.* **37**, 1951 (1962).
57. R. T. Hobgood, Jr., R. E. Mayo, and J. H. Goldstein, *J. Chem. Phys.* **39**, 2501 (1963).
58. W. A. Anderson, R. Freeman, and C. A. Reilly, *J. Chem. Phys.* **39**, 1518 (1963).
59. R. T. Hobgood, Jr., G. S. Reddy, and J. H. Goldstein, *J. Phys. Chem.* **67**, 110 (1963).
60. Y. Arata, H. Shimizu, and S. Fujiwara, *J. Chem. Phys.* **36**, 1951 (1962).
71. B. J. Nist, "A Nuclear Magnetic Resonance Summary of Small Ring Compounds" (unpublished) (1962).
72. H. M. Hutton and T. Schaefer, *Can. J. Chem.* **41**, 2774 (1963).
73. L. F. Johnson, V. Georgian, L. Georgian, and A. V. Robertson, *Tetrahedron* **19**, 1219 (1963).
74. E. Lustig, *J. Chem. Phys.* **37**, 2725 (1962).
75. L. D. Hall, *Chem. Ind. (London)* **1963**, 950.
76. F. A. L. Anet, R. A. B. Bannard, and L. D. Hall, *Can. J. Chem.* **41**, 2331 (1963).
77. R. U. Lemieux and J. W. Lown, *Tetrahedron Letters* **1963**, 1229.
78. N. S. Bhacca and D. H. Williams, *J. Am. Chem. Soc.* **86**, 2742 (1964).
79. A. Nickon, M. A. Castle, R. Harada, C. E. Berkoff, and R. D. Williams, *J. Am. Chem. Soc.* **85**, 2185 (1963).
80. R. U. Lemieux and J. Howard, *Can. J. Chem.* **41**, 308 (1963).
81. Y. Fujiwara and S. Fujiwara, *Bull. Chem. Soc. Japan* **37**, 1010 (1964).
82. J. Massicot and J. P. Marthe, *Bull. Soc. Chim. France* **1962**, 1962.
83. F. A. L. Anet, *Can. J. Chem.* **39**, 789 (1961).
84. P. Laszlo and P. von R. Schleyer, *J. Am. Chem. Soc.* **85**, 2017 (1963).
85. G. V. Smith and H. Kriloff, *J. Am. Chem. Soc.* **85**, 2016 (1963).
86. A. Saupe, *Z. Naturforsch.* **20b**, 572 (1965).
87. S. Castellano and C. Sun, *J. Am. Chem. Soc.* **88**, 4741 (1966).
88. J. Martin and B. P. Dailey, *J. Chem. Phys.* **37**, 2594 (1962).
89. H. M. Hutton, W. F. Reynolds, and T. Schaefer, *Can. J. Chem.* **40**, 1758 (1962).
90. N. Jonathan, S. Gordon, and B. P. Dailey, *J. Chem. Phys.* **36**, 2443 (1962).
91. V. G. Kowalewski and D. G. de Kowalewski, *J. Chem. Phys.* **36**, 266 (1962).
92. G. S. Reddy and J. H. Goldstein, *J. Am. Chem. Soc.* **84**, 583 (1962).
93. R. Freeman and D. H. Whiffen, *Mol. Phys.* **4**, 321 (1961).
94. R. Freeman and N. S. Bhacca, *J. Chem. Phys.* **38**, 293 (1963).
95. R. J. Abraham and H. J. Bernstein, *Can. J. Chem.* **39**, 905 (1961).

61. A. A. Bothner-By and C. Naar-Colin, *J. Am. Chem. Soc.* **83**, 231 (1961).
62. S. Castellano, private communication quoted by A. A. Bothner-By (see footnote 4, Chapter V).
63. H. Günther, private communication quoted by A. A. Bothner-By (see footnote 4, Chapter V).
64. A. A. Bothner-By, C. Naär-Colin, and H. Günther, *J. Am. Chem. Soc.* **84**, 2748 (1962); A. Bothner-By and H. Günther, *Discussions Faraday Soc.* **34**, 127 (1962).
65. S. Castellano and G. Caporiccio, *J. Chem. Phys.* **37**, 1951 (1962).
66. C. N. Banwell and N. Sheppard, *Discussions Faraday Soc.* **34**, 115 (1962); C. N. Banwell, N. Sheppard, and J. J. Turner, *Spectrochim. Acta* **11**, 794 (1960).
67. N. Muller, *J. Chem. Phys.* **37**, 2729 (1962).
68. R. K. Harris, private communication quoted by A. A. Bothner-By (see footnote 4, Chapter V).
69. R. T. Hobgood, Jr. and J. H. Goldstein, *J. Mol. Spectry.* **12**, 76 (1964).
70. S. Meiboom and L. C. Snyder, *J. Am. Chem. Soc.* **89** 1038 (1967).

APPENDIX D

CARBON−PROTON SPIN−SPIN COUPLING CONSTANTS

Table D.1. One-Bond Couplings (from G. E. Maciel, J. W. McIver, Jr., N. S. Ostlund, and J. A. Pople, *J. Am. Chem. Soc.* **92**, 1, 11 (1970) and references given there.) Carbon concerned is in bold type where necessary. Entries 1−23 represent sp^3 carbons; 23−48 sp^2 and sp carbons.

Entry Number	Compound	$^1J_{CH}$ (Hz)
1	CH_4	125.0
2	CH_3CH_3	124.9
3	$CH_3CH_2CH_3$	119.4
4	$(CH_3)_3CH$	114.2
5	$CH_3CH = CH_2$	122.4
6	$C_6H_5 \cdot CH_3$	129.4
7	$CH_3C \equiv CH$	132.0
8	CH_3F	149.1
9	CH_3Cl	150.0
10	CH_3Br	151.5
11	CH_3I	151.1
12	CH_2Cl_2	178.0
13	$CHCl_3$	209.0

14	CF_3H	239.1
15	CH_3OH	141.0
16	CH_3CH_2OH	140.3
17	CH_3CH_2OH	126.9
18	$(CH_3)_2CHOH$	142.8
19	$(CH_3)_2CHOH$	126.9
20	CH_3OCH_3	140.0
21	CH_3COOH	130.0
22	CH_3CN	136.1
23	CH_3NO_2	146.0

24	$CH_2 = CH_2$	156.2

25
$$\begin{array}{c} CH_3 \\ \end{array} C=C \begin{array}{c} H \\ CH_3 \end{array}$$
CH$_3$ / C=C / CH$_3$ with H — 148.4

26
t-Bu, H / t-Bu, H $C=C$ — 151.9

27
t-Bu, H / t-Bu, t-Bu $C=C$ — 143.3

28
H$_a$, H$_c$ / H$_b$, F $C=C$ —
(a) 159.2
(b) 162.2
(c) 200.2

29	$CH_2 = C = CH_2$	168.2

30	$\underset{H}{\overset{C_6H_5}{\diagdown}} C = C \underset{H}{\overset{C_6H_5}{\diagup}}$	155.0
31	$\underset{H}{\overset{C_6H_5}{\diagdown}} C = C \underset{C_6H_5}{\overset{H}{\diagup}}$	151.0
32	$\underset{H}{\overset{CH_3}{\diagdown}} C = N \diagdown_{OH}$	163.0
33	$\underset{H}{\overset{CH_3}{\diagdown}} C = N \diagup^{OH}$	177.0
34	CH_2O	172.0
35	NH_2CHO	188.3
36	CH_3CHO	172.4
37	$HCOO^-$ (aq.)	194.8
38	$HCOOH$	222.0
39	$HCOOCH_3$	226.2
40	$\underset{H}{\overset{F}{\diagdown}} C = O$	267.0
41	$HC \equiv CH$	249.0
42	$CH_3C \equiv CH$	248.0

43	$CH_2OH \cdot C \equiv CH$	248.0
44	$C_6H_5 \cdot C \equiv CH$	251.0
45	HCN	269.0

46

165

47

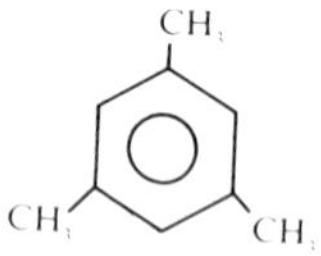

154

48

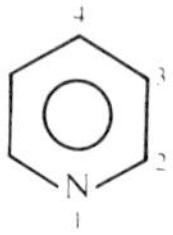

2,6 : 170
3,5 : 163
4 : 152

_____

49

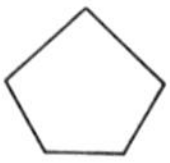

161

50

136

51

131

52

127

53	(cyclopropene structure)	$sp^2 : 220$
54	(tetrahydrofuran structure, positions labeled 1 and 2)	1 : 149 2 : 133
55	(1,4-dioxane structure)	145

Table D.2. Two-Bond Couplings adapted from Table 3.9 in "Carbon-13 NMR Spectroscopy" by E. Breitmaier and W. Voelter, 3rd edition, VCH, 1987, p. 141; coupled nuclei indicated by bold type.

Entry Number	Compound	$^2J_{CH}$ (Hz)	Ref.
56	**CH₃-CH₃**	-4.5	a
57	**CH₂Cl-CH₂Cl**	-3.4	a
58	**CHCl₂-CHCl₂**	1.2	a
59	**CH₂ = CH₂**	-2.4	a
60	(cis/trans dichloroethylene: H and Cl on one C=C, Cl and H on other)	0.8	a
61	(dichloroethylene: H and H, Cl and Cl)	16.0	a
62	$(CH_2)_n$ **C = CH₂**	$n = 4 : 4.2$ $n = 5 : 5.2$ $n = 6 : 5.5$	b b b

63	$(CH_3)_2C = O$	5.5	c
64	$H_2N-CH = CH-CHO$	6.0	c
65	CH_3-CHO	26.7	d
66	$(C_2H_5)_2CH-CHO$	22.1	d
67	$CH_2 = CH-CHO$	26.9	d
68	$NH_2-CH = CH-CHO$	20.0	c
69	$CH \equiv C-CHO$	33.2	d
70	$CH_2Cl-CHO$	32.5	d
71	$CHCl_2-CHO$	35.3	d
72	$CCl_3 - CHO$	46.3	d
73	$CH \equiv CH$	49.3	b
74	$C_6H_5O-C \equiv C-H$	61.0	b
75	$C_6H_5O-C \equiv C-CH_3$	10.8	b

[a] H. Yoder, R. H. Tuck, and R. F. Hess, *J. Am. Chem. Soc.* **91**, 539 (1969).

[b] E. F. Mooney and P. H. Winson, in *Ann. Rev. NMR Spectros.*, Vol. 2, E. F. Mooney (ed.), Academic Press, New York.

[c] "Carbon-13 NMR Spectroscopy," cited above.

[d] G. A. Olah and A. M. White, *J. Am. Chem. Soc.* **89**, 7072 (1967).

AUTHOR INDEX

A

Abraham, R. J., 189, *250*
Adams, J. Q., 98, 128, *142*
Ackerman, J. A., 480, 489, 519
Ackerman, J. L., 328, *355*
Ahmad, M., 300, 310, *323*
Alderman, D. W., 414, 415, *435*
Allerhand, A., 321, *324*
Amiel, Y., 113, *144*
Anderson, E. W., 449, 454, *518*
Anderson, W. A., 9, 36, *78, 86*, 155, *250*
Andrew, E. R., 20, *37*, 401, *435*
Anet, F. A. L., 117, *144*, 194, *250*, 251, 299, 300, 304, 305, 310, 314, 316, *322, 323*
Anet, R., 299, 305, *322, 323*
Armour, E. A. G., 110, *143*
Arnold, J. T., 30, *37*, 119, *144*, 315, *323*
Aue, W. P., 327, 334, *355, 356*

B

Bachmann, P., 334, *356*
Baldeschwieler, J. D., 218, *252*, 453, 454
Banwell, C. N., 454
Barfield, M., 192, 206, 208, 216, *250, 251*
Bartholdi, E., 327, *355*
Basus, V. J., 229, *252*
Baum, J., 505, *520*
Baumgarten, R. K., 128, *145*
Bax, A., 328, 353, *355*
Beaudet, R. A., 454
Becker, E. D., 113, 119, *144*
Behling, R. W., 388, *397*
Belford, G. G., 198, *251*
Bell, R. A., 209, *251*
Bendall, M. R., 245, *253*, 493, *520*
Ben-Ephraim, P. A., 113, *144*

Benoit, H., 422, *435*
Bel'ski, V. K., 414, *435*
Benz, F. W., 506, *521*
Bergland, G. D., 70, *86*
Berlin, A. J., 300, *322*
Bernstein, H. J., 112, *143*, 150, 159, *250*, 266, *322*
Bhacca, N. S., 118, *144*
Bickelhaupt, F., 112, *143*
Bielecki, K., 512, 514, 515, *521*
Bloch, F., 1, 9, 23, *36, 37*, 218, 251
Bloembergen, N., 16, 23, *36*, 259, 266, *321*
Bloom, A. L., 218, 231, *252, 253*
Bodenhausen, G., 331, 334, 337, 350, *355, 356*
Boekelheide, V., 113, *144*
Boicelli, C. A., 310, *323*
Booth, H., 201, *251*
Bothner-By, A. A., 113, *144*
Bottomley, P. A., 491, 494, 504, *520*
Bourn, A. J. R., 117, *144*, 316, *324*
Bovey, F. A., 104, 109, 117, 128, 130, 141, *143, 144, 145, 146*, 227, *252*, 265, 300, 311, *321, 323*, 344, *356*, 359, 361, 365, 367, 372, 376, 383, 388, *396, 397*, 420, *435*, 449, 454, 457, *518*
Bradbury, A., 401, *435*
Bradley, R. B., 113, *144*
Brady, T. J., 503, *520*
Brame, E. G., 378, *396*
Brany, M., 474, *519*
Brewster, A. I., 228, 367, *396*
Brey, W. S., 447, *518*
Briggs, J., 135
Brown, L. C., 444, 453
Brown, T. R., 474, *519*
Bruch, M. D., 359, 376, *396*
Brumlik, G. C., 128, *145*

Buckingham, A. D., 93, *142*, 225, *252*
Budinger, T. F., 502, *520*
Bunn, A., 419
Burfitt, I., 216, *251*
Burt, C. T., 474, *519*

C

Cais, R. E., 59, *86*, 374, 376, 378, *396*
Calder, J. C., 114, *144*
Cargioli, J. D., 140, *146*
Carhart, R. E., 98, *143*, 216, *251*
Carman, C. J., 98, 128, *143*
Carr, D. T., 453, 459
Carr, H. Y., 290, *322*
Carr, J. M., 501, *520*
Castellano, S., 203, *251*
Cavalli, L., 189, *250*, 440, *517*
Cavanaugh, J. R., 95, *142*
Chakrabarti, B., 206, 208, *251*
Chen, F., 424, *436*
Cheney, B. V., 99, *143*
Cheng, H. N., 308, *323*
Chernikova, N. Yu., 414, *435*
Cholli, A. L., 433, *436*
Christensen, K. A., 209, *251*
Ciula, R., 206, *251*
Cohen, A. D., 119, *144*
Cohn, M., 474, *519*
Connor, D. S., 206, *251*
Conover, W. W., 49, *86*
Corio, P. L., 161, 163, 174, *250*
Cotton, F. A., 124, *145*
Craig, N. C., 453
Crapo, L. M., 445
Crawford, B., Jr., 124, *145*, 445, 453
Cresp, T. M., 350, *356*
Creyghton, J. H. N., 505, *521*
Crook, R. A., 78, *86*
Crosby, R. C., 78, *86*
Cross, A. D., 455, *518*
Cudby, E. A., 419

D

Dailey, B. P., 94, *142*, 189, *250*
Dalling, D. K., 306, *323*
Damadian, R., 489, *520*
Darby, N., 350, *356*
Davis, D. R., 127, *145*, 454, *518*
de Kanter, F. J. J., 112, *143*

Delpuech, J.-J., 39, *85*, 290, *322*, 506, *521*
Derome, A. E., *39*, *85*
de Wolf, W. H., 112, *143*
Dharmatti, S. S., 30, *37*
Dickinson, W. C., 88, *142*
Ditchfield, 97, *142*
Djerassi, C., 118, *144*
Doddrell, D., 216, 245, *251*, *253*
Doering, W. von E., 317, *324*
Domenick, R. L., 141, *146*
Dorman, D. E., 98, 104, 117, *143*, *144*, 367, *396*
Douglass, D. C., 399, *435*
Drakenberg, T., 310, *323*
Drysdale, J. J., 124, *145*, 445
Dumais, J. J., 431, 433, *436*
Duncan, T. M., 402, 424, *435*, *436*
Dybowski, C., 424, *436*
Dyer, J. R., 114, *144*

E

Eades, R. G., 401, *435*
Earl, W. L., 78, *86*
Eaton, D. R., 131, *145*
Ech, G. W., 350, *356*
Edelman, R. R., 503, *520*
Edelstein, W. A., 495, *520*
Eikenberry, J. N., 136, *146*
Elleman, D. D., 444, 453
Ellet, J. D., 422, *436*
Ellis, P. D., 97, *142*, 229, *252*
Emid, S., 505, *521*
Emsley, J. W., 46, 440, 457, *518*
Engel, A. K., 431, 433, *436*
Englert, G., 225, *252*
Ernst, R. R., 9, *36*, 78, *86*, 228, *252*, 327, 328, 334, 337, 350, 351, 354, *355*, *356*, 495, 500, *520*
Ettinger, R., 94, 95, *142*

F

Fan, S., 509, 511, *521*
Farrar, T. C., 291
Feeney, J., 46, 506, *521*
Ferguson, R. C., 365, 378, *396*
Fermi, E., 131, *145*, 147, *250*
Ferrige, A. G., 63, *86*
Fessenden, R. W., 109, *143*
Fields, R., 440, *517*

Filipovich, G., 437, 440, *515*
Finegold, H. S., 127, *145*
Fischer, T., 379, *396*, *397*
Fleming, W. W., 419
Flynn, G. W., 453
Forsén, R.A., 231, *253*
Forsén, S., 218, 231, *251*, *253*, 319, *324*
Fossel, E. T., 501, *520*
Frank, L. R., 503, *520*
Frankiss, S. G., 459
Fraser, R. R., 136, *146*
Freed, J. H., 430, *436*
Freedman, M. H., 140, *146*
Freeman, P. K., 115, *144*
Freeman, R., 223, 224, 229, 241, *252*, *253*, 270, *322*, 328, 329, 331, 334, 337, *355*, *356*
Frost, G. H., 135
Fukushima, E., 39, *85*
Fyfe, C. A., 399, 419, *434*, *435*

G

Gadian, D. G., 474, *519*
Gaoni, Y., 113, *144*
Garroway, A. N., 505, *520*
Geller, S. C., 503, *520*
Gerstein, B. C., 399, 424, *435*, *436*
Gibby, M., 402, *435*
Gill, D., 290, *322*
Gillen, K. T., 262, *321*
Glonek, T., 474, *519*
Goering, H. L., 136, *146*
Goldstein, J. H., 98, 128, *143*
Gopinathan, M. S., 449, *518*
Gordon, R. E., 493, *520*
Gorenstein, D. G., 474, 475, *519*
Gouesnard, J.-P., 460, *518*
Govil, G., 449, *518*
Grant, D. M., 98, 99, *142*, *143*, 192, 209, *250*, *251*, 256, 306, *321*, *323*, 414, 415, *435*
Graves, R. E., 460
Gray, G. A., 399, 410, *434*
Green, D. K., 262, *321*
Griffin, R. G., 399, 430, *434*, *436*
Grostic, M. F., 115, *144*
Grunwald, E., 313, *323*
Grutzner, J. B., 229, *252*
Günther, H., 350, *356*
Günther, M.-E., 350, *356*

Gutowsky, H. S., 33, *37*, 119, 124, 141, *144*, *145*, *146*, 198, *251*, 292, 295, 321, *322*, *324*, 424, *436*, 440, 444, 454, *515*

H

Haeberlen, U., 399, 422, 424, *434*, *435*, *436*
Hahn, E. L., 33, *37*, 225, 232, *252*, *253*, 290, 407, *435*
Haigh, C. W., 110, 111, 112, *143*
Hall, F. H., 449, 454, *518*
Hall, G. E., 125, *145*
Hall, L. D., 300, *323*
Halstead, J. A., 289, *322*
Hansen, P. E., 216, *251*
Hansen, W. W., 1, *36*
Haq, M. A., 300, *323*
Harris, R. K., 190, *250*, 256, *321*, 399, 419, *434*, *435*, 445, 460, *518*
Hart, F. A., 135
Hartmann, S. R., 407, *435*
Havens, J. R., 399, *435*
Haw, J. F., 78, *86*
Heatley, F., 365, 388, *396*, *397*
Helfand, E., 433, *436*
Hellwege, K. H., 379, 397
Hester, R. K., 328, *355*
Hill, H. D. W., 229, *252*, 270, *322*, 329, *355*, 399, 410, *434*
Hinshaw, W. S., 493, 494, *520*
Hoffman, R. A., 218, *252*, 319, *324*
Holland, G. M., 494, *520*
Hollis, D. P., 474, *519*
Holm, C. H., 295, *322*, 454
Hood, F. P., 141, *146*, 300, 311, *323*, 441, 454
Hoult, D. I., 59, 61, *86*, 495, *520*
Howden, M. E. H., 114, *144*
Hoyland, J. R., 90, *142*
Hubbard, P. S., 262, *321*
Huber, L. M., 422, *435*
Hunsberger, I. M., 119, *144*
Hutchison, J. M. S., 495, *520*

I

Imbaud, J. P., 460

J

Jackman, L. M., 113, 119, *144*, 308, *323*
Jardetzky, O., 474, *519*
Jautelat, M., 228, *252*

Jeener, J., 327, *355*
Jelinski, L. W., 39, *85*, 131, *145*, 372, 388, *396*, *397*, 409, 429, 431, 433, *435*, *436*
Jenneskens, L. W., 112, *143*
Jennings, W. B., 125, *145*
Jensen, F. R., 300, *322*
Johnsen, U., 379, *397*
Johnson, C. E., Jr., 109, *143*
Johnson, G., 495, *520*
Johnson, L. F., 306, *323*
Jonas, J., 18, 37
Jones, K., 440, *516*
Josey, A. D., 131, *145*
Jost, R., 310, *323*

K

Kaiser, R., 460
Kaplan, F., 127, *145*
Karhan, J., 334, *356*
Karplus, M., 194, 216, *250*, *251*
Kearns, D. R., 388, *397*
Keeler, J., 229, *252*
Keith, H. D., 420
Kellogg, J. B. M., 1, *36*
Kempsell, S. P., 331, *355*
Kendrick, R. D., 419
Kinsinger, J. B., 379, *396*, *397*
Klose, G., 94, *142*
Knight, W. D., 30, *37*
Koenig, J. L., 399, *435*
Koermer, G. S., 136, *146*
Kolbe, K., 379, *397*
Kometani, J. M., 374, 378, *396*
Komoroski, R. A., 289, *322*, 399, *435*
Koolhaas, W. E., 112, *143*
Kornegay, R. L., 449, 454, *518*
Kosak, A. I., 128, *145*
Kozlowski, S. A., 359, *396*
Kraakman, P. A., 112, *143*
Kuhlmann, K. F., 256, *321*
Kumar, A., 328, 334, 351, *355*, *356*, 495, 500, *520*
Kuntz, S. D., 65, *86*, 506, *521*
Kurchinin, M. M., 414, *435*
Kurland, R. J., 225, *252*
Kusch, P., 1, *36*

L

Lacher, J. R., 114, *144*
Lamb, W. E., 87, *142*

Lambert, J. B., 445, 454
Lampman, G., 206, *251*
Landis, P. W., 455, *518*
Lauterbur, P. C., 127, *145*, 225, *252*, 445, 489, 494, 502, *520*
Lavanish, J., 206, *251*
Lee, J., 445, 453
Levitt, M. H., 229, *252*, 331, 344, 354, *355*, *356*
Levy, G. C., 140, *146*, 260, 277, 288, 289, *321*, *322*, 460, *519*
Lewin, A. H., 454
Lewis, A., 205, *251*
Lewis, D. W., 136, *146*
Li, Y.-F., 449, 454, *518*
Lichter, R. L., 460, *519*
Liddel, U., 119, *144*
Liepins, R., 378, *396*
Lin, Y. S., *316*, 324
Lindeman, L. P., 98, 128, *142*
Lindon, J. C., 63, *86*
Lindström, G., 31, *37*
Lowe, I. J., 401, *435*
Lowenstein, A., 313, *323*
Lunazzi, L., 310, *323*
Luz, Z., 321, *324*
Lyerla, J. R., 140, *146*, 260, 277, 288, *321*, *322*, 399, 419, *435*

M

Macciantelli, D., 310, *323*
MacFall, J. R., 494, *520*
Mallion, R. B., 110, 111, 112, *143*
Mann, B. E., 298, 306, 310, 316, *322*, *323*, *324*
Mansfield, P., 424, *436*, 495, *520*
Manus, M. M., 209, *251*
Marshall, J. L., 216, *251*
Martin, G. J., 39, *85*, 290, *322*, 460, 506, *518*, *521*
Martin, J., 189, *250*
Martin, M. L., 39, *85*, 290, *322*, 460, 506, *518*, *521*
Matsushima, M., 453
Maudsley, A. A., 337, *356*
Mavel, G., 474, *519*
Maxwell, D. E., 33, *37*
McBrierty, V. J., 399, *435*
McCall, D. W., 9, 21, 33, *36*, *37*, 292, *322*, 399, *435*, 440, *515*
McConnell, H. M., 106, *143*, 445, 453
McCrackin, F. L., 104, *143*

McDonagh, J., 501, *520*
McGarvey, B. R., 440, *515*
McIntyre, H. M., 288, *322*
McLauchlan, K. A., 223, 224, 225, *252*
McLean, A. D., 445, 453
McMahon, P. E., 198, *251*
Meese, M. T., 228, *252*
Mehring, M., 91, 429, 434
Meiboom, S., 290, 313, 321, *322, 323, 324*
Meinwald, J., 205, *251*
Meinwald, Y. C., 430, *436*
Meirovitch, E. R., 430, *436*
Meuli, R. A., *503*, 520
Meyer, L. H., 440, *515*
Middleton, W. J., 454
Miller, L., 334, 337, *356*
Millman, S., 1, *36*
Mirau, P. A., 222, *252*, 269, 333, 340, 344, 348, 349, *355, 356*, 509, 511, *521*
Mislow, K., 124, *144*
Mohanty, S., 460
Mondeshka, D., 350, *356*
Mooney, E. F., 440
Morosoff, N. 378, *396*
Morris, G. A., 241, *253*, 328, 337, *355, 356*
Moss, G. P., 135
Muller, N., 209, *251*, 453, 459
Müller, R., 460, *519*
Munowitz, M., 354, *356*, 505, *520*
Musher, J. I., 174, 223, 224, 250, *252*

N

Nagarajan, K., 127, *145*
Nagayama, K., 351, *356*
Nair, P. M., 124, *145*
Nakashima, T. T., 509, 511, *521*
Narasimhan, P. T., 449, *518*
Neff, B. L., 328, *355*
Nelson, G. L., 140, *146*, 277, *322*
Newell, G. F., 89, 90, *142*
Newmark, R. A., 311, *323*, 446, 448, 449, 460, *518*
Newton, M. D., 209, *251*
Newton, T. H., 494, *520*
Niedermeyer, R., 334, *356*
Noggle, J. H., 262, *321*
Noyce, D. S., 300, *322*

O

O'Neill, I. K., 474, *519*
Otocka, E. P., 105, *143*

P

Pachler, K. G. R., 189, 223, 224, *250, 252*
Packard, M. E., 1, 30, *36, 37*, 118, *144*
Packer, K. J., 399, 419, *434*
Page, T. F., 460
Pake, G. E., 19, *36*
Park, J. D., 114, *144*
Parr, R. G., 90, *142*
Patel, D. R., 114, *144*
Patt, S. L., 506, *521*
Paul, E. G., 98, *142*
Pauli, W., 1, *36*
Pauling, L., 108, *143*
Peat, I. R., 140, *146*
Pegg, D. T., 245, *253*
Petit, M. A., 136, *146*
Petrakis, L., 444
Phillips, J. B., 113, *144*
Phillips, L., 440, 457, *518*
Phillips, W. D., 124, 131, *145*, 445, 459
Pines, A., 354, *356*, 402, *435*, 505, 512, 515, *520*
Pirkle, W. H., 125, 126, *145*
Pollock, J. W., 114, *144*
Pople, J. A., 92, 97, 108, 109, 113, *142, 143, 144*, 150, 159, 225, *250, 252*, 266, *322*
Porte, A. L., 119, *144*
Pound, R. V., 1, 16, 23, *36*, 259, *321*
Powles, J. G., 262, *321*
Pritchard, J. G., 127, *145*
Pschorn, U., 429, *436*
Purcell, E. M., 1, 16, 23, *36*, 259, 290, *321, 322*
Pykett, I. L., 494, 495, *520*

R

Raban, M., 124, *144*
Rabenstein, D. L., 509, 511, *521*
Rabi, I. I., 1, *36*
Rabii, M., 422, *435*
Radda, G. K., 474, 504, *519, 520*
Ralph, E. K., 65, *86*, 506, *521*
Ramsey, K. C., 447, *518*
Ramsey, N. F., 1, *36*, 89, *142*
Randall, E. W., 218, *252*
Rao, B. D. N., 474, *519*
Raymond, F. A., 115, *144*
Redfield, A. G., 65, *86*, 506, *521*
Redpath, T., 495, *520*
Reeves, L. W., 445

Reich, H. J., 228, *252*
Reid, C., 119, *144*
Reilly, C. A., 444, 453
Reuben, J., 77, *86*
Reynolds, W. F., 140, *146*
Rice, D. M., 430, *436*
Richards, C. P., 474, *519*
Richards, R. E., 474, *519*
Roberts, G. C. K., 474, 506, *519, 521*
Roberts, J. D., 98, 114, 124, 127, *143, 144, 145,* 216, 217, 227, 228, *251, 252,* 445, 454, 460, 489, 490, *518*
Robinson, V. J., 460
Roeder, S. B. W., 39, 85
Rosen, B. R., 503, *520*
Rotaru, V. K., 414, *435*
Roth, W. R., 317, *324*

S

Saika, A., *22,* 142, 444, 454, 460
Salovey, R., 365, *396*
Santee, E. R., Jr., 374, 378, *396*
Santini, A. E., 229, *252*
Saunders, J. K. M., 136, *146*
Saunders, M., 119, *144, 317, 324*
Saupe, A., 225, *252*
Say, B. J., 419
Sayer, B. A., 209, *251*
Schaefer, J. F., 416, *435,* 469, *518, 519*
Schaefer, T., 138, *146,* 204, *454*
Scheraga, H. A., 430, *436*
Schertler, P., 206, *251*
Schilling, F. C., 59, *86,* 104, 128, 142, *145, 146,* 229, 231, 248, 249, *253,* 334, *355,* 359, 367, 369, 370, 373, 374, 383, *396, 397,* 414, 415, 420, *435,* 480, 489, 519
Schmicker, H., *350,* 356
Schneider, W. G., 95, 112, 138, 140, *142, 143, 146,* 150, 159, *250,* 266, *322*
Schröder, G., 317, *324*
Schulman, E. M., 209, *251*
Schulman, J. M., 209, *251*
Schwartz, M., 262, *321*
Sederholm, C. H., 300, 311, *322, 323,* 444, 445, 446, 448, 449, *518*
Seeley, P. J., 474, *519*
Sefcik, M. D., 431, *436*
Servis, K. L., 141, *146*
Shafer, P. R., 127, *145*
Shaka, A. J., 229, *252*

Sharkey, W. H., 454
Sheppard, N., 198, 213, *251,* 445, 454, *518*
Shoolery, J. N., 94, 112, 122, 124, *142, 144, 145,* 209, 231, *251, 253,* 274, *322,* 445, 453
Shulman, R. G., 474, *519*
Siddall, T. H. III, 308, *323*
Slichter, C. P., 92, *142*
Slichter, W. P., 9, 33, *36,* 292, *322*
Solomon, I., 256, *321*
Sommer, J., 310, *323*
Sondheimer, F., 113, 114, *144,* 350, *356*
Sorenson, W. O., 350, *356*
Spiesecke, H., 95, 140, *142, 146*
Spiess, H. W., 429, *436*
Stafford, S. L., 453
Staniforth, M. L.,
Stannett, V. T., 378, *396*
Stefaniak, L., 460, *518*
Steiner, R. E., 504, *520*
Stejskal, E. A., 416, *435*
Sternhell, S., 208, *251*
Stewart, W. E., 308, *323*
Stimson, E. R., 430, *436*
Strandberg, M. W. P., 225, *252*
Sun, C., 203, *251*
Surles, J. R., 378, *396*
Sütcliffe, L. H., 46, 445, 453
Swager, S., 449, 454, *518*
Sykes, B. D., 506, *521*

T

Taft, R. W., 440, *516*
Tailefer, R., 209, *251*
Takahashi, M., 454, *518*
Tarpley, A. R., Jr., 98, 128, *143*
Thayer, A. M., 512, 515, *521*
Thomas, H. A., 30, *37*
Thompson, D. S., 446, *518*
Tiers, G. V. D., 33, *37,* 130, 141, 142, *145, 146,* 227, *252,* 265, *321,* 361, *396,* 424, *436,* 437, 440, 445, 457, 460, *515, 518*
Timmons, M. L., 378, *396*
Tonelli, A. E., 59, *86,* 128, *145,* 334, *355,* 369, 370, 378, *396*
Torchia, D. A., 288, *322*
Torrey, H. C., 1, *36*
Tseng, S., 420, *435*
Turkenberg, L. A. M., 112, *143*
Turner, D. L., 331, 334, *355, 356*
Turner, J. J., 198, 213, *251*

U

Ugurbil, K., 474, *519*

V

VanderHart, D. L., 78, *86*
Van Gorkom, M., 125, *145*
Vanning, H., 419
Van Vleck, J. H., 20, *36*
Venkateswarlu, P., 460
Vogel, M., 127, *145*
von Philipsborn, W., 460, *519*

W

Wagner, G., 386, *397*
Wagner, H. L., 104, *143*
Walling, C., 209, *251*
Watnick, P. L., 431, *436*
Watson, C. J., 113, *144*
Waugh, J. S., 109, 124, *143*, *145*, 229, *252*, 328, *355*, 402, 422, *435*, *436*
Webb, G. A., 460, *518*
Wedeen, V. J., 503, *520*
Wehrli, F. W., 136, 141, *146*, 209, *251*, 494, *521*
Weigert, F. J., 217, 228, *252*, 460
Weitekamp, D. P., 512, 514, 515, *520*
Wells, E. J., 445
Welti, D. D., 328, *355*
Welti, I., 495, 500, *520*

Werstiuk, N. H., 209, *251*
Wirthlin, T., 136, 141, *146*, 209, *251*
Whitesides, G. M., 136, *146*
Wiberg, K. B., 206, *251*
Williams, D., 444, 453
Williams, D. H., 118, *144*
Williams, G. A., 454
Williamson, K. L., 449, 454, *518*
Wilson, C. W. III, 374, 378, 379, *396*, *397*
Winson, P. H., 440, *517*
Witanowski, M., 460, *518*
Wittebort, R. J., 430, *436*
Wolofsky, R., 113, *144*
Woodward, A. E., 420, *435*
Wortman, J. J., 378, *396*
Wray, V., 440, 457, *517*
Wright, D. A., 277, *322*
Wüthrich, K., 351, *356*, 386, *397*

Y

Yannoni, C. S., 399, 419, *434*, *435*

Z

Zacharias, J. P., 1, *36*
Zambelli, A., 365, 367, *396*
Zax, D. B., 512, 514, 515, *521*
Zilm, K. W., 414, 415, *435*, 512, 514, 515, *521*

SUBJECT INDEX

Specific compounds are indexed here only if they are mentioned in the text proper. Compounds listed in tables and appendices but not specifically named in the text are generally not indexed. References are given to text page, but tables are mentioned if present. Compounds are usually listed by name; by formula only if very simple.

Quantities denoted by Greek letters are listed before the corresponding Roman letters, e.g., δ before D, γ before G, and φ before P.

A

AA'BB' spectrum, 179–190
AA'XX' spectrum, 179–183
AB spectrum, 151–157
AB₂ spectrum, 159–162
AB₃ spectrum, 162–163
A₂B₂ spectrum, 184, 186
A₂B₃ spectrum (calc.), 602–603
ABₙ spectrum, 162–163
ABC spectrum, 175–179
ABX spectrum, 164–175
AMX spectrum, 163–165, 221–222
AX spectrum, 151–153
Acetaldehyde
  diethyl acetal, symmetry of, 128
  double resonance spectrum, 218–219
Acetamide, 227, 228
Acetic acid, geminal coupling in, 192
Acetone, geminal coupling in, 192
Acetophenone, 2-hydroxy-, proton exchange in, 319–321
N-Acetyl-L-alanyl-L-alanyl-L-alanyl methyl ester, NH resonance in, 227
Acetyl-L-proline, bond rotation in, 308
Acetylacetone
  hydrogen bonding in, 120-122
  keto-enol equilibrium in, 120–121

spectrum of, 121
Acetylene, 108, 217
Acetylenes
  ¹³C-¹³C J coupling in, 217
  ¹³C-¹H J coupling in, 210, 211
  J coupling in, 210, 211, 217
  shielding anisotropy in, 108
Acquisition time, 63–64
Acrylic esters, shielding in, 115, 116
Acrylyl fluoride, perfluoro-
  ¹⁹F-¹⁹F J coupling in, 447
Adamantane
  as reference, 79
  solid state spectrum, 415–416
Adenosine triphosphate, ³¹P spectrum of, 480
Aldehydes, shielding in, 95, 116
Aliasing, of frequencies in excess of Nyquist limit, 63
Alicyclic rings, shielding in, 118–119
Alkenes. See Olefinic double bonds
Alkyl halides
  chemical shifts in, 94
  geminal coupling in, 193
Alkyl nitrites, restricted rotation in, 307
Amides
  ¹⁴N, ¹⁵N-¹H J coupling in, 227–228
  ¹⁴N quadrupolar relaxation in, 227
  restricted rotation in, 307–310

Ammonia, as ^{14}N, ^{15}N chemical shift
  reference, 460, 461
Ammonium ion
  ^{14}N^{15}N-^{1}H J coupling in, 470–471
  ^{14}N^{15}N chemical shift in, 460, 461
  quadrupolar relaxation in, 30
Anisotropy. *See also* Shielding anisotropy
  diamagnetic. *See* Diamagnetic anisotropy,
    Shielding anisotropy, Paramagnetic
    anisotropy
Annulenes, shielding in, 113–114
Aromatic rings
  ^{13}C-^{1}H coupling in, 212–213 (table)
  ^{1}H shielding in, 138–141
  ^{1}H-^{1}H coupling in, 203–205
  ring currents in, 108–114
  shielding anisotropy of, 108–114
ATP. *See* Adenosine triphosphate
Axial protons. *See* Cyclohexane,
    Cyclohexanes, Vicinal coupling
Azoxybenzene, di-n-hexyloxy, as liquid
    crystal, 225

B

B_0, main polarizing magnetic field, 4–7,
    12–13, 19, 23–29, 41–50, 51, 61, 81, 87,
    237
B_1, magnetic component of resonant
    radiofrequency (rf) field, 4–11, 23–29,
    53, 407–409
B_2, magnetic component of decoupling rf
    field, 56, 59, 222, 231, 401
B_{loc}, local magneitc field, 21, 87, 400
^{10}B, ^{11}B, 3
Benzaldehyde, 4-N, N'-dimethylamino-d_2-,
    restricted rotation in, 310
Benzaldehyde, restricted rotation in, 310
Benzene, 72, 109–110, 140, 210, 225
  ^{13}C chemical shift, 140
  ^{13}C-^{1}H J coupling in, 210
  ^{1}H chemical shift, 139
  liquid crystal spectrum, 225
  shielding in, 109–110
  as solvent, 72
Benzenes,
  dinitro, ^{1}H spectra of, 123
  substituted
    as AA'BB' system, 180, 183, 186–189
    as ABC system, 175, 179
    ^{13}C chemical shifts, 140

^{13}C J coupling in, 210
  ^{1}H chemical shifts, 139
  ^{1}H J coupling in, 204–205
Bicyclocarboxamides, long-range, ^{1}H
    coupling in, 205–206
Bicyclo[1.1.1]pentane, long-range ^{1}H
    coupling in, 206
Bicyclo[3.1.0]hexene-2, 114–115
Biopolymers, NMR of, 381–395. *See also*
    Poly-γ-benzyl L-glutamate,
    Carboxypeptidase, Polynucleotides,
    Bovine pancreatic trypsin inhibitor
Bisphenol A, 410–414
  magic angle spectrum, 413
Bloch equations, 23–29, 292–299
Block averaging, 65
Boltzmann distribution of nuclear spins, 12
Boltzmann spin temperature, 12
Bond exchange, rate of
  in bullvalene, 317
  in cyclooctatetraene, 314
Bond hybridization
  effect on ^{13}C-^{13}C J couplings, 216
  effect on ^{13}C-^{1}H J couplings, 209
  effect on ^{15}N-^{1}H J couplings, 470
  effect on ^{31}P-^{1}H J couplings, 475
BPTI, bovine pancreatic trypsin inhibitor,
    389
Broadening, dipolar. *See* Dipolar
    broadening
1-Bromo-2-chloroethane, spectrum of,
    189–190
2-Bromo-5-chlorothiophene, as AB ^{1}H spin
    system, 156–157
Bullvalene
  bond exchange in, 317–318
  spectrum of, 318
Butane, ^{13}C chemical shift, 101
Butyric acid, 96, 121

C

^{12}C, 4
^{13}C
  abundance, 3
  chemical shift, 9, 79, 93–95, 97–106,
    129–130, 136, 140–142
  anisotropy of, 90, 402–406
  cross-polarization, 237, 407–411
  decoupling, 57, 401
  dipolar broadening, 400–401

^{13}C (*continued*)
  J coupling
    ^{13}C-^{13}C, 214–218
    ^{13}C-^{1}H, 208–214
  Nuclear Overhauser effect, 271–274
  references, 79, 71–74
  sensitivity, 237
  side-bands, 209–213
  spectra, 31, 58, 246, 248, 249, 368–369, 371,
    372, 374, 412, 413, 415, 417, 419, 421
  spin-lattice relaxation, 260–263, 267–271,
    278–287 (table)
  T$_1$, *see* spin-lattice relaxation
CCl$_3$F, as solvent and reference, 437, 439,
  438–439 (table)
CDCl$_3$
  shielding in, 72
  as solvent, 72
CH$_3$Br, shielding in, 94
CH$_3$Cl, shielding in, 94
CH$_3$F, shielding in, 94
CH$_3$I, shielding in, 94
C$_6$H$_6$. *See* Benzene
C$_6$D$_6$, as solvent, 72
cyclo-C$_6$HD$_{11}$, inversion of, 300–304
Carbon-13. *See* ^{13}C
Carbon-carbon double bond
  ^{1}H *J* coupling across, 176–178, 194, 207
  shielding anisotropy, 114–115
Carbon-carbon single bond, shielding
  anisotropy, 118–119
Carbonyl double bond
  effect on *J* coupling, 192
  shielding anisotropy, 116–118
Carboxypeptidase, 387
Carr-Purcell pulse sequence, 290–291, 321
Chemical shift, 30–33, 87–146, 150–190,
  375–378, 402–406, 437–445, 460–470,
  474–480
  ^{13}C, 31, 72–74, 79, 95 (figure), 97–106,
    402–406; *see individual compounds*
  ^{1}H(^{2}H), 72–74, 94 (figure), 87–146,
    150–190; *see individual compounds*
  ^{19}F, 165, 375–378, 437–441, 438–439
    (table)
  ^{15}N(^{14}N), 460–470, 461–468 (table)
  ^{31}P, 474–480, 475 (figure), 476–479
  anisotropy of, 83–93, 402–406
  effect of anisotropic groups, 106–119
  effect of hydrogen bonding, 119–122
  effect of molecular asymmetry, 122–128

  inductive effects in, 93–97
Chemical shift imaging, 504
3-Chloro-3-aminopyridine, spectrum of,
  171–173
p-Chlorobenzaldehyde, spectrum of,
  186–189
Conformers. *See also specific compounds*
  averaging of *J* couplings in, 193–203
  averaging of chemical shifts in,
    299–312
  in polymers, 368, 369
Contact term. *See* Fermi contact term
Coherent noise, 67
Cope rearrangement, in bullvalene, 317
Copolymers, structure of, 378–381
Coproporphyrin I methyl ester, shielding in,
  113
Correlated 2D NMR spectroscopy, 337–353
  heteronuclear, 337–341
  J-correlated, 349–353
  nuclear Overhauser correlated, 341–349
COSY form of 2D NMR, 349–353
Coupling. *See J* Coupling, Spin-spin
  coupling
Cryoshims, for superconducting magnets,
  46
Curvature, of magnetic field, 47
Cycling, of magnetic field, 45
Cyclobutanes
  inversion of, 200, 305
  vicinal *J* couplings in, 200–201
Cycloheptane, inversion of, 305
Cycloheptatriene, inversion of, 305–306
Cyclohexane
  ^{1}H shielding in, 118–119
  inversion of, 298–305
  J coupling in, 201
d_{11}-Cyclohexane, inversion of, 300–304
Cyclohexanes. *See also specific compound*
  inversion of,
  J coupling in, 193–195, 201–203
Cyclohexanol, 3,3,4,4,5,5-hexadeutero-,
  J coupling in, 193–195
Cyclopropane
  ^{13}C chemical shift, 95
  ^{1}H chemical shift, 94
  ^{1}H-^{13}C *J* coupling in, 210
Cyclopropanes, as AA'XX' ^{1}H spin
  systems, 183, 200
Cytidine, ^{1}H-^{13}C heterocorrelated 2D
  spectrum of, 340–341

D

δC, ^{13}C chemical shift reference, 72–74 (table)
δH, ^{1}H chemical shift reference, 72–74 (table)
D. *See specific deutero compounds*
DDS (2,2-Dimethyl-2-silapentane-sulfonate) as internal reference, 75, 76 (Table 2.3)
D$_2$O, as solvent, 72 (Table 2.2), 312, 381–382, 387, 389, 391
Debye correlation time, 16
cis-Decalin, inversion rate, 306–307
1,4-Decamethylenebenzene, shielding in, 111
Decoupling (double resonance), 55–59, 216
  ^{1}H-^{1}H, 218–224
  heteronuclear, 225 et seq.
    ^{1}H-^{13}C, 227–232
    ^{1}H-^{14}N, 225–227
    in spectral editing, 239–250
  homonuclear, 218–224
  and nuclear Overhauser effect, off resonance, 231–232
DEPT spectral editing sequence, 245–250
Detection of NMR signal, 7–11, 40, 62–67
Detector, 40, 61–67
Deuterated solvents, 72–74 (table)
Deuterium, 3, 30, 72–74
  effect on chemical shifts, 141–142
  spectral simplification by, 193–195, 301–305, 386–388
  quadrupolar NMR of, 424–434
Diamagnetic anisotropy, and shielding, 106–119
  in acetylenes, 108
  in aromatic ring, 108–114
  in double bonds, 114–118
  effect on chemical shift, 106–119
  in single bonds, 118–119
Diamagnetic susceptibility, 32, 106–119
2,3-Dibromopropionic acid,
  J-resolved 2D spectrum, 331–334
  relative signs of coupling in, 222–225
3,3-Dibenzylphthalide, asymmetry in, 128
2,3-Dibromobutyric acid, ^{1}H *J* coupling in, 222–225
1,2-Dibromo-1-chloro-2,2-difluoroethane, ^{19}F-^{19}F *J* coupling in, 449
1,1-Dibromo-1-fluoro-2,2-difluorobromoethane, ^{19}F-^{19}F *J* coupling in, 446
1,2-Dibromotetrafluoroethane, ^{19}F-^{19}F *J* coupling in, 446

o-Dichlorobenzene, ^{1}H spectrum, 186–187
Diethyl acetal, asymmetry in, 128
Diethyl fumarate, shielding in, 117
Diethyl maleate, shielding in, 117
Diethyl sulfite, asymmetry in, 127
Diethylmethyl ammonium ion, asymmetry in, 128
1,1-Difluoroallene, 158
1,1-Difluoroethylene, 158
1,2-Difluoro-tetrachloroethane ^{19}F chemical shifts, 311
Digital filtering, 68
Digitization rate, 63
Digital resolution, 63–65
p, *p'*-Di-*n*-Hexyloxyazoxybenzene, as liquid crystal, 225
Dimers, H-bonded, in alcohols, 120
N, *N*-Dimethylformamide
  bond rotation in, 307
  shielding in, 117
N, *N*-Dimethylnitrosamines, restricted rotation in, 307
Dinitrobenzenes, spectra of, 123
Dioxane
  geminal ^{1}H coupling in, 213
  spectrum of, 214
  vicinal ^{1}H coupling in, 213
Dipolar broadening, 18–23, 400
Dipolar decoupling, 401
Dipole, nuclear. *See* Nucleus, Nuclear spins
DNA, 386–388
Dome-dish correction, of magnetic field, 45–46
Double bond. *See* Carbon-carbon double bond, Carbonyl double bond
Double resonance, 55–59
Dwell time, 63–64
Dynamic range, 63–64

E

Editing, spectral, 232–250
Electron(s)
  circulation in aromatic rings, 108–114
  G-factor, 133
  magnetic moment of, 2, 15
  nuclear screening by, 30–33
  orbital motions, role in *J* coupling, 33–36, 147–150
  π, in benzene ring, 108–114
  unpaired, effect on T_1, 15, 133, 266
    T_1 of, 133

Electronegative groups (and atoms)
  effect on ^{1}H-^{19}F J coupling, 448
  effect on ^{1}H-^{1}H J coupling, 191–192
  in relation to chemical shifts
    ^{1}H, 31–32, 93–97, 120–122, 138–140
    ^{13}C, 31–32, 95, 98, 140
    ^{19}F, 437–441
    ^{14}N, 460–469
    ^{31}P, 474
Energy levels
  magnetic, 5–7
  of spin systems, 150–190 (*see also*
    particular spin systems)
Enols, O^{1}H shielding in, 120–122
Episulfides, gem ^{1}H J coupling in, 193
Epoxide rings
  as ABC ^{1}H systems, 175
  ^{1}H J coupling in, 200
Equatorial protons. *See* Cyclohexane(s),
  Vicinal coupling
Equivalence, of spins, 157–158
Ethanes
  1,2-disubstituted, as AA′BB′ system
  geminal ^{1}H coupling in, 189
  spin systems, 183, 189–190, 198–199
  1,1,2,2-tetrasubstituted, vicinal coupling
    ^{1}H in, 199
  1,1,2-trisubstituted, as AB$_2$ ^{1}H systems,
    198
  vicinal ^{1}H coupling in, 189–190, 198–199
Ethanol
  ^{1}H spectrum, 120
  O^{1}H shielding, 119–120
  proton exchange, 313
Ethyl group, ^{1}H spectrum, 120
  as A$_2$B$_3$, AA′B$_3$ ^{1}H systems, 127
Ethyl orthoformate
  ^{13}C spectrum, 31
  ^{1}H spectrum, 31
Ethylene, geminal ^{1}H-^{1}H J coupling in, 191
Ethylene imines, gem. ^{1}H J coupling in,
  193
Ethylenes, substituted
  as ABC ^{1}H systems, 175, 176–178
  geminal ^{1}H J coupling in, 175–178, 193
  vicinal ^{1}H J coupling in, 175, 176–178
Exchange rates, 291–321
  Bloch equations and, 292–299
  bond isomerization, 314–317
  bond rotation, 307–312
  of protons, 312–313, 315, 316
  ring inversions, 299–307

Excited electronic states
  in J coupling, 148
  in nuclear shielding, 89
External lock, 44–45
  for solids NMR, 78–79 (Table 2.4)
Eyring rate theory, 300

F

^{19}F, 3, 88, 158, 165, 174–175, 180, 437–459
  chemical shifts, 437–441
  J coupling
    ^{19}F-^{13}C, 457
    ^{19}F-^{19}F, 441–448
    ^{19}F-^{1}H, 449–456
  relative signs, 448, 456
Fast Fourier transform, 68–70
Fast passage, 267
Fermi contact term
  in chemical shifts, 131–133
  in J coupling, 147
Field-frequency lock. *See* Lock
Field sweep, 51–52
First order spectra, 151–153
Fluorocarbons
  ^{19}F-^{19}F couplings in, 441–448
  ^{19}F-^{1}H couplings in, 449–456
  chemical shifts in, 437–441
Folding-in (or aliasing) of frequencies
  beyond the Nyquist limit, 63
Formaldehyde, ^{1}H-^{1}H coupling in, 191
Formamide
  double resonance spectrum, 225–227
  shielding in, 117
Formamide, N,N'-dimethyl
  bond rotation in, 307, 309
  ^{1}H and ^{13}C shielding in, 117
  spectrum of, 309
Fourier transform, 7–11, 52–55, 59–70
Frequency domain, 52
Frequency sweep, 51–52
Fumarate esters, ^{1}H shielding in, 117
Furan, as AA′BB′ spin systems, 183
  ^{1}H shielding in, 141
Furan, tetrahydro-
  ^{13}C-^{13}C coupling in, 214–215
  ^{13}C spectrum of, 215
Furans, substituted, as ABC systems, 175

G

γ, magnetogyric ratio, 2
g, nuclear g factor, 2

Gauche conformers
  vicinal ^{19}F-^{19}F couplings in, 446–447
  vicinal ^{19}F-^{1}H couplings in, 448–454
  vicinal ^{1}H-^{1}H couplings in, 197–200
Gauche coupling. *See* Gauche conformers
Geminal coupling
  ^{19}F-^{19}F, 446
  ^{19}F-^{1}H, 449–455
  ^{1}H-^{1}H, 191–193, 195, 359
Glycyl-L-proline, bond rotation in, 308
Glycylsarcosine, 2D NOESY spectrum of,
  348–349

H

^{1}H, 2–3, 88
^{2}H. *See* Deuterium
^{3}H (tritium), 12
^{1}H$_2$, 89–90
^{1}H^{3}H molecule, J coupling in, 148
HDO (HOD), 348, 389, 507–509
Hartmann-Hahn condition, 407
Heisenberg uncertainty principle, 292
Helium, liquid, 42
Heptane, perfluoro-, 441–445
Heterocyclic rings
  ^{13}C-^{1}H J couplings in, 213
  ^{1}H chemical shifts in, 141
  ^{1}H-^{1}H J couplings in, 204
Homogeneity of magnetic field. *See*
  Magnetic field, homogeneity
Hybridization
  effect on ^{13}C-^{13}C J couplings, 216
  effect on ^{13}C-^{1}H J couplings, 209–210
  effect on ^{14}N-^{1}H, ^{15}N-^{1}H J couplings, 470
Hydrocarbons, ^{13}C chemical shifts, 97–106
Hydrogen bond, effect on ^{1}H chemical
  shift, 119–122
Hydrogen bonded dimers of alcohols,
  119–120
Hydrogen, molecular, 89–90
Hydrogen nucleus. *See* ^{1}H
3-Hydroxyacetophenone, proton exchange
  in, 319–320
Hydroxyl group, proton exchange in, 119,
  313, 315, 316, 318–321

I

I, nuclear spin quantum number, 1–2
Imaging, 489–505
INEPT spectral editing sequence, 241–245

Inductive effects
  in aromatic rings, 138–141
  ^{13}C, 95, 140
  in chemical shifts, 32, 93–97
  ^{19}F, 437
  ^{1}H, 94, 96–97, 138
  ^{14}N, ^{15}N, 460
  ^{31}P, 474–475
Internal lock, 44–45, 47
Inversion of rings, measurement of rate,
  299–307
Isobutane, ^{13}C chemical shift, 100
Isooctane, J-resolved ^{13}C 2D spectrum of,
  333–337
Isotactic, definition, 361

J

J, definition of, 34
J coupling, 33–36, 147–250
  ^{13}C-^{13}C, 214–216
  ^{13}C-^{1}H, 208–209
  ^{19}F-^{13}C, 457
  ^{19}F-^{19}F, 441–448
  ^{19}F-^{1}H, 448–456
  ^{1}H-^{1}H, in aryl rings, 203–205
    geminal, 191–193
    sign of, 151, 224–225
    vicinal, 193–203
  ^{14}N, ^{15}N, 470–474
  ^{31}P, 475–489
  virtual, 174
J-resolved 2D NMR spectroscopy, 328–337

K

Karplus function, in vicinal ^{1}H-^{1}H
  J coupling, 194–197
Keto-enol equilibrium, 120–122
11-Ketosteroids, shielding in, 118

L

Lamb term, in nuclear shielding, 87–89
Larmor precession, 4, 5, 7, 22, 87
Lattice, definition of, 13
Line broadening
  dipolar, 18–23, 400
  exchange, 291–317
Line shape
  Lorentzian, 27–28
  in rate measurement, 292–299

Line width, 18–23, 400
  of ^{13}C resonances, 400–401
  in relation to T_2, 22
  in solids, 400–409
Liquid crystal spectra, 225
Lock
  external, 44–45
  internal, 44–45
2,6-Lutidine, 159–162
Lysozyme, ^{1}H spectrum, 511

M

Macromolecules, NMR of, 357–395. *See also specific macromolecule*
Magnetic field, 2, 4–8, 30–33, 41–50, 87–88
  cycling, 45
  homogeneity, 45–50
  locking, 44–45
  shimming, 45–50
Magnetic moment
  of electron, 2
  of nucleus, 2–4
  of proton, 2–3
Magnetic shielding and chemical shift, 30–33, 87–142
Magnetogyric ratio
  in Bloch equations, 24
  definition, 2
  relation to J coupling, 149
  tables, 3, 524–528
Magneton
  Bohr, 2, 15, 133, 266
  nuclear, 2–3
Magnets
  cycling, 45
  electromagnets, 41–42
  homogeneity, 45–50
  permanent, 41
  shimming of, 45–50
Maleate esters, ^{1}H shielding in, 117
Methane
  ^{1}H chemical shift, 94
  ^{1}H J coupling in, 191
Methanol
  as AB_3 ^{1}H spin system, 162
  geminal coupling in, 192
Methyl acrylate, ^{1}H shielding in, 115
Methyl bromide, ^{1}H and ^{13}C chemical shift, 94
Methyl chloride, ^{1}H and ^{13}C chemical shift, 94
  ^{13}C satellites in ^{1}H spectrum, 210

Methyl fluoride, ^{1}H and ^{13}C chemical shifts, 94
Methyl halides
  ^{13}C-^{1}H coupling in, 209
  ^{1}H shielding in, 94
Methyl iodide, ^{1}H and ^{13}C chemical shifts, 94
  ^{13}C T_1 relaxation in, 262–263
Methyl mercaptan, spectrum of, 164
Methyl methacrylate
  ^{1}H spectrum of, 207
  polymer of, 359–365
Methylammonium ion, ^{1}H exchange in, 314, 317
Microcells, 75
Molecular motion, frequency spectrum of, 13
Molecular symmetry, NMR observation of, 122–131
Motional averaging, 21
Multiple quantum NMR spectroscopy, 354
Multiplet collapse, by ^{1}H exchange, 313, 315, 316
Multiplets. *See J coupling, Spin systems*

N

^{14}N,
  quadrupole moment of, 30, 225–228
  quadrupolar relaxation, 29–30, 263–265
^{14}N,^{15}N, 225–227, 457–474
  chemical shifts, 460–469
  heteronuclear decoupling, 225–227
  inherent sensitivity, 457
  J coupling, 225–227, 470–474
Neutron, magnetic moment of, 3
Nickel chelates, ^{1}H shielding in, 131
Nitrites, alkyl, restricted rotation in, 307
Nitrosamines, N,N-dimethyl, restricted rotation in, 307
p-Nitrotoluene, absolute sign of J coupling in, 225
NMR imaging, 489–505
NOE. *See Nuclear Overhauser effect*
NOESY form of 2D NMR, 341–349
Noise, coherent, 67
Nonequivalence of spins, 157–158
Nuclear g factor, 2
Nuclear induction, 7, 61
Nuclear magnetic moment, 1–3
  interactions in solids and liquids, 9–11, 18–23, 400–401
  macroscopic, 9–11, 23–28, 400–401
Nuclear magneton, 2

Nuclear Overhauser effect, 56, 271–289
  ^{13}C-^{1}H, 272–274
  ^{1}H-^{1}H, 272–274
  measurement of, 274–275, 278–287 (table)
Nuclear screening, 30–33, 87–93. *See also*
    specific nuclei and compounds
Nucleic acids, 386–388
Nucleus
  angular momentum of, 1
  g factor of, 2
  magnetic energy states, 5–6
  quantum numbers, 1–2
  spin, 1
Nyquist frequency, 63

O

^{16}O, 3. *See also* Oxygen
Off-resonance decoupling, 231–232
Olefinic double bonds
  ^{13}C-^{13}C ^{1}J coupling in, 217
  ^{1}H-^{1}H ^{3}J coupling in, 203–204
  ^{1}H-^{1}H ^{4}J coupling in, 207
  shielding anisotropy of, 114–117
Optically active solvent, splitting of *dl*
    isomer spectra by, 124–125
Overhauser effect, nuclear, 56, 271–289
  ^{13}C-^{1}H, 272–274
  ^{1}H-^{1}H, 272–274
  measurement of, 274–275, 278–287 (table)
Oximes, geminal ^{1}H-^{1}H coupling in, 193
  restricted rotation in, 308
Oxygen
  effect on T_1, 76, 266
  groups containing
    effect on aromatic ^{1}H-^{1}H coupling, 205
    effect on chemical shift, 93–98
    effect on geminal ^{1}H-^{1}H coupling, 192,
      204
    effect on vicinal ^{1}H-^{1}H coupling, 197,
      204

P

ϕ scale, 439
^{31}P, 3, 474–489
  chemical shifts, 474–480
  inherent sensitivity, 474
  J coupling, 475–489 (tables)
  sign of ^{31}P-^{1}H couplings, 475, 476–479,
    481–488 (tables)

Paraffinic hydrocarbons, ^{13}C chemical
    shifts, 97–106, 128–130
Paramagnetic anisotropy
  of electron *g* factor, 133
  of ring currents, 114
Paramagnetic compounds
  effect on T_1, 15, 131, 266
  shielding in, 131–138
Pascal triangle, 35, 241
Pentad configurational sequences, 360, 363,
    364 (table)
  in poly(methyl methacrylate), 360
  in polypropylene, 368–370
Perfluoroacrylyl fluoride, ^{19}F-^{19}F
    J couplings, 447
Perfluoroheptane, 441, 445
Perfluorocyclohexane, crystalline WAHUHA
    spectrum of, 422–425
Perfluoropropene, ^{19}F-^{19}F *J* couplings in, 447
Perturbation metehods, in quantum
    mechanical calculations, 149
Phase, of NMR signals, 26–28, 54–55,
    65–67
Phase cycling, 67
Phase memory, in relation to T_2, 23
Phase sensitive detection, 62, 65, 67
Phenanthrene, spectrum of, 112
α-Phenethylamine, as chiral solvent, 126
Phenyl group, anisotropic shielding by. *See*
    Benzene
Phosphonic esters, symmetry of, 128
Phosphoric esters, symmetry of, 128
4-Picoline, geminal ^{1}H coupling in, 192
Pinenes, α- and β-, shielding in, 116
Poly-γ-benzyl L-glutamate
  2D NOESY spectrum, 345–347
  ^{1}H spectrum, 345, 383–385
Polybutadiene, 370–373, 418–422
  solid state ^{13}C spectrum, 418–422
Polybutene-1, T_1 of 392, 394 (table)
Poly(*dA-dT*), 391
Polyethylene, T_1 of, 392, 394 (table)
Polyisoprene, 373–375
Polymers, NMR of, 357–395. *See also*
    specific polymer
Poly(methyl methacrylate), 359–363
  T_1 of, 392–394 (table)
Polynucleotides, 386–392
Polypeptides, 381–384
Poly(phenylene oxide), 416–418
  magic angle spectrum, 417

Polypropylene, 365–370, 418–419
  solid state ^{13}C spectrum, 418–419
  T_1 of, 393–394 (table)
Poly(propylene oxide), T_1 of, 393–394
    (table)
Polystyrene, T_1 of, 393–394 (table)
Poly(vinyl fluoride), 374–377
Poly(vinylidene chloride), T_1 of, 392, 394
    (table)
Poly(vinylidene fluoride), 378
Porphyrins, shielding anisotropy, 113
Precession, Larmor, 4, 5, 7, 22, 87
Precession, nuclear, 4, 5, 7, 22, 87
Probe, design of, 60–62
L-proline, L-alloproline, ^{1}H-^{1}H couplings in,
    201
Propane
  ^{13}C shielding in, 99
  ^{1}H-^{1}H vicinal coupling in, 201
n-Propanol, 95–96
Propene. See Propylene
Propene, perfluoro-, ^{19}F-^{19}F J coupling in,
    447
Propionic acid, 2,3-dibromo-, as AMX ^{1}H
    system, 222–224
n-Propylamine, 95–96
n-Propylammonium ion, 95–96
Propylene, polymer. See Polypropylene
Propylene oxide, double resonance
    spectrum, 220
Proteins, 384–386
Proton, 2, 3, 87–88, 93–128, 151–214,
    312–313, 315, 316
  dipolar decoupling, 401
  exchange and multiplet collapse, 312–313,
    315, 316
  J coupling, 151–214
  shielding of, 87–88, 93–128
Proton decoupling, dipolar, 401
Pulsed NMR, 7–11, 40, 52–55, 59–62,
    232–250, 325–354, 399–434
Pulse sequences, 10–11, 59–60
Pyrazole, ^{1}H and ^{13}C chemical shifts in, 141
Pyridazine, ^{1}H and ^{13}C chemical shifts in,
    141
Pyridine, ^{1}H and ^{13}C chemical shifts in, 141
Pyridines, substituted, as ABC systems, 175
Pyrimidine, ^{1}H and ^{13}C chemical shifts in,
    141
Pyrrole, ^{1}H and ^{13}C chemical shifts in, 141

Q

Q, probe quality factor, 62
Quadrature detection, 65
Quadrupolar relaxation, 263–265
  and ^{2}H(deuterium) NMR, 424–434
Quadrupole, nuclear electric, 15, 29–30,
    226–227, 457, 460
Quantum mechanics
  and calculation of spectra, 151
  and NMR phenomenon, 5

R

Ramsey's equation, 89
Radiofrequency (rf) field, 5–6, 23–28,
    28–29, 50–55
  Bloch equations and, 23–28
  saturation and, 28–29
Rate processes. See also specific compounds
  NMR measurement of, 291–321
  by saturation transfer, 318–320
  by spin echo, 321
  by two-dimensional NMR, 321, 348–349
Rates, of chemical exchange. See Exchange
  rates, Rate processes
Receiver, 7–11, 40–43, 61–67, 80
  in imaging, 489–500
Receiver coil, 61–62
Redfield pulse, 506–509
References, 31, 33, 57, 58, 75–79
  external, 79 (table)
  for ^{19}F, 439
  internal, 31, 33, 57, 58, 76 (table)
  for ^{14}N, ^{15}N, 460
  for ^{31}P, 474
Relaxation, nuclear, 12–18, 23–29, 263–265,
    267–289, 388–395, 400–401, 424–434
  in Bloch equations, 23–29
  ^{13}C, 260–266, 388–395
  ^{1}H, 12–18, 23–29, 267–289
  longitudinal (see Relaxation spin lattice)
  measurement, 267–289
  molecular motion and, 13–14
  in polymers, 388–395
  quadrupolar, 263–265, 424–434
  and rate processes, 289–321
  scalar relaxation, 265–267
  spin echo measurement of, 233–237,
    289–291

spin lattice, 12–18, 23–29, 267–289
spin-spin, 18–23 (*see also* Line
 Broadening, Line shape, Line width)
temperature dependence, 16–18
transverse (*see* Relaxation, spin-spin)
Residence time, τ, 293–299
Resolution, in NMR spectra, 79–85
Resonance equation, 5
Ring currents, 108–114
in alicyclic rings, 114
in annulenes, 113
in benzene ring, 109–113
in porphyrins, 113
Rotating frame, 9–11, 25–28, 232–244, 268,
 335, 339, 342, 423
in Bloch equations, 25–28
in pulse description, 9–11

S

^{32}S, 4
s Character of bonds
effect on ^{13}C-^{13}C couplings, 216
effect on ^{13}C-^{1}H couplings, 208–210
effect on J couplings, general, 147, 191
effect on ^{15}N-^{1}H couplings, 470, 471–473
effect on ^{15}N-^{13}C couplings, 474
effect on ^{31}P-^{1}H couplings, 474, 476–488
 (tables)
Salicylaldehyde, 121, 319–321
Sample, preparation of, 71–79
Sample tubes, 71, 75, 77–79, 404–406
magic angle spinner, 77–79, 404–406
microcell, 75
Satellites, from ^{13}C-^{1}H couplings, in ^{1}H
 spectra, 209–214
Saturation, 28–29, 271, 505–506
progressive, for T_1 measurement, 271
of solvent, 505–506
Scalar relaxation, 265–267
Screening, nuclear. *See* Nuclear screening
Screening constant, 32–33, 88–93, 106–110
SECSY form of 2D NMR, 351
Sensitivity, limits to NMR observation, 79–85
Shielding
anisotropy of. *See* Shielding anisotropy
atomic, 30, 87–88
inductive effects in, 93–97
isotope effects in, 141–142
Ramsey's equation for, 89

Shielding anisotropy
in acetylenes, 108
in alicyclic compounds, 114
in annulenes, 113
in aromatic rings, 108–112
in C = O bonds, 115–118
in olefinic double bonds, 114
in polynuclear hydrocarbons, 112
in porphyrins, 113
in single bonds, 118–119
Shift reagents, 131–138
Shimming of magnetic field, 45–50
Shoolery's rule, 94–95
^{29}Si, 3
Sidebands
from ^{13}C-^{1}H couplings, in ^{1}H spectra,
 209–214
from field modulation, 44
from sample spinning, 413
Signal. *See* Line shape, Line width
Signal/noise ratio, 35, 67, 79–85, 237
Single bonds, shielding anisotropy of,
 118–119
S/N, *See* Signal/noise ratio
Solvents
for NMR measurements, 71–77 (table)
optically active, 125–126
suppression of resonance, 505–511
Spectra, analysis of, 33–36, 150–190
deceptive, 173–175
first order, 34–35
strong coupled, definition, 149
Spectral width, 66–67
Spin. *See also* Nucleus
electron, 2, 15, 133, 266
nuclear, 1–4
 systems of (*see* Spin systems)
temperature, definition, 12
Spin echo, 232–237, 289–291, 320–321
Carr-Purcell echoes, 290–291
Carr-Purcell-Meiboom-Gill experiment,
 290–291
Hahn experiment, 232, 290
Spin-spin coupling
dipole-dipole, 18–29, 289–291, 341–349,
 400–401
electron-mediated, 33–36, 147–250,
 325–32X, 349–353
Spin systems
AA'BB', 185–190

Spin systems (*continued*)
  A_2B_2, 186
  AA'XX', 179–184
  AB_2, 159–162
  AB_3, AB_n, 162–164
  AB, AX, 151–158
  ABC, 175–179
  ABX, 164–175
  AMX, 163–164
  descriptions of, 150–190
  equivalent and non-equivalent, 157–158
Spin temperature, defined, 12
Spin lattice relaxation, 12–18, 23–29,
    267–289, 388–395, 400–401
  in Bloch equations, 23–29
  ^{13}C, 260–266, 388–395
  ^{1}H, 12–18, 23–29, 267–289
  measurement, 267–289
  molecular motion and, 13–14
  in polymers, 388–395
  in solids, 400–401
  temperature dependence, 16–18
Spin-spin relaxation, 18–23. *See also* Line
    broadening, Line shape, Line width
  and Bloch equations, 233–28
  and rate processes, 289–321
  spin echo measurement, 233–237, 289–291
Spinning, of NMR samples, 47
Steroids
  carbonyl shielding anisotropy in, 118
  ^{19}F-^{1}H coupling in, 455–456
Styrene, 176–177, 191
Sucrose, 410–412
  magic angle spectrum, 412
Sugar, furanose, ^{1}H-^{1}H coupling in, 201
Sulfinic esters, symmetry of, 128
Sulfite esters, symmetry of, 128
Sulfoxides, symmetry of, 128
Superconducting magnets, 42–45, 47
Symmetry, molecular, NMR observation of,
    122–131
Syndiotactic, definition of, 360–361
SWATTR pulse sequence, for solvent
    suppression, 509–511

T

τ, residence time, definition, 293
T_1, spin lattice relaxation time, 12–18,
    23–29, 267–289, 388–395, 400–401
  in Bloch equations, 23–29

^{13}C, 260–266, 388–395
^{1}H, 12–18, 23–29, 267–289
measurement, 267–289
molecular motion and, 13–14
in polymers, 388–395
in solids, 400–401
temperature dependence, 16–18
T_2, spin-spin relaxation time, 18–23, 23–28,
    233–237, 289–291, 289–321
  and Bloch equations, 23–28
  spin echo measurement, 233–237, 289–291
  and rate processes, 289–321
Temperature, of sample, 78
Tetrabromoethane, ^{1}H-^{1}H J coupling in, 199
1,2,3,4-Tetrachloronaphthalenebis(hexachloro-
    cyclopentadiene) adduct, zero field ^{1}H
    spectrum, 515–517
Tetrad configurational sequences, 363, 364
    (table)
Tetrahydrofuran, as AA'BB' ^{1}H spin
    system, 183
  ^{13}C-^{13}C coupling in, 215–216
Tetramethylsilane, TMS, as chemical shift
    reference, 30–33, 76, 79
Thiophene, as AA'BB' proton system, 183
Thiophenes, substituted, as ABC proton
    spin systems, 175
Threonine, solvent suppression and, 507
Time domain, 52
Toluene, geminal ^{1}H coupling in, 192
  α-d_1-, ^{1}H chemical shift of, 141
Toluene, *p*-nitro, sign of vicinal ^{1}H-^{1}H
    coupling, 225
Trans conformers
  vicinal ^{19}F-^{19}F couplings in, 446–447
  vicinal ^{19}F-^{1}H couplings in, 448–454
  vicinal ^{1}H-^{1}H couplings in, 195–203
Trans coupling. *See* Trans conformers
Trans ^{1}H-^{1}H couplings, 193–207
  in cyclobutanes, 200–201
  in cyclohexanol, 194–195
  in cyclopropanes, 200
  in olefins, 194, 206–207
  in six-membered rings, 202
  in substituted ethanes, 196–200
Transitions, between nuclear magnetic states
  and Bloch equations, 23–29
  and cross polarization, 237–241
  J coupling and 150–190
  and resonance, 4–7, 50–52
  and spin-lattice relaxation, 12–18

and spin-spin relaxation, 18–23
Transmitter, design of, 50–52, 60–62
Transmitter offset, 66–67
Triad configurational sequences, 360–370
Tribromotrifluoroethane, ^{19}F-^{19}F J coupling in, 445–447
1,2,4-Trichlorobenzene, as ABC ^{1}H spin system, 177, 179
1,1,2-Trichloroethane, J-resolved 2D spectrum, 328–329
Trichlorofluoromethane, as solvent and reference, 437–441
2,4,5-Trichloronitrobenzene, J-resolved 2D spectrum, 329–330
Trifluoroacetic acid, as solvent, 74, 344–347, 384–386
Trifluorochloroethylene, as ^{19}F AMX spin system, 164–165
Triple resonance, 55–59
Tritium. *See* ^{3}H
Tropilidene, inversion of, 305–306

U

u-Mode, dispersion signal, 26–28
Uncertainty principle, 292

V

Valence bond calculation, of J couplings, 194
L-valine, symmetry of, 125–126
Vicinal J coupling
^{13}C-^{1}H, 209
^{19}F-^{19}F, 441–448 (table)
^{19}F-^{1}H, 448–456 (table)
^{1}H-^{1}H, 193–205
^{15}N,^{14}N-^{13}C, 474
^{15}N,^{14}N-^{1}H, 470–474 (table)
^{31}P-^{1}H, 475–489 (table)
Vinyl chloride, ^{1}H spectrum, 176–177

Vinyl groups
as ABC ^{1}H systems, 175–178
geminal ^{1}H coupling in, 176–177, 191
vicinal coupling in, 203–204, 206–207
Virtual coupling, 174–175, 441–442
Viscosity
in polymer solutions, 389–395
in relation to T_1, 14–18, 260–261, 266, 277
in relation to T_2, 18–21
v-Mode, absorption spectrum, 26–28
Vortex plugs, 75

W

WAHUHA line narrowing, 422–424
WEFT pulse sequence, for solvent nulling, 506–507
WATR pulse sequence, for solvent suppression, 509–511
Water. *See also* D$_2$O, HDO
NMR of, 18
as solvent, 30, 72, 382, 384, 387, 389
T_1, 17–18

X

X-Gradient, correction of, 47–50

Y

Y-Gradient, correction of, 47–50

Z

Z, atomic number, 3–4
Z, saturation factor, 28–29
Zero field NMR, 512–517
Zero filling, 67
Z-Gradient, correction of, 47–50